BIODEGRADABLE GREEN COMPOSITES

BIODEGRADABLE GREEN COMPOSITES

Edited by

SUSHEEL KALIA
Army Cadet College Wing
Indian Military Academy, Dehradun
India

WILEY

Published by John Wiley & Sons, Inc., Hoboken, New Jersey
Published simultaneously in Canada

Library of Congress Cataloging-in-Publication Data:

9781118911099 [hardback]

Set in 10/12pt Times by SPi Global, Pondicherry, India

Printed in the United States of America

10 9 8 7 6 5 4 3 2

2 2016

CONTENTS

CONTRIBUTORS

Pramendra Kumar Bajpai, Division of Manufacturing Processes and Automation Engineering, Netaji Subhas Institute of Technology, Dwarka, New Delhi, India

Jaspreet Kaur Bhatia, Department of Chemistry, DR B R Ambedkar National Institute of Technology, Jalandhar, Punjab, India

Núria Butchosa, School of Biotechnology, Royal Institute of Technology (KTH), AlbaNova University, Stockholm, Sweden

Samuel Furtado, Department of Mechanical Engineering, Instituto Superior Tecnico, University of Lisbon, Lisbon, Portugal

Koronis Georgios, Department of Mechanical Engineering, Instituto Superior Tecnico, University of Lisbon, Lisbon, Portugal

Sreerag Gopi, Gandhigram Rural University—Deemed University, Dindigul, Tamil Nadu, India; International and Inter University Centre for Nanoscience and Nanotechnology, Mahatma Gandhi University, Kottayam, Kerala, India

Mahbub Hasan, Department of Materials and Metallurgical Engineering, Bangladesh University of Engineering and Technology, Dhaka, Bangladesh

Azman Hassan, Department of Polymer Engineering, Faculty of Chemical Engineering, Universiti Teknologi Malaysia, Johor, Malaysia

Balbir Singh Kaith, Department of Chemistry, DR B R Ambedkar National Institute of Technology, Jalandhar, Punjab, India

Susheel Kalia, Department of Chemistry, Army Cadet College Wing, Indian Military Academy, Dehradun, Uttarakhand, India

Yekta Karaduman, Akdagmadeni Vocational High School, Bozok University, Akdagmadeni, Yozgat, Turkey

M.V. Manjula, Department of Physics, Indian Institute of Science, Bengaluru, Karnataka, India

Reghu Menon, Department of Physics, Indian Institute of Science, Bengaluru, Karnataka, India

Levent Onal, Department of Textile Engineering, Faculty of Engineering, Erciyes University, Talas, Kayseri, Turkey

Anitha Pius, Gandhigram Rural University—Deemed University, Dindigul, Tamil Nadu, India

V.S. Prasad, Chemical Sciences & Technology Division, National Institute for Interdisciplinary Science & Technology (NIIST-CSIR), Thiruvananthapuram, Kerala, India

Anupama Rangan, Department of Pharmaceutical Chemistry, Vivekananda College of Pharmacy, Bengaluru, Karnataka, India

K.G. Satyanarayana, Poornaprajna Institute of Scientific Research (PPISR), Bengaluru, Karnataka, India

Arlindo Silva, Department of Mechanical Engineering, Instituto Superior Tecnico, University of Lisbon, Lisbon, Portugal

Danmei Sun, School of Textiles and Design, Heriot-Watt University, Netherdale, Galashiels, UK

Sabu Thomas, International and Inter University Centre for Nanoscience and Nanotechnology, Mahatma Gandhi University, Kottayam, Kerala, India

Zainoha Zakaria, Department of Chemistry, Faculty of Science, Universiti Teknologi Malaysia, Johor, Malaysia

Qi Zhou, School of Biotechnology, Royal Institute of Technology (KTH), AlbaNova University, Stockholm, Sweden; Wallenberg Wood Science Center, Stockholm, Sweden

PREFACE

In recent years, increased awareness of the environmental damage caused by plastic waste has led to research aimed at producing biodegradable green composite materials. Natural fibers and fiber-reinforced biodegradable polymer composites have received much attention because of low density, nonabrasive, combustible, nontoxic, low cost, and biodegradable properties. The lack of good interfacial adhesion between fibers and matrix, low melting point, and water sensitivity make the use of natural fiber-reinforced composites less attractive. Surface roughness can be increased and moisture absorption is decreased by surface modification of natural fibers. Various methods have been explored in order to improve the compatibility between hydrophilic natural fibers and hydrophobic polymer matrices. Most of the chemical surface treatments of natural fibers involve silylation, acetylation, benzoylation, isocyanate treatment, and polymer grafting. Although these treatments can alter the wettability of natural fibers, the appropriate handling and disposal of the large amounts of hazardous chemicals that is often involved is unattractive and an additional cost to the production. Surface modification of natural fibers by alternative green methods may surpass the chemical treatments. Efforts are now being focused on environmentally friendly methods to increase the compatibility between natural fibers and polymer matrices, such as enzymatic and plasma treatments. The main objective of this book is to elucidate some important aspects about chemical and green approaches for surface modification of natural fibers, nanocellulose- and nanolignin-based biocomposites, biodegradable polyesters, hybrid composite materials, and applications of green composite materials.

The first chapter of this book discusses biodegradable polymers and organic and inorganic nanofillers for composites. Processing and applications of green composites are also included in this chapter. The second and third chapters review the surface modification of natural fibers with green treatments, that is, plasma treatment and enzymatic treatment. Reinforcing potential of modified natural fibers is also discussed here. Recent developments in surface modification of natural fibers with chemical treatments such as silylation, noncovalent surface modifications, polymer grafting, TEMPO-mediated oxidation, and cationization are discussed in Chapter 4. Biodegradability of biocomposites is also discussed here. Nanocellulose-based biocomposites are discussed in Chapter 5. Thermal, mechanical, barrier, antimicrobial, and optical properties of biocomposites are also discussed in this chapter.

Chapter 6 highlights the manufacturing techniques of poly(lactic acid)-based hybrid green composites. Properties and application of hybrid green composites are also discussed here. Chapter 7 deals with lignin- and nanolignin-based green composite materials. Characterization, methods of preparation, and applications of nanolignin are also discussed. Synthesis, properties, and applications of starch-based polymer composites are reviewed in Chapter 8. Fabrication techniques for biodegradable polyester-based green composites, such as hand lay-up fabrication, compression molding, injection molding, resin transfer fabrication, and pultrusion fabrication, are discussed in Chapter 9. This chapter also includes the processing of green composites through microwave heating and applications of such green composite materials. Chapter 10 includes the applications of green composite materials such as biomedical applications, consumer products, packaging, transportation, construction, energy industry, and sports and leisure industry.

The various chapters in this book are contributed by prominent researchers from industry, academia, and research laboratories across the world. This book covers scientific, technological, and practical concepts concerning the research, development, and realization of green composite materials. So this book will prove to be a very useful tool for scientists, academicians, research scholars, material engineers, and industries. This book can also be supportive for undergraduate and postgraduate students in institutes of materials science and other technical institutes and technologists and researchers from R&D laboratories working in this area.

The editor would like to express his gratitude to all contributors of this book, who have provided excellent contributions. The editor would like to thank his research students, who helped him in the editorial work. Finally, the editor gratefully acknowledges permissions to reproduce copyrighted materials from a number of sources.

Susheel Kalia
Dehradun, India

1

BIODEGRADABLE GREEN COMPOSITES

SREERAG GOPI[1,2], ANITHA PIUS[1], AND SABU THOMAS[2]

[1] *Gandhigram Rural University—Deemed University, Dindigul, Tamil Nadu, India*
[2] *International and Inter University Centre for Nanoscience and Nanotechnology, Mahatma Gandhi University, Kottayam, Kerala, India*

Biodegradable Green Composites, First Edition. Edited by Susheel Kalia.

1.1 INTRODUCTION

Conventional polymers are usually made from petroleum resources such as polyolefins, and they are ideal for many applications such as packaging, building resources, commodities, and consumer goods. Polyolefin-based plastics have become a foundation of modern civilization and are low cost, durable, resistant to solvents, waterproof, and resistant to physical aging. The resistance of polyolefin materials to degradation by microorganisms is both an advantage and, in the long term, a problem. It was estimated in 2002 that some 41%·w/w of the total global plastic production was used by packaging industries, with 47% of that production being used to package foodstuffs [1]. Most oil-derived packaging is nonrecyclable, or economically impractical to recycle, and quickly becomes landfill, equating to a huge quantity of nondegradable waste. Microorganisms found in landfill soils are unable to degrade conventional plastics [2], and as a result, they remain in the environment for a very long time [3]. This in itself has not been a huge concern until recently. Landfills are unattractive to residents living near them, and new ones are costly and difficult to establish. Landfills are becoming filled to capacity with more waste generated every day due to continued expansion of human urban areas and population increases. Recycling plastics is one possible solution, and since the early 1990s, more and more plastic waste is subject to recycling across developed nations [4, 5]. Australia recycled 18.5% (282,032 t) of the total plastics collected in 2008, and 58.2% of that amount was recycled here with the remainder exported for reprocessing [6]. Despite this admirable effort, it still leaves a phenomenal amount of waste plastics. Recycling is not without its problems since often recycled polymers are contaminated, resulting in inferior mechanical properties to feedstock created ones [7]. This reduces recycled feedstock desirability and hence the economic benefit to recycling. Polymer waste can also be disposed of by incineration, but given the current political climate on greenhouse emissions, this is becoming unfeasible. Incineration also produces harmful gasses and emissions, for example, burning poly(vinyl chloride) (PVC) produces furans and dioxins [8]. In this context, green composites gain their importance.

Figure 1.1 shows a classification of biodegradable polymers mainly in two families. A large number of these biodegradable polymers (biopolymers) are commercially available. They show a large range of properties, and they can compete with nonbiodegradable polymers in different industrial fields (e.g., packaging).

1.2 BIODEGRADABLE POLYMERS

1.2.1 Starch

Starch is a widely used bioplastic that is actually a storage polysaccharide in plants. It is composed of both linear and branched polysaccharides known as amylose and amylopectin, respectively. The ratio of these polysaccharides varies with their botanical origin, and generally, native starches contain around 85–70% amylopectin and 15–30% amylose. Starch softening temperature is higher than its degradation

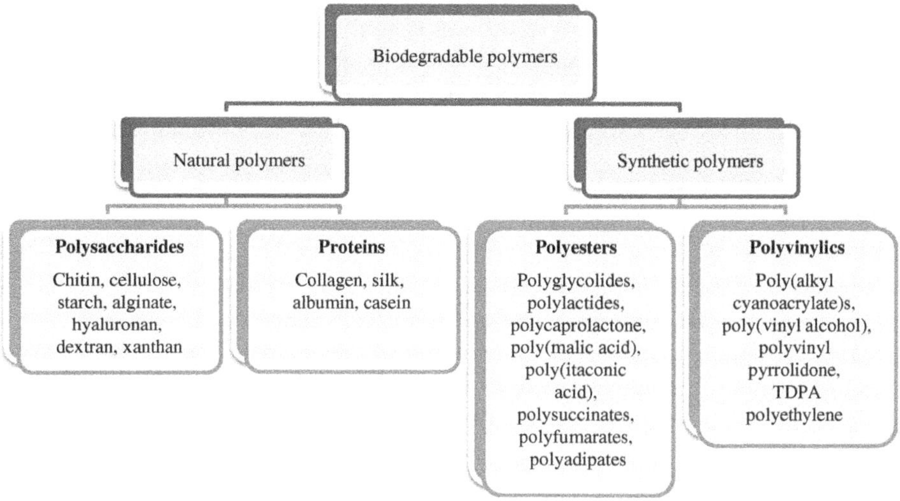

FIGURE 1.1 Classification of biodegradable polymers in four families. (*See insert for color representation of the figure.*)

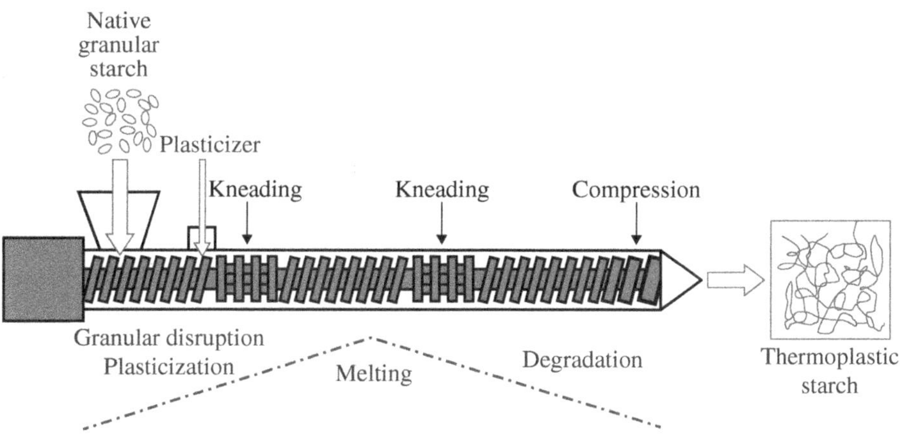

FIGURE 1.2 Thermoplastic starch formation. *Source*: Xie et al. [11], figure 37. Reproduced with permission from Elsevier. Copyright 2015. (*See insert for color representation of the figure.*)

temperature due to the presence of many intermolecular hydrogen bonds [9], which affects its processing. Plasticizers like water, glycerol, and sorbitol will help in increasing the free volume and thereby decreasing the glass transition and softening temperatures [10]. The schematic showing the process of obtaining TPS is shown in Figure 1.2. Traditional extrusion, injection molding, and compression molding can be used to process thermoplastic starch. The melt processing technique of obtaining thermoplastic starch is a complex operation that involves plasticization, devolatilization, melt–melt mixing, and morphology control. The final morphology of TPS depends on composition, mixing time, temperature, shear, and elongation rate of the operation. Although it is possible to make useful products from TPS alone, extreme moisture sensitivity of starch leads to limited practical application. Therefore, the reality in commercialization of starch-based plastics involves blending of TPS with other polymers and additives. Thermoplastic starch formation [11] is shown in Figure 1.2.

1.2.2 Cellulose

Cellulose is an abundant and ubiquitous natural polymer. It is the major structural component of plant cells and is found throughout nature. It is widely used in industrial applications in different forms. Cellulose is mostly obtained from wood and cotton at present for many applications; on the other hand, cellulose pulp is also being extracted from agricultural by-products such as bagasse, stalks, and crop straws. Currently, cellulose-based materials are used in two forms on an industrial scale [12]:

1. Regenerated cellulose is used for fiber and film production and cannot be melt processed.
2. Cellulose esters are used in a broad array of applications including coatings, biomedical uses, and other usual plastic applications.

Nonplant resources can also be used to produce cellulose, especially bacteria and tunicates. There is a considerable interest in obtaining cellulose from bacteria, popularly known as bacterial cellulose. *Acetobacter xylinum* produces this cellulose under unique culturing conditions to form a fibrous network [13]. *A. xylinum* produces cellulose with good mechanical strength and biodegradability. "Nanocellulose" comprises of fibrous or crystalline units of cellulose between 5 and 500 nm in diameter with a length of several micrometers. Nanocellulose is available in two forms, that is, microfibrils and nanowhiskers [14].

1.2.3 Chitin and Chitosan

Interest in these polymers is driven by their unique properties such as renewable, biocompatible, biodegradable, and nontoxic with excellent adsorption properties [15]. Chitin is an abundantly available natural polysaccharide and is the supporting material in many invertebrate animals such as insects and crustaceans. The monomers in chitin are 2-acetamido-2-deoxy-_-D-glucoses, which are attached to (1→4) linkages, and this polymer degrades by chitinase. The deacetylated chitin is known as chitosan; this deacetylation is more than 50% [16]. Chitosan is a semicrystalline polymer, and its crystallinity depends on the extent of deacetylation [17]. Chitosan is receiving more attention as a possible polysaccharide resource for biomedical applications [17].

The process of obtaining chitin from the shells of crab or shrimp starts with the extraction of proteins followed by treatment with calcium carbonate for dissolution of shells. The chitin obtained from this process is then deacetylated with 40% sodium hydroxide for 1–3 h at 120°C. This yields a 70% deacetylated chitosan [18]. The molecular weight of chitosan depends on the source, and it varies from 100 to 1100 kDa [19]. Commercial chitosan has around 50–90% deacetylation degree [17]. Chitosan has been extensively explored for films and fibers [16]. Fibers from these polymers are very useful, as wound dressing materials and absorbable sutures [19, 20]. They have generated interest in biomedical applications [17].

1.2.4 Proteins

A protein is a random copolymer of different amino acids. Based on the origin, proteins can be classified as plant proteins (e.g., soy, pea, canola, and wheat) and animal proteins (e.g., gelatin, whey, casein, and keratin). Generally, proteinaceous biomaterial can be defined as a stable 3D polymeric network that is strengthened by hydrophobic interactions and hydrogen bonding [21]. Unfolding and realigning of the proteins are necessary to produce useful biomaterials. Hydrophilic compounds and lipid compounds are used for plasticization of proteins. Water, glycerols, fatty acids, and oils are commonly used plasticizers for proteins. Plasticizers reduce the interactions between functional units and improve the polymeric chain mobility and intermolecular spacing. This also results in reduction in glass transition temperature of the proteins. Wet processing and dry processing are used to obtain biomaterials from proteins [22]. Wet processing involves solubilization and dispersion of proteins in solvent. The dissolved protein is casted to obtain films by drying the solvent. Dry processing is the more conventional method in which proteins are mixed with suitable additives followed by thermomechanical processing by conventional techniques such as extrusion and/or molding.

1.3 NANOFILLERS FOR COMPOSITES

Bionanocomposites are obtained from 100% biobased materials, in which the fillers and the matrix both are obtained from renewable resources. Nanofillers that are extensively used for the preparation of nanocomposites are listed below.

1.3.1 Cellulose-Based Nanofillers

Cellulose microfibrils and nanocrystalline celluloses or cellulose nanowhiskers (CNW) are the two types of nanoreinforcements obtained from cellulose [14]. Cellulose microfibrils consist of bundles of molecules that are elongated and stabilized through hydrogen bonding [14, 23]. The typical dimensions of these nanofibrils are 2–20 nm in diameters, while the lengths are in micrometer range. Also, these fibrils consist of both amorphous and crystalline regions. This is shown in Figure 1.3. The crystalline regions can be isolated by various techniques, and resultant material is known as whiskers. These whiskers are also known as nanorods and nanocrystals. The lengths of these whiskers typically

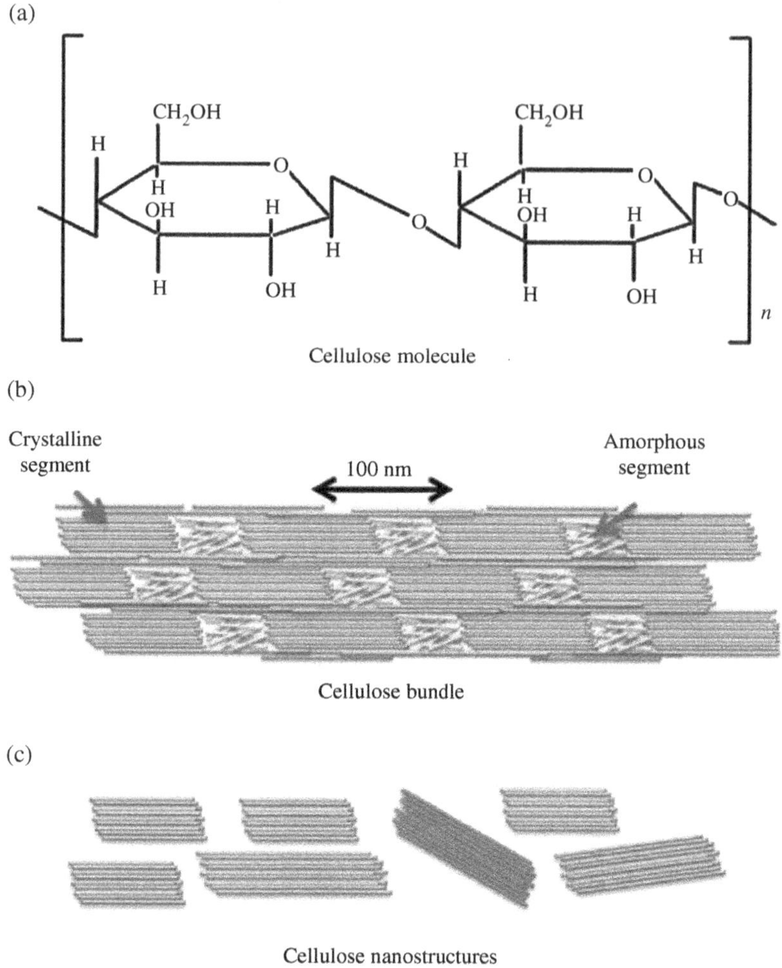

FIGURE 1.3 Schematics of (a) single cellulose chain repeat unit, showing the directionality of the 1–4 linkage and intrachain hydrogen bonding (dotted line), (b) idealized cellulose microfibril showing one of the suggested configurations of the crystalline and amorphous regions, and (c) cellulose nanocrystals after acid hydrolysis dissolved the disordered regions. *Source*: Reddy et al. [24], figure 6. Reproduced with permission from Elsevier. Copyright 2015. (*See insert for color representation of the figure.*)

range from 500 nm to 1–21 μm in length and diameter in the range of 8–20 nm [25]. Also, it was found that cellulose crystals have a modulus of around 150 GPa and a strength of 10 GPa [26]. This is a very interesting data as it suggests that cellulose can replace single-walled carbon nanotubes (SWCNTs) in many applications. Acid hydrolysis is the most widely used method for extracting CNW, which removes the amorphous regions while crystalline regions remain intact [27].

1.3.2 Carbon Nanotube

Carbon nanostructures including fullerene (buckyballs), carbon nanotubes (CNTs) (single walled and multiwalled), carbon nanofibers, carbon nanoparticles, and graphene nanosheets have been widely investigated due to their excellent physicochemical, mechanical, and electrical properties [28–30]. The allotropic behavior of carbon arises from different bonding states representing sp^3, sp^2, and sp hybridization. In general, the degree of carbon bond hybridization-*n* (sp^n) determines the structure of carbon nanomaterials and their functional properties [28]. CNTs have been synthesized using many methods; among them, (i) arc discharge, (ii) ablation using laser, (iii) chemical vapor deposition, and (iv) high pressure of carbon monoxide are found to be the most popular [31]. Especially for polymeric composites, CNTs impart several advantages:

1. Versatility as reinforcement in both thermoplastic and thermoset regimes [32]
2. Extremely high theoretical/experimental tensile strength (150–180 GPa) and modulus (640 GPa to 1 TPa) [32]
3. One-dimensional electronic structure, which enables significantly nonscattering electron transport [33]
4. Their compatibility with other chemical compounds, metal/metal oxide/chalcogenide nanoparticles, and polymeric materials [34] (reinforcement of CNTs into various polymer systems not only provides improvement in mechanical and thermal properties but also creates additional functional properties)

1.3.3 Clay

Layered silicates, also known as nanoclays, are most commonly utilized nanofillers in the synthesis of polymer layered silicate nanocomposites. Among these layered silicates, phylloscilicates (2 : 1) are extensively used in preparing clay-based nanocomposites. The crystal arrangement in the silicate layers is made up of two tetrahedrally coordinated atoms amalgamated to edge-shared octahedral sheets. The dispersibility of layered silicates into individual layers is governed by its own ability for surface modification via ion exchange reactions that can replace interlayer inorganic ions with organic cations. Renewable polyesters are mostly organophilic compounds, while the pristine silicate layers are miscible only with hydrophilic polymers. The silicate layers can be made miscible with hydrophobic polymer by introducing/exchanging interlayer cation galleries (Na^+, Ca^{2+}, etc.) of layered silicates with organic compounds.

1.3.4 Functional Fillers

Nowadays, medical implants are very common in practice that utilizes a wide range of biocompatible materials such as metals, alloys, ceramics, polymers, and composites [35]. Among them, bionanocomposites that are fabricated using the combination of biopolymers and various nanostructured inorganic/organic functional fillers receive extensive attention due to their diversified biomedical as well as biotechnological

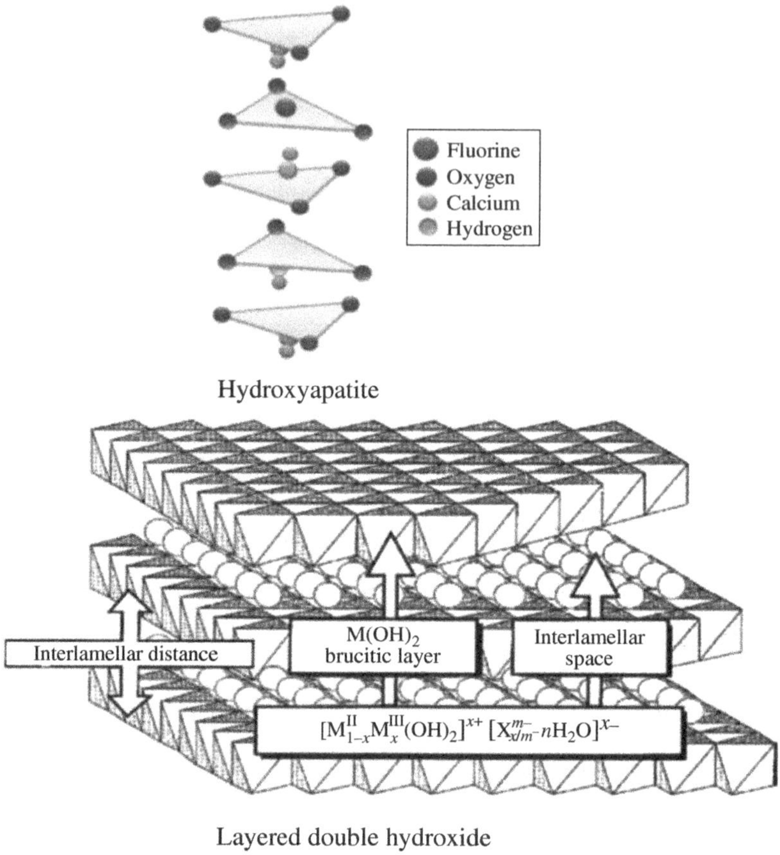

FIGURE 1.4 Schematic representation of hydroxyapatite and layered double hydroxide chemical structure. *Source*: Leroux and Besse [38], figure 13. Reprinted with permission from American Chemical Society. Copyright 2015. (*See insert for color representation of the figure.*)

application [36]. Nanostructured fillers play an important role in biocomposite fabrication, since they bring various desired functionalities to the composites [36]. Functional nanofillers such as cellulose nanofibers, hydroxyapatite (HAp), layered double hydroxides (LDH), silica nanoparticles, and polyhedral oligomeric silsesquioxanes (POSS) are mostly investigated for this proposes [36, 37]. Recently, HAp and LDH have received more attention due to their versatility in the fabrication of various nanocomposites for biomedical application [36] (Fig. 1.4).

1.4 NANOCOMPOSITES FROM RENEWABLE RESOURCES

Expansion of nanotechnology in recent years has influenced the scientific, technical, and economical competitiveness of renewable resource-based polymers in developing a range of high-performance engineering/consumer products [39]. Recently, scientists/engineers from academia and industries are investigating the use of nanostructures (cellulose nanostructures, CNTs, and nanoclays) as reinforcements in order to produce a new class of bionanocomposites [40, 41]. The inherent properties of these nanoparticles in enhancing the thermal, mechanical, dimensional stability, and other types of functional properties (electrical/electromagnetic shielding/barrier/fire retardant/triggered biodegradability/solvent resistance) of the composite materials with the added advantages of eco-friendliness were effectively utilized to create a new class of materials. They also provide additional advantages like ease to process, transparent, low density, and recyclable [14, 39].

1.4.1 Cellulose Nanocomposites

Cellulose-based nanostructures have been utilized for improving PLA's. Microcrystalline cellulose (MCC) is the best source for CNW; however, dispersion of MCC into individual fibers is difficult especially in melt processing. Mathew et al. reported that the reinforcement of MCC in PLA processed through twin-screw extrusion resulted in the retention of MCC as bundles and reduced in the mechanical properties [42]. It was found that the mechanical properties were not increased compared to virgin PLA, which is due to the presence of additives and unsuitable processing temperatures. Reinforcement of cellulose nanostructure into PLA creates a diversified impact that includes improvement in barrier properties, nucleation effects, and foam formation. Sanchez-Garcia and Lagaron [43] reported the barrier properties of PLA/CNW composites. Their research ensured that the addition of 3 wt% CNW into PLA was able to reduce the water and oxygen permeability by 82 and 90%, respectively. CNW acted as shield in PLA and caused the crystallinity development, which resulted in high barrier properties [25]. The effects of microfibrillated cellulose reinforcement in PLA on crystallization were studied by Suryanegara et al. [44]. They found that the microfibril acted as a nucleating agent and altered the crystallization behavior of PLA, which resulted in the enhancement of storage modulus up to 1 GPa [44].

1.4.2 CNT Nanocomposites

In starch-based plastic materials, utilization of CNTs as nanoreinforcement is limited by their effective distribution. One of the successful methods of dispersing CNTs into a starch matrix is functionalization. Fama et al. investigated the fabrication of starch/multiwalled CNT composites and reported their improved mechanical properties. They were successful in dispersing MWCNTs into a starch matrix by adopting new strategies to wrap the CNT surfaces with a starch–iodine complex. This also creased a strong adhesion between nanotubes and starch matrix, which caused an effective load transfer that improved their mechanical properties [45]. They were able to achieve a 70% increment in the stiffness with the MWCNT reinforcement of only 0.055 wt%. Another investigation of Fama et al. on starch-based nanocomposites using MWCNT modified with starch–iodine complex resulted in lower water permeability with high storage modulus [46]. These enhanced properties were obtained due to the uniform dispersion of MWCNT caused by the coating of same material that has been used for matrix.

1.4.3 Clay Nanocomposites

Bioplastic–clay nanocomposites have been receiving extensive attention due to their improved thermal, mechanical, and barrier properties as well as reduced flammability compared to their respective virgin polymers [47]. Other positive aspects of nanoclay reinforcement are that it does not hamper the biodegradation of biodegradable polymers [48]. These clay materials were classified into many types based on their chemical nature, structure, and unique properties of swelling as well as exfoliation [49]. In composite fabrication, montmorillonite, hectorite, and saponite are the three most commonly used nanoclay minerals, which belong to the smectites family [48]. Montmorillonite, saponite, and hectorite are the three most commonly used nanoclays in the synthesis of polymer nanocomposites; these nanoclays belong to smectites family [50]. Enhancement of polymer properties by individual clay layers can be obtained through their high aspect ratio and interfacial interactions with polymer networks.

1.4.4 Functional Nanocomposites

Biobased polymers with functional fillers such as HAp and LDH found a wide range of applications especially toward tissue engineering, drug deliver and gene therapy due to their compatibility and also noncytotoxic and noninflammatory toward with biological. Chen et al. [51] also demonstrated the precipitation technique for the fabrication of chitosan/HAp nanocomposites for the biomedical applications. Similarly, a wide range of literature available for the fabrication and characterization of HAp reinforced collagen nanocomposites for biomedical applications. Chang and Tanaka [52] reported the simultaneous titration method for the fabrication of HAp/collagen nanocomposites with glutaraldehyde cross-linking. Rhee and Tanaka [53] reported the precipitation method for the fabrication of HAp/collagen/chondroitin sulfate nanocomposites with the possible shaping and consolidation with mechanical pressing. Dagnon et al. [54] reported the solution casting fabrication of PLA/LDH nanocomposites and reported the increment of overall crystallinity. Chiang and Wu [55] investigated the fabrication of biodegradable PLA/LDH nanocomposites by solution mixing process and reported the significant increment of mechanical properties with 1.2% LDH reinforcement.

1.5 PROCESSING OF GREEN COMPOSITES

It is important to obtain uniform dispersion for two reasons: for effective stress transfer from matrix to the filler and also for the obtain network structure that can help in thermal and electrical conductivities. According to Thostenson et al. [56], uniform dispersion [57–59], wetting and orientation in the matrix [59, 60], and functionalization to improve the compatibility are the important issues that influence the processing of CNT polymer nanocomposites. Also, due to difference in size followed by larger surface found in SWCNTs leads to their aggregation compared to the MWCNTs in the polymer matrix. Among various approaches used in dispersion CNTs in polymer matrices include surface modification of nanotubes [61–63], polymer coating on the surface of CNTs [64], *in situ* polymerization technique to obtain CNT nanocomposite [65, 66], dispersion of CNT in polymer solutions using ultrasonic [67, 68], melt processing [69–71], surfactant chemistry [72, 73], electrospinning [74], electrode chemistry [75], and crystallization [76] as well as plasma treatment and/or chemical oxidation to attach functional group.

1.6 APPLICATIONS

1.6.1 Packaging

Narrow processing window, poor gas and water barrier properties, unbalanced mechanical properties, low softening temperature, and weak resistivity of the plastics have limited their use in a wide range of applications. As discussed in the previous sections, nanotechnology helps in overcoming these problems. Nanofillers help in improving the previously discussed properties of the bioplastics. Bionanocomposites exhibit remarkable improvement compared to the neat matrix and conventional composites due to nanoreinforcements. Polymers provide excellent balance in properties including mechanical, thermal, and barrier properties. It is well recognized that the incorporation of nanofillers especially nanoclay into the polymeric matrix can lead to significant enhancement in the barrier properties [47]. This improved barrier property in nanocomposites is explained on the basis of increased path length due to the presence of nanofillers that the same molecules need to traverse while diffusing through the matrix. Cellulose nanofibers were found to have the same reducing effect on barrier properties of TPS; this effect is attributed to the increase of the tortuosity induced by the presence of the nanofibers [77]. However, cellulose nanofibers are not as effective as that of nanoclay probably due to their shape that limits the increment in tortuous path.

1.6.2 Electronics, Sensor, and Energy Applications

Polymers reinforced with engineered/functional nanostructures provide additional electrical, optical, electromagnetic shielding, and magnetic properties and lead to the development of various advanced devices including light-emitting diodes, sensors, solar cells, display panels, and other medical devices [78, 79]. As the global demand for flexible electronic devices increases, polymer nanocomposites receive extensive attention in developing various devices [80]. Besides, ever-increasing uses of electrical and electronic equipment create environmental issues at the end of their life span and generate enormous waste products (e-waste) [81]. As a result, applications of bioplastics and their composites in electronic products are increasing due to their biodegradability and renewability with less environmental impact/carbon footprint [82]. Cellulose-based composite films (which are scientifically known as active paper) with active nanomaterials such as gold, silver, and CNTs have been used to construct strain, chemical, and biosensors. In recent years, a number of polymer composites reinforced with high conductive fillers have been reported as EMI shielding materials [83, 84]. Such materials found applications in high-performance shielded connectors, scientific/medical/consumer electronic devises, and military/security products.

1.6.3 Medicinal Applications

The versatility and adaptability of bionanocomposites enable these materials to be utilized for biomedical applications. An essential characteristic of medical biomaterials is biocompatibility, the ability to function appropriately in the human body to produce the desired clinical outcome, without causing adverse effects [85]. Bionanocomposites are an intuitive choice for medical applications, given that such materials are constructed from bioderived polymers, and such materials possess tunable mechanical properties. Biobased polymers are increasingly being recognized as biocompatible materials for clinical use. For example, plastics and films made from corn-derived PDO have been shown to be noncytotoxic and noninflammatory to clinically relevant cell lines [86]. Moreover, soy-derived polymers have been demonstrated to be useful as bone fillers [87].

1.7 CONCLUSION

Biodegradable green polymers are an emerging class of polymers that have many potential applications to replace the synthetic polymers. There is an immense opportunity in developing new biobased products, but the real challenge is to design sustainable biobased products. The major limitations of the present biodegradable polymers are their high cost. New environmental regulations and societal concerns have triggered the search for new products and processes that are compatible with the environment. Their unique balance of properties would open up new market development opportunities for biocomposites in the twenty-first-century green materials world. It can be concluded that biodegradable green composites are an essential guide for agricultural government, agricultural departments, automotive companies, composite producers and material scientists, and crop producers, all dedicated to the promotion and practice of eco-friendly materials and production methods. The biodegradable green composites can be effectively used as a material for structural, medical, automotive, and electronic applications. Nowadays, green composites are emerging as a viable alternative to glass fiber-reinforced composites especially in automotive and building product applications.

REFERENCES

[1] Ray SS, Bousmina M. "Biodegradable polymers and their layered silicate nano composites: in greening the 21st century materials world," *Progress in Materials Science*; **50**(8):962–1079 (2005).

[2] Mueller RJ. "Biological degradation of synthetic polyesters—enzymes as potential catalysts for polyester recycling," *Process Biochemistry*; **41**:21–6 (2006).

[3] Shimano M. "Biodegradation of plastics," *Current Opinion in Biotechnology*; **12**:5 (2001).

[4] Charles A. Hooper R. Potter AK Singh MM. "Diversion from landfill: quality products from valuable plastics," *Resources, Conservation and Recycling*; **36**:9 (2002).

[5] Subramanian PM. "Plastics recycling and waste management in the US," *Resources, Conservation and Recycling*; **28**:10 (2000).

[6] Oke M, Goldsworthy K. *National Plastics Recycling Survey*. Melbourne: Hyder Consulting Pty Ltd.; p. 53 (2009).

[7] Scott G. "Green' polymers," *Polymer Degradation and Stability*; **68**:7 (2000).

[8] Jayasekara R, Harding I, Bowater I, Lornergan G. "Biodegradability of selected range of polymers and polymer blends and standard methods for assessment of biodegradation," *Journal of Polymers and the Environment*; **13**(23):20 (2005).

[9] Halley PJ. "Thermoplastic starch biodegradable polymers," *Biodegradable Polymers for Industrial Applications*. Cambridge: CRC Press/Woodhead; p. 140–62 (2005).

[10] Coffin DR, Fishman ML. "Physical and mechanical properties of highly plasticized pectin/starch films," *Journal of Applied Polymer Science*; **54**:1311–20 (1994).

[11] Xie F, Halley PJ, Averous L. "Rheology to understand and optimize processibility, structures and properties of starch polymeric materials," *Progress in Polymer Science*; **37**:595–623 (2012).

[12] Amass W, Amass A, Tighe B. "A review of biodegradable polymers: uses, current developments in the synthesis and characterization of biodegradable polyesters, blends of biodegradable polymers and recent advances in biodegradation studies," *Polymer International*; **47**:89–144 (1998).

[13] Klemm D, Heublein B, Fink HP, Bohn A. "Cellulose: fascinating biopolymer and sustainable raw material," *Angewandte Chemie International Edition*; **44**:3358–93 (2005).

[14] Samir MASA, Alloin F, Dufresne A. "Review of recent research into cellulosic whiskers, their properties and their application in nanocomposite field," *Biomacromolecules*; **6**:612–26 (2005).

[15] Averous L. "Biodegradable multiphase systems based on plasticized starch: a review," *Journal of Macromolecular Science Part C: Polymer Reviews*; **44**:231–74 (2004).

[16] Ravi Kumar MNV. "A review of chitin and chitosan applications," *Reactive & Functional Polymers*; **46**:1–27 (2000).

[17] Francis Suh JK, Matthew HWT. "Application of chitosan-based polysaccharide biomaterials in cartilage tissue engineering: a review," *Biomaterials*; **21**:2589–98 (2000).

[18] Averous L. "Biodegradable multiphase systems based on plasticized starch: a review," *Journal of Macromolecular Science Part C: Polymer Reviews*; **44**:231–74 (2004).

[19] Ravi Kumar MNV. "Chitin and chitosan fibers: a review," *Bulletin of Materials Science*; **22**:905–15 (1999).

[20] Muzzarelli R, Muzzarelli C, Terbojevich M. "Chitin chemistry, upgrading a renewable resource," *Carbohydrates in Europe*; **19**:10–7 (1997).

[21] Verbeek CJR, Van Den Berg LE. "Extrusion processing and properties of protein-based thermoplastics," *Macromolecular Materials and Engineering*; **295**:10–21 (2010).

[22] Reddy M, Mohanty AK, Misra M. "Thermoplastics from soy protein: a review on processing. Blends and composites," *Journal of Biobased Materials and Bioenergy*; **4**:298–316 (2010).

[23] Wang B, Sain M. "Isolation of nanofibers from soybean source and their reinforcing capability on synthetic polymers," *Composites Science and Technology*; **67**:2521–7 (2007).

[24] Reddy M, Vivekanandhan S, Misra M, Bhatia SK, Mohanty AK. "Biobased plastics and bionanocomposites: current status and future opportunities," *Progress in Polymer Science*; **38**:1653–1689 Fig.6 (2013).

[25] de Souza Lima MM, Borsali R. "Rod like cellulose microcrystals: structure, properties, and applications," *Macromolecular Rapid Communications*; **25**:771–87 (2004).

[26] Helbert W, Cavaille J, Dufresne A. "Thermoplastic nanocomposites filled with wheat straw cellulose whiskers. Part I: processing and mechanical behavior," *Polymer Composites*; **17**:604–11 (1996).

[27] Gardner DJ, Oporto GS, Mills R, Samir MASA. "Adhesion and surface issues in cellulose and nanocellulose," *Journal of Adhesion Science and Technology*; **22**:545–67 (2008).

[28] Dai L. "From conventional technology to carbon nanotechnology: the fourth industrial revolution and the discoveries of C60, carbon nanotube and nanodiamond," *Carbon Nanotechnology: Recent Developments in Chemistry, Physics, Materials Science and Device Applications*. Amsterdam: Elsevier Science; p. 3–11 (2006).

[29] Tasis D, Tagmatarchis N, Bianco A, Prato M. "Chemistry of carbon nanotubes," *Chemical Reviews*; **106**:1105–36 (2006).

[30] Geim AK, Novoselov KS. "The rise of graphene," *Nature Materials*;**6**:183–91 (2007).

[31] Daenen M, De Fouw R, Hamers B, Janssen P, Schouteden K, Veld M. *The Wondrous World of Carbon Nanotubes*. Eindhoven: Eindhoven University Technology; p. 1–35 (2003).

[32] Moniruzzaman M, Winey KI. "Polymer nanocomposites containing carbon nanotubes," *Macromolecules*; **39**:5194–205 (2006).

[33] McEuen PL, Bockrath M, Cobden DH, Yoon YG, Louie SG. "Disorder, pseudospins, and backscattering in carbon nanotubes," *Physical Review Letters*; **83**:5098–101 (1999).

[34] Baur J, Silverman E. "Challenges and opportunities in multifunctional nanocomposite structures for aerospace applications," *MRS Bulletin*; **32**:328–34 (2007).

[35] Ramakrishna S, Mayer J, Wintermantel E, Leong KW. "Biomedical applications of polymer-composite materials: a review," *Composites Science and Technology*; **61**:1189–224 (2001).

[36] Hule RA, Pochan DJ. "Polymer nanocomposites for biomedical applications," *MRS Bulletin*; **32**:354–8 (2007).

[37] Millon L, Wan W. "The polyvinyl alcohol–bacterial cellulose system as a new nanocomposite for biomedical applications," *Journal of Biomedical Materials Research Part B: Applied Biomaterials*; **79**:245–53 (2006).

[38] Leroux F, Besse JP. "Polymer interleaved layered double hydroxide: a new emerging class of nanocomposites," *Chemistry of Materials*; **13**:3507–15 (2001).

[39] Pilla S, editor. *Handbook of Bioplastics and Biocomposites Engineering Applications*. Salem, MA: Wiley-Scrivener; p. 620 (2011).

[40] Lagaron JM, Lopez-Rubio A. "Nanotechnology for bioplastics: opportunities, challenges and strategies," *Trends in Food Science & Technology*; **22**:611–7 (2011).

[41] Petersson L, Oksman K. "Biopolymer based nanocomposites: comparing layered silicates and microcrystalline cellulose as nanoreinforcement," *Composites Science and Technology*; **66**:2187–96 (2006).

[42] Mathew AP, Oksman K, Sain M. "Mechanical properties of biodegradable composites from poly lactic acid (PLA) and microcrystalline cellulose (MCC)," *Journal of Applied Polymer Science*; **97**:2014–25 (2005).

[43] Sanchez-Garcia MD, Lagaron JM. "On the use of plant cellulose nanowhiskers to enhance the barrier properties of polylactic acid," *Cellulose*; **17**:987–1004 (2010).

[44] Suryanegara L, Nakagaito AN, Yano H. "The effect of crystallization of PLA on the thermal and mechanical properties of microfibrillated cellulose-reinforced PLA composites," *Composites Science and Technology*; **69**:1187–92 (2009).

[45] Fama LM, Pettarin V, Goyanes SN, Bernal CR. "Starch/multi-walled carbon nanotubes composites with improved mechanical properties," *Carbohydrate Polymers*; **83**:1226–31 (2011).

[46] Fama L, Rojo PG, Bernal C, Goyanes S. "Biodegradable starch based nanocomposites with low water vapor permeability and high storage modulus," *Carbohydrate Polymers*; **87**:1989–93 (2012).

[47] Sinha Ray S, Okamoto M. "Polymer/layered silicate nanocomposites: a review from preparation to processing," *Progress in Polymer Science*; **28**:1539–641 (2003).

[48] Nieddu E, Mazzucco L, Gentile P, Benko T, Balbo V, Mandrile R, Ciardelli G. "Preparation and biodegradation of clay composites of PLA," *Reactive & Functional Polymers*; **69**:371–9 (2009).

[49] Pavlidou S, Papaspyrides C. "A review on polymer-layered silicate nanocomposites," *Progress in Polymer Science*; **33**:1119–98 (2008).

[50] Usuki A, Hasegawa N, Kato M, Kobayashi S. "Polymer-clay nanocomposites", *Inorganic/Polymer Nanocomposite Membranes*. Heidelberg: Springer Berlin; p. 1–24 (2005).

[51] Chen F, Wang ZC, Lin CJ. "Preparation and characterization of nano-sized hydroxyapatite particles and hydroxyapatite/chitosan nano-composite for use in biomedical materials," *Materials Letters*; **57**:858–61 (2002).

[52] Chang MC, Tanaka J. "FT-IR study for hydroxyapatite/collagen nanocomposite cross-linked by glutaraldehyde," *Biomaterials*; **23**:4811–8 (2002).

[53] Rhee SH, Tanaka J. "Synthesis of a hydroxyapatite/collagen/chondroitin sulfate nanocomposite by a novel precipitation method," *Journal of the American Ceramic Society*; **84**:459–61 (2001).

[54] Dagnon KL, Ambadapadi S, Shaito A, Ogbomo SM, DeLeon V, Golden TD, Rahimi M, Nguyen K, Braterman PS, D'Souza NA. "Poly (L–lactic acid) nanocomposites with layered double hydroxides functionalized with ibuprofen," *Journal of Applied Polymer Science*; **113**:1905–15 (2009).

[55] Chiang MF, Wu TM. "Synthesis and characterization of biodegradable poly (l-lactide)/layered double hydroxide nanocomposites," *Composites Science and Technology*; **70**:110–5 (2010).

[56] Thostenson ET, Ren Z, Chou TW. "Advances in the science and technology of carbon nanotubes and their composites: a review," *Composites Science and Technology*; **61**:1899–912 (2001).

[57] Shaffer M, Windle A. "Analogies between polymer solutions and carbon nanotube dispersions," *Macromolecules*; **32**: 6864–6 (1999).

[58] Gong X, Liu J, Baskaran S, Voise RD, Young JS. "Surfactant-assisted processing of carbon nanotube/polymer composites," *Chemistry of Materials*; **12**:1049–52 (2000).

[59] Jin L, Bower C, Zhou O. "Alignment of carbon nanotubes in a polymer matrix by mechanical stretching," *Applied Physics Letters*; **73**:1197–9 (1998).

[60] Ajayan P, Stephan O, Colliex C, Trauth D. "Aligned carbon nanotube arrays formed by cutting a polymer resin—nanotube composite," *Science*; **265**:1212–4 (1994).

[61] Ma X, Yu J, Wang N. "Glycerol plasticized-starch/multiwall carbon nanotube composites for electroactive polymers," *Composites Science and Technology*; **68**:268–73 (2008).

[62] Chen J, Hamon MA, Hu H, Chen Y, Rao AM, Eklund PC, Haddon RC. "Solution properties of single-walled carbon nanotubes," *Science*; **282**:95–8 (1998).

[63] Mitchell CA, Bahr JL, Arepalli S, James M, Krishnamoorti R. "Dispersion of functionalized carbon nanotubes in polystyrene," *Macromolecules*; **35**:8825–30 (2002).

[64] Star A, Stoddart JF, Steuerman D, Diehl M, Boukai A, Wong EW, Yang X, Chung SW, Choi H, Heath JR. "Preparation and properties of polymer-wrapped single-walled carbon nanotubes," *Angewandte Chemie International Edition*; **40**:1721–5 (2001).

[65] Jia Z, Wang Z, Xu C, Liang J, Wei B, Wu D, Zhu S. "Study on poly (methyl methacrylate)/ carbon nanotube composites," *Materials Science and Engineering A*; **271**:395–400 (1999).

[66] Deng J, Ding X, Zhang W, Peng Y, Wang J, Long X, Li P, Chan ASC. "Carbon nanotube–polyaniline hybrid materials," *European Polymer Journal*; **38**:2497–501 (2002).

[67] Qian D, Dickey EC, Andrews R, Rantell T. "Load transfer and deformation mechanisms in carbon nanotube-polystyrene composites," *Applied Physics Letters*; **76**:2868–70 (2000).

[68] Safadi B, Andrews R, Grulke E. "Multiwalled carbon nanotube polymer composites: synthesis and characterization of thin films," *Journal of Applied Polymer Science*; **84**:2660–9 (2002).

[69] Siochi EJ, Working DC, Park C, Lillehei PT, Rouse JH, Topping CC, Bhattacharyya AR, Kumar S. "Melt processing of SWCNT polyimide nanocomposite fibers," *Composites Part B Engineering*; **35**:439–46 (2004).

[70] Tang W, Santare MH, Advani SG. "Melt processing and mechanical property characterization of multi-walled carbon nanotube/high density polyethylene (MWNT/HDPE) composite films," *Carbon*; **41**:2779–27785 (2003).

[71] Broda J. "Polymorphism in polypropylene fibers," *Journal of Applied Polymer Science*; **89**:3364–70 (2003).

[72] Gong X, Liu J, Baskaran S, Voise RD, Young JS. "Surfactant-assisted processing of carbon nanotube/polymer composites," *Chemistry of Materials*; **12**:1049–52 (2000).

[73] Schadler L, Giannaris S, Ajayan P. "Load transfer in carbon nanotube epoxy composites," *Applied Physics Letters*; **73**:3842–4 (1998).

[74] Dror Y, Salalha W, Khalfin RL, Cohen Y, Yarin AL, Zussman E. "Carbon nanotubes embedded in oriented polymer nanofibers by electrospinning," *Langmuir*; **19**:7012–20 (2003).

[75] Chen GZ, Shaffer MSP, Coleby D, Dixon G, Zhou W, Fray DJ, Windle AH. "Carbon nanotube and polypyrrole composites: coating and doping," *Advanced Materials*; **12**:522–6 (2000).

[76] Bin Y, Kitanaka M, Zhu D, Matsuo M. "Development of highly oriented polyethylene filled with aligned carbon nanotubes by gelation/crystallization from solutions," *Macromolecules*; **36**:6213–9 (2003).

[77] Tang X, Alavi S. "Recent advances in starch, polyvinyl alcohol based polymer blends, nanocomposites and their biodegradability," *Carbohydrate Polymers*; **85**:7–16 (2011).

[78] Godovsky D. "Device applications of polymer-nanocomposites," *Advances in Polymer Science*; **153**:163–205 (2000).

[79] Kim JY, Kim M, Kim HM, Joo J, Choi JH. "Electrical and optical studies of organic light emitting devices using SWCNTs-polymer nanocomposites," *Optical Materials*; **21**:147–51 (2003).

[80] Jung YJ, Kar S, Talapatra S, Soldano C, Viswanathan G, Li X, Yao Z, Ou FS, Avadhanula A, Vajtai R. "Aligned carbon nanotube-polymer hybrid architectures for diverse flexible electronic applications," *Nano Letters*; **6**:413–8 (2006).

[81] Widmer R, Oswald-Krapf H, Sinha-Khetriwal D, Schnellmann M, Boni H. "Global perspectives on e-waste," *Environmental Impact Assessment Review*; **25**:436–58 (2005).

[82] Nogi M, Yano H. "Transparent nanocomposites based on cellulose produced by bacteria offer potential innovation in the electronics device industry," *Advanced Materials*; **20**:1849–52 (2008).

[83] Huang JC. "EMI shielding plastics: a review," *Advances in Polymer Technology*; **14**:137–50 (1995).

[84] El-Shishtawy RM, Salam MA, Gabal M, Asiri AM. "Preparation, characterization and electromagnetic properties of polyaniline/carbon nanotubes/nickel ferrite nanocomposites," *Polymer Composites*; **33**:532–9 (2012).

[85] Ratner BD, Hoffman AS, Schoeri FJ, Lemons JE, editors. *Biomaterials Science: An Introduction to Materials in Medicine*. London: Elsevier Academic Press; p. 864 (2004).

[86] Bhatia SK, Kurian JV. "Biological characterization of Sorona polymer from corn-derived 1,3-propanediol," *Biotechnology Letters*; **30**:619–23 (2008).

[87] Giavaresi G, Fini M, Salvage J, Nicoli Aldini N, Giardino R, Ambrosio L, Nicolais L, Santin M. "Bone regeneration potential of a soybean-based filler: experimental study in a rabbit cancellous bone defects," *Journal of Materials Science: Materials in Medicine*; **21**:615–26 (2010).

2

SURFACE MODIFICATION OF NATURAL FIBERS USING PLASMA TREATMENT

Danmei Sun
School of Textiles and Design, Heriot-Watt University, Netherdale, Galashiels, UK

Biodegradable Green Composites, First Edition. Edited by Susheel Kalia.

2.1 INTRODUCTION

2.1.1 Natural Fiber Materials and their Properties

Natural fibers can be classified into two categories: vegetable (cellulose) and animal (protein) fibers. They have been noted for their renewable nature compared to synthetics, lack of irritation to the skin, and biodegradable nature. Natural fiber products are in consumer demand due to their physical properties such as high moisture regain in comparison to synthetic fibers, warmth and comfort of wool, softness of cotton, and general comfort in daily wear, making natural fibers the principal clothing fiber of the world.

There are various manufacturing processes between fibers and a ready-to-use product. Figure 2.1 shows the processes needed from fiber to finished fabric and factors affecting properties of the product for each process.

Among many of the properties, the surfaces of fiber, yarn, and fabric are very important; they affect the textile manufacturing processes and product properties, such as yarn spinning and yarn strength, coating processes and adhesive behavior of the coated layer, dyeing process, and dyed fabric properties.

2.1.2 Conventional Modification Methods and Drawbacks

Techniques have been developed to modify the surface of textiles for different purposes conventionally. Mercerization is a technique used to modify cotton surface luster. Cotton fabric is treated with strong caustic soda solution under tension at low temperature for 1–3 minutes, followed by washing while still under tension. If the cotton fabric is held under tension in the caustic solution, the fabric will maintain its original dimensions; fibers become more rounded in the cross section, which reflects light to

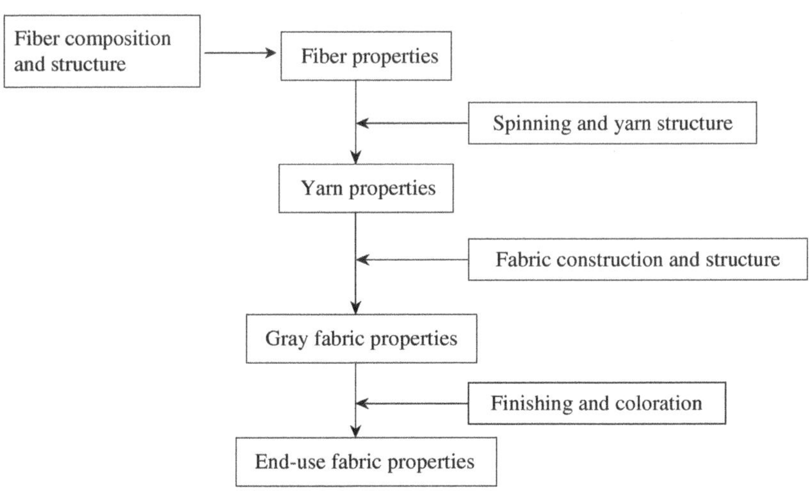

FIGURE 2.1 Relationship among fiber, structure of yarn and fabric, processing, and properties.

produce improved luster of the mercerized fabric. The lumen, a central canal running along the fiber axis, is fairly large in the unmercerized fiber and, having a different refractive index from cellulose, also scatters light at the lumen–cellulose interface. The lumen almost disappears in mercerizing and the fiber convolutions diminish. The cross section of the fiber is changed from approximately kidney shaped to circular. Mercerized cotton also undergoes a change in crystal structure as well as a reduction in the amount of crystalline material present. This relieves some of the stresses and increases the strength of the weak points in the fiber. The following processes have been used for dyeing and finishing of textiles traditionally: desizing using enzyme to remove the starch added in the warp preparation operation, scouring using detergent to remove all extraneous matter, such as grease, wax, spinning lubricant from the textile substrate, bleaching process using oxidizer to remove natural colorant, waterborne stain, and oil-borne soil.

In recent years, effluent treatment has become a major cost contributor driven by environmental legislation and the consumer. Legislation to tighten aqueous-based emissions is expected to increase; therefore there is a need to develop environmentally safe, effluent-free processes with reduced costs. There are also a number of problems associated with dyeing and finishing wool and cotton fabrics. The scales of wool fibers produce a surface barrier to finishing and dyestuff in wool processing. Another problem is that wool grease and cotton wax occur naturally in raw wool and cotton and other oils added to assist textile processing also affect the uptake of the finishing and dyeing solution of fabric. All these result in long processing time and high energy consumption when finishing and dyeing wool and cotton fabrics [1]. Environmental legislation now requires the textile industry to find alternatives to chemically intense methods in order to protect the environment.

2.1.3 Plasma Environment and the Advantages of Plasma Surface Modification

Plasma is often described as "the fourth state of matter," coined by Sir William Crookes in 1879. Langmuir and collaborators were the first to study phenomena in plasma in the early 1920s while working on the development of vacuum tubes for large currents, and it was Langmuir who in 1929 used the term "plasma" for the first time to describe ionized gases [2]. A specific characteristic of plasma is the visible glow discharge with colors ranging from blue-white to dark purple depending on the type of gas. Plasma is a mixture of electrons, negatively and positively charged particles, neutral atoms, and molecules [3]. Plasma can be divided into three major groups based on the mode of consumption of the gas used to create plasma, according to Yasuda [4]: (i) chemically nonreactive plasmas, which are mainly those of monatomic inert gases. Ar plasma can ionize other molecules or sputter materials but is not consumed in chemical reactions. (ii) Chemically reactive plasma, which are inorganic and organic molecular gases such as O_2, N_2, and CF_4, which are chemically reactive but do not form polymeric deposits in their pure gas plasma. (iii) Polymer-forming plasma. This type of plasma is obviously reactive but forms a polymeric solid deposit by themselves. Many organic and inorganic vapors are in the polymer-forming group. Low-pressure plasma, cold plasma, nonequilibrium plasma, and glow discharge are synonymously used terms to designate the same type of process. The technologies using these plasma-assisted processes are generally referred to as plasma processing and include such diverse approaches as plasma-assisted chemical vapor deposition, plasma-enhanced chemical vapor deposition, ionitriding, and plasma etching.

There is greater interest in plasma treatment of textiles [5]. Low-temperature plasma (LTP) has been used in textile dry finishing in treating different type of textile materials such as cotton, linen, wool, silk, and synthetic textile materials. In order to improve the efficiency of finishing and dyeing processes, plasma treatment has been employed as an environmentally friendly alternative to wet finishing of textiles to modify the surface properties of textiles. Compared to conventional finishing processes, the use of plasma treatment has the following advantages.

- It is a dry modification process, no further drying is required after the process.
- The dry production process is more reliable and safe.
- The chemical consumption of plasma treatment is very low.
- It does not destroy the bulk properties of the treated materials.
- Materials that are difficult to modify can be easily changed with plasma treatment.
- It produces no waste and the processing time is very short.

2.2 MECHANISMS OF PLASMA TREATMENT AND TYPES OF PLASMA MACHINES

2.2.1 Principle of Plasma Surface Modification

The plasma atmosphere consists of free electrons, radicals, ions, atoms, molecules, and different excited particles depending on the plasma gas used. Figure 2.2 describes the principle of plasma treatment. Free electrons gain energy from the imposed radio-frequency (RF) electric field, colliding with neutral gas molecules and transferring energy dissociating the molecules to form numerous reactive species. The primary process in the generation of activated species, especially ions, involves electron impact and photoionization [6]. Photoionization occurs when a photon is emitted, due to the energy falling of an electron from a higher level to a lower one, hitting a gas molecule with high enough energy [7]. The primary ion productions are shown as:

$$\mathrm{e} + \mathrm{M} \rightarrow \mathrm{M}^{+} + 2\mathrm{e}^{-} \tag{2.1}$$

$$h\gamma + \mathrm{M} \rightarrow \mathrm{M}^{+} + \mathrm{e}^{-} \tag{2.2}$$

where e is electron, M is plasma gas molecule, and $h\gamma$ is the energy of photons.

Ionization also takes place via collisions of metastable species. Free radicals may also be generated by electron impact, thermal effect, and photolysis. The impact with a monomer can lead to its excitation and dissociation, generation of free radicals.

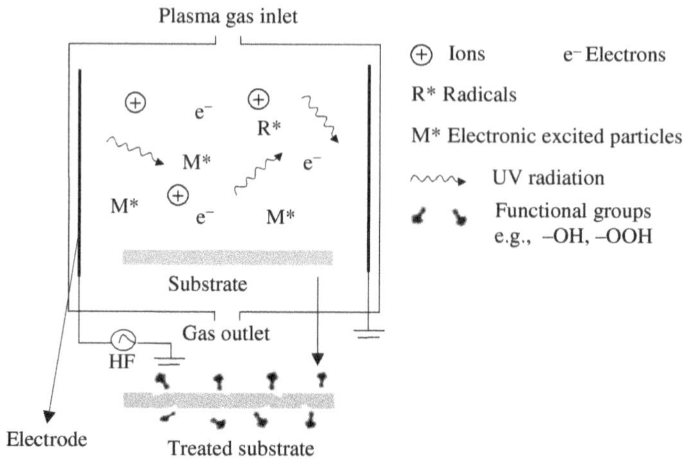

FIGURE 2.2 Mechanism of the surface treatment of substrate in a europlasma reactor.

It is the interaction of these excited species with solid surfaces placed in the plasma reactors, which results in the chemical and physical modification of the material surface. All of the active species react with the substrate surface. This creates chemical functionality on the substrate surface. The by-products of these reactions are readily removed by the vacuum system. Cleaning, modification, or coating occurs depending on the used plasma gases and treatment parameters. The formed reactive particles react in a direct way with the surface without damaging the bulk properties of the treated substrates. In fact the surface modification is limited to the outermost 10–1000Å of the substrate according to requirements the materials to be processed with the plasma.

2.2.2 Interactive Mechanisms between Plasma and Substrates

Ablation of materials by plasma can occur by two principal processes: one is physical sputtering, and the other is chemical etching. The sputtering of materials by chemically nonreactive plasma, such as inert gas plasma, is the typical example of physical sputtering. The physical sputtering process is a knock-on process by ions with high energy (more than a few tens of eV). Chemical etching occurs in chemically reactive types of plasma. This type of plasma gas includes inorganic and organic molecular gases, such as O_2, N_2, and CH_4, which are chemically reactive but do not form polymeric deposition in their pure gas plasma. During etching reaction, weight loss of the substrate occurs, and the topmost layer of the substrate is stripped off. The rate of weight loss is strongly dependent on the nature of the substrate, as well as the energy level of the plasma. Such weight loss is restricted at the topmost layer of the substrate, and at the inner layer, the weight loss process scarcely occurs. Substrate subjected to the plasma etching process possesses similar chemical and physical properties to the original substrate. The elemental composition, chemical structure, the degree of polymerization, and the crystallinity for the treated substrate are scarcely altered and are similar to those for the original substrate. The weight loss in the etching process is mainly due to bond scission of polymers and to reactions of the radicals generated in the polymer chains of the substrate by the plasma exposure [8]. Usually, etching is observed on ion-bombarded surfaces [9]. Ablation also causes changes in the surface morphology. The ionization reaction and dissociation of oxygen gas in plasma environment can be described in Equations 2.3 and 2.4:

$$e + O_2 \rightarrow O_2^- \tag{2.3}$$

$$e + O_2 \rightarrow O + O^- \tag{2.4}$$

Plasma polymerization occurs when an organic compound, a monomer, is introduced into plasma; the monomer gains energy from the plasma through inelastic collision and is activated and fragmented into activated small molecules. The small molecules are recombined to form a large molecule. Finally, polymers are deposited at the surface of substrates. The process is called plasma polymerization, and the product formed by plasma polymerization is termed as plasma polymer [10]. In many cases, polymers formed by plasma polymerization show distinguished chemical composition and chemical and physical properties from those formed by conventional polymerization. According to a US patent [11], hexafluoroethane undergoes bond cleavage to generate trifluoromethyl radical species when exposed to cold plasma conditions. The enhanced lubricity was claimed by hexafluoroethane-treated organic substrates in a trifluoromethyl reactive species:

$$R - CH_2. +. CF_3 \rightarrow R - CH_2 - CF_3 \tag{2.5}$$

$$R - CH_3 + CF_3. \rightarrow RCH_2. + HCF_3 \tag{2.6}$$

Once the polymers were deposited, they are constantly irradiated by the plasma until completing the polymerization process. In such a condition, even the deposited polymers are subjected to an interaction of the plasma. As a result, the polymers are degraded somewhat. More precisely, the polymers taken out after completing the plasma polymerization process are different in chemical structures than those deposited immediately. Therefore, polymerization and ablation of deposited polymers occur simultaneously. This is the conception of competitive ablation and polymerization (CAP) mechanism, which was proposed by Yasuda [4]. Plasma polymerization is in the balance of the polymer-forming process and ablation process, and the balance shifted to the polymer-forming process or the ablation process by changing the discharge power.

Plasma implantation is the treatment technique by introducing oxygen-containing functional groups on the treated substrate surface; it is one of the most useful and effective processes for surface modification. In general, oxygen plasma is used, but plasmas of other compounds involving carbon dioxide, carbon monoxide, nitrogen dioxide, and nitric oxide make the polymer surface hydrophilic. Gas molecules such as oxygen and nitrogen are activated by plasma. The activated oxygen and nitrogen species interact with polymer surfaces, and then special functions such as hydroxyl, carbonyl, carboxyl, amino, and amido groups are formed at the surface of the polymers. As a result, the implantation reactions lead to large changes in the surface properties of the polymer, for example, from hydrophobic to hydrophilic. Radical species, rather than ionic species, in the plasma may contribute mainly to the implantation reaction. Besides oxygen functionalities, chlorine functionalities, such as CCl_4, can also contribute to an increase in hydrophilicity [6]. In the implantation process, hydrogen will first be abstracted from polymer chains to create radicals at the midpoint of polymer chains, and the polymer radicals will combine with simple species.

Plasma-aided graft copolymerization can occur either by the creation of active species on the polymer surface followed by contact with monomer or direct grafting of the polymer with monomers under "monomer" plasma conditions [12]. In the former case, free radicals are formed on the polymer surface as a result of plasma treatment. These radicals can either directly initiate grafting or be converted into peroxide or hydroperoxides by the inclusion of an oxidative gas. These activated peroxides will also initiate grafting in the presence of the monomer species [13]. The latter case involves a combined plasma and monomer exposure in one step by the use of gaseous monomers in the working gas mixture [14]. Both of these techniques have shown great advantages over conventional grafting by offering a large range of chemical compounds to be used as monomers, varying thickness of monomer layers, and limited destruction [15].

Very rare details of reaction and experimental procedure of plasma treatment process have been given in published papers. This is due to the high dependence of the treatment result on reactor, uncertain operating parameters, and complexity of the plasma phenomena. The study of plasma reaction mechanisms is rather difficult. The chemistry involved is very complex, and only in rare cases is it possible to obtain information on reaction intermediates. Changes in treatment parameters may change the plasma reaction.

The results of plasma experiments strongly depend on the experimental parameters such as pressure, electrical field strength, gas velocity, and reaction chamber dimension. Since pressure, field strength, chamber diameter, and flow rate influence the average electron energy, the degree of ionization, gas temperature, and the rate of conversion, it is not possible to give accurate values of optimal working conditions valid for a variety of reactions [16].

It has to be noted that exposure to plasma can induce damage to the surface layers of sensitive materials. The main cause of this damage is the bombardment of the surface with ions and energetic neutral plasma particles; electron and UV bombardment might also contribute to the damage [8].

2.2.3 Types of Plasma Treatment Systems

There are two types of plasma treatment machines in general, namely, atmospheric pressure and low-pressure vacuum treatment machines. The former is to treat substrates under atmospheric pressure in an open environment, the latter in closed environment. Both types of plasma machines have been used to modify various types of textiles for different applications. Figure 2.3 shows the schematic diagram for the

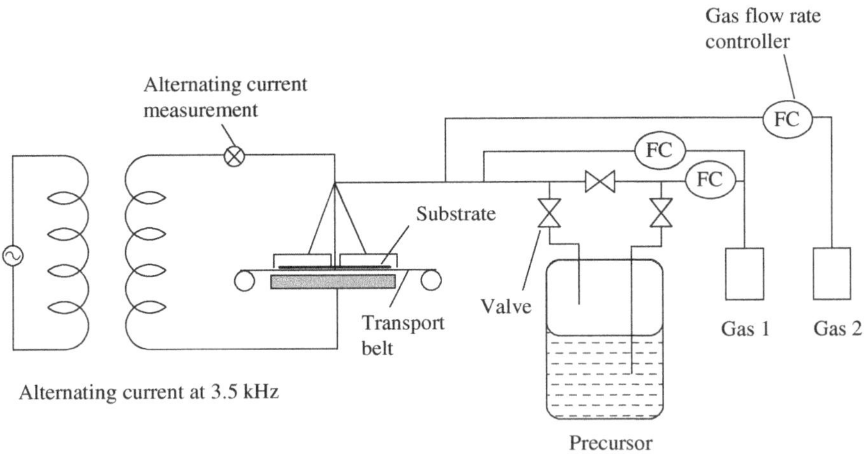

FIGURE 2.3 Schematic illustration of the alternating current plasma treatment system [18]. *Source*: Sun and Chen [20]. Reprinted with permission from Sage publications.

setup of an atmospheric pressure plasma-enhanced vapor deposition system, which was developed by Sheel and his colleagues used in the field of APCVD for application in environmental, construction, healthcare, and biotechnology sectors [17–19]. The plasma machine consists of five main components: reaction chamber, audio rate frequency generator, power distribution rack, measuring equipment, and gas supplier. The plasma reaction chamber containing cathode and anode is attached to an audio rate frequency generator. The samples to be treated are inserted between the two electrodes within the plasma reaction chamber or positioned on the sample holder statically after being wrapped on a thin and smooth tape, which allow the substrates to be transported from one side of the tape (untreated) to the other (treated). By opening and closing the valves and gas flow controllers, different gas and precursor chemical can be selected and introduced into plasma reactor. Sun and Chen [20] and Chu et al. [21] employed the plasma machine in treating Aramid fabrics and yarns to modify the surface for improved ballistic performance of the body armor material.

Gas at a metered rate is allowed to enter the chamber. Plasma will be generated once a low voltage between the electrodes is applied in the audio-frequency range. The whole process is controlled via a sophisticated system.

Sigma Technologies, an American company, produces atmospheric plasma equipment for treating films, fabrics, metals, and 3D parts. Enercon is another American company that developed commercial atmospheric plasma, which is highly effective at treating polymer films, nonwovens, textiles, foams, and other materials, which are processed in a roll-to-roll web format. Research work has been reported in using atmospheric pressure plasma machine made by Enercon for modification of various different textile materials [5, 22–24].

The other is vacuum plasma system. An example of this type of system is Europlasma. Sun and Stylios used a Europlasm 600 for their research on natural

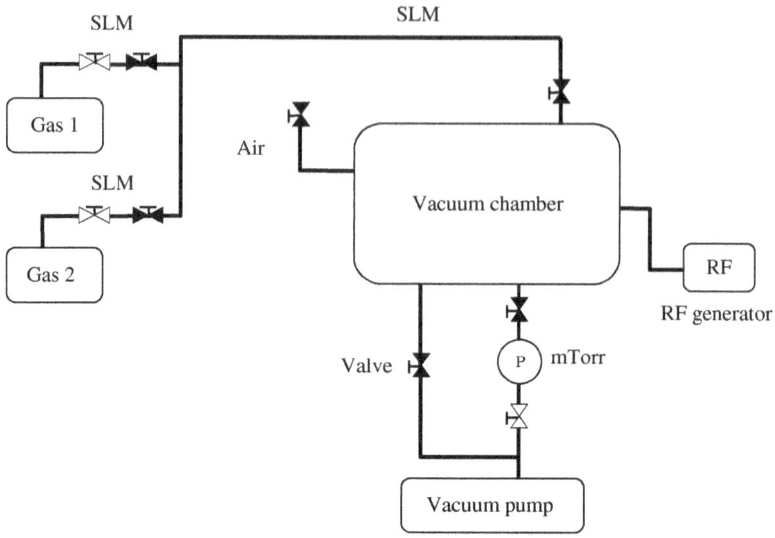

FIGURE 2.4 Working theory of vacuum plasma machine Europlasma CD600.

fibers [5, 25, 26]; its working theory is shown in Figure 2.4. It consists of six main parts: vacuum pump, high-frequency generator, power distribution rack, reaction chamber, PC rack, and measuring system. The reaction chamber containing positive and negative electrodes attached to an RF generator is evacuated from air and other gases by a pump. The substrates to be treated are placed between the positive and ground electrodes within the vacuum chamber. Once a certain pressure (base pressure) is reached, a gas inlet valve is opened and gas at a metered rate is allowed to enter the chamber. Once the flow rate and pressure in the chamber have stabilized, a high voltage between the ground and the positive electrode is applied in the RF range. The voltage creates matter composed of electrons, negatively and positively charged particles, neutral atoms, and molecules. The whole process is controlled via a sophisticated system of valves, pumps, and measuring equipment linked to a computer. At the bottom of the screen there is a menu bar with options to Logon menu, Start, Abort, Set points, and Report. The plasma process can be controlled manually or through a computer-controlled recipe.

Before treating the samples, the diagnostic menu was used to find out the parameters for proper plasma and then select recipe menu to create new recipe or change recipe if there is already one available to fill in the plasma parameters for the specimen, which will be treated. It includes the kind of gas or gases, reactor chamber base pressure (mTorr), treatment power W (Watt), treatment time (minutes), and reactor pressure (mTorr) or gas flow rate F (standard liter per minute (SLM)). Gas flow rate was used as one of the plasma parameters as the reactor pressure is not constant during specimen processing. Before the process started, the chamber door had to be closed. Air and old gases had to be pumped out by the vacuum pump; thus almost a vacuum level was created in the reaction chamber. When the base pressure or the evacuation time was reached, the plasma gas was introduced into the reaction chamber at the given parameters, and the RF started. RF energy is supplied into the process when the set power level is achieved, and then the plasma is produced. When the plasma treatment is finished, the old gas is cleaned out of the vacuum chamber. The vacuum relief breaks the vacuum stage by bringing air into the chamber to reach atmospheric pressure, and thus the chamber door could be opened and the treated sample could be taken out of the treatment chamber.

2.3 EFFECTS AND APPLICATIONS OF PLASMA TREATMENT

All finishing processes are designed to increase the attractiveness or serviceability of the textile product. Different fabric finishing methods can produce different fabric mechanical properties relating to aesthetic, garment manufacturing, and special application [27]. Surface properties sometimes play a more important role in the function of polymer materials than do the bulk of the polymer materials. Many functions of the polymer materials, for example, adhesion, printability, dyeability, antifogging, etc., are related to the topmost layer of the materials [28].

LTP treatment has been used to treat fabrics to improve shrinkage, wettability, and dyeability and reduce fabric surface resistivity [29, 30]. Theoretically, plasma treatment only affects the topmost surface of the substrate without having any effect on their internal structure and bulk characteristics. According to Kawabata [31], the surface property of the fabric relates to physical mechanical properties indirectly. Not only the physical properties but also the treated substrate surface chemical properties can be altered, depending on the used plasma gas(or gases) and treatment parameters.

2.3.1 Surface Morphology and Chemical Composition Change

The wettability of a polymer surface, defined as the contact angle between a drop of liquid and the surface [32], is an important characteristic. Plasma processes are available that can increase surface wettability. Changes in the wettability of a polymer are produced by adding oxygen- and nonoxygen-containing functional groups to the polymer surface. Oxygen-containing functional groups (carbonyl —C=O, hydroperoxide —OOH, hydroxyl —OH) produce larger wettability changes than those produced by nonoxygen-containing functional groups. Surface morphology changes can also cause wettability changes, but the wettability changes produced by this modification are small and, in general, are unimportant [33]. In general, oxygen plasma simultaneously produces wettability and molecular weight changes.

Sun and Stylios [5] reported plasma treatment on cotton and wool fabrics employing two different types of gases: polymerizing plasma gas hexafluoroethane (C_2F_6) and nonpolymerizing reactive plasma gas oxygen (O_2). The research showed that not only the topography but also the chemical composition of the surface of the fibers in the fabrics has been modified. Figure 2.5 shows the images from SEM examination results. It shows smoother surface for the C_2F_6-treated wool and cotton in comparison

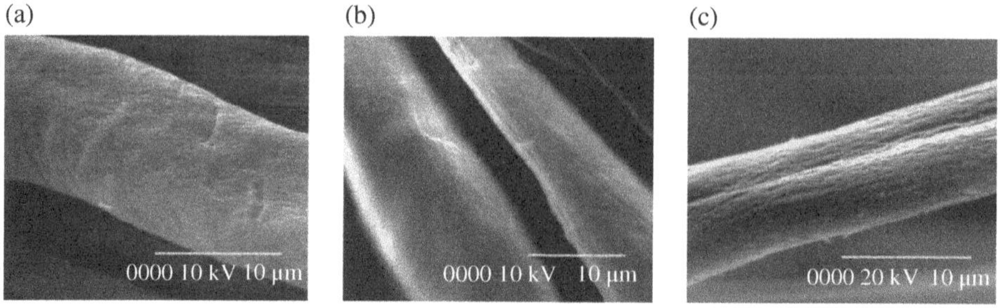

FIGURE 2.5 SEM images of (a) untreated cotton, (b) C_2F_6 plasma-treated cotton (0.10 SLM, 80 W), and (C) O_2 plasma-treated cotton (0.3 LSM, 300 W) [5]. *Source*: Sun and Stylios [5]. Reprinted with permission from Elsevier. Copyright 2014.

to the untreated fiber surface. On the other hand, holes are visible on the O_2 plasma-treated wool and cotton fiber surface. It was explained that the holes on the O_2 plasma-treated fiber surface can be attributed to the ablation effect caused by bombardment of O_2 plasma species on the fabric surface. As a result, the fabric surface became rougher. Plasma polymer might be formed on the treated cotton and wool surface in generating a smoother surface after C_2F_6 plasma treatment. Gibeop et al. and many other researchers [34–36] have examined the morphology of plasma-treated fiber surface using SEM technology.

Apart from surface morphological study of plasma treatment, the surface chemical composition of the treated wool and cotton fabrics has also been investigated using X-ray photoelectron spectroscopy (XPS). Figure 2.6 shows the results of cotton fabric treated using O_2 and C_2F_6 plasma. It is obvious that the intensity of the oxygen peaks from the O_2 plasma-treated surfaces is much stronger than untreated surfaces. Babai-Cline [37] gave more detailed analysis on this and indicated that alcohol, carbonyl, and ether groups are presented on the O_2-treated substrate surface. The scans of the C_2F_6-treated cotton fabric surface contain extra F_{1s} peaks compared with the untreated cotton and wool fabric surface. Sun and Stylios also reported that the O/C ratio of the treated samples increased significantly, indicating that oxygen containing group rich surface was produced after O_2 plasma treatment. The decrease in binding energy of plasma-treated cotton fabric indicates that plasma modification made some chain scission in the treated cotton molecules of the fibers. According to Inagaki [38], high-energy radiation of plasma treatment causes formation of radicals, and the formed radicals undergo chain scission.

2.3.2 Improved Hydrophilicity and Efficiency in Aqueous Processes

Further to the surface morphology and chemical change by plasma treatment, Sun and Stylios [26] studied the contact angles, wicking properties, scourability, and dyeability of wool and cotton fabrics affected by LTP treatments. It was reported that the

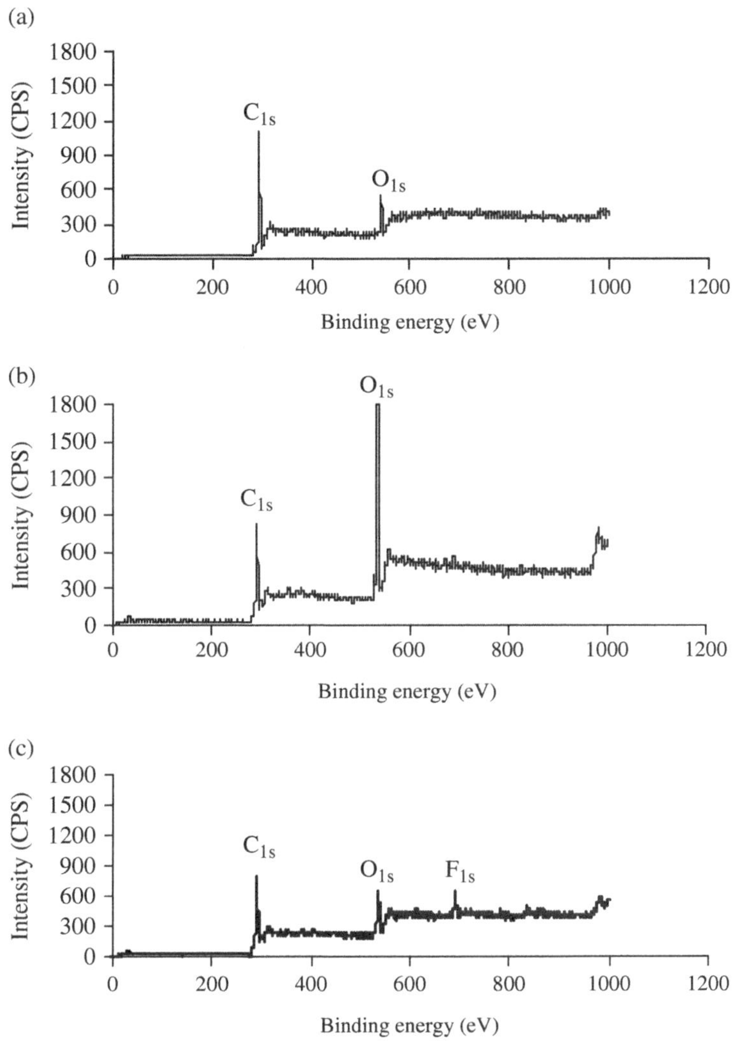

FIGURE 2.6 XPS images: overall scan of cotton fabric specimens: (a) untreated cotton fabric surface, (b) O_2 plasma treated (300 W, 0.3 SLM, 1 min), and (c) C_2F_6 treated (80 W, 0.1 SLM, 1 min) [5]. *Source*: Sun and Stylios [5]. Reprinted with permission from Elsevier.

contact angle decreased considerably after the O_2 plasma treatment compared with the untreated wool and cotton samples by using both the scouring and dyeing liquids.

Figure 2.7 shows the configuration of the drops on the untreated and O_2 plasma-treated cotton and wool fabric surface after the drops of dyeing solution were dropped on the fabric surface immediately. The drops were smaller and round above the fabric surface on the untreated cotton and wool fabric. Drops on the O_2 plasma-treated fabric surface spread out immediately after the drops were dropped on the fabric surface,

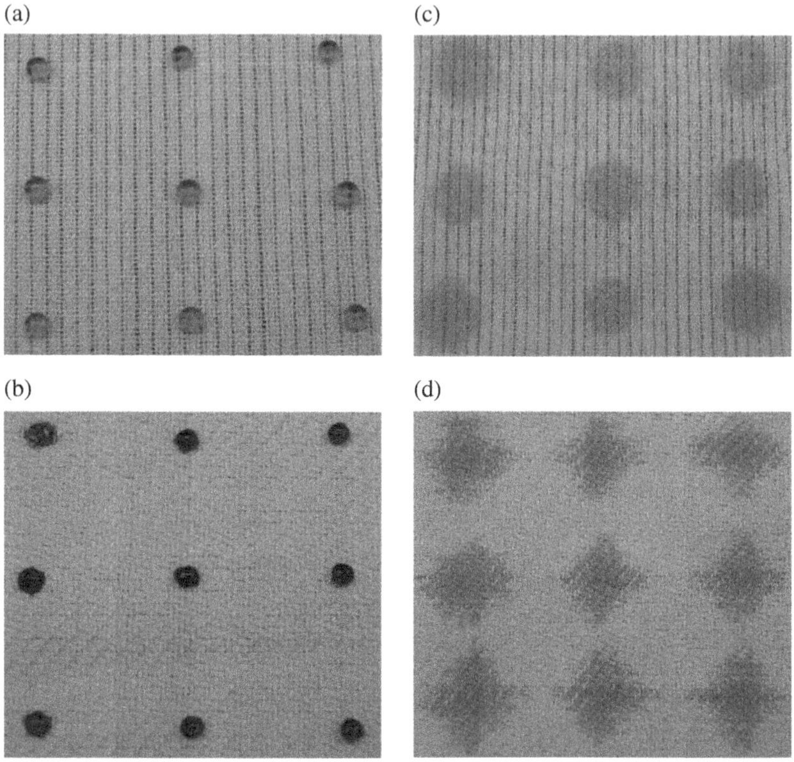

FIGURE 2.7 Comparison of the droplets on the untreated (a: cotton, b: wool) and O_2 plasma-treated (c: cotton, d: wool) fabric surface [1]. (*See insert for color representation of the figure.*)

that is, O_2 plasma-treated cotton and wool fabrics absorbed the drop liquid within a very short time. It was reported by Sun [1] that after 20 minutes, the shape of the liquid drops on the plasma-treated cotton and wool fabric surface was almost the same as that of the drops obtained immediately after the drops dropped on the fabric surface. This means that the dyeing liquids can pass through the surface fibers of the fabric and penetrate into internal fibers in a very short time. Similar phenomenon was observed for the drops from the scouring bath on both wool and cotton fabric surface.

The wickability of wool and cotton fabrics treated by O_2 plasma increased dramatically in both the scouring and dyeing baths, shown in Figure 2.8 for the results of cotton fabric. A relatively short time is required for the scouring and dyeing liquids to reach the same height for O_2 plasma-treated fabric comparing to the untreated. The dye solution reached a height of 100 mm in 7 minutes for the plasma-treated cotton samples compared to a height of 30 mm reached in 30 minutes for the untreated cotton in the dyeing bath. It took only 10 minutes for the plasma-treated cotton in comparison to 36 minutes for the untreated cotton to achieve a height of 100 mm in the scouring bath. It is believed that oxygen-containing polar groups and the holes on the wool and

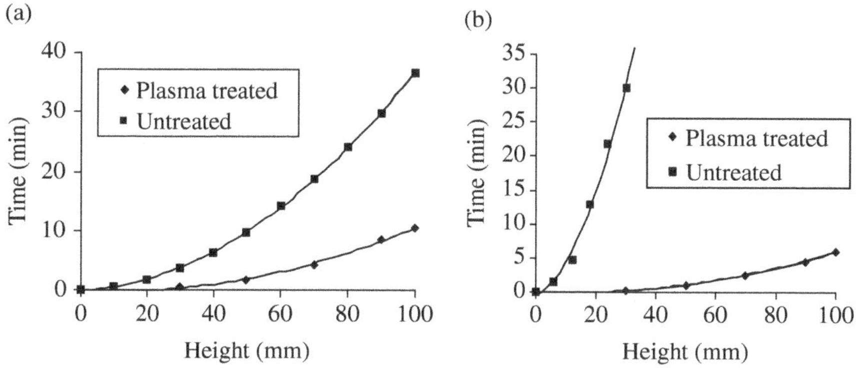

FIGURE 2.8 Comparison of wicking properties of O_2 plasma-treated and untreated cotton fabric in (a) scouring and (b) dyeing baths, respectively [25]. *Source*: Sun and Stylios [25]. Reprinted with permission from Sage Publications.

cotton fiber surface produced by O_2 plasma treatment increased the wettability of the O_2 plasma-treated fabrics. It is the increased wettability of fabrics caused by the changed surface properties during the plasma treatment results in the increased wicking rate of O_2 plasma-treated wool and cotton fabrics [24]. Moreover, according to K. K. Wong [39], the effective pore size presented in the plasma-treated fabric structure may be increased and adversely reduce capillary pressure, thus increasing wickability.

The dyeability of plasma treatment on cotton and wool fabrics revealed that O_2 plasma-treated fabrics have improved dyeability, indicated by the dyeing rate of the plasma-treated fabrics higher compared to the untreated, especially in the initial 30 minutes of dyeing wool fabrics using an acid dye, shown in Figure 2.9. It also shows that the time required to reach final exhaustion was shorter for the plasma-treated fabrics than the untreated fabrics. One reason behind the change is the holes produced by the nonpolymerizing reactive O_2 plasma gas. These holes provided new pathways for the dye to enter the fiber and hence increased the dyeing rate. The other and the more important reason for the enhanced dyeing rate of O_2 plasma treatment is the oxygen-containing polar groups generated on the O_2 plasma-treated fabric surface to allow the treated fabric absorbing dye solution much faster than the untreated fabric.

2.3.3 Improved Hydrophobicity

Kamlangkla et al. used sulfur hexafluoride (SF_6) that is a readily available and inexpensive fluoro-containing gas to treat cotton fabrics [40]. It was found that the hydrophobicity has been significantly improved, indicated by the massively increased contact angle of the droplets on SF_6-treated cotton fabric surface compared to the untreated cotton fabric. The XPS chemical analysis revealed that only the F• is the active species that graft onto the fabric surface producing C—F bonds after abstraction

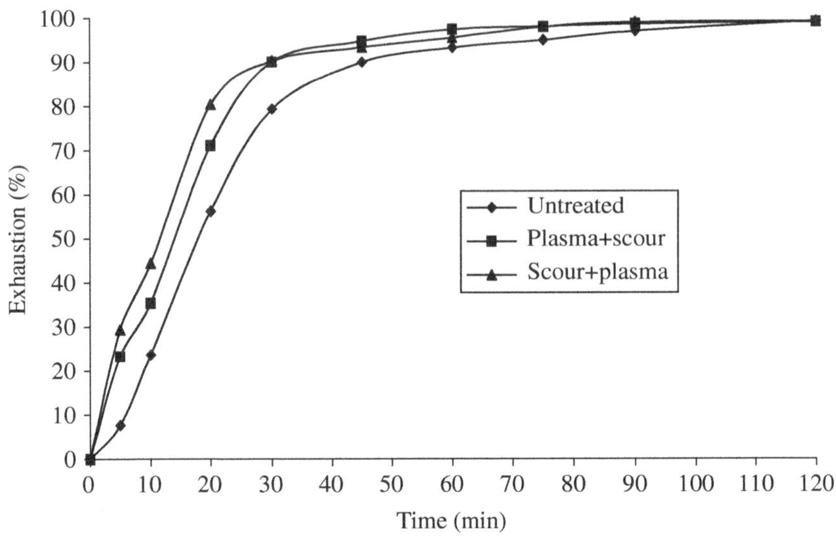

FIGURE 2.9 Exhaustion rate of dyestuff (1% Nelon Red P) on plasma-treated and untreated wool [24]. *Source*: Sun and Stylios [25]. Reprinted with permission from Sage Publications.

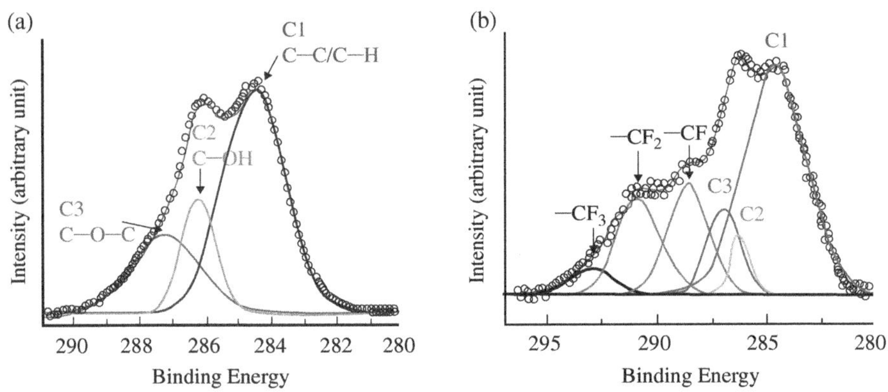

FIGURE 2.10 C1s XPS spectra of fabric surface: (a) untreated cotton and the molecular structure of cellulose; (b) cotton treated at the pressure of 0.05 Torr for 1 min [40]. *Source*: Kamlangkla et al. [40]. Reprinted with permission from Elsevier. Copyright 2014. (*See insert for color representation of the figure.*)

of H and O atoms by other plasma species (ions, radicals, electrons, and photons) [41]. Figure 2.10 shows that there are additional peaks becoming apparent at binding energies of 288, 291, and 293.5 eV. These signals were believed to be attributed to CF, CF_2, and CF_3 groups, respectively, after SF_6 plasma treatment. The research also concluded that the treatment pressure and treatment time are key parameters, and a compromise needs to be taken into consideration between the tensile strength and hydrophobic durability in terms of ageing and washing cycles. The same plasma gas was used by Selli et al. [42]; the hydrophobic properties after ageing of the treated silk fabric were studied.

It has been reported by Sun and Stylios that the surface smoothness and hydrophobicity of wool and cotton fabric have been improved after C_2F_6 plasma treatment [25]. It was believed that the increased contact angle of C_2F_6 plasma-treated samples was attributed to the fluorine-containing surface caused by the C_2F_6 plasma treatment, which produced a plasma polymer on the treated wool and cotton fabric surface. It was further suggested that fabrics that owned a smoother and hydrophobic surface can be used for antibacterial purposes.

2.3.4 Mechanical Properties Affected by Plasma Treatment

Sun and Stylios [5] reported that plasma treatment changed the surface properties and surface roughness leading to an increase in low-stress mechanical properties such as bending and shear rigidities. However the change in fabric hand value was not significant between treated wool and cotton fabrics and their untreated counterparts. Research made by Sinha and Panigrahi [43] revealed that the flexural strength of the composite prepared with fiber treated by plasma showed approximately 14% increase compared to raw fiber composites. Zhou et al. [44] investigated plasma treatment on ramie fibers, and the research showed up to 50% increase of the interfacial shear strength of the ramie fibers to polypropylene (PP) compared to the untreated due to the increased fiber surface roughness caused by plasma etching. Kim et al. [45] treated jute fiber using polymerizing plasma gas with the aid of a carrier; the tensile strength of plasma-treated jute/PP composite increased by 114% in comparison to the composite made by untreated jute and PP. Exposure of cotton yarn to the plasma of an electrical glow discharge resulted in an increase in breaking strength of the yarn [46]. Sobczyk-Guzendaa et al. studied plasma treatment on banana fibers [47]. The results show that the tensile strength increased dramatically compared to the untreated. The elastic modulus has also been increased by plasma treatment, meaning that the fiber tends to be stiffer. The elongation at breakage was decreased after plasma treatment. Ar plasma was introduced to a new process assigned as the pad-dry-plasma-cure process by Chen et al. [48]. The research showed that the addition of Ar plasma treatment could increase the cross-linking effect between cross-linking agent and cellulose molecules. The tensile strength retention of pad-dry-plasma-cure finished fabrics was found higher than that of traditional pad-dry-cure finished fabrics. Both maximum tensile strength and elongation at break were positively affected by the plasma treatment. The mechanical properties of natural fiber materials have been studied by many other researchers [49–52]. Ceria investigated mechanical properties of a wool fabric treated with N_2 using a roll-to-roll atmospheric plasma jet equipment. It was reported that the maximum tensile strength at break in both warp and weft directions of the woven wool fabric has been increased. Although the results show that extensibility was not affected by the plasma treatment along the warp direction of the fabric as a compensation of the feeding system, it was decreased along fabric weft and bias direction, implying an increase in fabric shear rigidity [47].

2.3.5 Medical Applications of Plasma Treatment

Investigations have also been made in the antimicrobial functions of plasma-treated natural fiber materials. Razic et al. [53] investigated low-pressure plasma in combination with silver containing agents to obtain antibacterial activities of cellulose. It was claimed that the antibacterial effects achieved do not lag behind commercial available product. Plasma treatment as a pretreatment method to enhance the absorption of antibacterial agent chitosan was studied by many researchers [54–56]. Plasma treatment added functional chemical groups on the cellulosic fabric surface, resulting in the increased adsorption of chitosan in which the amino groups are responsible for antimicrobial function of the treated fabric. The increased amino groups in the treated fabric surface increased the opportunity that a protonated amino group met the bioplasma of the bacteria contributing to the bacterial reduction probability.

Peršin et al. [57] investigated plasma treatment on cellulosic fabrics for wound dressing materials. Oxygen and ammonia plasma have been employed in their research as one-step and two-step approach. The antimicrobial property of the treated fabrics has been studied and comparison made to untreated fabric sample. It was found that NH_3 plasma treatment as a one-step approach significantly improved the hydrophilicity of the treated fabric but provided no antimicrobial activity on the used bacteria. On the other hand, fabrics treated through two-step approach (O_2 plasma treatment followed by silver chloride surface coating) exhibited superior antimicrobial performance. Such two-step treated fabric was capable of fully destroying all of the tested bacteria.

2.3.6 Plasma-Modified Fibers in Polymer Composites

Fiber-reinforced polymer composites made of natural fibers have attracted attention due to the many advantages such as biodegradable, nontoxic, etc. However, the main disadvantages of natural fibers in composites are the poor compatibility between fiber and matrix and the relative high moisture sorption. Therefore, chemical treatments are considered in modifying the fiber surface properties [58, 59].

Plasma treatment has been used to modify sisal fiber surface for improving the adhesion between sisal fibers as reinforcement and PP as the matrix of the fiber composite [60]. The single fiber pullout test showed that the maximum load needed to pull out fibers is much higher for the plasma treated in comparison to fiber modified by other methods for the same sample dimension. Bozaci et al. [61] modified flax fiber surface using plasma treatment under various treatment conditions to enhance the interfacial adhesion between flax fiber and matrix polyethylene and unsaturated polyester. Both the surface morphology and chemical change of the flax fiber surface were examined after plasma treatment. It was observed that a rough surface was obtained after argon and air plasma treatment. In the meantime, new polar groups such as O—C=O and C=O were found on the treated flax fiber surface. It was found that the interfacial shear strength increased for the composite containing plasma-treated flax. This is believed to be caused mainly by the surface roughness of the plasma surface modification other than the chemical reaction.

Li et al. studied [62] ramie fabrics treated by ethanol followed by an atmospheric helium plasma treatment for improvement in mechanical properties of ramie fabric-reinforced PP composites. It has been found that ethanol treatment alone had almost no influence on the adhesion of ramie fabrics with PP matrices, while ethanol treatment followed by plasma treatment of ramie fabrics provided the fabric composite much better mechanical properties: nearly 39% higher interlaminar shear strength, 28% higher flexural strength, and 20% higher tensile strength. It was argued that this was mainly caused by the hydrophobic surface of the ramie fibers by the reaction of cellulose and ethanol in plasma modification process. Secondly the increased surface roughness of the ramie fibers could also contribute to the improved mechanical properties due to the enhanced mechanical interlocking in the composite interface. It has been emphasized that plasma treatment time had great influence in the degree of the improvement in mechanical properties.

2.3.7 Other Areas of Applications

Shahidi et al. [63] reported that plasma-treated wool fabrics have antifelting behavior. Plasma-treated wool top [26] offers a number of processing benefits at different stages of the wool pipeline including significant improvements in spinning performance. The concept of rough surface produced by nonpolymerizing plasma gas can be used in many applications where a rough surface is needed to improve fiber cohesive force within yarn to achieve improved strength of yarn.

2.4 CONCLUSIONS AND INDUSTRIAL IMPLICATIONS

LTP treatment as an environmentally friendly method has significant impact on natural fibers and fabrics for various applications. The properties of the treated substrates are dependent on the plasma gas (or gases) used and other treatment parameters.

Both rougher and smoother surfaces can be produced using plasma treatment. Not only the surface morphology be changed but also the chemical composition of the treated substrate surface can be changed, by introducing various functional groups. The chemical change of the treated fiber and fabric surface plays a main role in further finishing processes of the textiles. Although the surface property alterations generated by plasma treatment are complex, there are lots of benefits in using plasma technology as a pretreatment to conventional finishing, dyeing, and other manufacturing processes and as an environmentally viable method. It is to be expected that this technology is being used in different branches of the industries and will conquer textile industry in the near future.

REFERENCES

[1] Sun, D., PhD thesis: The influence of conventional and plasma treatments on the handle and related properties of natural fabrics, Heriot-Watt University, Edinburgh, UK (2007).

[2] Langmuir, I., (1929). The interaction of electron and positive ion space charges in cathode sheaths, *Physical Review*, **33** (6), pp. 954–989.

[3] Inagaki, N., *Plasma surface modification and plasma polymerization*, Technomic Publishing Company Inc., New York, p. 1 (1996).

[4] Yasuda, H., *Plasma polymerization*, Academic Press, New York (1985).

[5] Sun, D. and Stylios, G., (2006). Fabric surface properties affected by low temperature plasma treatment, *Journal of Materials Processing Technology*, **173** (2), pp. 172–177.

[6] Sparavigna, A. C. and Wolf, R. A., *Atmospheric plasma treatments in converting and textile Industries*, Lulu Enterprises, Inc., Wisconsin (2008).

[7] Griem, H. R. and Lovberg, R. H., *Methods of experimental physics- plasma physics*, Academic Press Inc, New York (1970).

[8] Inagaki, N., *Plasma surface modification and plasma polymerization*, Technomic Publishing Company, Lancaster (1996).

[9] Grill, A., *Cold plasma in materials fabrication from fundamentals to applications*, IEEE Press, New York (1994).

[10] Grill, A., *Cold plasma in materials fabrication*, The Institute of Electrical and Electronics Engineers, Inc., New York, (1993).

[11] Auerbach, R. A., Cold plasma modification of organic and inorganic surfaces. US Patent, 4,188,426 (1980).

[12] Matthews, S. R., Master's thesis: Plasma aided finishing of textile materials, North Carolina State University, Raleigh (2005).

[13] Jagur-Grodzinski, J., *Heterogenous modification of polymers: matrix and surface reactions*, John Wiley & Sons, Ltd, Chichester, 1997.

[14] Simionescu, C. I. and Denes, F., (1980). The use of plasma-chemistry in the field of synthesis and modification of the natural macromolecular compounds, *Cellulose Chemistry and Technology*, **14**, p. 285.

[15] Simionescu, C. I.; Denes, F.; Macoveanu, M. M., and Negulescu, I., (1984). Surface modification and grafting of natural and synthetic fibers and fabrics under cold plasma conditions, *Makromolekulare Chemie, Supplement*, **8**, p. 17.

[16] Hollahan, J. R., *Techniques and applications of plasma chemistry*, A Wiley-Interscience Publication, New York (1974), p. 65.

[17] Shell, D. W.; Yates, H. M.; Evans, P.; Dagkaldiran, U.; Gordijn, A.; Finger, F.; Remes, Z., and Vanecek, M., (2009). Atmospheric pressure chemical vapour deposition of F doped SnO_2 for optimum performance solar cells, In 7th International Conference on Coatings on Glass and Plastics (ICCG7) Advanced Coatings for Large-Area or High-Volume Products, *Thin Solid Films*, **517** (10), pp. 3061–3065.

[18] Dagkaldiran, Ü.; Gordijn, A.; Finger, F.; Yates, H.M.; Evans, P.; Sheel, D. W.; Remes, Z., and Vanecek, M., (2009). Amorphous silicon solar cells made with SnO_2:F TCO films deposited by atmospheric pressure CVD, *Materials Science and Engineering B: Solid-State Materials for Advanced Technology*, **159**, pp. 6–9.

[19] Hodgkinson, J. L.; Thomson, M.; Cook, I., and Sheel D. W., (2011). Non-thermal atmospheric pressure plasma etching of F: SnO_2 for thin film photovoltaics, *Journal of Nanoscience and Nanotechnology*, **11** (9), pp. 8403–8407.

[20] Sun, D. and Chen, X., (2012). Plasma modification of kevlar fabrics for ballistic application. *Textile Research Journal*, **82**, pp. 1928–1934.

[21] Chu, Y.; Chen, X.; Sheel, D. W., and Hodgkinson, J. L., (2014). Surface modification of aramid fibres by atmospheric pressure plasma-enhanced vapour deposition, *Textile Research Journal*, **84** (12), pp. 1288–1297.

[22] Bhat, N. and Benjamin, Y., (1999). Surface resistivity behaviour of plasma treated and plasma grafted cotton and polyester fabrics, *Textile Research Journal*, **69**, pp. 38–42.

[23] McCord, M. G.; Hwang, Y. J., and Hauser, P. J. (2002). Modifying nylon and polypropylene fabrics with atmospheric pressure plasmas, *Textile Research Journal*, **72**, pp. 491–498.

[24] Sudheer, J.; Rachel, B.; Carol, K.; Vinitkumar, S.; Appachi, A.; Siva, P.; Gopal, C.; Ronald, K.; Rory W., and Seshadri, R. (2012). Atmospheric pressure plasma treatment and breathability of polypropylene nonwoven fabric, *Textile Research Journal*, **42** (4), pp. 501–514.

[25] Sun, D. and Stylios, G. (2005). Investigating the plasma modification of natural fibre fabrics—the effect on fabric surface and mechanical properties, *Textile Research Journal*, **75** (9), pp. 639–645.

[26] Sun, D. and Stylios, G. (2004). Effect of low temperature plasma treatment on the scouring and dyeing of natural fabrics, *Textile Research Journal*, **74** (9), pp. 751–757.

[27] Taylor, M. A., *Technology of textile properties*, Forbes Publications Ltd., London (1990).

[28] Byrne, K. M.; Ryder, A., and Rakowski, W. (1994). Plasma Treatment—A Multipurpose Finish for Wool, Proceedings of Aachen Textile Conference, Aachen, Germany, 10–11 November, pp. 197–211.

[29] Zuchairah, I. M.; Pailthorpe, M. T., and David, S. K. (1997). Effect of glow discharge-polymer treatments on the shrinkage behaviour and physical properties of wool fabric, *Textile Research Journal*, **67**, pp. 69–74.

[30] Kan, C. W.; Chan, K.; Yuen, C. W. M., and Miao, M. H. (1999). Low temperature plasma on wool substrates: the effect of the nature of the gas, *Textile Research Journal*, **69**, pp. 407–416.

[31] Kawabata, S., *The standardization and analysis of hand evaluation*, The Textile Machinery Society of Japan, OSAKA (1980).

[32] Adamson A. W., *Physical chemistry of surfaces*, Interscience, New York, p. 345 (1960).

[33] Grill, A., *Cold plasma in materials fabrication from fundamentals to applications*, IEEE Press, Piscataway, p. 122 (1994).

[34] Gibeop, N.; Lee, D. W.; Prasad, C. V.; Foru, F.; Kim, B. S., and Song, J. I. (2013). Effect of plasma treatment on mechanical properties of jute fibre/poly (lactic acid) biodegradable composites, *Advanced Composite Materials*, **22** (6), pp. 389–399.

[35] Boonla, K. and Saikrasun, S. (2013). Influence of silk surface modification via plasma treatments on adsorption kinetics of lac dyeing on silk, *Textile Research Journal*, **83** (3), pp. 288–297.

[36] Li, Y.; Sun, J.; Cheng, P.; Jiang, Y.; Zhou, Z.; Zhang, Q., and Qiu, Y. (2011). Aging of hydrophobized surface of ramie fibres induced by atmospheric pressure plasma treatment with ethanol pretreatment, *Journal of Adhesion Science and Technology*, **27** (22), pp. 2387–2397.

[37] Babai-Cline, M. and Wightman, J. P. (1995). Surface analysis and peel strength of aged, oxygen-plasma-modified unplasticised poly(vinyl chloride) films, *International Journal of Adhesion and Adhesives*, **15**, pp. 185–190.

[38] Inagaki, N., *Plasma surface modification and plasma polymerisation*, Technomic Publishing Company, Inc., Lancaster (1996).

[39] Wong, K. K. (2001). Wicking properties of linen treated with low temperature plasma, *Textile Research Journal*, **71**, pp. 49–56.

[40] Kamlangkla, K.; Paosawatyanyong, B.; Pavarajarn, V.; Hodak, Jose H., and Hodak, Satreerat K (2010). Mechanical strength and hydrophobicity of cotton fabric after SF6 plasma treatment, *Applied Surface Science*, **256** (20), pp. 5888–5897.

[41] Hodak, S. K.; Supasai, T.; Paosawatyanyong, B.; Kamlangkla, K., and Pavarajarn, V. (2008). Enhancement of the hydrophobicity of silk fabrics by SF6 plasma, *Applied Surface Science*, **254**, pp. 4744–4749.

[42] Selli, E.; Riccardi, C.; Massafra, M. R., and Marcandalli, B. (2001). Surface modifications of silk by cold SF6 plasma treatment, *Macromolecular Chemistry and Physics*, **202**, pp. 1672–1678.

[43] Sinha, E. and Panigrahi, S. (2009). Effect of plasma pretreatment on structural wettability of jute fibre and flexural strength of its composite, *Journal of Composite Materials*, **45**, pp. 1791–1802.

[44] Zhou, Z.; Liu, X.; Hu, B.; Wang, J.; Xin, D.; Wang, Z., and Qiu, Y. (2011). Hydrophobic surface modification of ramie fibres with ethanol pretreatment and atmospheric pressure plasma treatment, *Surface and Coatings Technology*, **205**, pp. 4205–4210.

[45] Kim, B. S.; Nguyen, M. H.; Hwang, B. S., and Lee, S. (2008). Effect of plasma treatment on the mechanical properties of natural fibre/PP composites, in: De Wilde, W.P. and Brebbia, C.A. (Eds.), *High performance structures and materials IV*. WIT transactions on built environment, WIT Press, Southampton, pp. 159–166.

[46] Stone, R. B.; John, R., and Barrett, J. (1962). Study reveals interesting effects of gas plasma radiations on cotton yarn, *Textile Bulletin*, **88**, pp. 65–69.

[47] Anna Sobczyk-Guzendaa, A.; Szymanowskia, H.; Jakubowskia, W., Błasińskab, A.; Kowalskia, J., and Gazicki-Lipmana, M. (2013). Morphology, photocleaning and water wetting properties of cotton fabrics, modified with titanium dioxide coatings synthesized with plasma enhanced chemical vapor deposition technique, *Surface and Coatings Technology*, **217**(25), pp. 51–57.

[48] Chen, C.; Chen, J. C., and Yao, W. H. (2010). Argon plasma treatment for improving the physical properties of crosslinked cotton fabrics with dimethyloldihydroxyethyleneurea-acrylic acid, *Textile Research Journal*, **80** (8), pp. 675–682.

[49] Ceria, A.; Rombaldonib, F.; Roveroa, G.; Mazzuchettib, G., and Sicardia, S. (2010). The effect of an innovative atmospheric plasma jet treatment on physical and mechanical properties of wool fabrics, *Journal of Materials Processing Technology*, **210** (5), pp. 720–726.

[50] Sinha, E. (2009). Effect of cold plasma treatment on macromolecular structure, thermal and mechanical behavior of jute fiber, *Journal of Industrial Textiles*, **38** (4), pp. 317–339.

[51] Goud, V. S. (2012). Influence of plasma processing parameters on mechanical properties of wool fabrics, *Indian Journal of Fibre & Textile Research*, **37** (3), pp. 292–298.

[52] Kan, C. W. (2008). KES-F analysis of low temperature plasma treated wool fabric, *Fibres & Textiles in Eastern Europe*, **16** (1), pp. 99–102.

[53] Razic, S. E.; Cunko, R.; Brkosek, V., and Matica, B. (2011). Antimicrobial modification of cellulose fabrics using low-pressure plasma and silver compounds, *Tekstil: Journal of Textile & Clothing Technology*, **60** (9), pp. 427–440.

[54] Strnad, J.; Sauperl, O., and Fras-Aemljic, L. (2010). Cellulose fibres functionalised by chitosan, in: Elnashar, M. (Ed.), *Characterisation and application biopolymers*, Sciyo Publication, Rijeka, pp. 181–200.

[55] Lim, S. H. and Hudson, S. M. (2004). Synthesis and antimicrobial activity of water-soluble chitosan derivative with a fiber-reactive group, *Carbohydrate Research*, **339** (2), pp. 313–319.

[56] Zhang, Z.; Chen, L., Ji, J.; Huang, Y., and Chen, D. (2003). Antibacterial properties of cotton fabrics treated with chitosan, *Textile Research Journal*, **73**, pp. 1103–1106.

[57] Peršin, Z.; Maver, U.; Pivec, T.; Maver, T.; Vesel, A.; Mozetič, M., and Stana-Kleinschek, K. (2014). Novel cellulose based materials for safe and efficient wound treatment, *Carbohydrate Polymers*, **100**, pp. 55–64.

[58] Li, X.; Tabil, L. G., and Panigrahi, S., (2007). Chemical treatments of natural fiber for use in natural fiber-reinforced composites: a review, *Journal of Polymers and the Environment*, **15** (1), pp. 25–33.

[59] Kalia, S.; Thakur, K.; Celli, A.; Kiechel, M. A., and Schauer, C. L., (2013). Surface modification of plant fibers using environment friendly methods for their application in polymer composites, textile industry and antimicrobial activities: a review, *Journal of Environmental Chemical Engineering*, **1**, pp. 97–112.

[60] Mukhopadhyay, S.; Pal, R.; Narula, V., and Mayank, M., (2013). A study of interface behavior in sisal fibre composites—single fibre pull out test. *Indian Journal of Fibre & Textile Research*, **38** (1), pp. 87–91.

[61] Bozaci, E.; Sever, K.; Sarikanat, M.; Seki, Y.; Demir, A.; Ozdogan, E., and Tavman, I. (2013). Effects of the atmospheric plasma treatments on surface and mechanical properties of flax fiber and adhesion between fiber–matrix for composite materials, *Composites: Part B*, **45**, pp. 565–572.

[62] Li, Y.; Moyo, S.; Ding, Z.; Shan, Z., and Qiu, Y. (2013). Helium plasma treatment of ethanol-pretreated ramie fabrics for improving the mechanical properties of ramie/polypropylene composites, *Industrial Crops and Products*, **51**, pp. 299–305.

[63] Shahidi, S.; Ghoranneviss, M., and Dalal, S. (2014). Effect of atmospheric pressure plasma treatment/followed by chitosan grafting on antifelting and dyeability of wool fabric, *Journal of Fusion Energy*, **33** (2), pp. 177–183.

3

REINFORCING POTENTIAL OF ENZYMATICALLY MODIFIED NATURAL FIBERS

LEVENT ONAL[1] AND YEKTA KARADUMAN[2]

[1]*Department of Textile Engineering, Faculty of Engineering, Erciyes University, Talas, Kayseri, Turkey*

[2]*Akdagmadeni Vocational High School, Bozok University, Akdagmadeni, Yozgat, Turkey*

Biodegradable Green Composites, First Edition. Edited by Susheel Kalia.

3.1 INTRODUCTION

In recent years, there is an increasing interest in composites reinforced with natural fibers as an alternative to glass fiber-reinforced plastics. Natural fibers are environmentally friendly, sustainable, renewable, and abundantly available materials. They are low cost and have low density in comparison to glass fibers. Composites reinforced with natural fibers have adequate specific mechanical properties (mechanical property/density) comparable with glass fiber composites on a per-weight basis [1, 2]. These properties make them ideal materials for the production of semistructural components in various industries like automotive, packaging, housing, and recreation [3, 4].

The properties and mechanical performance of natural fiber composites rely on the properties of fiber and resin, fiber volume fraction (overall and directional), fiber architecture, and fiber–matrix interface. The quality of adhesion between fibers and polymer matrix is especially important since it determines the load distribution capability of the material. Nevertheless, it is also the most problematical issue with natural fiber composites. There is a difference between the surface energies of inherently hydrophilic natural fibers and hydrophobic polymer resins resulting in a poor adhesion between the two. To date, several fiber and matrix modification methods are successfully implemented resulting in a relatively good adhesion at the interface. Some of these techniques are based on covering the surface of the fiber with a substance that is compatible with polymer resin to a certain extent (silane treatment, acetylation, coupling agents, etc.) [5–7] and some on removing particular fiber ingredients that are of no value in terms of composite reinforcement but have adverse effects on fiber–matrix bonding (alkali treatment) [8, 9]. Enzymatic catalyzation of the removal of these substances as a means of fiber pretreatment emerged as a relatively new and promising method that is effective, cost saving, and environmentally friendly in comparison to conventional chemical modification methods. The main principle of enzymatic treatment is to catalyze the removal of undesirable substances like pectin, hemicelluloses, lignin, fat, and waxes from the surface of the fibers in an effort to improve fiber–matrix adhesion. Enzymatic treatment has numerous advantages over traditional techniques. Traditional modification methods often require harsh acidic or alkaline conditions and take place at high temperatures usually greater than 100°C. Enzymes substantially lower the activation energy so that the reactions can occur in relatively mild conditions, that is, low temperature and a neutral pH. This results in a significantly lower energy requirement in case of enzymatic catalyzation. Furthermore, enzymes possess a high degree of specificity, that is, they act on only a particular substrate creating no damage on other fiber components. They do not cause water pollution unlike most chemicals. The use of enzymes can reduce the excessive usage of hazardous chemicals for fiber pretreatments, thus saving freshwater resources. These remarkable properties make enzymes very attractive for fiber pretreatment prior to composite manufacture.

In this chapter, the basic properties of enzymes, their classification, and some properties are introduced. Natural fibers and fiber components as substrates for enzymes are explained. The enzyme types used for natural fiber modification were presented. The present and future applications of enzymes in natural fiber pretreatment are reviewed.

3.2 ENZYMES

3.2.1 A Brief History

Enzymes are naturally occurring or synthetic macromolecular substances composed wholly or largely of protein that catalyze biochemical reactions by lowering the activation energy. The substances upon which enzymes act are known as substrates for which the enzyme possesses a specific binding or active site [10]. Among the earliest examples of the enzyme activity that is recognized for centuries is the fermentation of sugar to alcohol by yeast. The term "enzyme" (Greek: in yeast) was coined by German physiologist Wilhelm Kühne in the late ninetieth century to describe leavening process, and subsequently scientific research and understanding of enzymes started to develop. Since then, the structure and function of many enzymes have been fully discovered leading to their effective usage as catalyzers in various industrial applications.

The industrial use of enzymes begins with German chemist Otto Röhm who discovered the effectiveness of microbial proteases for the removal of protein-based stains from clothes. Today, nearly all commercial detergents contain these enzymes in order to catalyze the removal of stains like blood, milk, egg, etc. Enzymes found numerous applications in textile, food, paper, and biofuel industries due to both economic and environmental benefits. Enzymes not only accelerate the rate of chemical reactions and reduce the reaction time, but they can also help to reduce the energy requirement for a specific process, thus minimizing the process cost.

Although the usage of enzymes in natural fiber composite processing is a relatively new practice, enzymes have a long history with natural fibers in textile industry, especially in wet pretreatment and finishing processes. Amylases are used in many textile factories since the early twentieth century in desizing process to catalyze the degradation and removal of starch from woven fabrics. Since the late 1980s cellulases have been used extensively for the enzymatic depilling of cotton and to create a fading effect in denim fabrics. Catalases have been introduced in the early 1990s to destroy hydrogen peroxide after bleaching in an attempt to save freshwater resources. Pectinases, cellulases, and lipases have been used for many years in cotton scouring process in order to degrade and remove the pectin, waxes, and other hydrophobic materials in the outer layers of the fibers in order to facilitate the following treatments such as dyeing and wet finishing.

Enzymatic treatment was also used for the modification of bast fibers such as flax, jute, hemp, and ramie mostly for improving wettability and softness with less regard to strength properties. Kozlowski et al. [11] reviewed the use of enzymes for bast fiber processing mostly considering the textile properties such as whiteness, softness, water absorbency, fineness, and brightness. Textile experience can provide the composite industry with an important guidance since the substrate for the enzymes, that is, natural fibers, is common, despite the fact that the final aim of enzymatic treatment is totally different. In textile industry, the main goal of most enzymatic treatments is modification of the fiber surface either to prepare for the subsequent wet processing or to improve the hand and appearance of the textile product. In the case of composite pretreatment, however, the purpose is to remove the substances like pectin, lignin, hemicelluloses, fat, and waxes from fiber surface in an effort to create a rougher surface morphology. This increases the effective fiber surface area and results in a more efficient wetting of fibers, thus improving the mechanical interlocking between fiber and matrix. Therefore, a good fiber–matrix interface can be obtained, which is critical for a high mechanical performance.

3.2.2 Classification and Nomenclature

Enzymes are classified according to the chemical reaction they catalyze, since this is the specific property that distinguishes one enzyme from another. There are thousands of different types of enzymes in terms of types of reactions they catalyze and in terms of structure. Therefore, it is of a paramount importance to have a rational basis for classification and naming of enzymes. The Enzyme Commission of International Union of Biochemistry and Molecular Biology in its report of 1961 established a numerical classification scheme for enzymes based on the type of reaction they catalyze. The code number starts with the letters EC (i.e., Enzyme Commission), which is followed by four digits separated by points. The first digit denotes one of the six main classes of enzymes, that is, oxidoreductases, transferases, hydrolases, lyases, isomerases, and ligases. The second and third digits denote the subclass and the sub-subclass, respectively, representing a finer classification of the enzyme and giving further information about the reaction type. The fourth digit is the serial number of the enzyme in that sub-subclass, and it defines the actual substrate for the reaction. Enzymes that catalyze the same reaction have the same EC number even though they have different protein structures. These are called isoenzymes, and they carry the same four-digit EC classification number. They are distinguished from each other by specifying the source of that particular enzyme. Each enzyme is assigned two names, one being the recommended or trivial name, which is convenient for everyday use, and the other a systematic name. The systematic name is the name of substrate(s) followed by a word ending with *-ase* specifying the type of reaction the enzyme catalyzes [12].

3.2.3 Enzyme Structure

Enzymes are large protein macromolecules consisting of up to 20 different kinds of amino acids. The sequence of these amino acids determines the chemical and physical properties of proteins, their relationships to other proteins, and their reaction mechanism [13]. Amino acids are linked together by peptide bonds to form polypeptide chains. Polypeptide chains fold up to constitute a unique three-dimensional shape called "conformation." Conformation is critical in that, when it is disrupted, the enzyme loses its functionality (i.e., denaturation of the enzyme). The major factors that cause denaturation of an enzyme are pH, heat, and chemicals. In general, a protein containing more than 40 amino acid residues is called polypeptide. The majority of polypeptides contain between 100 and 1000 amino acid residues. The characteristics of an individual protein depend more on its amino acid sequence than which amino acids it contains.

Protein structure can be described in terms of its primary, secondary, tertiary, and quaternary structures. A protein's primary structure is the amino acid sequence of its polypeptide chain(s). Higher levels of protein structure—secondary, tertiary, and quaternary—refer to the three-dimensional shapes of folded polypeptide chains. Secondary structure is the local spatial arrangement of a polypeptide's backbone atoms without regards to the conformation of its side chains [13]. Tertiary structure refers to the three-dimensional structure of polypeptide including its side chains, and quaternary structure refers to the spatial arrangement of its subunits. Regions of these higher-level structures are stabilized by interactions between amino acids such as hydrophobic interactions, ionic bonds, hydrogen bonds, van der Waals forces, and covalent bonds.

Proteins are composed of amino acids. There are 20 commonly occurring amino acids. All has a central (α) carbon atom to which a primary amino group ($-NH_2$), a carboxyl group (—COOH), a hydrogen atom, and a side group or chain (R) are attached. The 20 standard amino acids differ in the structures of their side chains (R). Proline is the only exception because it contains a secondary amino group (—NH—) instead of a primary amino group.

Twenty standard amino acids are classified by the polarities of their side chains. According to this classification, there are three major types of amino acids: those with nonpolar R-groups, those with uncharged polar R-groups, and those with charged polar R-groups. Polar molecules display ionic character and are therefore hydrophilic and soluble in water. On the other hand, nonpolar molecules are generally insoluble in water but soluble in organic solvents.

With the exception of glycine, all of the common amino acids are optically active, that is, they rotate the plane of polarized light. Biochemistry commonly use the Fischer convention to describe different forms of amino acid molecules called spatial isomers or stereoisomers. These are two mirror image forms of an amino acid that cannot be superimposed by rotation of the molecule. The isomers are termed D- and L-isomers, the prefix D indicating rotation of polarized light to the right and L to the left. The amino acid residues of proteins all have the L-configuration.

3.2.4 Enzymatic Catalysis

Enzymes are catalysts that increase the rate of chemical reactions by lowering activation energy. Enzymes differ from ordinary chemical catalysts in several points [13]:

- The rates of enzymatically catalyzed reactions are several orders of magnitude greater than those of the corresponding uncatalyzed and chemically catalyzed reactions.

- Enzymatically catalyzed reactions take place under relatively mild conditions, that is, temperatures below 100°C, atmospheric pressure, and a neutral pH.
- Enzymes have a greater degree of reaction specificity with respect to their substrates.
- The catalytic activities of enzymes can be regulated by allosteric control, covalent modification of enzymes, and changing the amount of enzymes synthesized.

All enzymes have active sites consisting of an indentation that is complementary in shape to the substrate (geometric complementarity). Moreover, the amino acid residues on the active site are arranged in such a way that they attract the substrate (electronic complementarity). This complementarity between enzymes and their substrates called "lock and key model" is first proposed by Emil Fischer in 1894. In contrast to Fischer who assumed a rigid enzyme–substrate relation not allowing for changes in the structures of enzyme and its substrate during the reaction, Koshland, in 1958, proposed an alternative model for enzyme action in which the binding site of the enzyme undergoes some conformational changes as a result of substrate binding (a phenomenon called induced fit). When the complementarity conditions are met, enzymes and their substrates bind together through active sites by forces such as van der Waals, electrostatic, hydrogen bonding, and hydrophobic interactions.

Enzymes vary considerably in their degree of specificity. Some enzymes are highly specific, being capable of acting only on one compound (substrate), whereas others are not so specific, having the ability to act on closely related substrates but with different efficiencies.

The catalytic activity of some enzymes requires the presence of nonproteinaceous compounds known as "cofactors." Cofactors may be metal ions, such as Cu^{2+}, Fe^{3+}, or Zn^{2+}, or organic molecules known as "coenzymes." Enzymes that require cofactors are referred to as apoenzymes when not bound to the cofactor (catalytically inactive) and as holoenzymes when bound to the cofactor (catalytically active).

3.3 NATURAL FIBERS AS ENZYME SUBSTRATES

Natural fibers are divided into three main groups according to their origin: plant fibers, animal fibers, and mineral fibers (Fig. 3.1) [14]. Almost all natural fibers are possible candidates for composite reinforcement. However, in order to utilize a natural fiber effectively in a particular application, it must satisfy the cost and performance standards as well as the environmental requirements. There are a number of natural fibers that have been shown to satisfy such demands. Most of these fibers are plant based, and they are lignocellulosic in nature, that is, they contain lignin and cellulose as major components. The most popular lignocellulosic plant fibers for composite reinforcement are bast fibers such as flax, hemp, jute, ramie, and kenaf and a leaf fiber such as sisal [15]. These fibers are made up

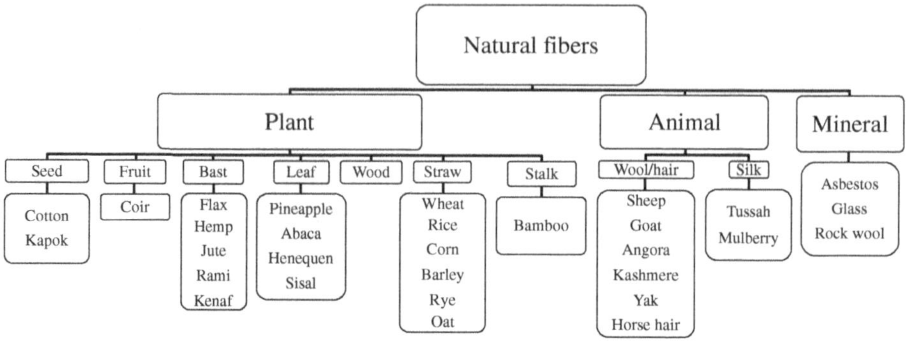

FIGURE 3.1 Classification of natural fibers. *Source*: Mohanty et al. [14], Figure 1.3. Reproduced with permission from Taylor and Francis Group LLC books.

of cellulose, lignin, hemicelluloses, and pectin polymers together with small amounts of waxes, ash, and other aromatic compounds. The amount of these polymers as well as their properties and orientation vary between different fiber types. The variation could even be according to the growing season, growing region, and location of the fiber on the plant. It is essential to know the basic physical and chemical properties of these fibers and their ingredients in order to understand the mechanism of enzymatic catalysis and establish its effective use for fiber pretreatment.

3.3.1 Physical Properties of Lignocellulosic Fibers

Lignocellulosic fibers are multicellular in structure. Most of them exist as fiber bundles (technical fiber) made up of several single cells, that is, single fibers [16]. Bast fibers like jute, flax, kenaf, and ramie are extracted from the outer layer of plant stem. The length of fiber bundles can be up to 300 mm. The number of single fibers in a bundle depends on the extraction method [16]. A single fiber has a laminar structure made up of lumen, primer wall, and seconder wall. Seconder wall can be further divided to three sublayers, namely, S1, S2, and S3. The strength of the fiber is determined by the amount and orientation of cellulose microfibrils within these layers as well as their degree of polymerization and crystallization. Other important polymers in the cell wall are hemicelluloses and lignins, which act as a matrix material between these cellulose microfibrils [16]. The primer wall is made up of randomly positioned microfibrils. S2 layer is much thicker in comparison to other layers making up approximately 80% of the total thickness and thus acts as the main load-bearing component [17]. Therefore, it is mainly the orientation of the cellulose microfibrils at S2 layer that determines the mechanical properties of the fiber [18, 19]. Fiber modulus increases with a decrement in microfibril angle (MFA) [20, 21]. As a general rule, fibers with a lower MFA and higher cellulose content possess higher strength and Young's modulus with lower breaking elongation. Physical properties of some well-known cellulosic fibers are given in Table 3.1 [16].

TABLE 3.1 Physical properties of natural fibers (single fiber)[a]

Fiber	Length *l* (mm)	Diameter *d* (μm)	Aspect ratio (*l*/*d*)	Microfibril angle (°)	Density (kg/m^3)	Moisture uptake (%)
Cotton	20–64	11.5–17	2752	20–30	1550	8.5
Flax	27–36	17.8–21.6	1258	5	1400–1500	12
Hemp	8.3–14	17–23	549	6.2	1400–1500	12
Jute	1.9–3.2	15.9–20.7	157	8.1	1300–1500	12
Kenaf	2–61	17.7–21.9	119	—	1220–1400	17
Rami	60–250	28.1–35	4639	—	1550	8.5
Sisal	1.8–3.1	18.3–23.7	115	10–22	1300–1500	11
Coir	0.9–1.2	16.2–19.5	64	39–49	1150–1250	13

[a] Ansell et al. [16], table 2.4. Reproduced with permission from Elsevier.

3.3.2 Chemical Properties and Composition of Lignocellulosic Fibers

Natural lignocellulosic fibers are cell walls that are composed of cellulose, hemicellulose, lignin, pectin, aromatics, waxes and other lipids, ash, and water-soluble compounds [22]. The relative proportion of these components depends on the geographical origin, growing conditions, and the degree of maturity of the plant as well as the location of the fiber on the plant. The relative ratio of these components, types of chemical and physical bonds that are present, and orientation of the macromolecules define the physical and chemical properties of the fibers [23]. Table 3.2 shows the properties and chemical composition of plant fibers [24].

3.3.2.1 Cellulose Cellulose is the most abundant polymer in nature. It is formed by biosynthesis in living cells. Plants use carbon dioxide (CO_2) and water to produce glucose molecules by utilizing light energy from the sun. These glucose units are then converted to cellulose by condensation polymerization reaction with the elimination of water [25].

Cellulose is a linear polymer of glucose. Glucose ($C_6H_{12}O_6$) has a ring structure made up of one oxygen atom (—O—) and five (—CH—) groups, four with pendant —OH groups and one with pendant —CH_2OH group. Glucose molecules add on successively through β-1,4 linkages with the elimination of water to form long cellulose chains [25]. Cellulose macromolecules are linked together by intramolecular hydrogen bonds, which occur between the pendant —OH and —CH_2OH groups creating microfibrils of 2–20 nm in diameter and 100–40,000 nm in length [22]. Figure 3.2 shows the chemical structure of cellulose macromolecule. The regions where the cellulose macromolecules are fairly parallel to each other forming close-packed units through intramolecular hydrogen bonds are called crystalline regions [26]. Crystalline regions are very hard to penetrate for most chemicals and enzymes. Therefore, enzyme activity generally takes place in the amorphous regions, that is, the regions with randomly orientated macromolecules that lack the highly oriented structure of crystalline regions resulting in a more open and flexible structure.

TABLE 3.2 Properties and chemical composition of plant fibers[a]

Fiber	Density (g/cm^3)	Diameter (μm)	Cellulose (wt.%)	Hemicellulose (wt.%)	Lignin (wt.%)	Pectin (wt.%)	Moisture (wt.%)	Wax/oil (wt.%)
Cotton	1.5–1.6	12–38	85–90	5.7	—	0–1	7.85–8.5	0.6
Jute	1.3–1.45	25–200	61–71.5	13.6–20.4	12–13	0.2	12.5–13.7	0.5
Flax	1.50	40–600	71	18.6–20.6	2.2	2.3	8–12	1.7
Hemp	1.47	25–500	70–74	17.9–22.4	3.7–5.7	0.9	6.2–12	0.8
Rami	1.50	—	68.6–76.2	13.1–16.7	0.6–0.7	1.9	7.5–17	0.3
Sisal	1.45	50–200	66–78	10–14	10–14	10	10–22	2
Pineapple	—	20–80	70–82	—	5–12.7	—	11.8	—
Coir	1.15	100–450	32–43	0.15–0.25	40–45	3–4	8	—

[a] Bismarck et al. [23] and Mohanty et al. [24]. Reproduced with permission from Wiley-VCH Verlag GmbH & Co. KGaA.

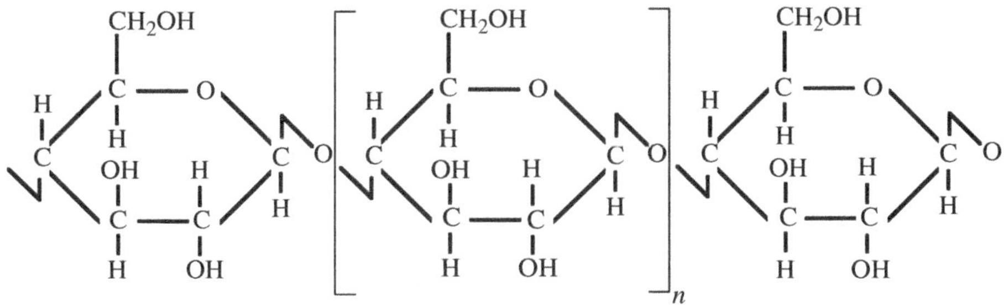

FIGURE 3.2 Chemical structure of cellulose.

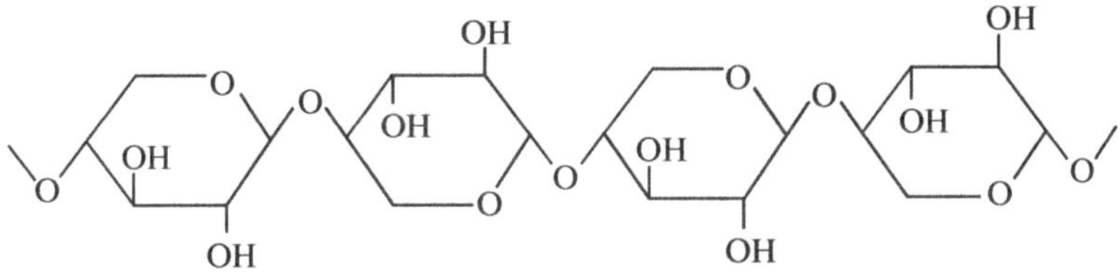

FIGURE 3.3 Hemicellulose-β-1-4-D-xylan.

3.3.2.2 ***Hemicellulose*** Hemicellulose is a type of carbohydrate that is also part of plant cell walls. It is associated with cellulose, lignin, pectin, and other aromatic substances within plant cell walls. Hemicelluloses are found in the middle lamellae, in the primary wall and in S2 layer [22, 27]. Hemicellulose is a common term for a highly heterogeneous group of polysaccharides, such as glucans (a polymer of D-glucose monomers—$C_6H_{12}O_6$), mannans (a polymer of the sugar mannose—$C_6H_{12}O_6$), galactans (a polymer of the sugar galactose—$C_6H_{12}O_6$), arabinans (a polymer of arabinose—$C_5H_{10}O_5$), and xylans (a polymer of the pentose sugar xylose—$C_5H_{10}O_5$) [22]. Hemicelluloses have amorphous structure with linear or branched macromolecules, and thus it is more susceptible to enzymatic action when compared to crystalline cellulose. Figure 3.3 shows the -β-1 → 4-D-xylan as a representative of hemicellulose [16]. Hemicellulose molecules contain approximately 200 sugar units, and they are much smaller in comparison to cellulose, which contains approximately 7,500–15,000 units.

3.3.2.3 ***Lignin*** Lignin is responsible for strength, rigidity, and protection against microbial pathogens of cell walls. Covalent linkages occur between lignin and hemicelluloses, and together they act as a cement material between cellulose macromolecules [22]. The aromatic ring structure is the primary constituent in lignin. Lignin is composed of oxyphenyl propan units from one or more of the following alcohols: *p*-coumaryl, coniferyl, or sinapyl [16, 22, 28]. Lignin has a hydrophobic structure unlike hemicelluloses [16, 29]. Lignin has a branched structure expanding in all possible directions unlike cellulose, which generally tends to form a long chain

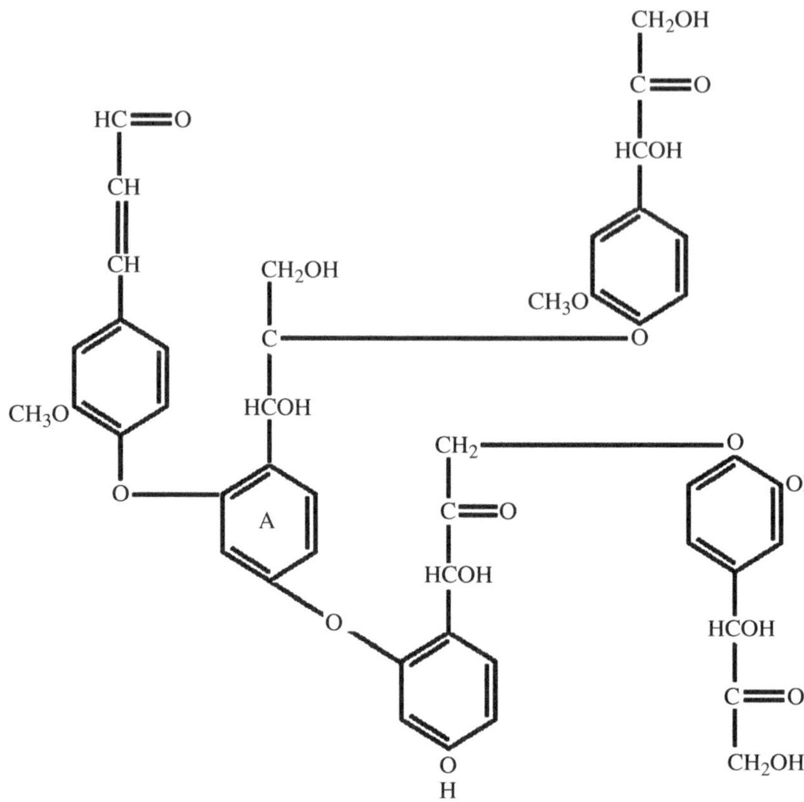

FIGURE 3.4 Lignin polymer.

(Fig. 3.4) [16]. Hence lignin polymer forms a more open structure when compared to cellulose creating more available sites for enzyme penetration.

3.3.2.4 Pectin Pectin and other complex pectic polysaccharides are known as pectins. They can be considered as matrix materials that bind cellulose and hemicellulose macromolecules together. Pectin amounts are often low in lignocellulosic fibers. Pectins have an extremely heterogeneous structure comprising a diverse group of substances [30]. Galacturonic acid residues linked through *α* bonds (poly-α-(1-4)-galacturonic acid) are the major component of most pectins. Pectins can be represented with rhamnose and galactose units (Fig. 3.5) [16, 22, 31].

Pectic substances are grouped into two types: homogalacturonan in which the main polymer chain consists of α-D-galacturonate units linked by 1,4-glycosidic bonds and heterogalacturonan (rhamnogalacturonan) in which the primary chain consists of 1,4-linked α-D-galacturonates and with about 2–4% L-rhamnose units that are -β-1,2- and -β-1,4-linked to D-galacturonate units [32]. The side chains of rhamnogalacturonans are generally substituted by L-arabinose or D-galacturonic acid units.

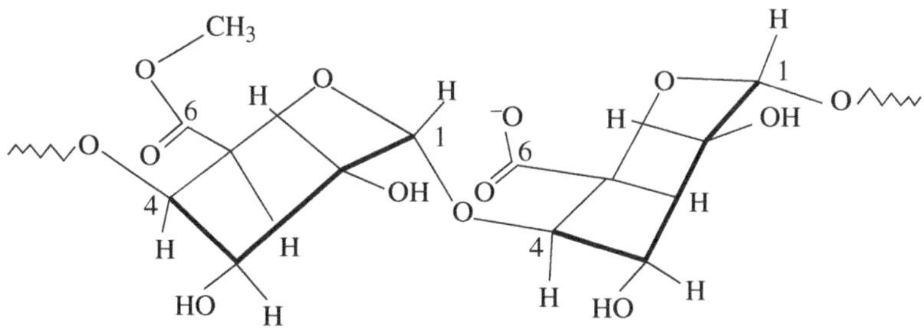

FIGURE 3.5 Poly-α-(1-4)-galacturonic acid.

In plant tissues, about 60–70% of the galacturonate units are esterified with methanol and occasionally with ethanol. Based on the degree of esterification, pectic substances are classified into protopectin, pectinic acid, pectin, and polygalacturonic acid [33]. Protopectin is a parent pectic substance that yields pectin or pectinic acid upon hydrolysis. Pectic acids are the galacturonans that contain negligible amounts of pectic acid. Pectinic acids are galacturonans that contain various amounts of methoxy groups. Pectin is a generic name for mixtures of different pectic substances containing pectinic acid as the major component [34].

3.3.2.5 Other Aromatic Compounds Aromatic compounds in plant cell walls are condensed tannins, which are phenolic complexes arising from the condensation of leucoanthocyanidin and catechin [35] and low molecular weight phenolic acids, for example, ferulic and *p*-coumaric acids.

3.3.2.6 Fats, Waxes, and Lipids Fats, waxes, and lipids are hydrocarbons that are insoluble in water. The amounts of these compounds are relatively low in bast fibers. Waxes are esters of long-chain alcohols. The waxy cuticle influences the properties of natural fibers. In bast fibers, the cuticle is outside the stem, and little wax occurs on the fiber surface. In retting, the cuticular layer of stem is separated from the bast fibers. Lipid compounds in natural fibers include long-chain fatty acids (carboxylic acid), fatty alcohols, and wax esters (esters of long-chain fatty alcohols with long-chain fatty acids) [36].

3.4 TYPES OF ENZYMES USED IN NATURAL FIBER MODIFICATION

3.4.1 Cellulases

Cellulases belong to the group of hydrolytic enzymes. They are derived from a wide range of bacteria and fungi. They catalyze the degradation of cellulose to smaller oligosaccharides and eventually to glucose. Cellulases can be used solely, that is, only one type of cellulase and also as enzyme mixtures containing two or more cellulase types. The processes where cellulases are effectively used include denim biostoning for achieving a worn effect similar to that of stonewashed jeans and cotton biofinishing for removing small fibers and obtaining a clean and smooth fabric surface. Cellulases are also used for scouring of cellulose-containing fibers such as cotton, flax, jute, hemp, etc. Successful application of cellulase depends on the parameters such as fiber type, enzyme composition, pH, temperature, enzyme concentration, treatment time, liquor ratio, and mechanical agitation.

There are three types of cellulases with different functions. **Endoglucanases** (EG: 1,4-β-D-glucan glucanohydrolase; EC 3.2.1.4) act on amorphous regions and cleave the bonds in cellulose chains and create smaller oligosaccharides reducing the degree of polymerization. **Cellobiohydrolases** (also called exoglucanases, CBH: 1,4-β-D-glucan cellobiohydrolase; EC 3.2.1.91) act on the crystalline parts, and they also catalyze the breakdown of cellulose macromolecules to smaller oligosaccharides but at a much slower rate in comparison to endoglucanases. **β-glucosidases** (BGL: cellobiase or β-D-glucoside glucohydrolase, EC 3.2.1.21) hydrolyze the oligosaccharides to glucose [37].

Cellulases with other enzymes can be used in retting of bast plants as an alternative to conventional retting methods such as water or dew retting, in order to separate fibers from the nonfiber substances. Enzymatic retting has been studied for many years in order to obtain a more controlled way of isolating fibers. Cellulases were employed in enzymatic retting with promising results [38].

3.4.2 Xylanases

Hemicelluloses in the plant cell walls are an extremely heterogeneous mixture of several branched polymers. The most common types of these polymers are xylan and galactoglucomannan [39], and other less abundant hemicelluloses are glucomannan, xyloglucan, arabinogalactan, and arabinan [40]. Xylans are generally used as representatives of hemicelluloses. Xylan is a complex polysaccharide composed of β-1,4-linked xylopyranosyl residues. Depending on the plant source, it can contain side chains of arabinosyl, glucuronosyl, methylglucuronosyl, acetyl, feruloyl, and *p*-coumaroyl residues.

Xylanases can be considered as the primary degrading enzyme of the hemicelluloses. Nonetheless, several other enzymes such as cellulases, arabinases, mannanases, and noncellulosic glucanases are also effective on hemicellulases due to its complex structure with different kinds of branched structures [41].

The enzymes that catalyze the breakdown of xylan can be classified into two main groups, that is, xylanases and β-1,4-xylosidases. Xylanases destroy the xylose backbone by cleaving bonds within the polymer. Xylanases are further divided into three distinct groups, namely, exoxylanases, endoxylanases, and xylosidases. β-1,4-Xylosidases attack on the side chains and hydrolyze xylobiose. They also degrade short-chain xylooligosaccharides generated by the action of xylanases, thus releasing xylose units from xylobiose and xylooligomers. Other enzymes that catalyze the degradation of xylan side chains are [39]:

- **α-L-Arabinofuranosidases (EC 3.2.1.55)** catalyze the cleavage of terminal arabinose residues from the side chains of xylan and other arabinose-containing polysaccharides [42].

- **α-D-Glucuronidases (EC 3.2.1.131)** hydrolyze the linkages between 4-*O*-methylglucuronic/glucuronic acid and xylose residues in glucuronoxylan.
- **Acetyl xylan esterases (EC 3.1.1.72)** remove the acetyl groups from acetylated xylan that hinder the approach of xylanases that cleave xylan backbone and facilitate the action of xylanases [43].
- **Ferulic acid and *p*-coumaric acid esterases** cleave the ester bonds between arabinose side chains and feruloyl or *p*-coumaroyl residues, respectively [44].

Xylanases can be used in conjunction with the pectinolytic enzyme systems for the degumming of bast fibers such as jute, hemp, flax, and ramie [45, 46].

3.4.3 Pectinases

Pectinases are a diverse group of enzymes whose primary task is to degrade pectic substances. They are classified depending upon their substrate and mode of enzymatic reaction. Industrially important pectinases and their actions are summarized below [33]:

- **Polymethylgalacturonases (PMG)** degrade the highly esterified pectin.
- **Polygalacturonases** hydrolyze glycosidic linkages of polygalacturonates (pectates) and convert them to oligogalacturonates.
- **Pectin lyases** convert pectic substances to unsaturated oligomethylgalacturonates.
- **Polygalacturonate lyases** degrade pectate to unsaturated oligogalacturonates.
- **Pectinesterases** hydrolyze the methoxy groups from 6-carboxyl group of galacturonan backbone of pectin.

Regarding industrial applications, pectinases are classified into two groups as acidic and alkaline pectinases. Alkaline pectinase is used for the retting and degumming of fibers such as jute, flax, hemp, ramie, kenaf, and coir, whereas acidic pectinases are mostly used in the fruit juice industries. Alkaline pectinases come mostly from bacterial sources. In the industrial sector, alkaline pectinases, mainly from *Bacillus* spp., are utilized [34]. Species of *Cladosporium*, *Penicillium*, *Aspergillus*, and *Rhodotorula* have been isolated from dew-retted plants [47], whereas *Bacillus*, *Pseudomonas*, and *Micrococcus* spp. were reported to be the major retting species in the case of water retting of flax [48], jute [49], sisal, and coir [50]. Pectinases were reported to be the most effective among the retting enzymes especially acting on the pectin in the cell walls [51].

3.4.4 Laccases

Laccases (benzenediol: oxygen oxidoreductase; EC 1.10.3.2) use molecular oxygen to oxidize various aromatic and nonaromatic compounds by a radical-catalyzed reaction mechanism [52]. The substrates for laccases include *o*- and *p*-diphenols, aminophenols, benzenethiols, polyphenols, methoxy-substituted phenols, polyamines, hydroxyindols, some aryl diamines, and other compounds [53].

The reactivity of laccases is related to their redox potential (E°), which is the energy required to capture one electron from a reducing substrate. E° value depends upon the origin of the enzyme, that is, plant, fungi, or bacteria. As the E° value increases, the rate of substrate oxidation increases [54].

Laccases contain four copper atoms per monomer as cofactors that are bound to three redox sites (i.e., T1, T2, and T3). Laccases oxidize the reducing substrate at T1 in the presence of Cu^{2+}, whereas the reduction of molecular oxygen to water takes place at the T2 and T3 sites [53].

The main role of laccase is the depolymerization of lignin. It can be used solely or in association with other lignin-degrading enzymes such as high redox potential peroxidases, cellobiose:quinone oxidoreductase, cellobiose dehydrogenase, glyoxylate oxidase, glucose oxidases, veratryl alcohol oxidase, and some esterases [53]. In pulp and paper industry, the removal of lignin from woody tissues by employing laccase from *T. versicolor* has been found to be very efficient [55, 56].

3.5 EFFECT OF ENZYMATIC TREATMENT ON THE STRUCTURE AND PROPERTIES OF NATURAL FIBERS

The effect of enzymatic treatment on bast and other natural fibers is mainly on the chemical composition of the fiber, which consequently affects related physical and mechanical properties. Enzymes have proved to be very effective in the destruction and removal of particular fiber constituents like lignin, pectin, hemicellulose, and other aromatic compounds depending on the type of enzyme(s) used and reaction conditions. Therefore, enzymatic treatment can be very versatile depending upon fiber properties required for particular applications. In this context, textile and composite applications often demand different fiber properties. In textile applications, enzymatic treatment is employed to improve fineness, comfort, and appearance of fabric, whereas composite industry is more interested in antibacterial, moisture absorption, strength, and stiffness properties of fibers and their composites.

Processing of bast fibers with cellulases to remove hemicellulases, pectins, waxes, and impurities is a promising technique as an alternative to alkaline treatment of fibers prior to composite production. Pectinases, proteases, xylanases, and lipases were employed together with cellulases on flax fibers, and improved wettability was obtained [57].

It was reported that *Trichoderma* endoglucanases (cellulases) improve the pilling properties of linen fabric and the bending of flax fibers. Surface fibrils were eliminated by a 6 h treatment resulting in weight loss of fibers. Yarn strength decreased as a consequence of weight loss in all type of fabrics such as cotton, linen, viscose, ramie, and their blends. Retention of strength after 48 h of incubation followed the order of viscose rayon $\ll$ cotton $<$ ramie $<$ linen, whereas weight loss increased in the order ramie $<$ linen $<$ cotton $<$ viscose rayon. It has been reported that X-ray crystallinity and moisture absorption of the samples did not change after the treatments, which indicated cellulase activity only on the surfaces of accessible cellulose chains [58].

Mooney et al. [59] analyzed retted and nonretted flax fibers by chemical and enzymatic means in order to determine the compositional changes brought about by retting process. They used the enzymes endoglucanase V from *Trichoderma viride* and xylanase I from *Aspergillus awamori*. Treatment with two different enzymes showed that nonretted extract contained a relatively high molecular weight xylan not found in the retted flax indicating that xylan macromolecules degraded to some extent with enzymatic treatment. It was shown that endoglucanase V from *T. viride* solubilized cellulose as well as xylan and xyloglucan oligomers from the extract. However, xylan and xyloglucan polymers in the fibers entrapped by cellulose were not accessible to the enzyme. These polymers could be released by the hydrolysis of the cellulose and then can be degraded by the action of enzymes.

Saikia et al. [60] investigated microbial degumming of ramie fibers. They observed that the gum content decreased in 3–8 days of treatment. Tensile strength reduced with an increase in treatment time. The optimum conditions for *Bacillus subtilis*, *Aspergillus* sp., and *Curvularia* sp. bacterial treatments were recorded as pH 7.5 and at 35°C.

Kalia and Sheoran [61] treated ramie fibers with cellulase from *Streptomyces albaduncus*. It was shown that the fiber morphology changes remarkably in comparison to untreated and chemically treated fibers. Fiber surface became soft and bright with a reduction in thermal stability and crystalline structure after cellulase treatment. Biologically modified ramie fibers showed enhanced softness and smooth appearance, which was attributed to surface protuberances of extracellular cellulose. On the other hand, thermal stability of fibers was not altered by the modification. Percentage of crystallinity of the fibers reduced from 80.14% (untreated ramie fiber) to 76.73% (*S. albaduncus*-treated ramie fiber) with enzymatic treatment. The authors attributed this decrement to the production of extracellular cellulose, which hydrolyzes cellulose by exocleavage of β-1,4-glycosidic linkage. The crystallinity index of ramie fiber and biologically modified ramie fiber recorded as 0.7522 and 0.6968. Lower crystallinity index indicated poor orientation of cellulase crystals in the fiber.

Kalia and Vashishta [62] modified the surface of sisal fiber using bacterial cellulase and methyl methacrylate. They obtained a bright, smooth fiber due to presence of extracellular protuberant structures of cellulase, which resulted in removal of gummy substances and small fibrils protruding from the fiber surface. It has been found that thermal stability of sisal fibers enhanced with cellulase treatment. Maximum thermal stability (425°C) was observed in the case of sisal fibers modified with both bacterial cellulase and poly(MMA). In case of bacterial cellulase treatment, thermal stability enhanced because of the removal of amorphous and gel-like polysaccharide layer on the surface together with small fibrils protruding from the surface.

Joko et al. [63] investigated the properties of jute fibers treated with various enzymes such as cellulase, pectinase, hemicellulase, xylanase, and laccase. It was reported that lignin and hemicellulose constituents were removed from the surface of the fiber by enzymatic treatment. Breaking strength of enzyme-treated fibers reduced by 15–25%.

Jayapriya and Vigneswaran [64] used white rot fungi cellulase and a mixture of enzymes (cellulase, xylanase, and pectinase) to treat jute fibers. Flexural rigidity and tenacity reduced with enzymatic treatment. The lignin polymers were degraded by enzymes. Lignin content of raw jute was recorded as 11.50%. Significant reductions in lignin content were observed, that is, 44.16 and 25.91% for *Phanerochaete chrysosporium* and *Ceriporiopsis subvermispora*, respectively, after a period of 30 days with accompanying reductions in cellulose contents. Treatment with *P. chrysosporium* and *C. subvermispora* led to a decrease of 12.26 and 16.86% in tenacity of raw jute fibers, respectively. The authors attributed this decrement to selective removal of lignin, resulting in the compaction of microfibrils in biosoftened jute. Another possible reason for the strength loss was thought to be partial surface hydrolysis of the fibers. It was observed that the strength loss showed higher rate in the mixed enzyme systems than solely cellulase. This was attributed to higher enzymatic surface hydrolysis in the presence of pectinase and xylanase in the mixed enzyme system. Elongation to break values increased with enzymatic treatments. Stiffness of the raw jute fiber as indicated by flexural rigidity reduced substantially in case of cellulase and mixed enzyme systems. SEM studies revealed that the smoothness and evenness of the fiber surface increased with enzymatic treatments.

Park et al. [65] modified pine fibers with cellulase enzyme from *Trichoderma reesei*. They observed a significant increase in crystallinity index. It was observed that enzymatic hydrolysis of fibers resulted in changes to the surface and pore structure. Three independent measurements, that is, nonfreezing bound water, polymer adsorption, and crystallinity index, confirmed that the amorphous regions of fibers are found to be more readily available as enzyme substrate when compared to crystalline regions. For the low-dosage enzymatic treatments, the average fiber length and fines content did not change, whereas substantial decrement in fiber length was observed for the high-dosage treatment. After 240 min of high-dosage treatment, the fiber length decreased from 2.56 to 0.32 mm and fines content increased from 3.02 to 33.4%. The average fiber width was not reduced after 240 min of high-dosage treatment. The authors claimed that the significant decrease in fiber length without any decrement in the fiber width shows that enzymatic treatment does not cause cleavage in the fiber axial direction. It was also demonstrated that the fiber length decrement is due to the cuts in the cross direction of the fibers rather than cleavage in axial direction (Fig. 3.6).

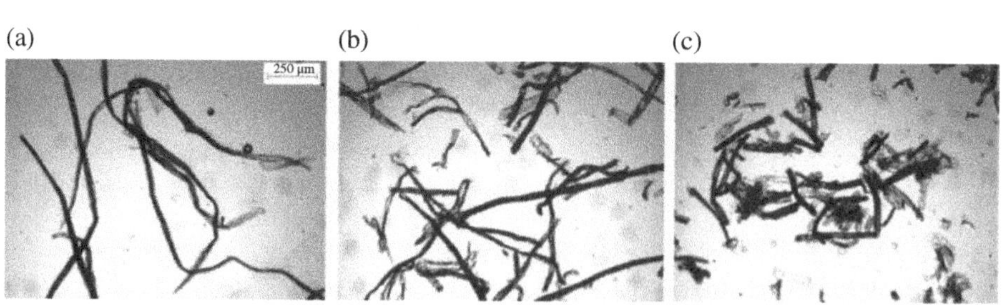

FIGURE 3.6 Microscopic images for (a) no treatment, (b) 60 min, and (c) 240 min high-dosage cellulase treatments. Scale bar shows 250 μm in (a). *Source*: Park et al. [65], figure 1. Reprinted with permission from John Wiley & Sons. Copyright 2006.

Azelee et al. [66] investigated the efficient removal of lignin from kenaf by using xylanase from *T. reesei*. A two-stage pretreatment route was adopted using calcium hydroxide, $Ca(OH)_2$, and peracetic acid, PPA, to break the recalcitrant lignin layer from other polysaccharides. It was shown that the pretreatment provides 59.25% lignin removal while maintaining 87.72% hemicellulose and 96.17% cellulose.

Sharma [46] investigated the enzymatic degradation of residual noncellulosic polysaccharides present on dew-retted flax fibers. A mixture of pectinase and Ultrazyme and an enzyme from *Ceraceomyces sublaevis* were found to be the most effective enzymes for depolymerization of noncellulosic material.

Enzymatic degumming of ramie fibers was investigated by using pectate lyase enzyme [67]. Culture supernatants from the *Amycolata* sp. with high pectate lyase activities were most efficient in fiber separation reducing the gum content of ramie fibers by 30% within 15 h. On the other hand, xylanase from *Amycolata* sp. contributed little to degumming.

Akin et al. [68] investigated enzyme retting of flax with enzyme formulations consisting of Viscozyme L, a pectinase-rich commercial multienzyme complex prepared from the fungus *Aspergillus* and ethylenediaminetetraacetic acid (EDTA) chelator. Enzyme levels of 0.3% (v/v) plus 25 mM EDTA gave the best yarn properties. The usage of chelator EDTA causes the removal of Ca^{++} from pectin bridges and destabilizes the cell wall. Longer incubation times increased separation of bast fibers. Addition of chelator agent (EDTA) at 50 mM improved the retting further with a greater separation of the fibers from cuticle-/epidermis-based bast region (Figs. 3.7 and 3.8). Table 3.3 shows the properties of enzyme-retted and commercially cleaned and cottonized flax fiber. Enzyme retting of fibers with 0.3% Viscozyme L plus chelator resulted in finer fibers, with the addition of 50 mM EDTA in the formulation producing the finest fibers. Lower levels of chelator resulted in coarser fibers. Fiber elongation reduced with enzyme retting. Fiber strength was lowest with 0.3% Viscozyme L treatments. Typically, fiber strength reduced with an increase in enzyme concentration. Fiber length and uniformity were similar for all treatments.

Ossola and Galante [69] tested the scouring of flax fibers with enzymes. They showed the effectiveness of enzymes such that pectinase > xylanase > galactomannanase = protease > lipase > or = laccase. They also showed that pectinase treatment increases yarn strength and decreases the number of imperfections when compared to traditional chemical scouring. Figure 3.9 shows the results of yarn count and strength of control (no enzyme) and enzymatically treated yarns. "No-enzyme control" is the baseline value, and experimental results are as percentages above or below the baseline. All enzymatic treatments improved yarn count (thinner yarns) without negatively affecting its strength. Pectinase plus chelators gave the highest yarn count of all enzymatic treatments and the best overall strength value. Other enzymes like xylanase, galactomannanase, and protease gave good strength parameters without improving yarn count.

Liu et al. [70] treated bamboo fibers with a number of enzymes, that is, xylanase, cellulase, pectin lyase, and laccase. Fiber fineness improved as a result of the removal of hemicellulosic substances.

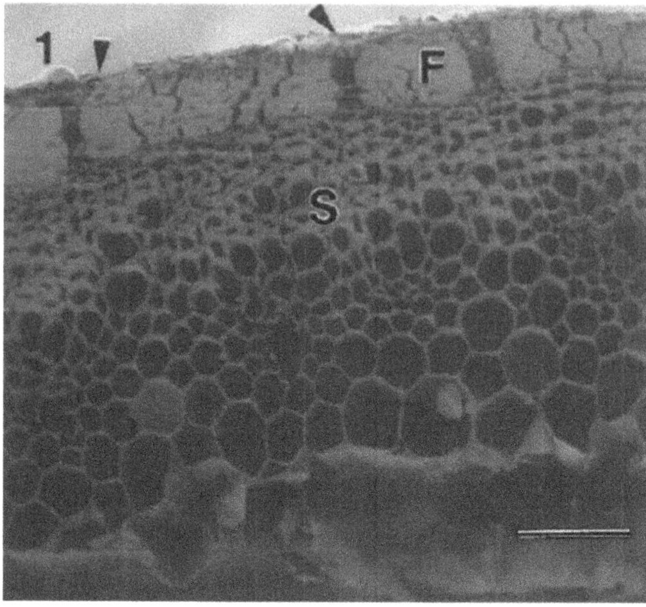

FIGURE 3.7 Scanning electron micrograph of control, unretted Ariane fiber flax stem showing fiber bundles (F) between shive (S) and the cuticle/parenchyma barrier (arrows). Bar = 50 μm. *Source*: Akin et al. [68]. Reprinted with permission from Elsevier. Copyright 2001.

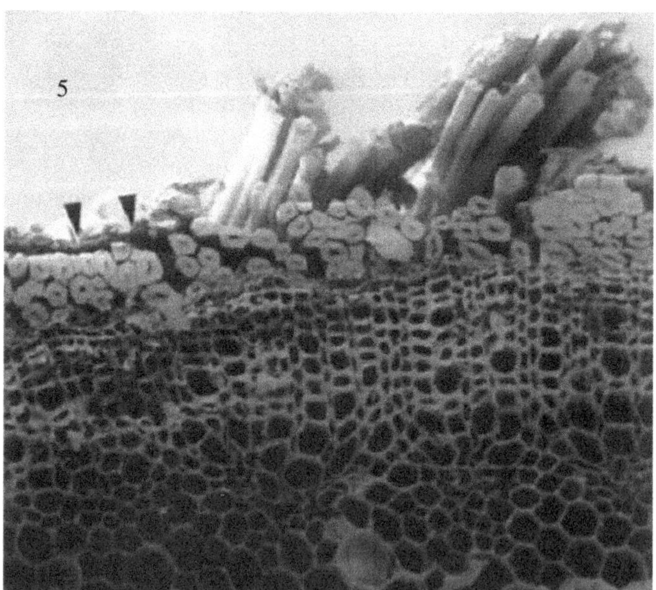

FIGURE 3.8 Scanning electron micrograph of Ariane fiber flax stem retted in Viscozyme L plus 50 mM EDTA at pH 5.0 for 8 h showing significant separation from the epidermis/cuticle (arrows) and loosening of fibers and fiber bundles within the bast region. Bar = 50 μm. *Source*: Akin et al. [68]. Reprinted with permission from Elsevier. Copyright 2001.

TABLE 3.3 Properties of enzyme-retted and commercially cleaned and cottonized flax fiber[a]

Sample	Retting treatment		Micronaire	Strength (g tex^{-1})[b]	Length		Shirley-cleaned yield (%)
	EDTA (mM)	Viscozyme (%)			UQL	Uniformity	
Seed flax straw	50	0.05	$5.9 \pm 0.1^{c,d}$	$25.9 \pm 2.9^{c,d}$	1.355 ± 0.009	78.3 ± 0.6	$25.3 \pm 1.0^{e,f}$
Seed flax straw	25	0.05	6.0 ± 0.2^{c}	$19.6 \pm 1.1^{g,e}$	1.200 ± 0.075	73.7 ± 1.6	23.6 ± 1.0^{f}
Ariane (early)	50	0.05	$5.8 \pm 0.1^{d,g}$	24.0 ± 2.0^{d}	1.346 ± 0.114	74.6 ± 9.7	$30.7 \pm 8.8^{g,e}$
Ariane (early)	25	0.05	5.7 ± 0.1^{g}	20.9 ± 1.3^{g}	1.428 ± 0.024	78.4 ± 2.1	$37.9 \pm 0.3^{c,d}$
Ariane (early)	50	0.3	3.9 ± 0.1^{h}	13.0 ± 1.3^{h}	1.247 ± 0.025	71.3 ± 1.9	61.4 ± 0.7^{b}
Ariane (early)	25	0.3	4.6 ± 0.1^{f}	15.8 ± 1.8^{f}	1.247 ± 0.037	73.0 ± 2.0	58.7 ± 1.1^{b}
Ariane (early)	Dew retted	3 weeks	5.3 ± 0.1^{e}	36.2 ± 2.3^{b}	1.311 ± 0.082	73.8 ± 0.5	43.0 ± 1.1^{c}
Ariane (late)	50	0.05	6.7 ± 0.1^{b}	26.8 ± 3.4^{c}	1.428 ± 0.065	79.9 ± 2.5	$32.3 \pm 0.3^{d,g}$

[a] Akin et al. [68], Table 1. Reprinted with permission from Elsevier. Copyright 2001.
[b–h] Within columns, values with different letters differ at $P < 0.05$.

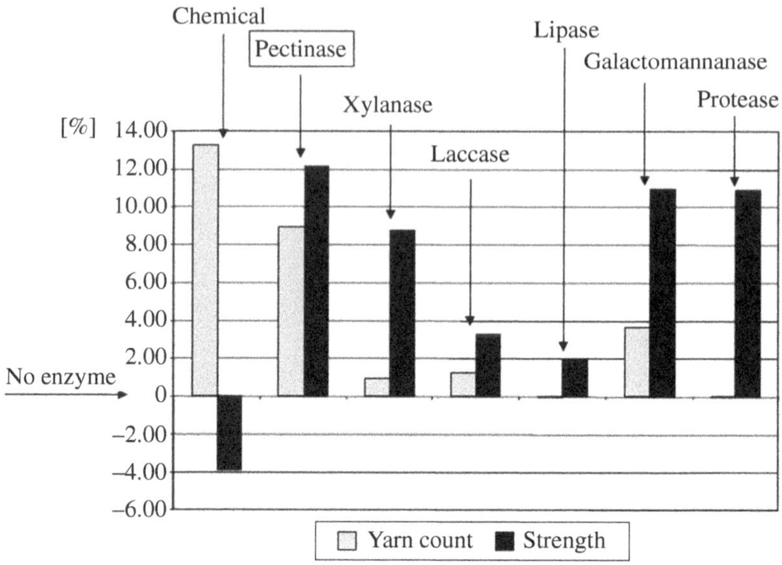

FIGURE 3.9 Yarn count and strength. A high yarn count value indicates a thin yarn, thus greater removal of material from the fiber. *Source*: Ossola and Galante [69], figure 4. Reprinted with permission from Elsevier. Copyright 2003.

Ouajai and Shanks [71] investigated the morphology and structure of hemp fiber after bioscouring using pectate lyase and Scourzyme L. The reaction was performed at 55°C and pH 8.5 without agitation. Greater enzyme concentration and a longer treatment time improved the removal of the low-methoxy pectin component. It has also been reported that removal of pectate caused no crystalline transformation in the fibers, except for a slight decline in the crystallinity order index. They determined that a linear relationship exists at a low enzyme concentration, indicating that the gradual removal of pectate from fiber is directly proportional to the enzyme concentration. However, the further increase of enzyme concentration (2.5–10% owf) did not improve the efficiency simply because of the limitation of the substrate to bind with the enzyme. Another important finding of this study was that the IR lateral crystallinity index decreased as a function of Scourzyme concentration and treatment time. This suggested that the enzymatic treatment resulted in a less ordered crystalline structure. Surface composition of fibers after the treatment was also observed with FTIR. Results indicated that after treatment, pectin still existed but with a degree of methyl ester content. This suggested that the enzyme specifically attacks and removes the nonesterified fraction from pectin. The tensile modulus of the fibers gradually decreased with Scourzyme L concentration. The authors attributed this result to reduction in the noncellulosic content that binds the cellulose fibers together. The tensile stress showed similar behavior to the tensile modulus, whereas elongation at the break did not display any reasonable trend. This indicated that enzymatic treatment of fibers causes some degradation in the mechanical properties of hemp fibers. Pectin removed in the early stage did not affect much the strength of the fibers. In contrast, prolonged enzymatic treatment led to the removal of pectin deposited in cell walls, which provide the strength of the fibers. SEM images of

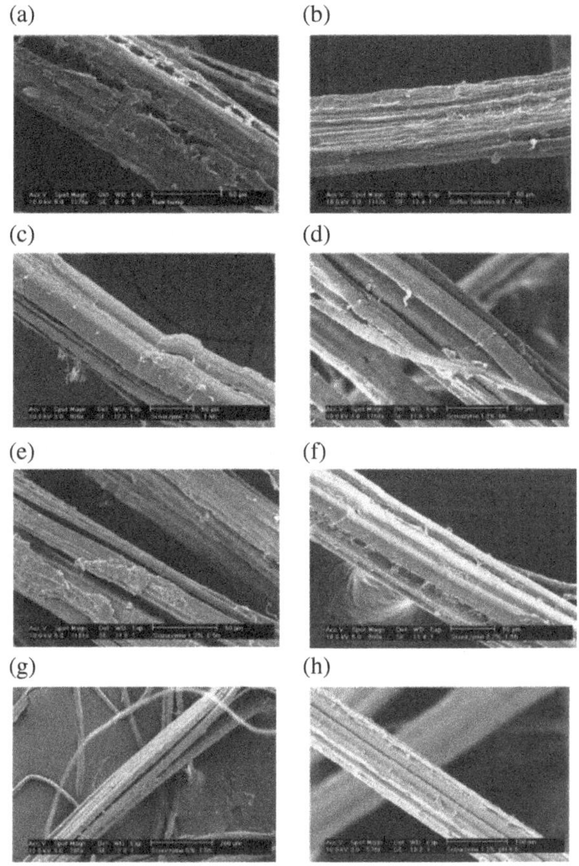

FIGURE 3.10 SEM micrographs of Scourzyme-treated hemp fibers (a–b) untreated, (c) 1.2% Scourzyme L-treated, (d) 1.2% Scourzyme L with extended treatment time (6 h), (e) 1.2% Scourzyme L with a short treatment time (30 min), (f) the treatment at low (0.2%) enzyme concentration, (g) the treatment at high (5%) enzyme concentration (h) 1.2% Scourzyme L at pH 6.5. *Source*: Ouajai and Shanks [71]. Reprinted with permission from Wiley-VCH Verlag GmbH & Co. KGaA. Copyright 2005.

control and enzymatically treated fibers are shown in (Fig. 3.10). Figure 3.10a, b shows the fiber bundles of untreated hemp fibers covered by noncellulosic materials. The surface of 1.2% Scourzyme L-treated fibers (Fig. 3.10c) is smooth, but the fibers are still assembled in bundles. Extended treatment time (6 h) causes further elimination of pectin and deeper interfibrillar disintegration of the bundle (Fig. 3.10d). Figure 3.10f, g shows the treatment at low (0.2%) and high (5%) enzyme concentrations. Low enzyme concentration was sufficient to expose the fiber surface but not enough to proceed further and break down the calcified pectin between the fibers. The best result was obtained with 1.2% enzyme concentration at a pH of 8.5 (Fig. 3.10g). Scouring at an enzyme concentration of 1.2% at pH 6.5 (Fig. 3.10h) caused a lower change in fiber surface when compared to pH 8.5 treated fiber indicating that the optimum pH range for enzymatic treatment is between pH 8 and 9.

3.6 POLYMER COMPOSITES REINFORCED WITH ENZYMATICALLY MODIFIED NATURAL FIBERS

Enzymatic treatment of fibers to obtain a better fiber–matrix interface in reinforced composites has been the subject of several studies. In general, enzymatic treatment of fibers improves fiber–matrix adhesion and the ultimate mechanical properties of the composites. Table 3.4 shows the property improvements obtained with enzymatically modified natural fiber composites.

Foulk et al. [72] analyzed flax and cotton fiber fabric blends and recycled polyethylene composites. They compared the dew-retted and enzyme-retted (with Viscozyme

TABLE 3.4 Property enhancements of enzymatically modified natural fiber composites

Composite type	Enzyme type(s) used	Improved composite properties after enzymatic modification	Reference
Flax–cotton/PE	Pectinase-rich multienzyme product (Viscozyme L) + chelator	Interfacial adhesion Strength and stiffness	[72]
Jute/PES	Cellulase, pectinase, laccase, xylanase	Interfacial adhesion Young's modulus Flexural modulus	[73]
Jute/PES	Cellulase, pectinase, laccase, xylanase	Storage modulus	[74]
Hemp/PP	Pectinase, laccase, chelator	Interfacial adhesion Tensile strength	[75]
Abaca/PP	Fungamix	Moisture absorption Tensile strength Flexural strength Charpy impact strength	[76]
Hemp/PP	Pectinase	Tensile strength Flexural strength Elastic moduli Wettability	[77]
Hemp/cellulose acetate butyrate	Pectate lyase	Tensile modulus	[78]
Kenaf/PP	Laccase	Moisture absorption Tensile, flexural, and impact properties Fiber crystallinity Thermal properties Fiber–matrix adhesion	[79]
Kenaf/PP	Laccase	Tensile strength Fiber crystallinity Thermal properties	[80]

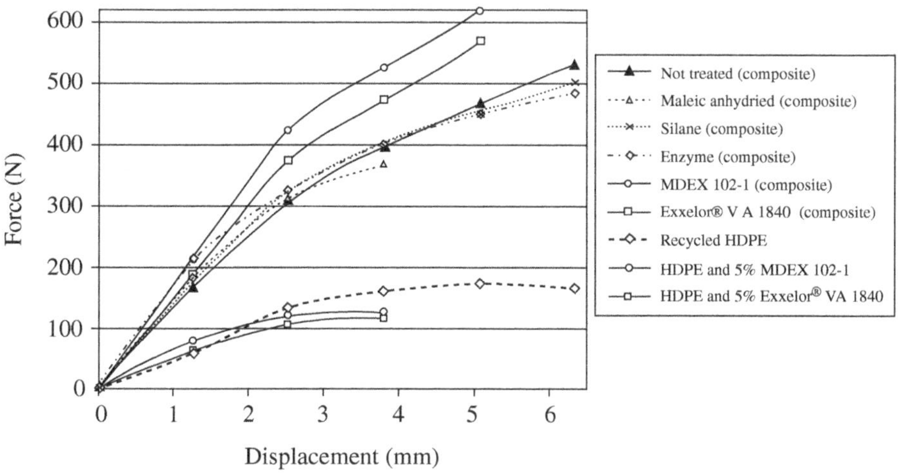

FIGURE 3.11 Typical force displacement curves for various composite materials. *Source*: Foulk et al. [72]. Reprinted with permission from Springer Science+Business Media. Copyright 2006.

L and a chelator, Mayoquest 200) flax fibers reinforced high-density polyethylene (HDPE) composites. Enzymatic treatment removed pectin, imparting enhanced interfacial adhesion between fibers and the matrix. Figure 3.11 shows the comparison between the force–elongation curves of various materials. It is noteworthy that enzyme-treated composite demonstrated higher initial stiffness behavior but similar strength to other samples. This result could be associated with better wettability of enzyme-treated fibers and improved fiber–matrix bonding. Composite percent elongation (12.1%) was largest for the enzyme-treated composite.

Karaduman et al. [73] investigated the effect of enzymatic treatment on the properties of jute fiber-reinforced polyester composites. Jute fabrics were treated with cellulase, pectinase, laccase, and xylanase enzyme solutions with varying enzyme mixtures and treatment times. The fiber–matrix interface and mechanical properties of enzyme-treated composites improved. SEM images of fiber–matrix interfaces of jute composites are given in Figure 3.12. It is clearly shown that enzyme-treated composite samples have better fiber–matrix adhesion when compared to untreated fiber composites. Enzyme acted on hemicellulose polymers reducing them to smaller polysaccharides. Enzymes catalyzed the removal of amorphous and disordered polymers that prevent α-cellulose chains from rearranging themselves in the direction of load. This also led to an increment in the Young's modulus of the composites reinforced with enzyme-treated fibers.

Karaduman and Onal [74] investigated dynamic mechanical and thermal properties of enzyme and NaOH-modified jute fiber polyester composites. Storage modulus of the composites increased by using treated fibers when compared to composites reinforced with untreated fibers. Thermal stability of the composites decreased due to the removal of thermally more stable lignin polymer.

(a) (b) (c)

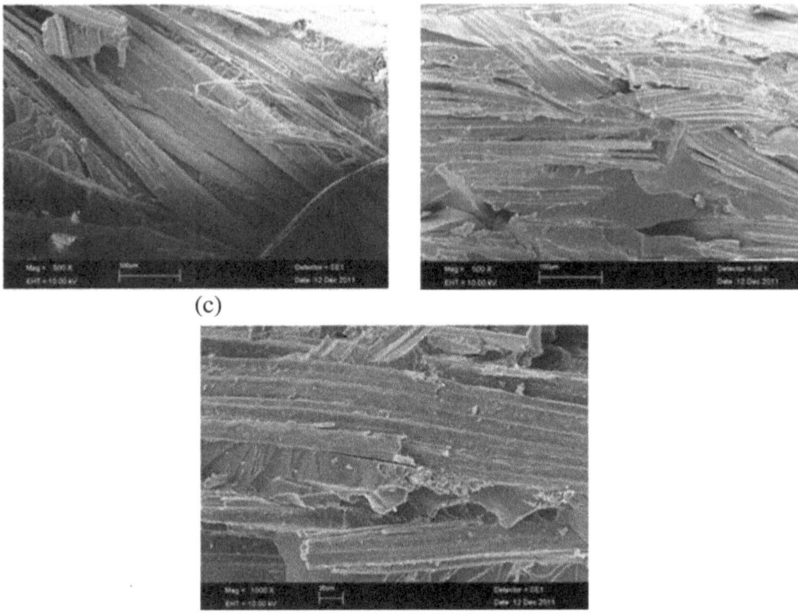

FIGURE 3.12 SEM images of fiber–matrix interfaces of jute composites. (a) Untreated. (b) Pectinase + laccase + cellulase treated. (c) Laccase treated [73].

George et al. [81] investigated surface and thermal characterization of enzyme-treated natural fibers (hemp and flax) for composite applications. They studied the effect of five enzymatic systems (xylanase, pectinmethylesterase, polygalacturonase, and laccase) on the chemical, morphological, and thermal properties of natural fibers. They treated flax and hemp fibers with hemicellulases, pectinases, and oxidoreductases. Each treatment changed the topography of fibers exposing the individual fiber bundles. Treatment with hemicellulase and pectinase improved the thermal properties of both fibers. Fiber treatment with xylanase and pectinases led to a reduction in hemicelluloses content of both fibers. They also reported that the removal of amorphous hemicellulosic material from the fiber surface led to the exposure of crystalline cellulose network and resulted in a lower contact angle for all the treated samples. The results indicated that enzymes could be used to remove the least stable pectic and hemicellulosic content and increase the thermal stability of natural fibers. Table 3.5 shows the chemical compositions of the treated samples with respect to cellulose, hemicellulose, and lignin. It was shown that, for both fiber types, the cellulose content remained unaltered in all the treatments except for xylanase + cellulase treatment, which clearly indicates the role of cellulase enzyme in the degradation of cellulose. On the other hand, hemicellulose content in the fibers substantially decreased after xylanase and pectinase (polygalacturonase and pectinmethylesterase) treatments. The authors attributed this result to the removal of pectic matter bonded to hemicellulosic content, which leads to disintegration and rupture of the

TABLE 3.5 Chemical compositions of the treated samples with respect to cellulose, hemicellulose, and lignin[a]

		Component (% of dry weight)		
Fiber	System	Cellulose	Hemicellulose	Lignin
Hemp	Untreated	67.77 ± 2.25	12.14 ± 3.51	7.24 ± 0.64
	Xylanase	67.18 ± 1.46	7.89 ± 0.81	7.04 ± 0.61
	Xylanase + cellulase	57.42 ± 2.12	11.80 ± 1.02	6.48 ± 1.35
	Laccase	68.01 ± 2.48	10.97 ± 0.86	6.88 ± 0.05
	Polygalacturonase	68.63 ± 4.58	6.40 ± 2.21	5.09 ± 1.10
	Pectinmethylesterase	68.35 ± 1.31	7.45 ± 1.81	6.14 ± 1.17
Flax	Untreated	64.08 ± 3.09	16.10 ± 3.02	4.14 ± 0.42
	Xylanase	63.56 ± 2.65	11.37 ± 1.25	3.53 ± 0.09
	Xylanase + cellulase	54.50 ± 1.31	8.69 ± 1.96	4.18 ± 0.89
	Laccase	63.06 ± 2.09	13.35 ± 2.64	1.45 ± 0.12
	Polygalacturonase	61.31 ± 1.99	2.55 ± 0.84	1.43 ± 0.28
	Pectinmethylesterase	61.12 ± 1.72	4.15 ± 1.82	2.81 ± 0.06

[a] George et al. [81], Table 3. Reprinted with permission from Elsevier. Copyright 2014.

network. Lignin content of hemp fiber seems little affected by enzymes showing values well between the experimental ranges. Flax fiber showed different behavior in terms of its lignin content where notable reductions were recorded especially for laccase and polygalacturonase treatments. This was probably because of different structures of the two fibers governing the accessibility of enzymes through deeper sites of lignin deposition.

Li and Pickering [75] used chelators and enzymes to separate hemp fiber from its bundles as well as remove noncellulosic materials in an attempt to improve interfacial bonding in the composites. They used pectinase (P), laccase (L), 5 g/l EDTMP.Na5 (ethylene diamine tetra methylene phosphonic acid pentasodium salt) (E), 10 g/l EDTMP.Na5 (E2), E + P, and E + L solutions. They reported that the crystallinity and thermal properties improved after enzymatic treatment. Composites with treated hemp fiber showed an increase of 19% in tensile strength when compared to composites with untreated fibers. This result was attributed to improved fiber–matrix adhesion as a result of enzymatic treatment. Thermal stability of the treated fibers was increased due to the reduced amount of noncellulosic material and increased crystallinity. The crystallinity index of untreated fibers recorded as 66.3%, whereas the crystallinity index of treated fibers recorded in the range of 85.8–86.3 for chelator and enzymatic treatments. The tensile strength of fibers reduced slightly for all treatments when compared to untreated fiber (Fig. 3.13). This result was associated with cellulose degradation during the treatments imparting larger effect on the composite properties than increased crystallinity index values. Composite tensile strength values were compared in (Fig. 3.14). All treatments produced an improvement in composite strength. Untreated fibers are bundled together and covered with noncellulosic compounds such as wax, pectin, and lignin, which would limit the effective surface area for the reach of matrix and hinder chemical fiber–matrix bonding.

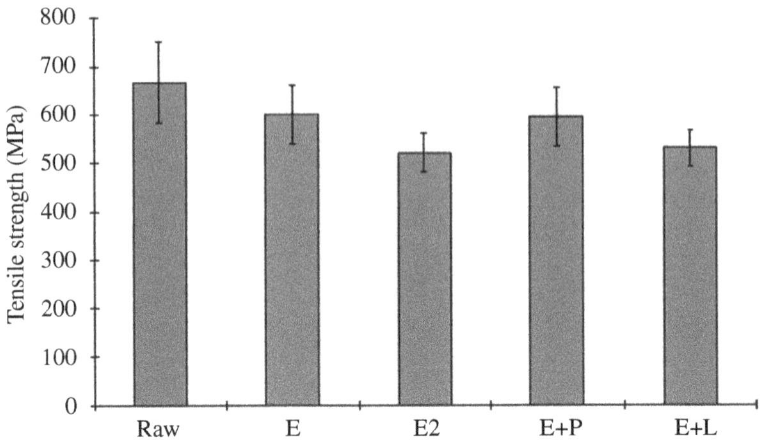

FIGURE 3.13 Single fiber tensile strength of untreated and treated fibers. *Source*: Li and Pickering [75]. Reprinted with permission from Elsevier. Copyright 2008.

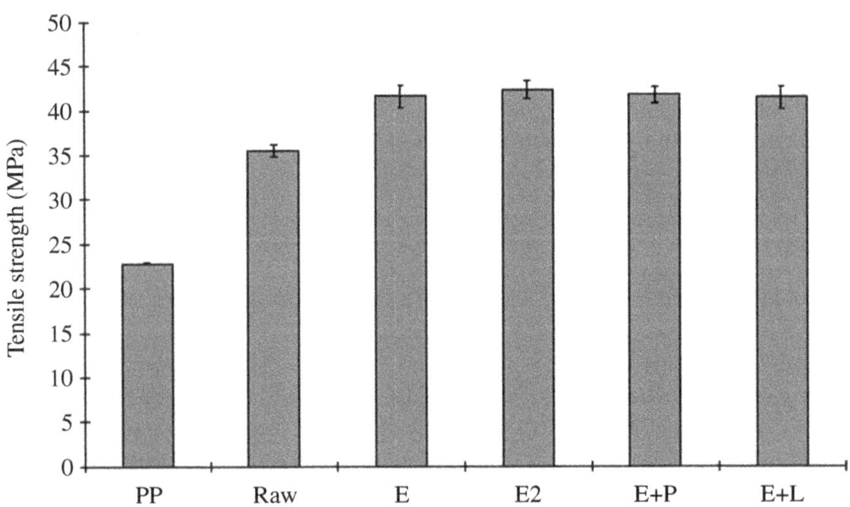

FIGURE 3.14 Composite tensile strength of untreated and treated fibers. *Source*: Li and Pickering [75]. Reprinted with permission from Elsevier. Copyright 2008.

Removal of these substances during treatment separates fiber bundles and exposes hydroxyl groups on the fiber surface. This would increase the interfacial bonding between fiber and matrix resulting in improved composite strength despite the slightly lower single fiber strength for treated fibers. SEM photographs of untreated and treated fiber composite fracture surfaces are shown in Figure 3.15. Fracture surfaces for untreated fiber composite showed debonding between fiber and matrix, whereas the fiber–matrix adhesion is far better for treated sample.

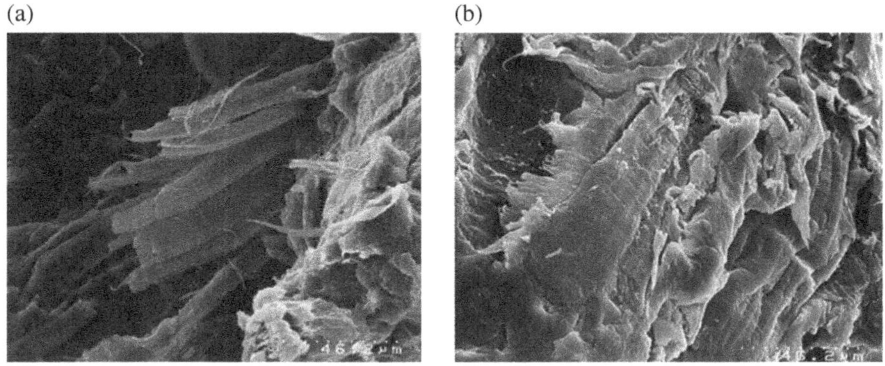

FIGURE 3.15 SEM photomicrographs of composite fracture surfaces (a) untreated (b) E2 treated. *Source*: Li and Pickering [75]. Reprinted with permission from Elsevier. Copyright 2008.

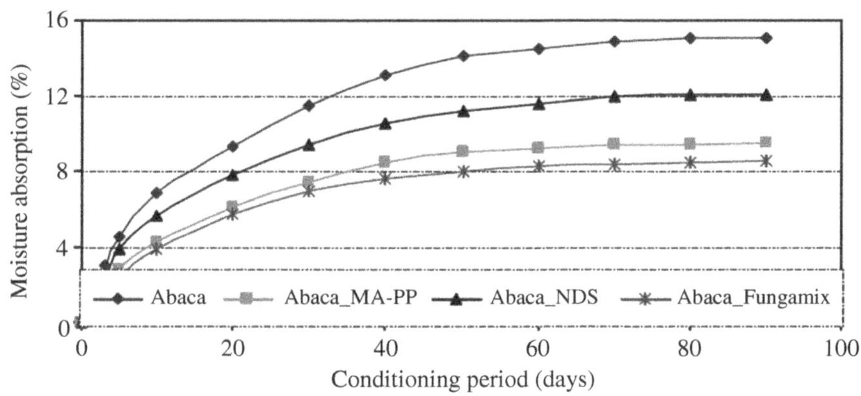

FIGURE 3.16 Moisture absorption of modified and unmodified abaca/PP composites. *Source*: Bledzki et al. [76]. Reprinted with permission from Elsevier. Copyright 2010. (*See insert for color representation of the figure.*)

Bledzki et al. [76] produced polypropylene composites with enzyme-modified abaca fiber. Prior to composite production, they modified the fibers with fungamix (Novozyme) and natural enzyme. They reported that the moisture absorption of the composites reduced 20–45% due to modification (Fig. 3.16). Tensile strength 5–45% and flexural strength 10–35% increased due to modification (Fig. 3.17). They attributed these improvements to selective removal of surface extractives, fat, protein, lignin, and hemicellulose without destroying fiber surface, which leads to a stronger fiber–matrix adhesion. Charpy impact strength of composites also improved with natural digestive system (NDS) and fungamix treatments. It is noteworthy that the fungamix-modified fiber composite was even superior to MAPP modified one

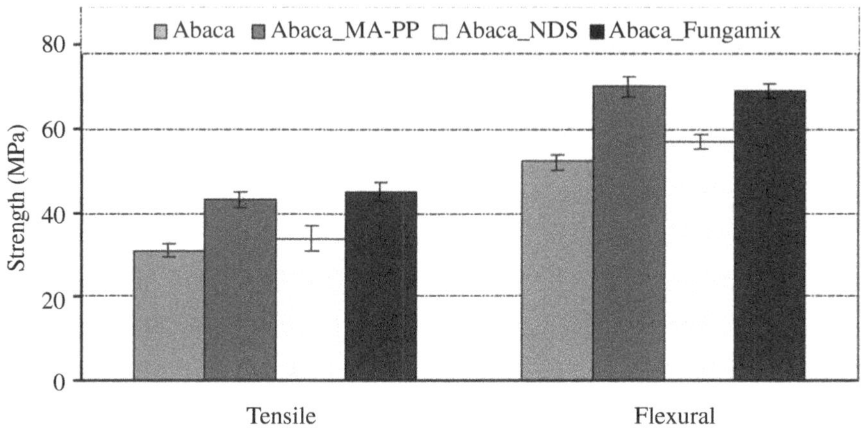

FIGURE 3.17 Tensile and flexural strength of modified and unmodified abaca/PP composites. *Source*: Bledzki et al. [76]. Reprinted with permission from Elsevier. Copyright 2010. (*See insert for color representation of the figure.*)

in terms of moisture absorption, mechanical, and rheological properties. Enzyme-modified composites also demonstrated better resistance in acid and alkaline medium. This study is a good example of enzymatic treatment to first separate the fiber bundles into fiber ultimates as well as fiber surface modification with an aim of improving interaction between fibers and resin by removing loosely bonded materials.

Saleem et al. [77] treated hemp fibers with pectinase to determine its effect on the mechanical properties of thermoplastic composites. Incubating technical fibers with enzyme reduced their cross-sectional area and tensile strength. They reported that the tensile strength, flexural strength, and elastic moduli of the composites improve after 8% enzymatic treatment in maleic anhydride medium. They attributed this improvement to increased surface area of fibers after enzymatic treatment as a consequence of the separation of technical fibers into smaller bundles and single fiber cells. This resulted in better wettability with polypropylene matrix.

Ouajai and Shanks [78] prepared cellulose acetate butyrate biocomposites reinforced with modified hemp. Composite containing pectate lyase enzyme-treated fibers showed higher modulus in comparison to untreated or alkali-treated fibers at a similar fiber length of 100 μm.

Islam et al. [79] used raw and laccase-treated kenaf fiber to produce recycled polypropylene composites. Enzyme concentration and treatment time were considered as the treatment parameters. Density and crystallinity of the treated fibers increased. Enzymatic treatment improved composite tensile strength by 40%. It was also reported that flexural, impact, and thermal properties of the composites improved due to fiber treatment. Morphological images of the composites indicated better adhesion. Moisture absorption of the composites also reduced as a consequence of fiber treatment.

Islam et al. [80] produced composites using untreated and treated kenaf fibers with recycled polypropylene. Surface modifications such as ultrasound, enzyme, and alkali pretreatments were carried out in order to improve interfacial bonding between fibers and matrix. Results revealed that ultrasound was able to remove the highest amount of lignin (32%). Tensile strength of the composites increased by 57, 58, and 40% due to alkali, ultrasound, and enzymatic treatments, respectively. Optimum enzyme concentration and soaking time were found as 1.26% and 3.89h, respectively. A correlation among crystallinity, surface morphology, and tensile and thermal properties of the composites with the fiber–matrix interaction was established.

3.7 ENZYME-ASSISTED BIOGRAFTING METHODS

Enzyme-assisted biografting is a relatively new and very promising modification method that involves grafting of organic molecules onto lignocellulosic fibers with the aid of enzymes. The final aim of biografting is to increase the performance of lignocellulosics by improving mechanical properties such as strength and stiffness, hydrophobicity, resistance to moisture, and microbial attack. Laccases, lipases, and peroxidases are among the enzymes being investigated for biografting of organic molecules onto lignin for improved properties of lignocellulosic fibers [82].

Lignocellulosic fibers usually have a very poor microbial and moisture resistance, and they are hydrophilic in nature. These negative properties limit their usage in composite industry despite their sufficient mechanical properties and low price. Enzyme-assisted modification of lignocellulosic fibers is a way to overcome these problems and add functional properties to natural fibers [61, 62, 83–85].

Laccases and peroxidases are main enzymes studied for the surface functionalization of lignocellulosics. These enzymes can oxidize a wide range of natural and synthetic polymers and generate reactive species such as phenoxy radicals increasing the reactivity of polymers. Laccases are usually preferred since they are more readily available and easy to manipulate in comparison to peroxidases. Furthermore, peroxidases require hydrogen peroxide as cofactor, which increases the process cost [82]. Laccase can catalyze the oxidation of a range of aromatic compounds especially phenols. Laccase can oxidize only phenolic groups of lignin to phenoxy radicals due to the random polymer structure of lignin and redox potential limitations. These phenoxy radicals can be exploited to graft functional molecules onto lignocellulosic fibers (Fig. 3.18).

Laccase-assisted biografting of phenolics on lignocellulosic materials enhances hydrophobicity and helps to reduce the chances of microorganism growth [86]. It also has possibility to reduce incompatibility between fiber and matrix.

Laccase grafting is used in papermaking to improve the properties of the pulp. Laccase is used to copolymerize low molecular weight phenols with the pulp before papermaking through a radical coupling reaction [87]. The addition of methyl syringate (MS) gave a twice wet tensile index of unbleached kraft pulp but had little effect on the dry tensile index. The radical concentration in laccase-/MS-treated fibers increased substantially.

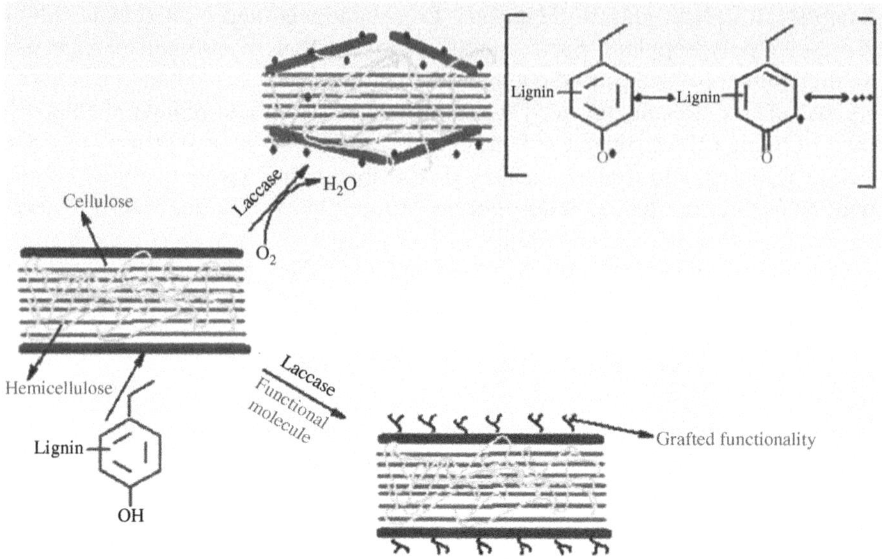

FIGURE 3.18 Laccase-assisted grafting of functional molecule on lignocellulosics. *Source*: Kalia et al. [82], figure 2. Reprinted with permission from Elsevier. Copyright 2014. (*See insert for color representation of the figure.*)

Laccase was reacted with 4-hydroxybenzoic acid (4-hba) and high-kappa (91) kraft pulp in order to obtain grafting of 4-hba. Laccase was shown to increase the molecular weight of 4-hba through a polymerization reaction [88].

Schroeder et al. [89] investigated enzymatic coating of lignocellulosics with polyphenols. They used laccase from *Trametes hirsuta* to conduct laccase-induced coating of flax fibers. A screening was carried out with different phenols for their potential as monomers for enzyme-catalyzed polymerization resulting in a coating with an improved antimicrobial performance. Bacterial growth of *B. subtilis* and *Staphylococcus aureus* was reduced significantly using ferulic acid and hydroquinone. Using laccase-induced coupling and polymerization, coloration and antimicrobial performance were achieved.

Aracri et al. [90] studied the effects of laccase-catalyzed grafting of ferulic acid on sisal specialty pulp fibers. Refining the pulp prior to the enzymatic treatment resulted in increased grafting, which in turn led to the handsheets with improved strength properties.

Enzymatic grafting of simple phenols on flax and sisal pulp fibers was carried out [91]. Two laccases were used in the presence of different phenolic compounds. In most cases the enzymatic treatments resulted in increased kappa number of pulps suggesting the incorporation of phenols into fibers with covalent bonds.

Fillat et al. [92] studied the grafting of natural phenols to flax fibers. Unbleached flax fibers for paper production were treated with laccase from *Pycnoporus cinnabarinus* and low molecular weight phenols. Covalent binding of the phenoxy radicals on fiber was established. Paper handsheets from treated pulps showed antimicrobial activity against bacteria.

Ubasart et al. [93] investigated the hydrophobization of cellulose fibers from hardwood kraft pulp using laccase and hydrophobic phenolic compound (lauryl gallate). The treatment resulted in significantly lower water penetration in the handsheets and wettability of the paper surface. SEM images of the fiber network revealed the presence of the sizing agent, which is a product of the reaction between laccase and lauryl gallate.

Chandra et al. [94] modified high lignin content kraft pulps with laccase in the presence of gallic acid. The result was a modified pulp with 34, 20, and 72% improvements in burst, tensile, and wet tensile strength compared to untreated control samples. They found that fully bleached pulps were not responsive to the laccase treatment indicating that lignin was the major target for laccase treatment. The strength increases were a combined effect of improvements of hydrogen bonding between fibers and creation of phenoxy radical cross-links within the sheet.

Kudanga et al. [95] used alkylamines instead of phenolics in a laccase-catalyzed coupling of long-chain alkylamines to lignin model molecules and lignocellulose. They used a wide range of alkylamines and lignin molecules. The grafting of alkylamines onto beech veneers resulted substantial increase (up to 84.2%) in hydrophobicity with respect to contact angle measurements. There are numerous other studies that showed laccase-assisted grafting of various groups onto lignocellulosics improves the strength, antimicrobial resistance, and hydrophobicity.

Chemoenzymatic grafting is a relatively new and nonselective way of enzyme-assisted biografting that targets the modification of not only lignin but all substituents in lignocellulosic fibers. Kudanga et al. [96] used this method for functionalization of lignocellulose materials. They used oxiranes coupled onto long-chain alkylamines, and oxiranes were in turn coupled onto phenolic lignin model compounds. Treatment of beech veneers with dodecylamine using this process led to a 90% increase in hydrophobicity.

Dong et al. [97] investigated hydrophobic modification of jute fiber via laccase-assisted grafting and its use for composite reinforcement. Dodecyl gallate (DG) was used for coupling reaction as described in Figure 3.19. The hydrophobicity of jute fabrics was estimated by means of contact angle and wetting time measurements. The mechanical properties of modified and unmodified jute fiber/polypropylene composites were studied. The results showed that covalent coupling of DG to the jute substrates occurred. The surface hydrophobicity of the jute fabric increased dramatically after the enzymatic modification. The breaking strength of the modified jute fiber/PP composite increased with modification. SEM studies revealed a better fiber–matrix adhesion at the composite interface. The time-dependent contact angle of a water drop on jute fabric samples is shown in Figure 3.20. For the control fabric the contact angle was recorded as $29.75° \pm 3.75°$ at the beginning, and the water droplet disappeared within 3–5 s. On the other hand, the laccase-/DG-treated sample showed larger contact angle $111.49° \pm 2.90°$ due to the hydrophobic DG grafted on the surface. The contact angle of this sample remained unchanged in 300 s. Photographs of colored water droplet on the surface further confirm the increase in surface

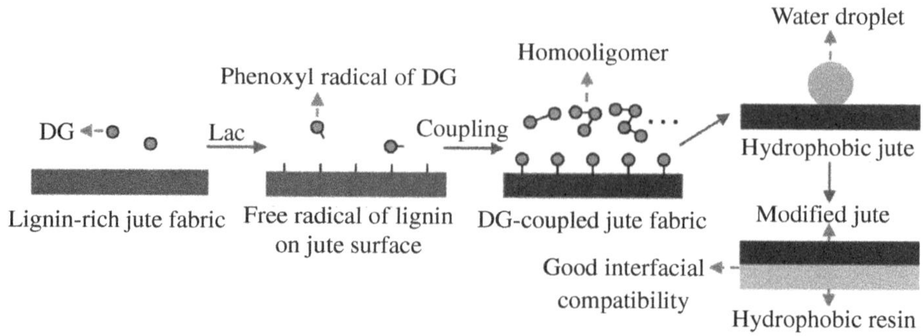

FIGURE 3.19 Schematic illustration of the grafting reaction of dodecyl gallate (DG) on the lignin-rich jute fabric catalyzed by laccase. *Source*: Dong et al. [97]. Reprinted with permission from Elsevier. Copyright 2014. (*See insert for color representation of the figure.*)

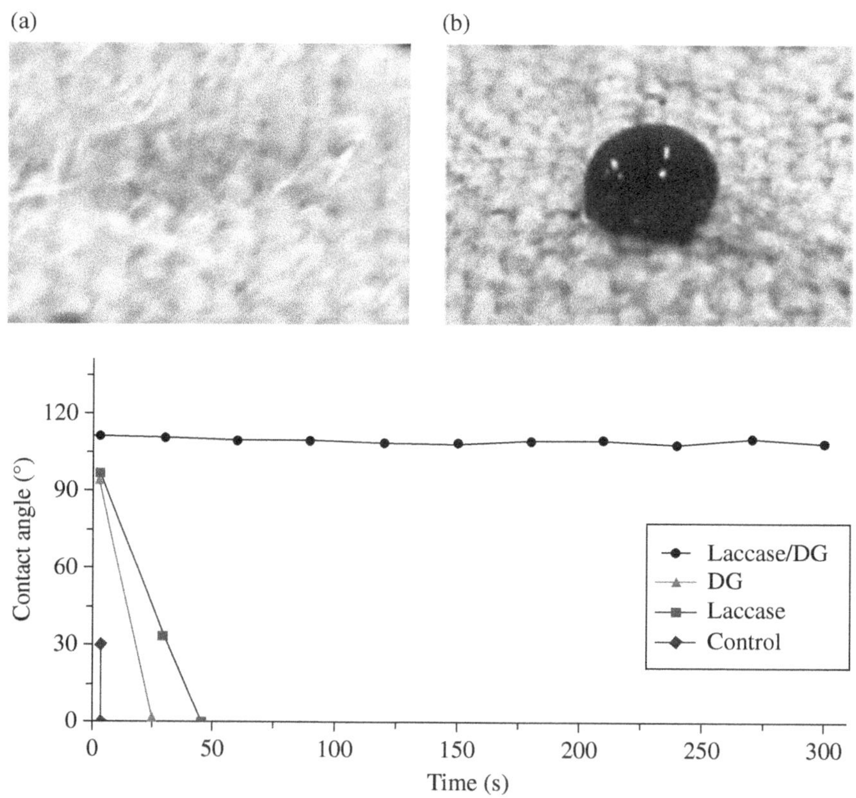

FIGURE 3.20 Contact angles over the time for the control jute (control), the laccase-treated jute (laccase), the DG-alone-treated jute (DG), and the laccase-/DG-treated jute (laccase/DG). Insert: digital photographs of water drop captured on the surface of control jute after 5 s (a) and of treated jute with laccase and DG after 30 min (b). *Source*: Dong et al. [97]. Reprinted with permission from Elsevier. Copyright 2014. (*See insert for color representation of the figure.*)

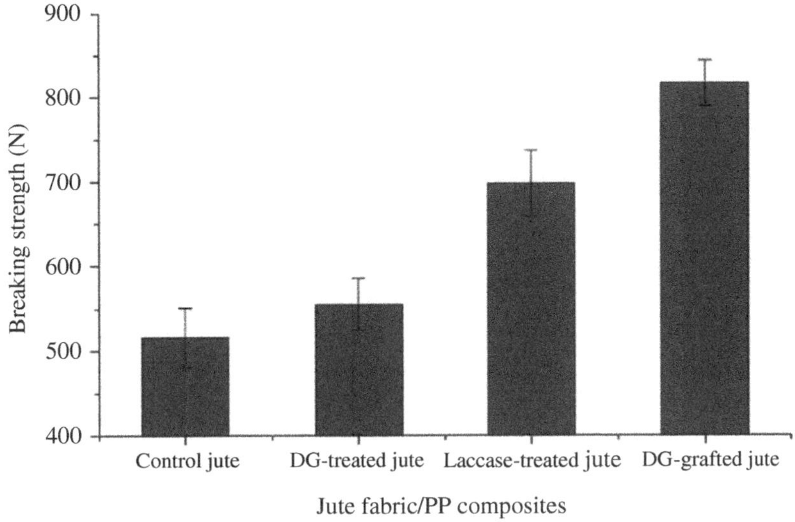

FIGURE 3.21 Breaking strength of polypropylene (PP) composites reinforced by dodecyl gallate (DG)-grafted jute fabric, laccase-treated jute fabric, DG-treated jute fabric, and control jute fabric. *Source*: Dong et al. [97]. Reprinted with permission from Elsevier. Copyright 2014.

hydrophobicity (Fig. 3.20). A comparison between the breaking strengths of various jute/PP composite samples was given in Figure 3.21. The grafted jute fabric-reinforced composites yielded the highest breaking strength followed by the laccase-treated jute and the DG-treated jute. This result indicated that the compatibility between jute fibers and PP matrix improved as the hydrophobicity of jute fibers improved by grafting of DG.

Although enzyme-assisted biografting processes for paper- and pulp-related applications are very common, its research with the direct aim of composite industry is yet scarce. Enzyme-assisted biografting yielded many achievements for natural fiber-based materials used in paper-related products such as improved strength, hydrophobicity, and resistance to microbial attack. Especially, hydrophobicity improvement of natural fibers is very important for composite applications since it reduces the incompatibility between fiber and matrix and yields higher mechanical properties accompanied by lower moisture absorption. Therefore, enzyme-assisted biografting of natural fibers for composite applications is a new area of research that is full of potential and needs to attract more attention from scientists and composite industry.

3.8 CONCLUSIONS

Enzymatic pretreatment of natural fibers to obtain a better fiber–matrix interface in natural fiber composites emerged as a cost-effective and environmentally friendly method. Enzymes are highly selective in terms of their substrates resulting in a very productive technique of fiber modification. Enzymatic treatment is environmentally friendly unlike conventional chemical methods, which commonly employ hazardous chemicals and use excessive amounts of freshwater. Enzymatic catalyzation of the removal of substances like hemicelluloses, pectin, lignin, and other compounds results in a purified and cellulose-rich fiber that is more compatible with polymer resin. Enzymes can be used solely or in association with other enzymes for fiber modification. To date, the most popular enzymes used for natural fiber modification are cellulases, laccases, pectinases, and xylanases and their mixtures having different enzyme ratios. Recent studies showed that enzymatic treatment of fibers leads to an improvement in fiber–matrix adhesion in natural fiber composites resulting in enhanced mechanical properties. Although enzymatic treatment of natural fibers for textile purposes has a long history, its usage in composite industry is at an early stage of development and needs to be further investigated for its effective usage.

REFERENCES

[1] Joshi, S.V., Drzal, L.T., Mohanty, A.K., Arora, S. (2004). Are natural fiber composites environmentally superior to glass fiber reinforced composites? *Composites Part A: Applied Science and Manufacturing*, **35**, 371–376.

[2] Wambua, P., Ivens, J., Verpoest, I. (2003). Natural fibres: can they replace glass in fibre reinforced plastics? *Composites Science and Technology*, **63**, 1259–1264.

[3] Bledzki, A.K., Faruk, O., Sperber, V.E. (2006). Cars from bio-fibres. *Macromolecular Materials and Engineering*, **291**, 449–457.

[4] Mehta, G., Mohanty, A.K., Thayer, K., Misra, M. (2005). Novel biocomposites sheet molding compounds for low cost housing panel applications. *Journal of Polymers and the Environment*, **13**, 169–175.

[5] Xie, Y., Hill, C.A.S., Xiao, Z., Militz, H., Mai, C. (2010). Silane coupling agents used for natural fiber/polymer composites: a review. *Composites Part A: Applied Science and Manufacturing*, **41**, 806–819.

[6] Rowell, R.M. (2004). Acetylation of natural fibers to improve performance. *Molecular Crystals and Liquid Crystals*, **418**, 153–164.

[7] Mohanty, S., Nayak, S.K., Verma, S.K., Tripathy, S.S. (2004). Effect of MAPP as a coupling agent on the performance of jute–pp composites. *Journal of Reinforced Plastics and Composites*, **23**, 625–637.

[8] Onal, L., Karaduman, Y. (2009). Mechanical characterization of carpet waste natural fiber-reinforced polymer composites. *Journal of Composite Materials*, **43**, 1751–1768.

[9] Gassan, J., Bledzki, A.K. (1999). Possibilities for improving the mechanical properties of jute/epoxy composites by alkali treatment of fibres. *Composites Science and Technology*, **59**, 1303–1309.

[10] Smith, A.D., *Oxford Dictionary of Biochemistry and Molecular Biology*, Oxford University Press, Oxford, 2000, pp. 209, 210.

[11] Kozlowski, R., Batog, J., Konczewicz, W., Mackiewicz-Talarczyk, M., Muzyczek, M., Sedelnik, N., Tanska, B. (2006). Enzymes in bast fibrous plant processing. *Biotechnology Letters*, **28**, 761–765.

[12] Jenkins, R.O., Enzymes, in A. Cavaco-Paulo, G.M. Gübitz, ed., *Textile Processing with Enzymes*, Woodhead Publishing Limited, Cambridge, 2003, p. 1.

[13] Voet, D., Voet, J.G., Pratt, C.W., *Fundamentals of Biochemistry: Life at the Molecular Level*, Fourth Edition, John Wiley & Sons, Inc., Hoboken, 2013, pp. 315–395.

[14] Mohanty, A.K., Misra, M., Drzal, L.T., Selke, S.E., Harte, B.R., Hinrichsen, G., Natural fibers, biopolymers and biocomposites, in A.K. Mohanty, M. Misra, L.T. Drzal, ed., *Natural Fibers, Biopolymers and Biocomposites*, CRC press, Boca Raton, 2005, p. 1.

[15] Niska, K.O., Sain, M., *Wood-Polymer Composites*, Woodhead Publishing Limited, Cambridge, 2008.

[16] Ansell, M.P. Mwaikambo, L.Y., The structure of cotton and other plant fibres, in S.J. Eichhorn, J.W.S. Hearle, M. Jaffe, T. Kikutani, ed., *Handbook of Textile Fibre Structure, Volume 2: Natural, Regenerated, Inorganic and Specialist Fibres*, Woodhead Publishing Limited, Cambridge, 2009, p. 62.

[17] Dinwoodie, J.M., *Timber: Its Nature and Behaviour*, E & FN Spon, London, 2000.

[18] McLaughlin, E.C., Tait, R.A. (1980). Fracture mechanism of plant fibres. *Journal of Materials Science*, **15**, 89–95.

[19] Aziz, S., Ansell, M., Optimising the properties of green composites, in C. Baillie, ed., *Green Composites—Polymer Composites and the Environment*, Woodhead Publishing Ltd., Cambridge, 2004, p. 154.

[20] Mwaikambo, L.Y. (2006). Review of the history, properties and application of plant fibres. *African Journal of Science and Technology*, **7**, 120–133.

[21] Satyanarayana, K.G., Arizaga, G.G.C., Wypych, F. (2009). Biodegradable composites based on lignocellulosic fibers—an overview. *Progress in Polymer Science*, **34**, 982–1021.

[22] Akin, D.E., in Chemistry of plant fibres, J. Müssig, ed., *Industrial Applications of Natural Fibres: Structure, Properties and Technical Applications*, John Wiley & Sons, Ltd., Chichester, 2010, p. 13.

[23] Bismarck, A., Mishra, S., Lampke, T., Plant fibers as reinforcement for green composites, in A.K. Mohanty, M. Misra, L.T. Drzal, ed., *Natural Fibers, Biopolymers, and Biocomposites*, Taylor & Francis Group, Boca Raton, 2005, p. 37.

[24] Mohanty, A.K., Misra, M., Hinrichsen, G. (2000). Biofibres, biodegradable polymers and biocomposites: an overview. *Macromolecular Materials and Engineering*, **276–277**, 1–24.

[25] Ciechanska, D., Wesolowska, E., Wawro, D., An introduction to cellulose fibres, in S.J. Eichhorn, J.W.S. Hearle, M. Jaffe, T. Kikutani, ed., *Handbook of Textile Fibre Structure, Volume 2: Natural, Regenerated, Inorganic and Specialist Fibres*, Woodhead Publishing Limited, Cambridge, 2009, p. 3.

[26] Nevell, T.P., Zeronian, S.H., Chapter 1, in T.P. Nevell, S.H. Zeronian, ed., *Cellulose Chemistry and Its Applications*, Ellis Horwood Ltd, Chichester, 1985, p. 15.

[27] Focher, B., Physical characteristics of flax fibre, in H.S.S. Sharma, C.F. van Sumere, ed., *The Biology and Processing of Flax*, M Publications, Belfast, 1992, p. 11.

[28] Sarkanen, K.V., Ludwig, C.H., *Lignins: Occurrence, Formation, Structure and Reactions*, Wiley-Interscience, New York, 1971.

[29] Thielemans, W., Can, E., Morye, S.S., Wool, R.P. (2002). Novel applications of lignin in composite materials. *Journal of Applied Polymer Science*, **83**, 323–331.

[30] Sakai, T., Sakamoto, T., Hallaert, J., Vandamme, E.J., Pectin, pectinase, and protopectinase: production, properties, and applications, in S. Neidleman, A.I. Laskin, ed., *Advances in Applied Microbiology*, Academic Press, New York, 1993, p. 213.

[31] E. Beleski-Carneiro, Y. Alquini, F. Reicher, "Pectins from the fruit of myrtaceae," Proceedings of the Third International Symposium on Natural Polymers and Composites, Sao Pedro, Brazil, May 14–17, 2000, pp. 162–170.

[32] Whitaker, J.R., Microbial pectolytic enzymes, in M. Fogarty, C.T. Kelly, ed., *Microbial Enzymes and Biotechnology*, Second Edition, Elsevier Applied Science, London, 1991, p. 133.

[33] Gummadi, S.N., Manoj, N., Kumar, D.S., Structural and biochemical properties of pectinases, in J. Polaina, A.P. MacCabe, ed., *Industrial Enzymes: Structure, Function and Applications*, Springer, Dordrecht, 2007, p. 99.

[34] Kashyap, D.R., Vohra, P.K., Chopra, S., Tewari, R. (2001). Applications of pectinases in the commercial sector: a review. *Bioresource Technology*, **77**, 215–227.

[35] Windham, W.R., Petersen, J.C., Terrill, T.H., Tannins as anti-quality factors in forage, in D.E. Akin, L.G. Ljungdahl, J.R. Wilson, P.J. Harris, ed., *Microbial and Plant Opportunities to Improve Lignocellulose Utilization by Ruminants*, Elsevier, New York, 1990, p. 127.

[36] Morrison, I.I.I. W.H., Holser, R., Akin, D.E. (2006). Cuticular wax from flax processing waste with hexane and super critical carbon dioxide extractions. *Industrial Crops and Products*, **24**, 119–122.

[37] Oinonen, A.M., Cellulases in the textile industry, in J. Polaina, A.P. MacCabe, ed., *Industrial Enzymes: Structure, Function and Applications*, Springer, Dordrecht, 2007, p. 51.

[38] Akin, D., Morrison, W., Gamble, G., Rigsby, L. (1997). Effect of retting enzymes on the structure and composition of flax cell walls. *Textile Research Journal*, **67**, 279–287.

[39] Pastor, F.I.J., Gallardo, O., Sanz-Aparicio, J., Diaz, P., Xylanases: molecular properties and applications, in J. Polaina, A.P. MacCabe, ed., *Industrial Enzymes: Structure, Function and Applications*, Springer, Dordrecht, 2007, p. 65.

[40] De Vries, R.P., Visser, J. (2001). Aspergillus enzymes involved in degradation of plant cell wall polysaccharides. *Microbiology and Molecular Biology Reviews*, **65**, 497–522.

[41] Deblois, S., Wiegel, J., Hemicellulases in lignocellulose degradation, in D.E. Akin, L.G. Ljungdahl, J.R. Wilson, P.J. Harris, ed., *Microbial and Plant Opportunities to Improve Lignocellulose Utilization by Ruminants*, Elsevier, New York, 1990, p. 275.

[42] Saha, B.C. (2000). α-l-arabinofuranosidases: biochemistry, molecular biology and application in biotechnology. *Biotechnology Advances*, **18**, 403–423.

[43] Coughlan, M.P., Hazlewood, G.P. (1993). β-1,4-d-xylan-degrading enzyme systems: biochemistry, molecular biology and applications. *Biotechnology and Applied Biochemistry*, **17**, 259–289.

[44] Williamson, G., Kroon, P.A., Faulds, C.B. (1998). Hairy plant polysaccharides: a close shave with microbial esterases. *Microbiology*, **144**, 2011–2023.

[45] Puchart, V., Katapodis, P., Biely, P., Kremnicky, L., Christakopoulos, P., Vrsanska, M., Kekos, D., Marcis, B., Bhat, M. (1999). Production of xylanases, mannanases, and pectinases by the thermophilic fungus Thermomyces lanuginosus. *Enzyme and Microbial Technology*, **24**, 355–361.

[46] Sharma, H.S.S. (1987). Enzymatic degradation of residual non-cellulosic polysaccharides present on dew retted flax fibers. *Applied Microbiology and Biotechnology*, **26**, 358–362.

[47] Fogarty, M.W., Ward, O.P. (1972). Pectic substances and pectolytic enzymes. *Process Biochemistry*, **7**, 17–31.

[48] Rosemberg, J.A., de Franca, F.P. (1967). Importance of galacturonic acid in controlling the retting of flax. *Applied Microbiology*, **15**, 484–486.

[49] Ahmed, M. (1963). Study on jute retting bacteria. *Journal of Applied Bacteriology*, **26**, 117–126.

[50] Jayasankar, N.P., Agate, A.D., Bate, J.B. (1967). Microbial decomposition of pectic substances. V. Evidence for the role of *Micrococcus* species in the retting of sisal and coconut husks. *Journal of the Indian Institute of Science*, **49**, 10–17.

[51] Adamsen, A., Akin, D., Rigsby, L. (2002). Chelating agents and enzyme retting of flax. *Textile Research Journal*, **72**, 296–302.

[52] Thurston, C.F. (1994). The structure and function of fungal laccases. *Microbiology*, **140**, 19–26.

[53] Alcalde, M., *Laccases: Biological Functions, Molecular Structure, and Industrial Applications*, Springer, Dordrecht, 2007.

[54] Xu, F., Berka, R.M., Wahleithner, J.A., Nelson, B.A., Shuster, J.R., Brown, S.H., Palmer, A.E., Solomon, E.I. (1998). Site-directed mutations in fungal laccase: effect on redox potential, activity and pH profile. *Biochemical Journal*, **334**, 63–70.

[55] Bourbonnais, R., Paice, M.G. (1990). Oxidation of nonphenolic substrates—an expanded role for laccase in lignin biodegradation. *FEBS Letters*, **267**, 99–102.

[56] Bajpai, P. (2004). Biological bleaching of chemical pulps. *Critical Reviews in Biotechnology*, **24**, 1–58.

[57] Aehle, W., *Enzymes in Industry: Production and Applications*, Wiley-VCH, Weinheim, 2004.

[58] Buschle-Diller, G., Zeronian, S., Pan, N., Yoon, M.Y. (1994). Enzymatic hydrolysis of cotton, linen, ramie and viscose rayon fabrics. *Textile Research Journal*, **64**, 270–279.

[59] Mooney, C., Stolle-Smits, T., Schols, H., de Jong, E. (2001). Analysis of retted and non retted flax fibres by chemical and enzymatic means. *Journal of Biotechnology*, **89**, 205–216.

[60] Saikia, R., Boruah, P., Samanta, R. (2009). Microbial degumming of decorticated ramie and its fiber characteristics. *Indian Journal of Fiber and Textile Research*, **34**, 187–190.

[61] Kalia, S., Sheoran, R. (2009). Modification of ramie fibers using microwave-assisted grafting and cellulase enzyme-assisted biopolishing: a comparative study of morphology, thermal stability, and crystallinity. *International Journal of Polymer Analysis and Characterization*, **16**, 307–318.

[62] Kalia, S., Vashishta, S. (2012). Surface modification of sisal fibers (*Agave sisalana*) using bacterial cellulase and methyl methacrylate. *Journal of Polymers and the Environment*, **20**, 142–152.

[63] Joko, K., Tokuda, S., Kikumoto, N., Sugai, J., Hayashi, T., Arai, M. (2002). Enzymatic function on jute fiber with various commercial enzymes. *Sen'i Gakkai shi*, **58**, 22–28.

[64] Jayapriya, J., Vigneswaran, C. (2010). Process optimization for biosoftening of ligno-cellulosic fiber with white rot fungi and specific enzymatic systems. *Journal of Natural Fibers*, **7**, 17–33.

[65] Park, S., Venditti, R.A., Abrecht, D.G., Jameel, H., Pawlak, J.J., Lee, J.M. (2007). Surface and pore structure modification of cellulose fibers through cellulase treatment. *Journal of Applied Polymer Science*, **103**, 3833–3839.

[66] Azelee, N.I.W., Jahim, J.M., Rabu, A., Murad, A.M.A., Bakar, F.D.A., Illias, R.M. (2014). Efficient removal of lignin with the maintenance of hemicellulose from kenaf by two-stage pretreatment process. *Carbohydrate Polymers*, **99**, 447–453.

[67] Brühlmann, F., Leupin, M., Erismann, K.H., Fiechter, A. (2000). Enzymatic degumming of ramie bast fibers. *Journal of Biotechnology*, **76**, 43–50.

[68] Akin, D.E., Foulk, J.A., Dodd, R.B., McAlister III, D.D. (2001). Enzyme-retting of flax and characterization of processed fibers. *Journal of Biotechnology*, **89**, 193–203.

[69] Ossola, M., Galante, Y. (2004). Scouring of flax rove with the aid of enzymes. *Enzyme and Microbial Technology*, **34**, 177–186.

[70] Liu, L., Cheng, L., Huang, L., Yu, J. (2012). Enzymatic treatment of mechanochemical modified natural bamboo fibers. *Fibers and Polymers*, **13**, 600–605.

[71] Ouajai, S., Shanks, R. (2005). Morphology and structure of hemp fibre after bioscouring. *Macromolecular Bioscience*, **5**, 124–134.

[72] Foulk, J.A., Chao, W.Y., Akin, D.E., Dodd, R.B., Layton, P.A. (2006). Analysis of flax and cotton fiber fabric blends and recycled polyethylene composites. *Journal of Polymers and the Environment*, **14**, 15–25.

[73] Karaduman, Y., Gokcan, D., Onal, L. (2013). Effect of enzymatic pretreatment on the mechanical properties of jute fiber-reinforced polyester composites. *Journal of Composite Materials*, **47**, 1293–1302.

[74] Karaduman, Y., Onal, L. (2013). Dynamic mechanical and thermal properties of enzyme-treated jute/polyester composites. *Journal of Composite Materials*, **47**, 2361–2370.

[75] Li, Y., Pickering, K.L. (2008). Hemp fibre reinforced composites using chelator and enzyme treatments. *Composites Science and Technology*, **68**, 3293–3298.

[76] Bledzki, A.K., Mamun, A.A., Jaszkiewicz, A., Erdmann, K. (2010). Polypropylene composites with enzyme modified abaca fibre. *Composites Science and Technology*, **70**, 854–860.

[77] Saleem, Z., Rennebaum, H., Pudel, F., Grimm, E. (2007). Treating bast fibers with pectinase improves mechanical characteristics of reinforced thermoplastic composites. *Composites Science and Technology*, **68**, 471–490.

[78] Ouajai, S., Shanks, R.A. (2009). Biocomposites of cellulose acetate butyrate with modified hemp cellulose fibres. *Macromolecular Materials and Engineering*, **294**, 213–221.

[79] Islam, M.R., Beg, M.D.H., Gupta, A. (2013). Characterization of laccase-treated kenaf fibre reinforced recycled polypropylene composites. *Bioresources*, **8**, 3753–3770.

[80] Islam, M.R., Beg, M.D.H., Mina, M.F. (2014). Fibre surface modifications through different treatments with the help of design expert software for natural fibre-based biocomposites. *Journal of Composite Materials*, **48**, 1887–1899.

[81] George, M., Mussone, P.G., Bressler, D.C. (2014). Surface and thermal characterization of natural fibres treated with enzymes. *Industrial Crops and Products*, **53**, 365–373.

[82] Kalia, S., Thakur, K., Kumar, A., Celli, A. (2014). Laccase-assisted surface functionalization of lignocellulosics. *Journal of Molecular Catalysis B: Enzymatic*, **102**, 48–58.

[83] Kalia, S., Vashistha, S. (2012). Surface modification of sisal fibers using cellulase and microwave-assisted grafting: a study of morphology, crystallinity, and thermal stability. *International Journal of Polymeric Materials*, **61**, 1130–1141.

[84] Kalia, S., Sheoran, R., Mittal, H., Kumar, A. (2013). Surface modification of ramie fibers using microwave assisted graft copolymerization followed by Brevibacillus parabrevis pretreatment. *Advanced Materials Letters*, **4**, 742–748.

[85] Kalia, S., Thakur, K., Celli, A., Kiechel, M.A., Schauer, C.L. (2013). Surface modification of plant fibers using environment friendly methods for their application in polymer composites, textile industry and antimicrobial activities: a review. *Journal of Environmental Chemical Engineering*, **1**, 97–112.

[86] Nyanhongo, G.S., Kudanga, T., Prasetyo, E.N., Gübitz, G.M. (2011). Enzymatic polymer functionalisation: advances in laccase and peroxidase derived lignocellulose functional polymers. *Advances in Biochemical Engineering/Biotechnology*, **125**, 47–68.

[87] Liu, N., Shi, S., Gao, Y., Qin, M. (2009). Fiber modification of kraft pulp with laccase in presence of methyl syringate. *Enzyme and Microbial Technology*, **44**, 89–95.

[88] Chandra, R.P., Felby, C., Ragauskas, A.J. (2005). Improving laccase-facilitated grafting of 4-hydroxybenzoic acid to high-kappa kraft pulps. *Journal of Wood Chemistry and Technology*, **24**, 69–81.

[89] Schroeder, M., Aichernig, N., Guebitz, G.M., Kokol, V. (2007). Enzymatic coating of lignocellulosic surfaces with polyphenols. *Biotechnology Journal*, **2**, 334–341.

[90] Aracri, E., Roncero, M.B., Vidal, T. (2011). Studying the effects of laccase-catalysed grafting of ferulic acid on sisal pulp fibers. *Bioresource Technology*, **102**, 7555–7560.

[91] Aracri, E., Fillat, A., Colom, J.F., Gutierrez, A., Rio, J.C., Martinez, A.T., Vidal, T. (2010). Enzymatic grafting of simple phenols on flax and sisal pulp fibres using laccases. *Bioresource Technology*, **101**, 8211–8216.

[92] Fillat, A., Gallardo, O. Vidal, T., Pastor, F.I.J., Diaz, P., Roncero, M.B. (2012). Enzymatic grafting of natural phenols to flax fibres: development of antimicrobial properties. *Carbohydrate Polymers*, **87**, 146–152.

[93] Ubasart, J.G., Colom, J.F., Vila, C., Hernandez, N.G., Roncero, M.B., Vidal, T. (2012). A new procedure for the hydrophobization of cellulose fibre using laccase and a hydrophobic phenolic compound. *Bioresource Technology*, **112**, 341–344.

[94] Chandra, R.P., Lehtonen, L.K., Ragauskas, A.J. (2004). Modification of high lignin content kraft pulps with laccase to improve paper strength properties. 1: laccase treatment in the presence of gallic acid. *Biotechnology Progress*, **20**, 255–261.

[95] Kudanga, T., Prasetyo, E.N., Sipila, J., Guebitz, G.M., Nyanhongo, G.S. (2010). Reactivity of long chain alkylamines to lignin moieties: implications on hydrophobicity of lignocellulose materials. *Journal of Biotechnology*, **149**, 81–87.

[96] Kudanga, T., Prasetyo, E.N., Sipila, J., Nyanhongo, G.S., Guebitz, G.M. (2010). Chemo-enzymatic functionalisation of lignocellulose materials using oxiranes. *Process Biochemistry*, **45**, 1557–1562.

[97] Dong, A., Yu, Y., Yuan, J., Wang, Q., Fan, X. (2014). Hydrophobic modification of jute fiber used for composite reinforcement via laccase-mediated grafting. *Applied Surface Science*, **301**, 418–427.

4

RECENT DEVELOPMENTS IN SURFACE MODIFICATION OF NATURAL FIBERS FOR THEIR USE IN BIOCOMPOSITES

JASPREET KAUR BHATIA[1], BALBIR SINGH KAITH[1], AND SUSHEEL KALIA[2]

[1] *Department of Chemistry, DR B R Ambedkar National Institute of Technology, Jalandhar, Punjab, India*

[2] *Department of Chemistry, Army Cadet College Wing, Indian Military Academy, Dehradun, Uttarakhand, India*

Biodegradable Green Composites, First Edition. Edited by Susheel Kalia.

4.1 INTRODUCTION

With rising interest in environmentally friendly biopolymers as substitute for petroleum-based synthetic materials, immense efforts are made to produce eco-friendly and eco-efficient products, which can meet out the demands of consumers. The depletion of petroleum resources and environmental legislations are the main driving forces behind the effort of searching for new materials. Efforts are being made on all fronts including industries, government, and researchers for development and utilization of greener materials. Biodegradable plastics and biopolymers are the only solution to the complex problem of environmental issues. Renewable plant resources and agricultural stocks are the basis of new generation bioplastics, which can be tailored to develop environmentally friendly materials. Vegetable oils used for lubricants, soy and corn oil for preparation of thermoset and thermoplastic composites, and corn for polylactic acid synthesis are the few examples of utilization of bioresources for polymeric material development [1, 2]. Enhanced performance along with cost-effectiveness is the demand of today's market, which can be achieved by blending of both petroleum-based and biobased plastics. Bioplastics may or may not contain 100% biomaterials, they may contain high, medium, or low content of biobased components, but at all levels these materials can contribute toward the solution to complex problem of the environment [3].

The idea of composite materials and use of bioplastics is not new. Paper and silk are the biopolymers that are used by mankind for centuries. In ancient times composites made up of mud and straw were used for building applications. Wattle and daub, the oldest composite material, is composed of a mixture of lattice of wooden strips with sticky material of soil, clay, sand, animal dung, and straw [4]. Before the 1990s, major thrust was on the development of new materials for industrial as well as for general applications. Being the first choice for the development of synthetic polymers and reinforced composites, low-cost petroleum raw material played an important role. However, after the 1990s, adverse environmental impact of synthetic components, depletion of petroleum resources, and development of advanced techniques shifted the interest of industries and researchers toward biobased materials [5, 6]. In the light of need of cost-effective engineered materials and environmental consciousness, research continued on the development of natural fiber composites for different applications like automotive parts, packaging material, building material, and many more [7–10]. In the 1940s, Henry Ford used soy protein plastic-based natural fiber-reinforced biocomposites for making of different parts of cars. In 1908, Ford released Model-T in which side panels were made up of hemp fiber-reinforced composites. These side panels showed 10 times higher strength than steel panel [11]. Natural fibers are the potential ecological alternatives of conventional fibers in composites due to their higher specific properties at low cost. In addition to cost-effectiveness, the nonabrasive nature of natural fibers toward mixing and molding equipment and less health hazards guaranteed their acquiescence in market [12, 13]. Table 4.1 shows properties of certain natural fibers.

The utilization of natural fibers as reinforcing materials in composites has few restrictions due to certain properties of natural fibers. High moisture absorption due

TABLE 4.1 Mechanical properties of some natural fibers[a]

Fiber	Density (g/cm^3)	Tensile strength (MPa)	% Elongation	Young's modulus (GPa)
Sisal	1.3–1.5	400–700	2.0–1.4	9.0–38.0
Cotton	1.5–1.6	287–597	3.0–10.0	5.5–12.6
Ramie	1.5	220–940	2.0–3.8	44–128
Flax	1.4–1.5	345–1500	1.2–3.2	27.6–80.0
Jute	1.3–1.46	393–800	1.5–1.8	10–30
Hemp	1.48	550–900	1.6	70
Bamboo	0.8	391–1000	—	48–89
Coir	1.2	175–220	15.0–30.0	4.0–6.0
Sugarcane bagasse	—	222	1.1	17.9–27.1
Banana	—	700–800	2.5–3.7	27–32

[a] From Bogoeva-Gaceva et al. [12] Table 4. Reprinted with permission from John Wiley & Sons.

to hydrophilic nature, poor microbial resistance, and low thermal resistance are the major hurdles for their success as reinforcing material. Moreover, nonuniformity of their characteristics along with local and seasonal quality variations is the foremost challenge for their use in highly specific applications [14].

Regardless of these limitations, natural fibers are intensely used for more and more applications. Automobile industries, building products, flower pots, marine piers, and packaging industry are the emerging market areas exploring natural fiber composite-based new products. Studies showed that the market of natural fiber composites has expected a growth rate of 15–20% per year. The global market for composites report estimated the increase of composite market to nearly 17.2 billion pounds by 2019. This market is expected to show a compound annual growth rate (CAGR) of 3.7% through the forecast period of 2014–2019 [15].

4.2 BIOCOMPOSITES

Biocomposites are the composite materials consisting of one or more phases derived from biopolymers or renewable resources. Biocomposites mainly consist of natural fibers as reinforcement and biopolymers as matrix component [5]. Biocomposites are the materials with combination of properties, which are not present in nature. These materials show high performance but at affordable prices. Awareness of environment bent area of research toward materials associated with renewable, sustained, recyclable, and biodegradable properties. These composite materials are renewable materials, easy to degrade, and eco-friendly in nature and are categorized under biocomposites. The development and use of biocomposites is not a new technique to mankind. Examples like use of bows constructed from adhesively bonded laminates of horns and tendons of animals are of centuries back. Development of biocomposites has gone through significant changes. Choice of materials for development of biocomposites is important in terms of properties required for special applications

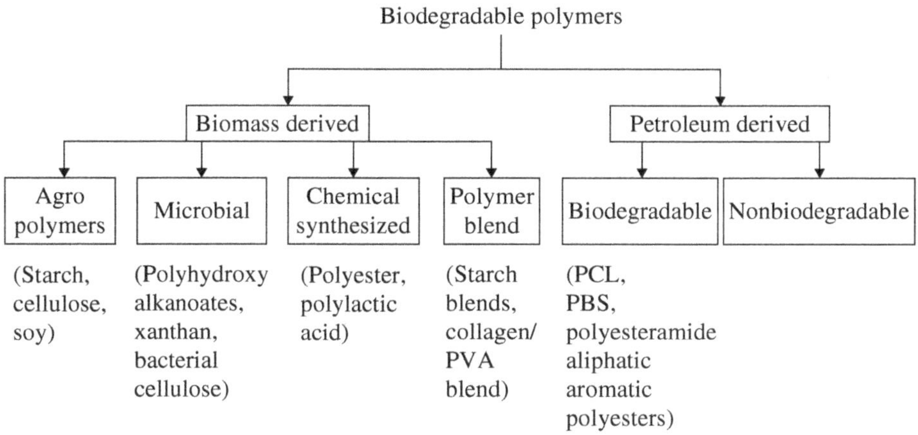

FIGURE 4.1 Classification of matrices used in biocomposites.

along with maintenance of biodegradable and eco-friendly nature. Over the last decade, biocomposites developed are competitive with synthetic one like glass–epoxide and glass–polypropylene (PP) composites. Hybrid composites involving reinforcement of natural along with synthetic materials also resulted in good results [16].

4.2.1 Classification: Biomass Derived and Petroleum-Derived Matrix

Biodegradable polymeric materials have been classified as natural or synthetic depending upon the origin. Classification of matrices used is represented in Figure 4.1.

Renewable resources like plants and bacteria are the sources of large number of polymeric matrices. The utilization of biomass for development of composites provides a route to utilize huge raw materials available in nature with the advantage of reduction in the volume of waste material, reduction of released atmospheric carbon dioxide, and complete biodegradability. The use of biomass-derived matrices results in composite material obtained from completely renewable resources and may be categorized as green composites. The biodegradable polymers can be classified depending upon their origin from agropolymers (starch, cellulose, soy, etc.), microbial (polyhydroxyalkanoates, xanthan, or bacterial cellulose (BC)), chemically synthesized from agro-based products (polyester, polylactic acid, etc.), and polymer blend (starch blends collagen/PVA blends) [17]. Starch and polyhydroxybutyrate (PHB) are most extensively studied biodegradable polymers for composites [18–20]. Substitution of synthetic polymers with biomass-derived matrices faces major challenges in terms of limited performance and high cost along with lack of on-shelf stability and short lifetime. Moreover, most natural polymers used as matrices are water sensitive. Naturally available biomass needs to be developed with different approaches to make it suitable as matrix polymer for different composite applications. Their properties can be altered and improved by use of plasticizers, blending with hydrophobic and thermoplastic polymers, and through chemical modifications. The modifications result in new products with properties suitable to act as a matrix for biocomposites.

Starch has been extensively studied as the matrix in its modified and unmodified form to develop different biocomposite systems. Unmodified starch-based thermoplastics are sensitive to humid conditions and are used in place of polystyrene. The presence of plasticizer may cause fragmentation of starch granule, hydrogen bond breakage, and partial depolymerization, which convert starch to starch plastics [21]. Blended acetylated starch acetate has been used as matrix to develop NaOH-treated corn stalk-reinforced composites. The composites are processed by extrusion at 150°C with 2–14% fiber content. The composites formed have increased density and shear modulus up to 10% fiber content [22]. Starch-filled PP composites were prepared by extrusion of PP resin with 5, 10, 15, and 20 wt % of potato starch. The increase in starch content enhanced tensile modulus, flexural strength, and flexural modulus of the PP composites along with the increase in moisture, water absorption, and density [23].

Cellulose from plants is a versatile material to develop composite matrices. Different cellulose esters like cellulose acetate, cellulose acetate butyrate, and cellulose acetate propionate are now used in different biocomposites. Cellulose matrices result in attractive combinations of mechanical and thermal properties. Cellulose esters can be extruded, injection molded, and blow molded into sheets and films. Their mechanical properties depend upon the type and nature of plasticizers [24–26].

Proteins like casein, zein, and gelatin are the viable alternatives of synthetic polymers. Soy protein plastics are good substitutes for petroleum products. Soy-based plastics have been developed as eco-friendly plastics. Compression molding and extrusion are the two usual techniques to process soy plastics [27]. The preparation and use of soy plastics can be traced back in 1930s. Since such plastics are brittle and moisture sensitive, therefore, it is difficult to prepare soy plastics with good mechanical properties. However, the use of plasticizers and cross linker assists in the incorporation of desired properties in the soy protein plastics [28]. Biodegradable and environmentally friendly green composites were prepared using natural fibers and soy protein-based resins. Nam and Netravali [29] prepared fully biodegradable green composites using ramie fiber and soy protein concentrate (SPC). SEM images of tensile fracture surfaces of green composites showed good interfacial bonding between fibers and SPC resin. Water resistance and modulus were enhanced by reinforcing acrylated soy protein isolate with ramie fibers aligned in vertical, horizontal, as well as in both directions. Vertical alignment leads to highest water resistance and mechanical properties [30]. Wang et al. [31] investigated hemp fiber whiskers reinforced soy protein composites by using thiodiglycol. Huang and Netarvali developed biodegradable green composites made up of microsized bamboo fibrils and SPC resin. The incorporation of desired modifications in soy protein could improve physicochemical properties of composites, but at the same time, it affects the biodegradation of composites [32].

TABLE 4.2 Some applications of biodegradable polymers[a]

Product	Society	Composition	Application
Mater-Bi®	Novamont (Italy)	Starch and polyester	Collection bags for green waste, agricultural films, disposable items
Polynat®	Roverc'h (France)	Rye flower (80%)	Disposable items, flower containers
Ecofoam®	American Excelsior Company (USA)	Starch	Wrapping plastics
Biopol®	Goodfellow (Great Britain)	PHB/PHV	Razors, bottles
Eco-pla®	Cargill Dow (USA)	PLA	Sanitary products, sport clothes, conditioning and packaging
Bio-D®	Cirad (France)	Proteins extracted from cotton seed	Agricultural films
Ecoflex®	BASF (Germany)	Copolyester	Agricultural films
Eastar Bio®	Eastman (Great Britain)	Copolyester	Agricultural films
BAK 1095®	Bayer (Germany)	Polyester amide	Disposable items, flower containers

[a] From Vroman and Tighzert [7]. Public domain.

Polyhydroxyalkanoates (PHA) matrices have been prepared by different processing techniques. Hot pressing and compression molding are found to be most promising techniques to prepare PHA-based composites [33, 34]. Gatenholm [35] reported the processing and properties of biodegradable composites of bacteria produced polyesters reinforced with wood cellulose. PHA-based composites basically have low strength so these can be strengthened by incorporation of reinforcing agents to develop useful biodegradable composites [36, 37].

Polycaprolactone (PCL), polybutylene succinate (PBS), polyester amide (PEA), and aliphatic aromatic polyesters are petroleum-based biodegradable polymers, which have been studied as matrices of natural fiber composites. PCL is a crystalline polymer having modulus between low-density and high-density polyethenes. PCL has been found to degrade by enzymes like lipase [38]. It is used as resin for adhesives, surface coatings, and fabrics due to low m.pt, high biodegradability, and biocompatibility. It has been used in starch-based formulations to enhance water resistance. Due to its biocompatibility it has potential biomedical uses like bone repair and bone fixation [39]. Polybutylene succinate (PBS) can be produced by condensation of glycol with aliphatic dicarboxylic acid. It is commercially available under the trade name "Bionolle." Properties like melt processing, thermal and chemical stability, and biodegradability make it suitable matrix material for composites [40]. Coir fiber [41], bamboo fiber [42], jute fiber [43], abaca fiber [44], and kenaf fiber [45] have been used as reinforcements in PBS matrices to enhance properties for desired applications. Different biopolymers have been commercially used for various applications (Table 4.2).

4.2.2 Advantage over Traditional Composites

Biocomposites involving biodegradable matrices either obtained from biomass or prepared from petroleum resources reinforced with natural fiber are of great interest because of their diversified applications. Increased awareness toward the environment renewed the demand of development and use of sustainable and ecologically strong naturally derived materials [46]. Along with the awareness drive, composites based upon natural polymers predominated the use of traditional nonbiodegradable matrices and reinforced materials due to highly specific and favorable properties of biocomposites. Natural fiber-reinforced composites have low specific weight than glass fiber, which results in high specific strength and stiffness. Biocomposites processing and production requires less energy and has low health hazards than traditional composites [47]. Low-cost production, easy thermal recycling, and favorable thermal and acoustic properties make biocomposites today's material of choice. Its low density, good mechanical properties, favorable processing environment, high stability, health benefits, and cost advantages make biocomposites a preferred choice in industries especially automotive and building industry [48–50].

4.3 NATURAL FIBER: STRUCTURE AND COMPOSITION

Natural fibers consist of hollow cellulose fibrils suspended in the matrix of hemicellulose and lignin. These cellulose microfibrils bound helically around amorphous lignin matrix. Microfibrils have a diameter of about 10–30 nm and each consist of 30–100 cellulose molecules. Lignin provides protection and acts as stiffener for stem, whereas hemicellulose is a compatibilizer between cellulose and lignin [51]. The cell wall is not homogeneous. Each fiber has a complex structure consisting of layers. A thin primary layer encircles a secondary wall, which is again consisting of three layers. A thick middle layer has a series of helically wound cellular microfibrils. The middle layer determines the mechanical properties of fiber [52]. The angle between fiber axis and microfibril is called microfibril angle. Each fiber has a characteristic value of microfibril angle. The efficiency of fiber to act as reinforcing agent depends upon the nature of cellulose and its percentage crystallinity. Celluloses, hemicelluloses, lignins, pectins, and waxes are major chemical constituents of natural fibers (Figs. 4.2 and 4.3).

Cellulose consists of long chains of β-D-anhydroglucopyranose units joined together by β-1,4-glycosidic linkages. Anhydroglucose units are of reducing end having free hemiacetal group at C-1, nonreducing one with free hydroxyl group at C-4, and internal rings joined at C-1 and C-4. C-2, C-3, and C-6 hydroxyl group are the active sites. These active sites direct the crystalline and physical properties of cellulose. Hydroxyl groups are involved in H-bonding, which causes the chain to group together in highly ordered structure. These ordered crystalline regions have area of disorders between them. It has a degree of polymerization of about 10,000. The crystalline regions provide fiber strength and thermoplastic behavior. On the other hand amorphous regions are responsible for hygroscopic nature of lignocellulosic fiber. Hemicellulose forming supporting matrix for cellulose microfibrils is

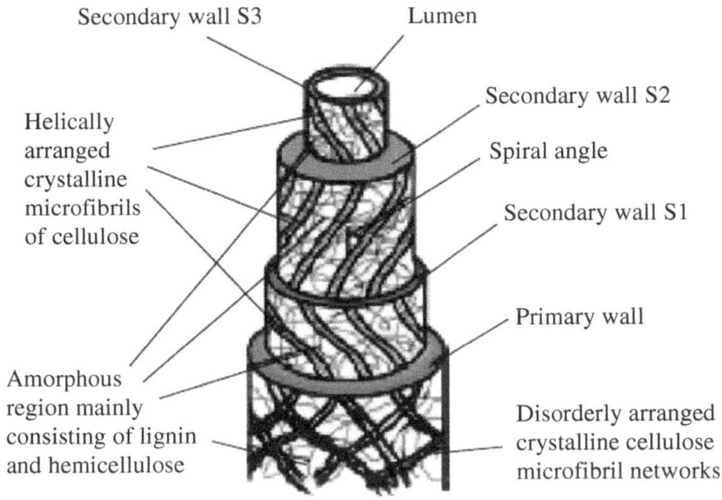

FIGURE 4.2 Structure of plant fiber (cell). *Source*: Rong et al. [51]. Reprinted with permission from Elsevier.

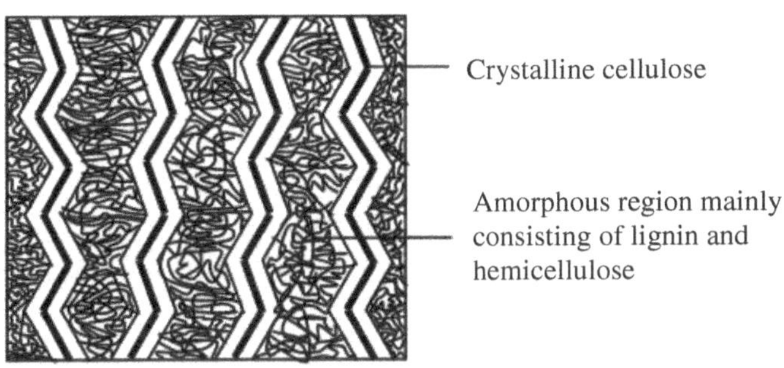

FIGURE 4.3 Schematic view of secondary layer of fiber cell wall. *Source*: Bledzki and Gassan [52]. Reprinted with permission from Elsevier.

a group of polysaccharides composed of combination of 5- and 6-membered ring sugars. Hemicellulose is a branched polymer having low degree of polymerization and is noncrystalline in nature. Hemicellulose is a hydrophilic component of natural fiber and sensitive toward acids and bases. Lignin is thought to be three-dimensional complex copolymer containing both aliphatic and aromatic components having high molecular weights. It is a thermoplastic containing hydroxyl, methoxyl, and carbonyl groups. It was found to contain five hydroxyl and five methoxyl groups per unit and considered as derivative of 4-hydroxy-3-methoxyphenyl propane. It is resistant to attack of acids but soluble in hot alkali. Lignin provides rigidity to plants [53]. Pectin provides flexibility to plants and consists of different polysaccharides. Waxes contain different types of alcohols.

TABLE 4.3 Chemical compositions of natural fibers[a]

Fiber	Cellulose (wt %)	Hemicellulose (wt %)	Lignin (wt %)	Waxes (wt %)
Abaca	56–63	20–25	7–9	3
Alfa	45.4	38.5	14.9	2
Bagasse	55.2	16.8	25.3	—
Bamboo	26–43	30	21–31	—
Banana	63–64	19	5	—
Coir	32–43	0.15–0.25	40–45	—
Cotton	85–90	5.7	—	0.6
Curaua	73.6	9.9	7.5	—
Flax	71	18.6–20.6	2.2	1.5
Hemp	68	15	10	0.8
Henequen	60	28	8	0.5
Isora	74	—	23	1.09
Jute	61–71	14–20	12–13	0.5
Kenaf	72	20.3	9	
Kudzu	33	11.6	14	—
Nettle	86	10	—	4
Oil palm	65	—	29	—
Piassava	28.6	25.8	45	—
Pineapple	81	—	12.7	—
Ramie	68.6–76.2	13–16	0.6–0.7	0.3
Sisal	65	12	9.9	2
Sponge gourd	63	19.4	11.2	3
Straw (Wheat)	38–45	15–31	12–20	—
Sun hemp	41–48	8.3–13	22.7	—

[a] From John and Anandjiwala [55], Table 2. Reprinted with permissions from John Wiley & Sons.

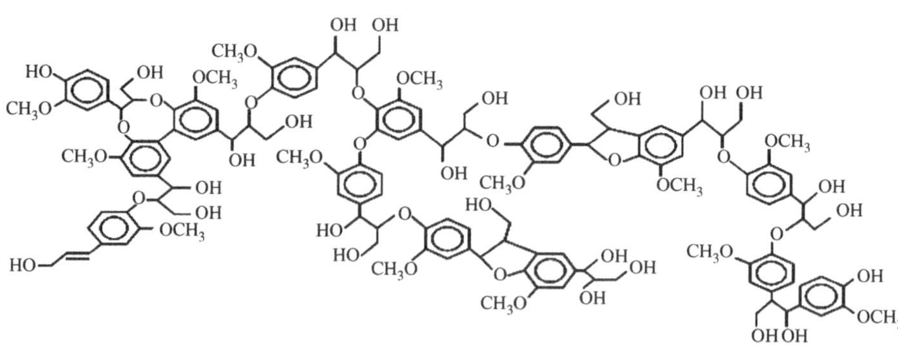

FIGURE 4.4 Lignin structure. Sette [54]. Reprinted with permissions from John Wiley & Sons.

The properties of fibers depend upon their structure, chemical composition, cell dimensions, and microfibril angle. Composition of natural fibers depends upon their origin as well as their geographical conditions. Table 4.3 represents chemical composition of few commercially important natural fibers (Fig. 4.4).

4.4 SURFACE MODIFICATION OF NATURAL FIBERS

Chemical treatment of natural fiber involves —OH and —H groups of cellulose backbone. It results in change of surface energy and polarity of fiber. Different chemical treatments like silane treatment, acylation, benzoylation, peroxide treatment, alkali treatment, isocyanate treatment, and permanganate treatment have been applied to fibers to improve their strength and fiber–matrix adhesion in fiber-reinforced composites [51, 56–60]. The mechanism of these methods depends upon the chemical structure of reagent.

4.4.1 Silylation, Esterification, and other Surface Chemical Modifications

Silane treatment involves the coupling of lignocellulosic fiber with silane coupling agent. It improves the degree of cross linking in the interface region resulting in excellent bonding. Silane treatment was applied on lignocellulosic fiber pretreated with NaOH. Pretreatment with NaOH increases the efficiency of fiber toward silane coupling agent. Silane treatment improves the interfacial properties [61–63]. It facilitates the interaction between filler and the polymer and lowers the adhesion factor [64]. John et al. [55] had proposed the mechanism for different silane coupling agents. Silane coupling agents react with —OH group of fiber to form stable covalent bond [65].

$$\text{Fiber–OH} + \text{R-Si(OH)}_3 \longrightarrow \text{Fiber-O–Si(OH)}_2\text{–R} \quad (1)$$

For the effective coupling between natural fiber and matrices, silane coupling agent should be bifunctional so that it can form bridge between two phases. Different silanes have been studied as coupling agents for natural fiber-reinforced composites. The most successful one is trialkoxysilane. Silane functional groups interact with polymer matrices involving different type of interactions depending upon the nature of functional group and compatibility toward polymer. The functional group may be glycidoxy, vinyl, amine, mercapto, or methacrylate. The type of silane reagent used for modification of natural fiber depends upon type and nature of matrix involve in composites [66]. The surface of pulp paper fiber was modified with silane coupling agents and surface treatment was investigated with the help of inverse gas chromatography [67]. Treatment of sisal fiber with aminoalkoxy silane coupling reagent showed formation of thin layer of siloxane and polysiloxane on fiber surface. It ultimately affects decomposition properties of sisal fiber [68]. The effect of silane treatment on fiber strength gives variable results. Alkali–silane treatment with 3-aminopropyltriethoxy silane showed significantly improved tensile and flexural properties of short fiber nonwoven kenaf PP composites [69]. Silane coupling of fiber is a promising process of improving durability and adhesion of fiber. Bilba and Arsene gave effect of silane treatment in combination with alkyltrialkoxy and dialkyldialkoxy silane.

These coupling agents affect porosity, dimension morphology, and hygroscopic characteristics of fiber. For composites reinforced with silane-treated fiber setting time get increased on treatment with 6% silane [70]. Organosilane treatment on surface properties of flax fiber showed an increase in hydrophobicity of fiber, and it also improved the mechanical properties of biocomposites significantly [71]. Mechanical properties of soy protein-based biocomposites reinforced with silane-modified cornhusk were found to be superior to biocomposites reinforced with unmodified fiber. Significant improvement of mechanical properties has been observed with silane-treated fiber at 5 wt % fiber loading [72].

Acetylation is a well-known esterification method to improve moisture resistance and dimensional stability. It involves the reaction of hydroxyl group of fiber with acetic or propionic anhydride at high temperature [12].

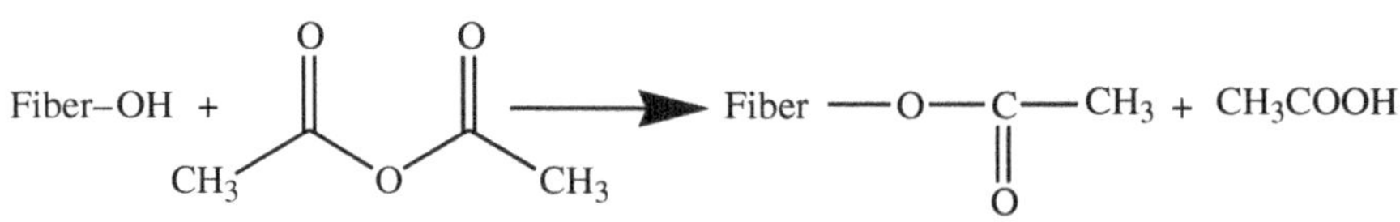

(2)

Acetylation of fiber significantly improved hydrophobic properties of fiber. Acetylation of jute fiber showed enhanced oil sorption properties on acetylation with acetic anhydride using *N*-bromosuccinimide (NBS) as a catalyst in a solvent-free system [73]. Esterification of flax fiber was done to change the surface properties. The chemically modified fiber showed increased thermal stability as compared to that of unmodified fiber [74]. Modification of lignocellulosic fiber obtained from different sugarcane bagasse pulping process, that is, kraft, was studied using octadecanoyl and dodecanoyl chloride. The modification resulted in the incorporation of nonpolar character to the fiber surface, which made it a suitable reinforcing agent in polymer matrix-based composites [75]. Freire et al. [76] studied the effect of reaction conditions and solvent type on the extent of esterification and fiber properties during controlled heterogeneous modification of lignocellulosic fibers with fatty acids. Heterogeneous cellulose/trifluoroacetic acid (TFAA)/pyridine/toluene system was used to modify lignocellulosic fiber with TFAA. The degree of substitution ranged between 0.04 and 0.30. It resulted in decreased hydrophobicity and lipophobicity of fiber surface [77]. Way et al. [78] studied the biodegradation behavior of maple fiber-reinforced polylactic acid composites. Treatment of maple fiber with NaOH followed by acetylation increased the rate of degradation, which was evident by gel permeation chromatography. It has been found that treatment of fiber with alkali degrades acetyl hemicellulose faster than cellulose and lignin. Thus, cellulose shows better chemical stability (Fig. 4.5).

The physical properties of lignocellulosic fiber can be altered by treatment with hydrogen peroxide. Hydrogen peroxide solubilizes a small part of lignin, ruptures hydrogen bonds, and decreases the cellulose crystallinity [80]. Hydrogen peroxide acts as a bleaching agent. Free radicals formed by peroxides react with lignocellulosic fiber. Sreekala et al. [81] proposed the scheme of peroxide treatment of cellulose.

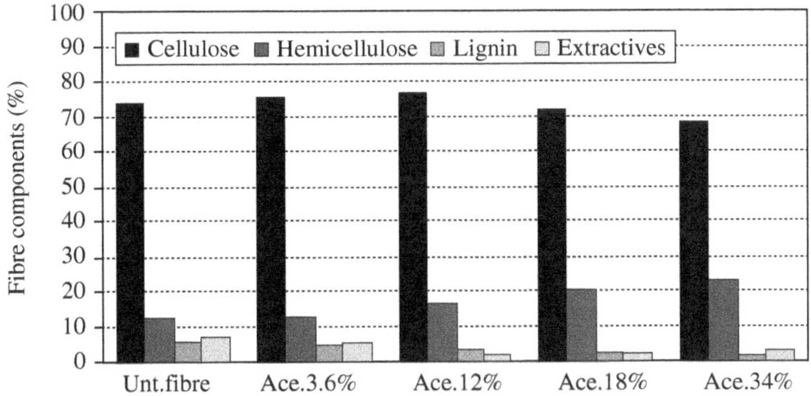

FIGURE 4.5 Effect of acetylation on fiber content. *Source*: Bledzki and Mamun [79]. Public Domain.

$$\text{RO-OR} \longrightarrow \text{RO}^* \quad (3)$$

$$\text{RO}^* + \text{Fiber–H} \longrightarrow \text{ROH} + \text{Fiber}^* \quad (4)$$

Benzoylation is another important technique for the surface modification of lignocellulosic fiber. Benzoyl chloride is the most commonly used reagent for the incorporation of benzoyl group on the cellulose. Reaction between lignocellulosic fiber and benzoyl chloride can be shown as [82]:

(5)

Benzoylation resulted in improved fiber–matrix adhesion and decreased water absorption. Joseph et al. [82] and Nair et al. [83] employed benzoyl chloride and NaOH for surface modification of sisal fiber. Benzoylation of oil palm fiber showed effect on dielectric properties of rubber composites. Introduction of modified fiber enhanced the permittivity of rubber from 0.5 to 2 [84]. Benzoylation of plain woven jute fiber in poly(L-lactic acid) reinforced composites enhanced tensile strength, tensile modulus, flexural strength, and impact strength of composites [85].

Treatment of lignocellulosic fiber with alkali is a most common method to improve its properties and is called mercerization [86]. Mercerization improves the fiber surface interaction, decreases fiber diameter, and increases aspect ratio. It ultimately improves the mechanical properties of lignocellulosic fiber composites (Fig. 4.6).

Alkali treatment of woven centaury fiber polyester composites exhibited enhanced tensile, thermal, and water absorption properties, whereas flexural strength and impact strength decreased on alkali treatment [88]. Mokaloba and

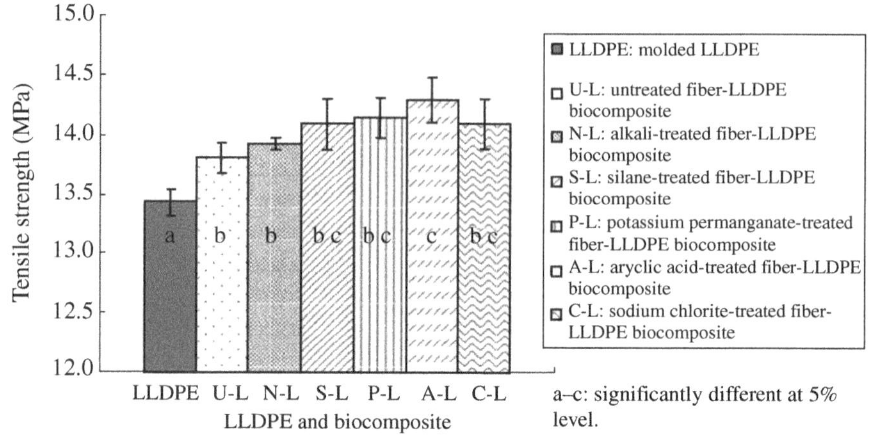

FIGURE 4.6 Effect of surface treatment on tensile strength of LLDPE/fiber-reinforced composites. *Source*: Li et al. [87]. Reprinted with permission from ASABE.

Batane investigated the influence of mercization on the surface morphology along with structure and interfacial adhesion characteristics of sisal fiber within PP matrix. Interfacial adhesion shear strength for alkali treated fiber improved by 173% as compared to untreated fiber [89]. Alkali treatment of flax yarn decreased the mechanical properties of fiber but enhanced the mechanical properties of composites [90]. Alkali treatment was reported to increase the amorphous nature of lignocellulosics by disturbing the hydrogen bonds in network structure [91]. A lot of work has been reported on the mercerization of lignocellulosic fibers [92, 93]. The effect of different mercerization media on mechanical properties of lignocellulosic fibers has been investigated, and 5–10% NaOH medium was found to give best results [94]. Jute fiber on treatment with 5% alkali solution showed an increase in flexural strength of red mud-filled jute fiber epoxy composites [95]. Alkali treatment enhanced fiber surface adhesion, which allowed an effective stress transfer from matrix to fiber. Mercerization of coir fiber improved the thermal stability of fiber. Tensile strength of composites with mercerized fiber was found to increase by 53% as compared to composites with untreated fiber [96].

Combination of different chemical treatments may be employed to get desired results. Surface modification of jute fibers was done by performing a series of chemical treatments, which involved detergent washing, dew axing, alkali, and acetic acid treatment. Morphology of modified jute fiber was studied using scanning electron microscopy (SEM) and Fourier transform infrared spectroscopy (FTIR), which showed marked morphological changes for better adhesion with matrix. Surfaces of treated and untreated fibers were examined using SEM micrographs. Figure 4.7 showed relatively rougher surfaces in treated fibers as compared to untreated fibers. Removal of surface impurities, noncellulosic substances, inorganic materials, and waxes resulted in cleaner and rougher surfaces in chemically modified jute fiber [97].

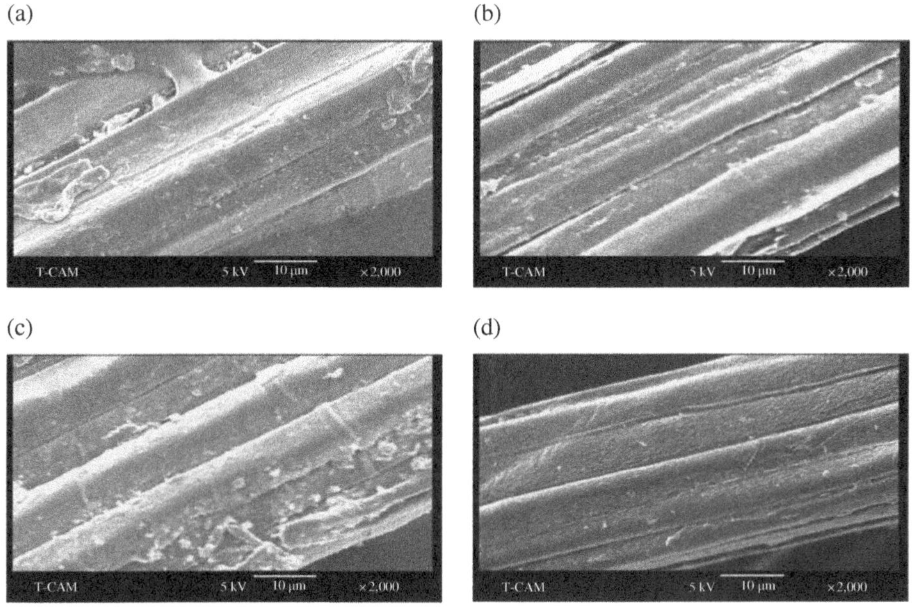

FIGURE 4.7 SEM micrographs of different phase surface treated jute fibers. (a) Untreated jute fiber, (b) detergent-washed jute fiber, (c) ethanol-washed jute fiber, and (d) final treated jute fiber. *Source*: Hossain et al. [97]. Reprinted with permission from Elsevier.

4.4.2 Noncovalent Surface Chemical Modifications

Plasma treatment and corona treatment provide more environmentally friendly approach for the surface modification of lignocellulosic fibers. Plasma is an ionized gas containing mixture of ions, electrons, neutral and excited molecules, and photons [98]. Gases like air, O_2, N_2, and helium have been used for the plasma treatment resulting in enhanced surface wettability, reduced surface friction, and removal of weakly bound polymer layers and etching, which ultimately influence their surface chemistry, morphology, and mechanical properties. CO_2 and O_2 treatment causes high degree of chain scission and degradation than any other active gases [99]. Low-pressure plasma treatment of lignocellulosic fiber has been reported for cleaning surfaces from contaminations such as air pollutants, oxide layers, and other surface additives. Contaminations can be removed by simple plasma sputtering with the help of noble gases, by oxidation of organic contaminants with oxygen plasma, or by reduction of oxides or sulfides using hydrogen plasma [100]. Plasma etching is another method to remove unwanted materials from surfaces. Four processes—sputtering, chemical etching, energetic ion etching, and ion inhibitor etching—are used in combination to remove materials from the surface [101, 102]. Atmospheric air pressure plasma treatment has found to show impact on mechanical properties of flax, hemp, and sisal fibers. A short atmospheric air pressure plasma treatment increased interfacial shear strength as surface roughness get enhanced due to cleaning of contaminant substances that hindered the adhesive process thus favors the

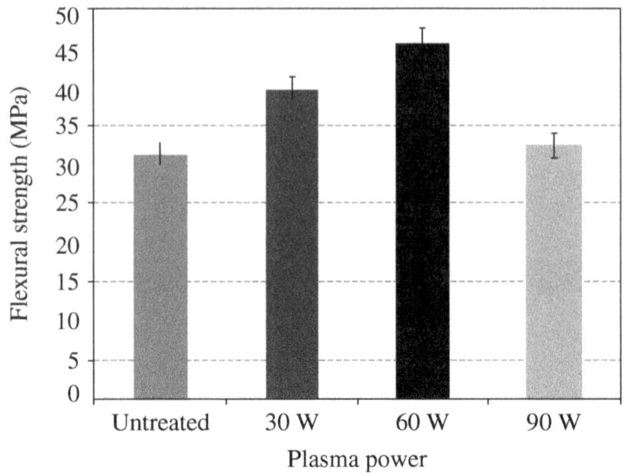

FIGURE 4.8 Effect of plasma power on flexural properties of jute/HDPE composites. *Source*: Seki et al. [106]. Public Domain. (*See insert for color representation of the figure.*)

mechanical interlocking between fibers and matrix [103]. Inbakumar et al. [104] reported that plasma treatment made the lignocellulosic fiber surface rough due to formation of significant amount of tiny grooves and cracks. Low-temperature atmospheric oxygen plasma treatment improved the interlaminar fracture behavior of basalt/epoxy composites by 16% along with the increase in wettability of fiber due to physical etching and formation of functional groups on surface containing oxygen and nitrogen [105]. Seki et al. [106] investigated the effect of power of plasma treatment on flexural strength of jute fiber composites. An increase in plasma power to 60W resulted in maximum flexural strength (Fig. 4.8).

Helium atmospheric pressure plasma jet treatment along with alcohol pretreatment improved ramie fiber for better adhesion to PP and polybutylene succinate matrix [107, 108]. Applied plasma voltage selection is critical for optimal improvement of bonding between fiber and matrix [109]. The existence of moisture on the surface of fiber may influence the effect of atmospheric pressure plasma treatment. Moisture regain decreases the surface hydrophobicity of fiber and thus have negative influence on the adhesion property of fiber with matrix. Ramie fiber with different moisture regains of 2.5, 6.1, and 23.5% showed that 2.5% moisture regain group has most significant plasma etching effect [110]. Dielectric barrier discharge (DBD) treatment at atmospheric pressure improved the wettability behavior of aramid fiber due to generation of new oxygen-containing groups like C—O, C=O, and O=C—O on the surface of fiber [111, 112]. Exposure time, treated power, and argon flux are important parameters for DBD plasma. Exorbital power, overlarge flux, and long-lasting exposure could partly destroy the effect of treatment and damage mechanical properties of fiber [113].

Activation of lignocellulosic fiber surface by corona discharge treatment is an eco-friendly solvent-free route, which acts on fiber surface without modifying bulk properties [114]. Corona discharge treatment improves adhesion, wettability, and hydrophilic properties of lignocellulosic fibers [115–117]. The term "corona" is used for the condition of air or other gases between charged electrodes. Air break down in a strong electric field and its ionized molecules conduct electricity, which can be seen as a soft bluish discharge between electrodes and is called corona [118]. Corona treatment affects the morphology of lignocellulosic fiber, which results in improved mechanical properties. Corona treatment improves the interfacial adhesion between lignocellulosic fibers and matrix [119].

4.4.3 Cationization

Cationization of lignocellulosic fibers can be done by three methods: (i) by direct reaction of epichlorohydrin and tertiary amine, (ii) by coupling of oligo-ionomers, and (iii) by grafting of cationic monomers [120].

The direct reaction of tertiary amine and epichlorohydrin is catalyzed by tertiary amine. This method results in the wide range of products with different chemical groups having different polarities, accessibility, and charge density. *n*-Alkyldimethyl-3-chloro-2-hydroxypropyl ammonium chloride having alkyl groups of different lengths were prepared by reaction of *n*-alkyldimethyl amine hydrochlorides with epichlorohydrin and used for the cationization of the lignocellulosic fiber [121]. Cationized lignocellulosic fibers are not sensitive toward pH changes and are more economical. Cationized lignocellulosic fibers show antibacterial activities, which depends upon the length of alkyl chains on the ammonium cation. Antimicrobial agents also have cationizing effect on cellulose, and thus cellulose pretreated with these agents shows an increased uptake of anionic dyes. Cationization of lignocellulosic fiber is an important technique to improve the uptake of dyes. Neutral lignocellulosic fiber has a small negative charge (−11 mV) due to the presence of some carboxylic acid groups from oxidation at primary hydroxyl sites. Negative charge is still higher in basic medium due to ionization of hydroxyl groups on hydroxymethyl side chains [122]. The presence of negative charges on the surface of lignocellulosic fiber reduces the efficiency of fixation of dye on the fiber. The introduction of cationic sites increases the interaction of anionic dyes [123, 124]. Cationization of lignocellulosic fiber by binding polyallylamine to aldehyde groups introduced by partial oxidation of cellulose resulted in a novel anion exchange material. The amount of cationization was found to depend upon molecular weight of polyallylamine, its concentration, and pH of the reaction medium [125].

Cationization of lignocellulosic fiber by coupling ionomers results in higher charge densities but poor selectivity. Grafting of cationic monomers resulted in higher surface selectivity. Zghida et al. [126] reported the grafting of quaternary ammonium group onto cellulosic materials. Grafted products were found to absorb oxyanions containing chromium atoms.

4.4.4 Polymer Grafting

Graft copolymerization onto lignocellulosic fiber is well known and most extensively studied technique for the modification. Graft copolymerization results in incorporation of covalently bonded polymer chains onto lignocellulosic fibers.

The presence of polymer chains may develop new properties or enhance the existing ones depending on the nature of polymeric chains grafted [127]. Ushako reported the first graft copolymer of cellulose by copolymerizing allyl ester with vinyl esters of cellulose [128]. Synthesis, properties, and applications of lignocellulosic fiber graft copolymers have been extensively studied [129–131]. Different methods have been developed for the synthesis of graft copolymers onto lignocellulosic fibers. Graft copolymers can be synthesized by free radical grafting, ionic grafting, and living radical polymerization [132]. Grafting involves the generation of active sites on the lignocellulosic backbone by different means using chemical initiators, radiations, plasma ion beam, and enzymes. Polymer chains get covalently bonded to these active sites and reaction propagates till dead polymer is obtained [133].

Initiation

$$\text{I-I} \longrightarrow 2\text{I}^* \quad (6)$$

$$\text{Fiber-OH} + \text{I}^* \longrightarrow \text{Fiber-O}^* + \text{I-H} \quad (7)$$

$$\text{M} + \text{I}^* \longrightarrow {}^*\text{M-I} \quad (8)$$

Propagation

$$\text{Fiber-OH} + {}^*\text{M-I} \longrightarrow \text{Fiber-O-M}^* + \text{I-H} \quad (9)$$

$$\text{Fiber-O-M}^* + n\text{M} \longrightarrow \text{Fiber-O-(M)}_n\text{-M}^* \quad (10)$$

$$\text{Fiber-O}^* + n\text{M} \longrightarrow \text{Fiber-O-(M)}_{n-1}\text{-M}^* \quad (11)$$

$$\text{M-I} + n\text{M} \longrightarrow \text{I-(M)}_n\text{-M}^* \quad (12)$$

Termination

$$\text{Fiber-O-(M)}_n\text{-M}^* + \text{M}^*\text{-(M)}_n\text{-O-Fiber} \longrightarrow \text{Fiber-O-(M)}_n\text{-M}_2\text{-(M)}_n\text{-O-Fiber} \quad (13)$$

$$\text{Fiber-O-(M)}_{n-1}\text{M}^* + {}^*\text{M-(M)}_{n-1}\text{-O-Fiber} \longrightarrow \text{Fiber-O-(M)}_{n-1}\text{-M}_2\text{-(M)}_{n-1}\text{-O-Fiber} \quad (14)$$

$$\text{Fiber-O-(M)}_n\text{-M}^* + \text{I}^* \longrightarrow \text{Fiber-O-(M)}_{n+1}\text{-I} \quad (15)$$

$$\text{I-(M)}_n\text{-M}^* + {}^*\text{M-(M)}_n\text{-I} \longrightarrow \text{I-(M)}_n\text{-M}_2\text{-(M)}_n\text{-I} \quad (16)$$

Where, M = Monomer, I = Initiator

Azo-bis-isobutyl nitrile (AIBN), benzoyl peroxide, hydrogen peroxide, potassium persulfate, sodium sulfite, and sodium thiosulfate are the common initiators used for grafting of polymer chains onto lignocellulosic fibers [134–136]. Radiations involved are either high-energy gamma radiations, low-energy UV radiations, or microwave radiations [137]. UV radiation-induced graft copolymerization involves the use of photo initiator to accelerate the initiation process. UV radiation-induced grafting causes less degradation of backbone, which results in better control over grafting [138].

Khan et al. studied the UV radiation-induced preirradiation and simultaneous grafting of MMA and 2-hydroxyethyl methacrylate onto jute fiber in aqueous medium [138, 139]. The use of microwave radiation as an initiator source is a greener way for graft copolymerization of lignocellulosic fibers. The microwave radiation-induced grafting occurs with or without chemical initiator. Kaith et al. carried out graft copolymerization of different vinyl monomers onto different lignocellulosic fibers under the influence of microwave radiations [140, 141] (Fig. 4.9).

(a) (b)

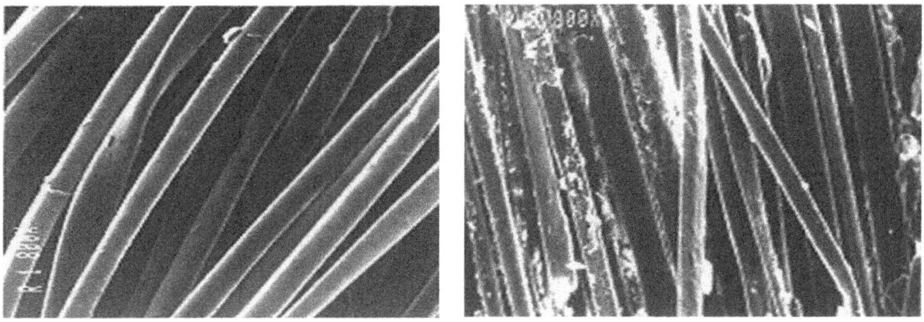

FIGURE 4.9 (a) SEM of *Saccharum spontaneum* (Ss) fiber and (b) SEM of grafted *Saccharum spontaneum* fiber prepared under microwave radiations. *Source*: Kaith et al. [141]. Reproduced with permission from Taylor & Francis.

Ionic grafting is less explored technique of graft copolymerization due to requirement of particular reaction conditions like low temperature, high purity of reagents, inert atmosphere, and anhydrous conditions [142]. This approach proceeds by forming cations and anions onto cellulose backbone, which initiates graft copolymerization. Cationic graft copolymerization involves the use of Lewis acids and bases as initiators [143, 144]. Graft copolymerization of lignocellulosic and its fibers provide a tool to overcome the drawbacks in its physical properties and chemical reactivity. Graft copolymerization alters the mechanical properties, thermal stability, dyeability, and ion exchange properties of lignocellulosic fibers [145]. The properties of grafted lignocellulosic depend upon chemical structure of monomer, length of grafted segment, and grafting percentage. Moisture resistance of lignocellulosic fibers can be enhanced by grafting of hydrophobic monomers. Grafted polymers block the sites sensitive for the moisture absorbance. UV radiation-induced preirradiation grafting of methylacrylate onto lignocellulosic fiber causes 61% decrease in hydrophilicity [139]. Mechanical properties of lignocellulosic and its fibers are significantly affected by grafting due to alteration of crystallinity of the backbone. Incorporation of specific properties through graft copolymerization onto natural fiber has widened its scope and applicability in different fields. Grafted fiber received increased interest due to their applicability as antibacterial surfaces, metal ion sorption materials, reinforcing agents in composite materials, thermoresponsive smart materials, photographic and motion picture film, LCD flat screens, and controlled drug delivery devices. The use of cellulosic materials as reinforcement of polymer matrix has been increased largely. Grafted fibers act as better reinforcing agents as they showed better fiber–matrix adhesion and better dispersion of fiber during composite processing [146–149]. The grafting of vinyl monomers is an important method for replacing hydrophilic groups on the surface of natural fibers by hydrophobic polymers. It improves the compatibility of natural fibers with matrices in natural fiber-reinforced polymer composites, which ultimately affects the mechanical properties. *Cannabis indica* fiber was graft copolymerized with acrylonitrile using redox initiator system to improve mechanical properties of polyester composites [150].

Reinforcement of poly(vinyl chloride) composites with poly(MMA) grafted oil palm fruit bunch fiber enhanced the tensile strength and elongation at break, whereas Young's modulus was found to decrease [151].

4.4.5 TEMPO-Mediated Oxidation

2,2,6,6-Tetramethylpiperidine-1-oxy radical (TEMPO)-mediated oxidation is a promising route to convert surface hydroxyl groups of natural fibers into charged carboxyl entities. TEMPO-mediated oxidation has an advantage over traditional hydrolysis with HCl and H_2SO_4, as in these cases, resulting sulfate and chlorate moieties are rather labile and get readily removed with mild alkaline conditions, whereas, charged carboxyl moieties generated during TEMO-mediated oxidation do not have that much liability. First report of TEMPO-mediated oxidation was given by De Nooy et al., which showed the oxidation of hydroxymethyl groups of polysaccharides while secondary hydroxyl groups remain unaffected [152]. TEMPO-mediated oxidation has been reported to different cellulose samples like cotton linters, wood pulp, rayon, parenchyma cellulose, and cellulose III [153–155]. 2,2,6,6-Tetramethylpiperidine-1-oxy radical/sodium hypochlorite/NaBr system at pH 10.0 is used normally for oxidation of cellulose. TEMPO-mediated oxidation is applicable not only to introduce carboxylate groups but the other functional groups like aldehyde can also be introduced to lignocellulosic fiber. Saito and Isogai [156] introduced aldehyde groups up to 0.225 mmol/g in native lignocellulosic fiber by oxidizing fiber with sodium hypochlorite, small amount of 2,2,6,6-tetramethylpiperidine-1-oxy radical, and sodium bromide at pH 10.0. The introduction of aldehyde group through TEMPO-mediated oxidation was found to increase the wet strength of fiber by forming hemiacetal linkages with cellular hydroxyl groups [157]. The amount of carboxyl group generated on the fiber surface by TEMPO-mediated oxidation depends upon the amount of sodium hypochlorite used. The increase in carboxyl content has been found to improve the tensile strength and reduce the tear index [158]. The effect of TEMPO-mediated oxidation on lyocell fibers was studied in terms of functional group content and functional group profile in relation to the molecular weight of lignocellulosic fiber through GPC using multiple detection and group-selective fluorescence labeling as per CCOA and FDAM methodology. Analysis of CHO and COOH profiles and molecular weight alterations showed a decrease in the average molecular weight during the first hour of TEMPO-mediated oxidation; no strong additional chain scission was obtained with prolonged oxidation time [159]. Sorption properties of lyocell fibers were improved by the selective TEMPO-mediated oxidation. TEMPO-oxidized lyocell fibers showed antifungal properties with silver ions against fungi from the *Candida* family, *Candida albicans* (ATCC 24433), and antibacterial activity against two strains: *Staphylococcus aureus* (ATCC 25923) and *Escherichia coli* (ATCC 25922) [160].

Electro-mediated oxidation with 4-acetamide-TEMPO catalyst at pH 6.8 without sodium hypochlorite was found to be an environmentally friendly modification of lignocellulosic fibers to introduce ion-exchangeable carboxyl and aldehyde groups. It resulted in the incorporation of 1.1 mmol/g of carboxylate groups and 0.6 mmol/g of aldehyde groups in the oxidized lignocellulosic fiber [162]. Praskalo et al. studied

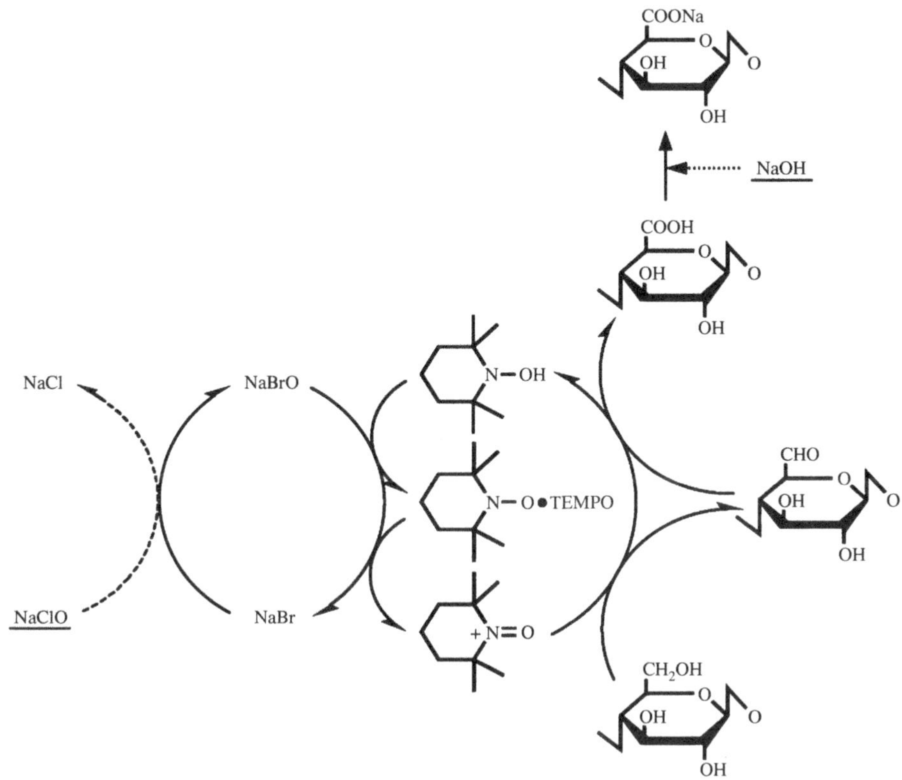

FIGURE 4.10 Schematic representation of the regioselective oxidation of cellulose by 2,2,6,6 tetramethyl-1-piperidinyloxy (TEMPO) process. *Source*: Missoum et al. [161]. Public domain.

the moisture sorption properties of oxidized lignocellulosic fibers. The oxidized lignocellulosic fiber with TEMPO-mediated oxidation was found to give 1.55 times higher moisture sorption than unoxidized fiber [163]. TEMPO-mediated oxidized lignocellulosic fibers were found to show better ion exchange behavior. TEMPO-oxidized lignocellulosic fibers have more ion exchange capacity especially for lead, calcium, and silver ions than those of carboxymethyl cellulose [164]. Fukuzumi et al. studied the thermal stability of TEMPO-oxidized cellulose. The decomposition temperature of oxidized cellulose was found to be lesser than that of native cellulose. Thermal stability of TEMPO-oxidized cellulose has been found to increase on methyl esterification of carboxyl groups with trimethylsilyldiazomethane or on ion exchange treatments with some metal solutions [165]. Oxidation of sisal pulp fibers by laccase–TEMPO system had been examined to study the effect of oxidative treatment on strength-related properties, water absorbency capabilities, and fiber morphology. Oxidation of sisal pulp by laccase–TEMPO system resulted in improvement of wet strength due to formation of a substantial amount of aldehyde groups in cellulose chains providing interfiber bonding through hemiacetal linkages [166] (Fig. 4.10).

4.4.6 Green Modification

Many species of bacteria, such as those in the genera *Gluconacetobacter, Agrobacterium, Aerobacter, Achromobacter, Azotobacter, Rhizobium, Sarcina, Salmonella,* and *Escherichia,* have been reported to produce extracellular cellulose [167]. Cellulose produced by *Gluconacetobacter* is called BC. The molecular formula of BC $(C_6H_{10}O_5)_n$ is the same as that of plant cellulose but has different physical and chemical properties [168]. The use of bacteria to modify the lignocellulosic fiber is a way to enhance their adhesion with renewable polymers. The use of bacteria has been recently reported to deposit sticky coatings of nanosized cellulose particles on the surface of natural fibers. Microbial cellulose synthesized in abundance by *Acetobacter xylinum* showed high mechanical strength and good physical properties [169]. Juntaro et al. reported the growth of BC nanofibers on the surface of micrometer-scale natural fibers, which resulted in a new class of renewable and degradable composites [170]. An integrated manufacturing approach for fiber-reinforced thermoplastic nanocomposites has been reported by Brown and Laborie [171]. They prepared BC/poly(ethylene oxide) nanocomposites by adding PEO to the growth medium of *A. xylinum.* Pommet et al. [172] described the process of modifying large quantity of natural fibers with BC through their use as substrate for bacteria during fermentation. The modified fibers were characterized to determined surface and mechanical properties by SEM, single fiber tensile test, X-ray photoelectron spectroscopy, and inverse gas chromatography. Quero et al. [173] first time reported the use of Raman spectroscopy to investigate micromechanical properties of BC/poly(L-lactic acid) resin composites.

Enzymatic retting of flax, enzymatic hemp separation, and enzymatic cottonization of bast fiber are groups of new technologies, which create suitable conditions to utilize traditional and renewable fibers for biocomposites. Enzymatic treatment affects the mechanical properties of fiber due to enzymatic decomposition of middle lamellae. Pectinase treatment of hemp fiber increased tensile strength, flexural strength, and modulus of elasticity of fiber-reinforced PP composites [174]. Tensile strength of cassava starch fiber was almost found to be double on treatment with *Trichoderma reesei* endoglucanases for 60 min with dose of 10 mg of enzyme per gram of substrate [175]. Treatment of fiber with different enzymes like pectinase, lactase, cellulose, and xylanase destroyed pectin, hemicellulose, and lignin from fiber bundle interface, which reduced the technical fiber diameter and increased the aspect ratio, thereby enhancing the fiber–matrix adhesion and improved mechanical properties [176]. Removal of thermally liable pectin and hemicellulose resulted in homogeneous surface, which improved the thermal stability. Removal of hemicellulose exposes the crystalline cellulose network, which consequently resulted in lower contact angle for fiber [177].

4.5 BIOCOMPOSITES: RECENT TRENDS AND OPPORTUNITIES FOR THE FUTURE

All over the world during the last three centuries, composite materials played the role of emerging materials for different applications. The increase in demand of composites for large number of applications has steadily grown the new market for modern composite materials. Bioplastic-based composites not only exhibit significance in day-to-day utilization but also constituent significant proportion of sophisticated and specified applications. Bioplastic-reinforced composites have four to six times greater tensile strength, 30–40% less weight, lower embedded energy, less noise, and high fatigue endurance limit in comparison to materials like steel and aluminum. Such materials could prove their worth as versatile materials and can be tailored according to required performance. Automobile industries, packaging materials, and household products are the major areas of application for biocomposites reinforced with natural fibers. The incorporation of stability both in terms of functional and structural aspects and durable shelf life along with pollution-free environmental degradation characteristics after their end use are the major challenges in replacing glass-reinforced composites with fiber-reinforced biocomposites.

Automobile companies remain in search of lightweight materials to increase fuel efficiency and sound absorbing materials for different parts of car and other vehicles. Natural fibers have excellent sound absorbing properties, and lightweight materials can be obtained. Natural fiber-reinforced composites are reported for both interior and exterior automobile parts [178]. Automotive manufacturers like Audi, BMW, Daimler Chrysler, Ford, Fiat, and Volkswagen are already major users of natural fiber composites in their cars. The use of natural fibers in automotive industry has risen exponentially in recent years. The European automobile industry has increased its usage from 70,000t in 2005 to more than 100,000t by 2010 [179]. Front door liners, boot liners, parcel shelves, seat backs, headrest, and sunroof interior shields are major interior parts utilizing natural fiber. Flax- and hemp-reinforced composites are of major interest for exterior automobile parts [180]. In addition to reduction in weight, other properties like reduced cost, good acoustic properties, and better thermal insulating properties are the benefits of using natural fiber in place of glass fiber. Ecological sustainability and low-energy requirements at the end disposal are main reasons of enhanced interest in natural fiber-based composites for automotive applications [181].

The use of biobased composites for building applications has already proved their worth. The technology road map for plants and crop-based renewable resources 2020 sponsored by the US Department of Energy has targeted to achieve goal of utilization of plant-derived resources up to 10% of basic chemical building blocks by 2020, which is further planned to be increased up to 50% by 2050 [49, 182]. Various natural fibers like sisal, palm, jute, and wood fiber in different polymer matrices like polyesters, epoxy, and phenolics have been explored for development of different building materials like laminates, door frames, door shutters, and partitions. Laminates and panel formed using sisal, jute, and coir fibers in polyester, phenolic, and polyurethane resins are comparable with the synthetic alternatives available. Natural fiber-reinforced composites are the promising alternative for wood in construction materials.

4.6 BIODEGRADABILITY OF BIOCOMPOSITES

Concern for the environment issues and waste management awaken the interest of researcher toward biodegradability of materials prepared. American society for testing of materials defines biodegradable plastics as materials that undergo change in

TABLE 4.4 Standard test methods for biodegradation[a]

Standard	Description
ASTM D 6400-99	Standard specifications for compostable plastics
ASTM D 6954-04	Standard guide for exposing and testing plastics that degrade in the environment by a combination of oxidation and biodegradation
ASTM D 5338-98	Standard test method for determining aerobic biodegradation of plastic materials under controlled composting conditions
ASTM D 6002-96	Standard guide for assessing the compostability of environmentally degradable plastics
ASTM D 6340-98	Standard test methods for determining aerobic biodegradation of radio labeled plastic materials in an aqueous or compost environment
ASTM D 5988-03	Standard test method for determining aerobic biodegradation in soil of plastic materials or residual plastic material after composting
ISO 14855-99	Determination of the ultimate aerobic biodegradability and disintegration of plastic materials under controlled composting conditions—method by analysis of evolved carbon dioxide
ISO 15314-04	Plastic—methods for marine exposure
ISO 16221-01	Water quality—guidance for the determination of biodegradability in the marine environment
EN 13432-2000	Requirements for packaging recoverable through composting and biodegradation—test scheme and evaluation criteria for the final acceptance of packaging

[a] From Baruah [183]. Public Domain.

chemical structure under specific environmental conditions. Factors affecting the degradation of material as well as mechanism of degradation depend upon the nature of method used for degradation. Degradation may be induced by UV, heat, or microorganism termed as photooxidative degradation, thermal degradation, and microbial degradation. Biodegradation may occur under aerobic or anaerobic conditions. Aerobic degradation produces carbon dioxide and water, whereas, anaerobic degradation results in methane along with carbon dioxide and water. Several methods have been explored for biodegradation of polymeric materials (Table 4.4).

Soil burial method is one of the most common methods used for determination of biodegradability of plastics and composites. Biodegradability was assessed in terms of loss of weight and change in physical appearance of material under consideration. Soil burial test was performed by burring sample of specific weight and dimensions in specific depth of soil for specific time period [184–187]. Composting method is another one of the frequently used methods for determination of biodegradability. Dry plastic or composite is placed in mixture of manure compost at 58°C with moisture content of 65% for specific time interval. Extent of biodegradation is measured by determining release of carbon dioxide [188, 189]. Pure culture method involves treatment of material with specific bacteria and fungi. Culture medium is inoculated with strain of specific bacteria or fungi and incubated for 4 weeks at optimal temperature and different films are compared for the growth of microorganisms [190]. Solubilization, ionization, hydrolysis, and enzymatic degradation are the different processes involved in biodegradation of materials depending upon the nature of natural fiber and matrices used [191]. The extent of biodegradation in different methods can be assessed by measuring weight, molecular weight distribution, tensile strength, biochemical oxygen demand, and carbon dioxide product [192–194]. Biodegradation studies of flax fiber-reinforced PP composites were done by burring samples in farm soil, and % weight loss was monitored for different intervals of time [195]. Loss in weight was also influenced by the presence of different additives (Fig. 4.11).

Biodegradation of composites is greatly influenced by the nature of polymers used as matrices. Both physical as well as chemical properties of matrix polymer influence the extent and mechanism of degradation. Addition of sawdust and wheat flour to PP enhanced biodegradability. PP–sawdust show greater biodegradability than PP–wheat flour due to higher water absorption ability of sawdust in comparison to wheat flour. The presence of higher moisture content in sawdust facilitate rate of degradation by microbial attack and hydrolysis. The total rate of degradation of hemp fiber-reinforced wheat gluten, hemp-gliadin, and glutenin composites was determined by calculating amount of CO_2 after 90 days. Wheat gluten and gliadin composites showed higher rate of degradation as compared to gluten-based hemp fiber-reinforced composites [196].

Surface treatment of fiber subsequentially affects the rate of degradation of composites. Modifications of natural fiber may result in positive or negative impact on rate of degradation. Maple fiber was subjected to different sequential treatment as NaOH treatment, NaOH followed by acetylation and NaOH followed by silane treatment. Composites reinforced with alkali alone showed earlier onset of degradation and faster rate of degradation than acetylated and silane-treated fiber. Increased rate of degradation is attributed to removal of wax and hemicellulose content from fiber surface [78]. Hydrophobic and hydrophilic nature of polymer, molecular weight distribution, and crystal structure also played an important role in the degradation process.

4.7 CONCLUSIONS

Enhanced environmental concern and legal reforms regarding utilization of petroleum resources have promoted the research in the direction of environmentally friendly materials. Biocomposites are the material of choice to reduce dependency on petroleum resources and to develop eco-friendly products. Development of commercially important natural fiber-reinforced bioplastics is greatly influenced by hydrophilic nature of fiber and properties of matrix used. Variation in fiber variety according to different geographical conditions is the other issue that must be handled carefully before commercial acceptance of natural fiber-based biocomposites. Surface treatments of natural fibers have been used to overcome the hydrophilic nature and also

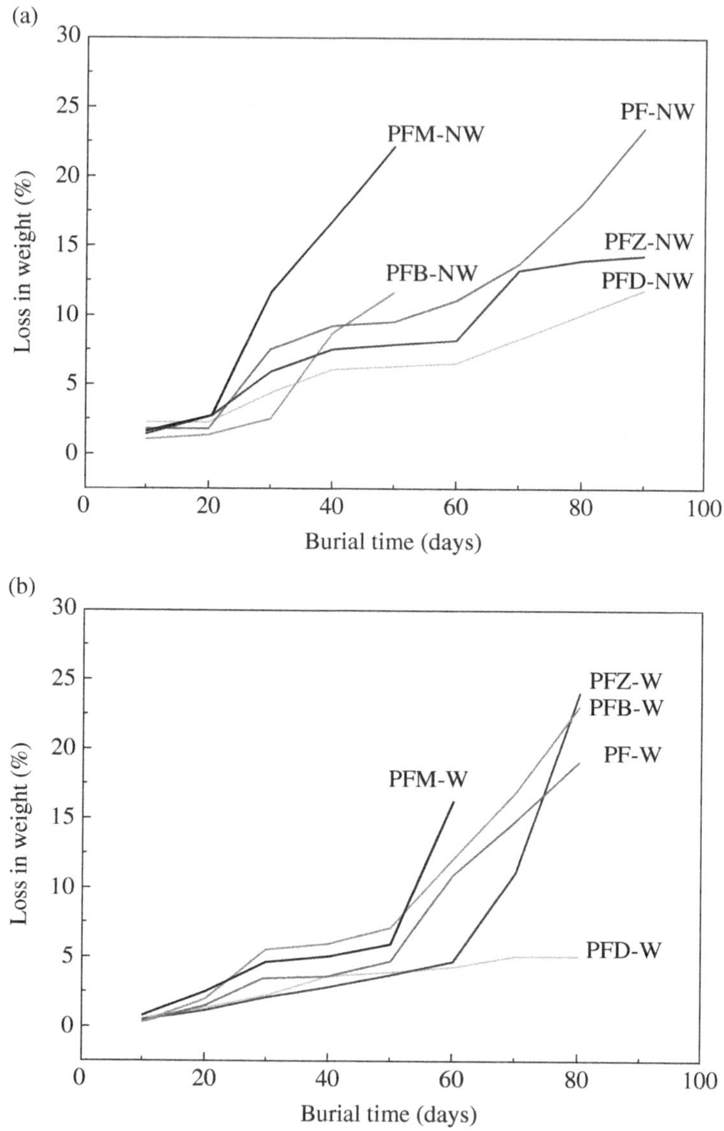

FIGURE 4.11 Loss of weight of (a) flax fiber reinforced (PF) and (b) nonwoven fiber reinforced (NW) poly(Lactic acid) composites with different additives: B, benzilic acid; D, dicumyl peroxide; M, mandelic acid; W, woven flax fiber reinforced poly(lactic acid) composites; and Z, zein protein. *Source*: Kumar et al. [195]. Public Domain.

to enhance compatibility between reinforcement and matrix. Treatment of natural fibers may involve chemical treatments like alkalization, silane treatment, acetylation, benzoylation, or through greener modifications like plasma, corona, and bacterial modifications, which can modify structural constituents and surface morphology.

These modifications can enhance thermal and mechanical properties of natural fibers. Reinforcement of composites with modified natural fibers can result in some specific composites and can be utilized for different specific applications. Recent trends involve development of composite materials, which can not only be used in diversified fields but also offers solution to the problems related to environmental issues. Researchers all over the world are indulged in the development of new methods for surface modifications of fibers so that desired properties could be incorporated along with sustained biodegradable characteristics. Biodegradable green composites are the promising substituent of petroleum products. However, issues related to commercialization of biodegradable composites need to be addressed globally.

REFERENCES

[1] Jamshidian, M., Tehrany, E.A., Imran, M., Jacquot, M., Desobry, S. (2010). Poly-lactic acid: production, applications, nanocomposites and release studies. *Comprehensive Reviews in Food Science and Food Safety*, **9**, 552–571.

[2] Ralston, B.E., Osswald, T.A. (2008). The history of tomorrow's materials: Protein-based biopolymer. *Plastics Engineering*, **64**, 36–40.

[3] Jawaid, M., Abdul Khalil, H.P.S. (2011). Cellulosic/synthetic fibre reinforced polymer hybrid composites: A review. *Carbohydrate Polymers*, **86**, 1–18.

[4] Shaffer, G.D. (1993). An archaeomagnetic study of a wattle and daub building collapse. *Journal of Field Archaeology*, **20**, 59–75.

[5] Fowler, P.A., Hughes, J.M., Elias, R.M. (2006). Biocomposites: Technology, environmental, credentials and market forces. *Journal of the Science of Food and Agriculture*, **86**, 1781–1789.

[6] Stewart, R. (2007). Biopolymers: Greater consumer awareness drives growth of biobased-resins market. *Plastics Engineering*, **63**, 24–30.

[7] Vroman, I., Tighzert, L. (2009). Biodegradable polymers. *Materials*, **2**, 307–344.

[8] Holbery, J., Houston, D. (2006). Natural-fiber-reinforced polymer composites in automotive applications. *JOM*, **58**, 80–86.

[9] Bakis, C.E., Bank, L.C., Brown, V.L., Cosenza, E., Davalos, J.F., Lesko, J.J., Machida, A., Rizkalla, S.H., Triantafillou, T.C. (2002). Fiber-reinforced polymer composites for construction—State-of-the-art review. *Journal of Composites for Construction*, **6**, 73–87.

[10] Davis, G., Song, J.H. (2006). Biodegradable packaging based on raw materials from crops and their impact on waste management. *Industrial Crops and Products*, **23**, 147–161.

[11] Suddell, B.C., Evans, W.J. Natural fiber composites in automotive applications, In: *Natural Fibers, Biopolymers, and Biocomposites*, Mohanty, A.K., Misra, M., Drzal, L.T. (Eds.), Taylor & Francis, New York, 2005.

[12] Bogoeva-Gaceva, G., Avella, M., Malinconico, M., Buzarovska, A., Grozdanov, A., Gentile, G., Errico, M.E. (2007). Natural fiber eco-composites. *Polymer Composites*, **28**, 98–107.

[13] Li, X., Tabil, L.G., Panigrahi, S. (2007). Chemical treatments of natural fiber for use in natural fiber-reinforced composites: A review. *Journal of Polymer and the Environment*, **15**, 25–33.

[14] Bismarck, A., Jimenez, A.B., Sarikakis, K. (2006). Green composites as panacea? Socio-economic aspects of green materials. *Environment, Development and Sustainability*, **8**, 445–463.

[15] Schlechter, M. (2014). The global market for composites: resins, fillers, reinforcements, natural fiber & nanocomposites, BCC Research. Available at https://www.reportbuyer.com/product/2009163/The-Global-Market-for-Composites-Resins-Fillers-Reinforcements-Natural-Fibers--Nanocomposites.html. Accessed on September 15, 2014.

[16] Jacob, M., Thomas, S., Varughese, K.T. (2004). Mechanical properties of sisal/oil palm hybrid fiber reinforced natural rubber composites. *Composites Science and Technology*, **64**, 955–965.

[17] Flieger, M., Kantorová, M., Prell, A., Řezanka, T., Votruba, J. (2003). Biodegradable plastics from renewable sources. *Folia Microbiologica*, **48**, 27–44.

[18] Wong, S., Shanks, R., Hodzic, A. (2002). Properties of poly(3-hydroxybutyric acid) composites with flax modified by plasticizer absorption. *Macromolecular Materials and Engineering*, **287**, 647–655.

[19] Carvalho, A., Zambon, M., Agnelli, J. (2003). Thermoplastic starch composites: 1-Influence of plasticizer and fiber content. *Polymer Degradation and Stability*, **79**, 133–138.

[20] Averous, L., Boquillon, N. (2004). Biocomposites based on plasticized starch: Thermal and mechanical behaviours. *Carbohydrate Polymers*, **56**, 111–122.

[21] Chiou, B., Glenn, G.M., Imam, S.H., Inglesby, M.K., Wood, D.F., Orts, W.J. Starch polymers: chemistry, engineering, and novel products, In: *Natural Fibers, Biopolymers, and Biocomposites*, Mohanty, A.K., Misra, M., Drzal, L.T. (Eds.), Taylor & Francis, New York, 2005.

[22] Ganjyal, G.M., Reddy, N., Yang, Y.Q., Hanna, M.A. (2004). Biodegradable packaging foams of starch acetate blended with corn stalk fibers. *Journal of Applied Polymer Science*, **93**, 2627–2633.

[23] Roy, S.B., Ramaraj, B., Shit, S.C., Nayak, S.K. (2011). Polypropylene and potato starch biocomposites: Physicomechanical and thermal properties. *Journal of Applied Polymer Science*, **120**, 3078–3086.

[24] Lu, J., Drzal, L.T. (2010). Microfibrillated cellulose/cellulose acetate composites: Effect of surface treatment. *Journal of Polymer Science Part B: Polymer Physics*, **48**, 153–161.

[25] Christian, S.J., Billington, S.L. (2011). Mechanical response of PHB- and cellulose acetate natural fiber-reinforced composites for construction applications. *Composites: Part B*, **42**, 1920–1928.

[26] Kunanopparat, T., Menut, P., Morel, M.H., Guilbert, S. (2008). Reinforcement of plasticized wheat gluten with natural fibers: From mechanical improvement to deplasticizing effect. *Composites Part A: Applied Science and Manufacturing*, **39**, 777–785.

[27] Guerrero, P., Retegi, A., Gabilondo, N., Caba, K.D.L. (2010). Mechanical and thermal properties of soy protein films processed by casting and compression. *Journal of Food Engineering*, **100**, 145–151.

[28] Kumar, R., Choudhary, V., Mishra, S., Varma, I.K., Mattison, B. (2002). Adhesives and plastics based on protein products. *Industrial Crops and Products*, **16**, 155–172.

[29] Nam, S., Netravali, A.N. (2006). Green composites. II Environment-friendly, biodegradable composites using ramie fibres and soy protein concentrate (SPC) Resin. *Fibers and Polymers*, **7**, 380–388.

[30] Kumar, R., Zhang, L. (2009). Aligned ramie fiber reinforced arylated soy protein composites. *Composites Science and Technology*, **69**, 555–560.

[31] Wang, L., Kumar, R., Zhang, L. (2009). Investigation into hemp fiber- and whisker-reinforced soy protein composites. *Frontiers of Chemistry in China*, **4**, 313–320.

[32] Huang, X., Netravali, A. (2009). Biodegradable green composites made using bamboo micro/nano-fibrils and chemically modified soy protein resin. *Composites Science and Technology*, **69**, 1009–1015.

[33] Khan, M.A., Kopp, C., Hinrichsen, G. (2001). Effect of vinyl and silicon monomers on mechanical and degradation properties of biodegradable Jute–Biopol® composite. *Journal of Reinforced Plastics and Composites*, **20**, 1414–1429.

[34] Avella, M., Martuscelli, E., Raimo, M. (2000). Review—Properties of blends and composites based on poly(3-hydroxy)butyrate (PHB) and poly(3-hydroxybutyrate-hydroxyvalerate) (PHVP) copolymers. *Journal of Materials Science*, **35**, 523–545.

[35] Gatenholm, P., Kubat, J., Mathiasson, A. (1992). Biodegradable natural composites- I. Processing and properties. *Journal of Applied Polymer Science*, **45**, 1667–1677.

[36] Singh, S., Mohanty, A.K., Sugie, T., Takai, Y., Hamada, H. (2008). Renewable resource based biocomposites from natural fiber and polyhydroxybutyrate-co-valerate (PHBV) bioplastic. *Composites: Part A*, **39**, 875–886.

[37] Shanks, R.A., Hodzic A., Wong S. (2004). Thermoplastic biopolyester natural fiber composites. *Journal of Applied Polymer Science*, **91**, 2114–2121.

[38] Chandra, R., Rustgi, R. (1998). Biodegradable polymers. *Progress in Polymer Science*, **23**, 1273–1335.

[39] Bloor, D., Flemings, M.C., Brook, R.J., Mahajan, S. Biodegradable polymers, In: *The Encyclopedia of Advanced Materials*, Bloor, D., Cahn, R.W. (Eds.), VCH Publishers, New York, Vol. **1**, 1994, p. 238.

[40] Zhang, Y., Yu, C., Chu, P.K., Lv, F., Zhang, C., Ji, J., Zhang, R., Wang, H. (2012). Mechanical and thermal properties of basalt fiber reinforced poly(butylene succinate) composites. *Materials Chemistry and Physics*, **133**, 845–849.

[41] Nam, T.H., Ogihara, S., Tung, N.H., Kobayashi, S. (2011). Effect of alkali treatment on interfacial and mechanical properties of coir fiber reinforced poly(butylene succinate) biodegradable composites. *Composites Part B Engineering*, **42**, 1648–1656.

[42] Lee, S.H., Wang, S. (2006). Biodegradable polymers/bamboo fiber biocomposite with bio-based coupling agent. *Composites: Part A*, **37**, 80–91.

[43] Liu, L., Yu, J., Cheng, L., Yang, X. (2009). Biodegradability of poly(butylene succinate) (PBS) composite reinforced with jute fibre. *Polymer Degradation and Stability*, **94**, 90–94.

[44] Teramoto, N., Urata, K., Ozawa, K., Shibata, M. (2004). Biodegradation of aliphatic polyester composites reinforced by abaca fiber. *Polymer Degradation and Stability*, **86**, 401–409.

[45] Liang, Z., Pan, P., Zhu, B., Dong, T., Inoue, Y. (2010). Mechanical and thermal properties of poly(butylene succinate)/plant fiber biodegradable composite. *Journal of Applied Polymer Science*, **115**, 3559–3567.

[46] Satyanarayana, K.G., Arizaga, G.G.C., Wypych, F. (2009). Biodegradable composites based on lignocellulosic fibers—An overview. *Progress in Polymer Science*, **34**, 982–1021.

[47] Mohanty, A.K., Misra, M., Hinrichsen, G. (2000). Biofibres, biodegradable polymers and biocomposites: An overview. *Macromolecular Materials and Engineering*, **276/277**, 1–24.

[48] Netravali, A.N., Chabba, S. (2003). Composites get greener. *Materials Today*, **6**, 22–29.

[49] Mohanty, A.K., Misra, M., Drzal, L.T. (2002). Sustainable bio-composites from renewable resources: Opportunities and challenges in the green materials world. *Journal of Polymers and the Environment*, **10**, 19–26.

[50] Misnon, M.I., Islama, M.M., Epaarachchi, J.A., Lau, K.T. (2014). Potentiality of utilising natural textile materials for engineering composites applications. *Materials and Design*, **59**, 359–368.

[51] Rong, M.Z., Zhang, M.Q., Liu, Y., Yang, G.C., Zeng, H.M. (2001). The effect of fiber treatment on the mechanical properties of unidirectional sisal-reinforced epoxy composites. *Composites Science and Technology*, **61**, 1437–1447.

[52] Bledzki, A.K., Gassan, J. (1999). Composites reinforced with cellulose based fibres. *Progress in Polymer Science*, **24**, 221–274.

[53] Kalia, S., Dufresne, A., Cherian, B.M., Kaith, B.S., Averous, L., Njuguna, J., Nassiopoulos, E. (2011). Cellulose-based bio- and nanocomposites: A review. *International Journal of Polymer Science*, **2011**, 1–35.

[54] Sette, M., Wechselberger, R., Crestini, C. (2011). Elucidation of lignin structure by quantitative 2D-NMR. *Chemistry-A European Journal*, **17**, 9529–9535.

[55] John, M.J., Anandjiwala, R.D. (2008). Recent developments in chemical modification and characterization of natural fiber-reinforced composites. *Polymer Composites*, **29**, 187–207.

[56] Andersson, M., Tillman, A.M. (1989). Acetylation of jute: Effects on strength, rot resistance and hydrophobicity. *Journal of Applied Polymer Science*, **37**, 3437–3447.

[57] Khan, M.A., Hassan, M.M., Drzal, L.T. (2005). Effect of 2-hydroxyethyl methacrylate (HEMA) on the mechanical and thermal properties of jute-polycarbonate composite. *Composites Part A: Applied Science and Manufacturing*, **36**, 71–81.

[58] Maiti, S.N., Subbaro, R., Ibrahim, M.N. (2004). Effect of wood fibers on the rheological properties of PP/wood fiber composites. *Journal of Applied Polymer Science*, **91**, 644–650.

[59] Paul, A., Joseph, K., Thomas, S. (1997). Effect of surface treatments on the electrical properties of low-density polyethylene composites reinforced with short sisal fibres. *Composites Science and Technology*, **57**, 67–79.

[60] Kittikorn, T., Stormberg, E., Ek, M., Karlsson, S. (2013). Comparison of water uptake as a function of surface modification of empty fruit bunch oil palm fibers in PP biocomposites. *Bioresources*, **8**, 2998–3016.

[61] Gonzalez, L., Rodnguez, A., de Benito, J.L., Marcos-Fernandez, A. (1997). Applications of an azide sulfonyl silane as elastomer crosslinking and coupling agent. *Journal of Applied Polymer Science*, **63**, 1353–1359.

[62] Culler, S.R., Ishida, H., Koenig, J.L. (1986). The silane interphase of composites: Effects of process conditions on γ-aminopropyltriethoxysilane. *Polymer Composites*, **7**, 231–238.

[63] Ghatge, N.D., Khisti, R.S. (1989). Performance of new silane coupling agents along with phenolic nobake binder for sand core. *Journal of Polymer Materials*, **6**, 145–149.

[64] Ghasemi, I., Farsi, M. (2010). Interfacial behaviour of wood plastic composite: Effect of chemical treatment on wood fibres. *Iranian Polymer Journal*, **19**, 811–818.

[65] Agrawal, R., Saxena, N.S., Sharma, K.B., Thomas, S., Sreekala, M.S. (2000). Activation energy and crystallization kinetics of untreated and treated oil palm fibre reinforced phenol formaldehyde composites. *Materials Science and Engineering A*, **277**, 77–82.

[66] Xie, Y., Hill, C.A.S., Xiao, Z., Militz, H., Mai, C. (2010). Silane coupling agents used for natural fiber/polymer composites: A review. *Composites: Part A*, **41**, 806–819.

[67] Shakeri, A., Haidar, K.T. (2004). Surface characterization of pulp paper fibres using inverse gas chromatography. *Iranian Polymer Journal*, **13**, 471–478.

[68] Zhou, F., Cheng, G., Jiang, B. (2014). Effect of silane treatment on microstructure of sisal fibres. *Applied Surface Science*, **292**, 806–812

[69] Asumani, O.M.L., Reid, R.G., Paskaramoorthy, R. (2012). The effects of alkali–silane treatment on the tensile and flexural properties of short fibre non-woven kenaf reinforced polypropylene composites. *Composites Part A: Applied Science and Manufacturing*, **43**, 1431–1440.

[70] Bilba, K., Arsene, M.A. (2008). Silane treatment of bagasse fiber for reinforcement of cementitious composites. *Composites Part A: Applied Science and Manufacturing*, **39**, 1488–1495.

[71] Moigne, N.L., Longerey, M., Taulemesse, J.M., Bénézet, J.C., Bergeret, A. (2014). Study of the interface in natural fibres reinforced poly(lactic acid) biocomposites modified by optimized organosilane treatments. *Industrial Crops and Products*, **52**, 481–494.

[72] Saenghirunwattana, P., Noomhorm, A., Rungsardthong, V. (2014). Mechanical properties of soy protein based "green" composites reinforced with surface modified cornhusk fibre. *Industrial Crops and Products*, **60**, 144–150.

[73] Teli, M.D., Valia, S.P. (2013). Acetylation of jute fiber to improve oil absorbency. *Fibers and Polymers*, **14**, 915–919.

[74] Baiardo, M., Frisoni, G., Scandola, M., Licciardello, A. (2002). Surface chemical modification of natural cellulose fibres. *Journal of Applied Polymer Science*, **83**, 38–45.

[75] Pasquini, D., Belgacem, M.N., Gandini, A., da Silva Curvelo A.A. (2006). Surface esterification of cellulose fibers: Characterization by DRIFT and contact angle measurements. *Journal of Colloid and Interface Science*, **295**, 79–83.

[76] Freire, C.S.R., Silvestre, A.J.D., Neto, C.P., Belgacem, M.N., Gandini, A. (2006). Controlled heterogeneous modification of cellulose fibers with fatty acids: Effect of reaction conditions on the extent of esterification and fiber properties. *Journal of Applied Polymer Science*, **100**, 1093–1102.

[77] Cunha, A.G., Freire, C.S.R., Silvestre, A.J.D., Neto, C.P., Gandini, A. (2006). Reversible hydrophobization and lipophobization of cellulose fibers via trifluoroacetylation. *Journal of Colloid and Interface Science*, **301**, 333–336.

[78] Way, C., Dean, K., Wuc, D.Y., Palombob, E. (2012). Biodegradation of sequentially surface treated lignocellulose reinforced polylactic acid composites: Carbon dioxide evolution and morphology. *Polymer Degradation and Stability*, **97**, 430–438.

[79] Bledzki, A.K., Mamun, A.A., Lucka-Gabor, M.M., Gutowski, V.S. (2008). The effects of acetylation on properties of flax fibre and its polypropylene composites. *Express Polymer Letters*, **2**, 413–422.

[80] Galdeano, M.C., Grossmann, M.V.E. (2005). Effect of treatment with alkaline hydrogen peroxide associated with extrusion on color and hydration properties of oat hulls. *Brazilian Archives of Biology and Technology*, **48**, 63–72.

[81] Sreekala, M.S., Kumaran, M.G., Joseph, S., Jacob, M., Thomas, S. (2000). Oil palm fibre reinforced phenol formaldehyde composites: Influence of fibre surface modifications on the mechanical performance. *Applied Composite Materials*, **7**, 295–329.

[82] Joseph, K., Mattoso, L.H.C., Toledo, R.D., Thomas, S., de Carvalho, L., Pothen, L., Kala, S., James, B. Natural fiber reinforced thermoplastic composite, In *Natural Polymers and Agrofibers Composites*, Frollini, E., Leao, A.L., Mattoso, L.H.C. (Eds), Embrapa Instrumentacao Agropecuaria, Sao Carlos, 2000, pp. 159–201.

[83] Nair, K.C.M., Thomas, S., Groeninckx, G. (2001). Fracture mechanics of composites with residual stresses, imperfect interfaces and traction-loaded cracks, *Composites Science and Technology*, **61**, 2159–2167.

[84] Santulli, C., Marzinotto, M., Mazzetti, C., Thomas, S. (2009). Influence of content and diameter of fibres and chemical treatment on the dielectric properties of oil palm fibres—Rubber composites. *Science and Engineering of Composite Materials*, **16**, 77–88.

[85] Khan, G.M.A., Terano, M., Gafur, M.A., Alam, M.S. (in press). Studies on the mechanical properties of woven jute fabric reinforced poly(l-lactic acid) composites, *Journal of King Saud University-Engineering Sciences*, DOI: 10.1016/j.jksues.2013.12.002.

[86] Ray, D., Sarkar, B.K., Rana, A.K., Bose, N.R. (2001). Effect of alkali treated jute fibres on composite properties. *Bulletin of Materials Science*, **24**, 129–134.

[87] Li, X., Panigrahi, S., Tabil, L. G. (2009). A study on flax fiber-reinforced polyethylene biocomposites. *Applied Engineering in Agriculture*, **25**, 525–531.

[88] Reddy, K.R.N., Rao, D.N.K., Rao, K.G.K., Kar, K.K. (2012). Studies on woven century fiber polyester composites. *Journal of Composite Materials*, **46**, 2919–2933.

[89] Mokaloba, N., Batane, R. (2014). The effects of mercerization and acetylation treatments on the properties of sisal fiber and its interfacial adhesion characteristics on polypropylene. *International Journal of Engineering, Science and Technology*, **6**, 83–97.

[90] Xue, D., Hub, H. (2013). Mechanical properties of biaxial weft-knitted flax composites. *Materials and Design*, **46**, 264–269.

[91] Lee, S.G., Choi, S.S, Park, W.H., Cho, D. (2003). Characterization of surface modified flax fibers and their biocomposites with PHB. *Macromolecular Symposium*, **197**, 89–100.

[92] Kaith, B.S., Singha, A.S., Kumar, S., Kalia, S. (2008). Mercerization of flax fiber improves the mechanical properties of fiber-reinforced composites. *International Journal of Polymeric Materials*, **57**, 54–72.

[93] Srinivasababu, N., Murali, K., Rao, M., Kumar, J.S. (2008). Tensile properties characterization of okra woven fiber reinforced polyester composites. *International Journal of Engineering*, **3**, 403–412.

[94] Modibbo, U.U., Aliyu, B.A., Nkafamiya, I.I. (2009). The effect of mercerization media on the physical properties of local plant bast fibres. *International Journal of Physical Sciences*, **4**, 698–704.

[95] Patel, B.C., Acharya, S.K., Mishra, D. (2012). Environmental effect of water absorption and flexural strength of red mud filled jute fiber/polymer composite. *International Journal of Engineering, Science and Technology*, **4**, 49–59.

[96] Rosa, M.F., Chiou, B., Medeiros, E.S., Wood, D.F., Williams, T.G., Mattoso, L.H.C., Orts, W.J., Imam, S.J. (2009). Effect of fiber treatments on tensile and thermal properties of starch/ethylene vinyl alcohol copolymers/coir biocomposites. *Bioresource Technology*, **100**, 5196–5202.

[97] Hossain, M.K., Dewan, M.W., Hosur, M., Jeelani, S. (2011). Mechanical performances of surface modified jute fiber reinforced biopol nano phased green composites. *Composites: Part B*, **42**, 1701–1707.

[98] Shenton, M.J., Lovell-Hoare, M.C., Stevens, G.C. (2001). Adhesion enhancement of polymer surfaces by atmospheric plasma treatment. *Journal of Physics D: Applied Physics*, **34**, 2754–2760.

[99] Zaldivar, R.J., Nokes, J.P., Patel, D.N., Morgan, B.A., Steckel, G. Kim, H.I. (2012). Effect of using oxygen, carbon dioxide, and carbon monoxide as active gases in the atmospheric plasma treatment of fiber-reinforced polycyanurate composites. *Journal of Applied Polymer Science*, **125**, 2510–2520.

[100] Collaud Coen, M., Dietler, G., Kasas, S., Gröning, P. (1996). AFM measurements of the topography and the roughness of ECR plasma treated polypropylene. *Applied Surface Science*, **103**, 27–34.

[101] Harth, K., Hibst, H. (1993). Surface modification of polypropylene in oxygen and nitrogen plasmas. *Surface and Coating Technology*, **59**, 350–355.

[102] Samukawa, S. (2007). High-performance and damage-free neutral-beam etching processes using negative ions in pulse-time-modulated plasma. *Applied Surface Science*, **253**, 6681–6689.

[103] Baltazar-y-Jimenez, A., Bistritz, M., Schulz, E., Bismarck, A. (2008). Atmospheric air pressure plasma treatment of lignocellulosic fibres: Impact on mechanical properties and adhesion to cellulose acetate butyrate. *Composites Science and Technology*, **68**, 215–227.

[104] Inbakumar, S., Morent, R., De Geyter, N., Demet, T., Anukaliani, A., Dubruel, P., Leys, C. (2010). Chemical and physical analysis of cotton fabric plasma-treated with a low pressure DC glow discharge. *Cellulose*, **17**, 417–423.

[105] Kim, M.T., Kim, M.H., Rhee, K.Y., Park, S.J. (2011). Study on an oxygen plasma treatment of a basalt fiber and its effect on the interlaminar fracture property of basalt/epoxy woven composites. *Composites Part B: Engineering*, **42**, 499–504.

[106] Seki, Y., Sever, K., Sarikanat, M., Güleç, H.A., Tavman, I.H. (2009). The influence of oxygen plasma treatment of jute fibers on mechanical properties of jute fiber reinforced thermoplastic composites. 5th International Advanced Technologies Symposium (IATS'09), May 13–15, 2009, Karabuk, Turkey.

[107] Zhou, Z., Liu, X., Hu, B., Wang, J., Xin, D., Wang, Z., Qiu, Y. (2011). Hydrophobic surface modification of ramie fibers with ethanol pretreatment and atmospheric pressure plasma treatment. *Surface and Coatings Technology*, **205**, 4205–4210.

[108] Wang, J., Zhou, Z., Huang, X., Zhang, L., Hu, B., Moyo, S., Sun, J., Qiu, Y. (2013). Effect of alcohol pretreatment in conjunction with atmospheric pressure plasmas on hydrophobizing ramie fiber surfaces. *Journal of Adhesion Science and Technology*, **27**, 1278–1288.

[109] Li, Y., Zhang, J., Chenga, P., Shic, J., Yaoa, L., Qiua, Y. (2014). Helium plasma treatment voltage effect on adhesion of ramie fibers to polybutylene succinate. *Industrial Crops and Products*, **61**, 16–22.

[110] Zhou, Z., Jilong, W., Xiao, H., Liwen, Z., Senelisile, M., Shiyuan, S., Yiping, Q. (2012). Influence of absorbed moisture on surface hydrophobization of ethanol pretreated and plasma treated ramie fibers. *Applied Surface Science*, **258**, 4411–4416.

[111] Jia, C., Chen, P., Liu, W., Li, B., Wang, Q. (2011). Surface treatment of aramid fiber by air dielectric barrier discharge plasma at atmospheric pressure. *Applied Surface Science*, **257**, 4165–4170.

[112] Xi, M., Li, Y., Shang, S., Li, D., Yin, Y.X., Dai, X.Y. (2008). Surface modification of aramid fiber by air DBD plasma at atmospheric pressure with continuous on-line processing. *Surface and Coatings Technology*, **202**, 6029–6033.

[113] Gua, R., Yua, J., Hua, C., Chena, L., Zhua, J., Hub, Z. (2012). Surface treatment of para-aramid fiber by argon dielectric barrier discharge plasma at atmospheric pressure. *Applied Surface Science*, **258**, 10168–10174.

[114] Lei, J., Shi, M., Zhang, J. (2000). Surface graft copolymerisation of hydrogen silicone fluid onto fabrics through Corona discharge and water repellency of grafted fabrics. *European Polymer Journal*, **36**, 1277–1281.

[115] Tu, X., Young, R.A., Denes, F. (1994). Improvement of bonding between cellulose and polypropylene by plasma treatment. *Cellulose*, **1**, 87–106.

[116] Ooi, S.K., Vanderhoek, N., Morrison, R., Bauer, A. (2004). Atmospheric plasma treatment (APT). Understanding the Basics. 58th Appita Annual Conference, 19–21 April, Canberra, Australia, pp. 407–412.

[117] Dong, S., Sapieha, S., Schreiber, H.P. (1993). Mechanical properties of Corona-modified cellulose/polyethylene composites. *Polymer Engineering and Science*, **33**, 343–346.

[118] Uner, B. (2009). The effect of support media on corona treated paper sorption properties. *Scientific Research and Essay*, **4**, 1024–1030.

[119] Amirou, S., Zerizer, A., Haddadou, I., Merlin, A. (2013). Effects of corona discharge treatment on the mechanical properties of biocomposites from polylactic acid and Algerian date palm fibres. *Scientific Research and Essays*, **8**, 946–952.

[120] Gruber, E., Granzow, C., Ott, T. Cellulose derivatives, In: *Cationization of Cellulose Fibers in View of Applications in the Paper Industry Cellulose Derivatives*, ACS Symposium Series, ACS Publication, Columbus, OH, Vol. **688**, 1998, pp. 94–106.

[121] Sawa, Y., Yamabayashi, C., Hoten, M. (2001). Cationization of cellulosic fibers by *n*-alkyldimethy-l,3-chloro-2-hydroxypropyl ammonium chlorides and their antibacterial activities. *Bulletin of Mukogawa Women's University National Science*, **48**, 19–26.

[122] Stan-Kleinschek, K., Ribitsch, V. (1998). Electrokinetic properties of processed cellulose fibers. *Colloids and Surfaces A: Physicochemical and Engineering Aspects*, **40**, 127–138.

[123] Kin, T.K., Yoon, S.H., Son, Y.A. (2004). Effect of reactive anionic agent on dyeing of cellulosic fibers with a berberine colorant. *Dyes and Pigments*, **60**, 121–127.

[124] Burkinshaw, S.M., Mignanelli, M., Froehling, P.E., Bide, M.J. (2000). The use of dendrimers to modify the dyeing behaviour of reactive dyes on cotton. *Dyes and Pigments*, **47**, 259–267.

[125] Wu, M., Shigenori, K. (2006). Cationization of cellulose fabrics by polyallylamine binding. *Journal of Applied Polymer Science*, **100**, 1666–1672.

[126] Zghida, H., Baouab, M.H.V., Gauthier, R. (2003). Sorption of chromium oxy-anions onto cationized ligno-cellulosic material. *Journal of Applied Polymer Science*, **87**, 1660–1665.

[127] Chauhan, G.S, Dhiman, S.K., Guleria, L.K., Kaur, I. (2002). Polymers from renewable resources: Kinetics studies of the radiochemical graft copolymerization of styrene onto cellulose extracted from pine needles and the effect of some additives on the grafting parameters in an aqueous medium. *Journal of Applied Polymer Science*, **83**, 1490–1500.

[128] Hebeish, A., Guthrie, J.T. *The Chemistry and Technology of Cellulosic Copolymers*, Springer Verlag, Heidelberg, 1998.

[129] Nasef, M.M., El-Sayed, A.H. (2004). Preparation and applications of ion exchange membranes by radiation-induced graft copolymerisation of polar monomers onto non-polar films. *Progress in Polymer Science*, **29**, 499–561.

[130] Thakur, V.K., Thakur, M.K., Gupta, R.K. (2014). Graft copolymers of natural fibers for green composites. *Carbohydrate Polymers*, **104**, 87–93.

[131] Thakur, V.K., Thakur, M.K., Singha, A.S. (2013). Free radical–induced graft copolymerization onto natural fibers. *International Journal of Polymer Analysis and Characterization*, **18**, 430–438.

[132] Bhattacharya, A., Misra, B.N. (2004). Grafting: a versatile means to modify polymers: Techniques, factors and applications. *Progress in Polymer Science*, **29**, 767–814.

[133] Jenkins, D.W., Hudson, S.M. (2001). Review of vinyl graft copolymerization featuring recent advances toward controlled radical-based reactions and illustrated with chitin/chitosan trunk polymers. *Chemical Reviews*, **101**, 3245–3273.

[134] Toledano-Thompson, T., Loria-Bastarrachea, M.I., Aguilar-Vega, M.J. (2005). Characterization of henequen cellulose microfibers treated with an epoxide and grafted with poly(acrylic acid). *Carbohydrate Polymers*, **62**, 67–73.

[135] Ouajai, S., Hodzic, A., Shanks, R.A. (2004). Morphological and grafting modification of natural cellulose fibres. *Journal of Applied Polymer Science*, **94**, 2456–2465.

[136] Mondal, I.H., Haque, M. (2007). Effect of grafting methacrylate monomers onto jute constituents with a potassium persulfate initiator catalyzed by Fe(II). *Journal of Applied Polymer Science*, **103**, 2369–2375.

[137] Aly, A.S., Sokker, H.H., Hoshem, A., Hebeish, A. (2006). Preparation of cellulosic membrane containing pyrrolidone moiety via radiation induced grafting and its waste water treatment. *American Journal of Applied Sciences*, **2**, 508–513.

[138] Khan, F., Ahmad, S.R. (2006). Graft copolymerization and characterization of 2-hydroxyethyl methacrylate onto jute fiber by photoirradiation. *Journal of Applied Polymer Science*, **101**, 2898–2910.

[139] Khan, F., Ahmad, S.R., Kronfli, E. (2004). UV-radiation–induced preirradiation grafting of methyl methacrylate onto lignocellulose fiber in an aqueous medium and characterization. *Journal of Applied Polymer Science*, **91**, 1667–1675.

[140] Maiti, M., Kaith, B.S., Jindal, R., Jana, A.K. (2010). Synthesis and characterization of corn starch based green composites reinforced with *Saccharum spontaneum* L graft copolymers prepared under micro-wave and their effect on thermal, physio-chemical and mechanical properties. *Polymer Degradation and Stability*, **95**, 1694–1703.

[141] Kaith, B.S., Jindal, R., Jana, A.K., Maiti, M. (2009). Rapid synthesis of graft copolymer of MMA onto Saccharum spontaneum under microwave irradiation for enhanced thermal modifications. *International Journal of Polymer Analysis and Characterization*, **9**, 364–387.

[142] Gupta, K.C., Sahoo, S. (2001). Graft copolymerization of 4-vinylpyridine onto cellulose using Co (III) acetylacetonate complex in aqueous medium. *Cellulose*, **8**, 233–242.

[143] Ikeda, I., Kurushima, Y., Takashima, H., Suzuki, K. (1988). Cationic graft polymerization of 2-oxazolines on cellulose derivatives. *Polymer Journal*, **3**, 243–250.

[144] Feit, B.A., Bar-Nun, A., Lahav, M., Zilkha, A. (1964). Anionic graft polymerization of vinyl monomers on cellulose and polyvinyl alcohol. *Journal of Applied Polymer Science*, **8**, 1869–1888.

[145] Princi, E., Vicini, S., Pedemonte, E., Mulas, A., Franceschi, E., Luciano, G., Trefiletti, V. (2005). Thermal analysis and characterisation of cellulose grafted with acrylic monomers. *Thermochimica Acta*, **425**, 173–179.

[146] Tiana, Y., Wuc, M., Liua, R., Wanga, D., Lina, X., Liu, W., Maa, L., Li, Y., Huanga, Y. (2011). Modified native cellulose fibers—A novel efficient adsorbent for both fluoride and arsenic. *Journal of Hazardous Materials*, **185**, 93–100.

[147] Maldas, D., Kokta, B.V. (1990). Effect of recycling on the mechanical properties of wood fibre polystyrene composites. Part 1: Chemithermo mechanical pulp as reinforcing filler. *Polymer Composites*, **11**, 77–83.

[148] Huang, Z., Liang, X., Hu, H., Gao, L., Chen, Y., Tong, Z. (2009). Influence of mechanical activation on the graft copolymerization of sugarcane bagasse and acrylic acid. *Polymer Degradation and Stability*, **94**, 1737–1745.

[149] Dong, H., Xu, Q., Li, Y., Mo, S., Cai, S., Liu, L. (2008). The synthesis of biodegradable graft copolymer cellulose-graft-poly(l-lactide) and the study of its controlled drug release. *Colloids and Surfaces B: Biointerfaces*, **66**, 26–33.

[150] Singha, A.S., Rana, A.K. (2012). Preparation and characterization of graft copolymerized *Cannabis Indica* L. fiber-reinforced unsaturated polyester matrix-based biocomposites. *Journal of Reinforced Plastics and Composites*, **31**, 1538–1553.

[151] Ratnam, C.T., Raju, G., Ibrahim, N.A., Ab. Rahman, M.Z., Yunus, M.Z.W. (2008). Characterization of oil palm empty fruit bunch (OPEFB) fiber reinforced PVC/ENR blend. *Journal of Composite Materials*, **42**, 2195–2204.

[152] de Nooy, A.E.J., Besemer, A.C., Van Bekkum, H. (1994). Highly selective TEMPO mediated oxidation of primary alcohol groups in polysaccharides. *Recueil des Travaux Chimiques des Pays-Bas*, **113**, 165–166.

[153] Tahiri, C., Vignon, M.R. (2000). TEMPO-oxidation of cellulose: Synthesis and characterization of polyglucuronans. *Cellulose*, **7**, 177–188.

[154] Saito, T., Isogai, A. (2004). TEMPO-mediated oxidation of native cellulose- The effect of oxidation conditions on chemical and crystal structures of the water-insoluble fractions. *Biomacromolecules*, **5**, 1983–1989.

[155] Fujisawa, S., Ikeuchi, T., Takeuchi, M., Saito, T., Isogai, A. (2012). Superior reinforcement effect of tempo-oxidized cellulose nanofibrils in polystyrene matrix: Optical, thermal, and mechanical studies. *Biomacromolecules*, **13**, 2188–2194.

[156] Saito, T., Isogai, A. (2006). Introduction of aldehyde groups on surfaces of native cellulose fibers by TEMPO-mediated oxidation. *Colloids and Surfaces A: Physicochemical and Engineering Aspects*, **289**, 219–225.

[157] Saito, T., Isogai, A. (2007). Wet strength improvement of TEMPO-oxidized cellulose sheets prepared with cationic polymers. *Industrial and Engineering Chemical Research*, **46**, 773–780.

[158] Mao, L., Law, K., Claude, D., Francois, B. (2008). Effects of carboxyl content on the characteristics of TMP long fibers. *Industrial and Engineering Chemical Research*, **47**, 3809–3812.

[159] Milanovic, J., Schiehser, S., Milanovic, P., Potthast, A., Kostic, M. (2013). Molecular weight distribution and functional group profiles of TEMPO-oxidized lyocell fibers. *Carbohydrate Polymers*, **98**, 444–450.

[160] Praskalo-Milanovic, J.Z., Kostic, M.M., Dimitrijevic-Brankovic, S.I., Skundric, P.D. (2010). Silver-loaded lyocell fibers modified by tempo-mediated oxidation. *Journal of Applied Polymer Science*, **117**, 1772–1779.

[161] Missoum, K., Belgacem, N.M., Bras, J. (2013). Nanofibrillated cellulose surface modification: A review. *Materials*, **6**, 1745–1766.

[162] Takuya, I., Saito, T., Isogai, A. (2010). TEMPO electromediated oxidation of some polysaccharides including regenerated cellulose fibre. *Biomacromolecules*, **11**, 1593–1599.

[163] Praskalo, J., Kostica, M., Potthast, A., Popov, G., Pejic, B., Skundric, P. (2009). Sorption properties of TEMPO-oxidized natural and man-made cellulose fibers. *Carbohydrate Polymers*, **77**, 791–798.

[164] Saito, T., Isogai, A. (2005). Ion-exchange behavior of carboxylate groups in fibrous cellulose oxidized by the TEMPO-mediated system. *Carbohydrate Polymers*, **61**, 183–190.

[165] Fukuzumi, H., Saito, T., Okita, Y., Isogai, A. (2010). Thermal stabilization of TEMPO-oxidized cellulose. *Polymer Degradation and Stability*, **95**, 1502–1508.

[166] Aracria, E., Vidala, T., Ragauskas, A.J. (2011). Wet strength development in sisal cellulose fibers by effect of a laccase–TEMPO treatment, *Carbohydrate Polymers*, **84**, 1384–1390.

[167] Ross, P., Mayer, R., Benziman, M. (1991). Cellulose biosynthesis and function in bacteria. *Microbiological Reviews*, **55**, 35–58.

[168] Römling, U. (2002). Molecular biology of cellulose production in bacteria. *Research in Microbiology*, **153**, 205–212.

[169] Czaja, W., Krystynowicz, A., Bielecki, S., Brown, R.M. (2006). Microbial cellulose-the natural power to heal wounds. *Biomaterials*, **27**, 145–151.

[170] Juntaro, J., Pommet, M., Kalinka, G., Mantalaris, A., Shaffer, M.S.P., Bismarck, A. (2008). Creating hierarchical structures in renewable composites by attaching bacterial cellulose onto sisal fibres. *Advanced Materials*, **20**, 3122–3126.

[171] Brown, E.E., Laborie, M.G. (2007). Bioengineering bacterial cellulose/poly(ethylene oxide) nanocomposites. *Biomacromolecules*, **10**, 3074–3081.

[172] Pommet, M., Juntaro, J., Heng, J.Y.Y., Mantalaris, A., Lee, A., Wilson, K., Kalinka, G., Shaffer, M.S.P., Bismarck, A. (2008). Surface modification of natural fibers using bacteria: Depositing bacterial cellulose onto natural fibers to create hierarchical fiber reinforced nanocomposites. *Biomacromolecules*, **9**, 1643–1651.

[173] Quero, F., Nogi, M., Yano, H., Abdulsalami, K., Holmes, S.M., Sakakini, B.J., Eichhorn, S.J. (2010). Optimization of the mechanical performance of bacterial cellulose/poly(l-lactic) acid composites. *ACS Applied Materials & Interfaces*, **2**, 321–330.

[174] Saleem, Z., Rennebaum, H., Pudel, F., Grimm, E. (2008). Treating bast fibres with pectinase improves mechanical characteristics of reinforced thermoplastic composites. *Composites Science and Technology*, **68**, 471–476.

[175] Woehl, M.A., Canestraro, C.D., Mikowski, A., Sierakowski, M.A., Ramosa, L.P., Wypych, F. (2010). Bionanocomposites of thermoplastic starch reinforced with bacterial cellulose nanofibres: Effect of enzymatic treatment on mechanical properties. *Carbohydrate Polymers*, **80**, 866–873.

[176] Karaduman, Y., Gokcan, D., Onal, L. (2013). Effect of enzymatic pretreatment on the mechanical properties of jute fiber-reinforced polyester composites. *Journal of Composite Materials*, **47**, 1293–1302.

[177] George, M., Mussone, P.G., Bressler, D.C. (2014). Surface and thermal characterization of natural fibres treated with enzymes. *Industrial Crops and Products*, **53**, 365–373.

[178] Ahmad, F., Choi, H.S., Park, M.K. (2014). A review: Natural fiber composites selection in view of mechanical, light weight, and economic properties. *Macromolecular Materials and Engineering*, **300**, 10–24.

[179] Thomas, S. (2002). Effect of chemical modification on the properties of natural fiber reinforced composites. Proceedings of the 10th European Conference on Composite Materials (ECCM-10), 3–7 June, Brugge, Belgium.

[180] Sexena, M., Pappu, A., Haque, R., Sharma, A. Sisal fiber based composites and their applications, In: *Cellulose Fibers: Bio- and Nano-Polymer Composites: Green Chemistry and Technology*, Kalia, S., Kaith, B.S., Kaur, I. (Eds.), Springer, Berlin/Heidelberg/New York, 2011, p. 589.

[181] Huda, M.S., Darzal, L.T. Natural Fiber composites for automotive sectors, In: *Properties and Performance of Natural-Fibre Composites*, Pickering, K. (Ed.), Woodhead Publishing Limited, CRC Press, Cambridge, 2008.

[182] Mohanty, A.K., Misra, M., Drzal, L.T., Selke, S.E., Harte, B.R., Hinrichsen, G. Natural fibers, biopolymers and biocomposites: An introduction, In: *Natural Fibers, Biopolymers and Biocomposites*, Mohanty A.K., Misra, M., Drzal, L.T. (Eds.), CRC Press, Boca Raton, FL, 2005, pp. 3.

[183] Baruah, S.D. (2011). Biodegradable polymer: The promises and the problems. *Science and Culture*, **77**, 466–470.

[184] Behera, A.K., Avancha, S., Basak, R.K., Sen, R., Adhikari, B. (2012). Fabrication and characterizations of biodegradable jute reinforced soy based green composites. *Carbohydrate Polymers*, **88**, 329–335.

[185] Salim, Y.S., Sharon, A., Vigneswari, S., Ibrahim, S.M.M., Amirul, A.A. (2012). Environmental degradation of microbial polyhydroxyalkanoates and oil palm-based composites. *Applied Biochemistry and Biotechnology*, **167**, 314–326.

[186] Rana, A.K., Singha, A.S. (2014). Studies on the performance of polyester composites reinforced with functionalized grewia optiva short fibers. *Advances in Polymer Technology*, **33**, 4.

[187] Jayaramudu, J., Reddy, G.S.M., Varaprasad, K., Sadiku, E.R., Ray, S.S., Rajulu, A.V. (2013). Preparation and properties of biodegradable films from Sterculia urens short fiber/cellulose green composites. *Carbohydrate Polymers*, **93**, 622–627.

[188] Gunning, M.A., Geever, L.M., Killion, J.A., Lyons, J.G., Higginbotham, C.L. (2013). Mechanical and biodegradation performance of short natural fibre polyhydroxybutyrate composites. *Polymer Testing*, **32**, 1603–1611.

[189] Abraham, E., Elbi, P.A., Deepa, B., Jyotishkumar, P., Pothen, L.A., Narine, S.S., Thomas, S. (2012). X-ray diffraction and biodegradation analysis of green composites of natural rubber/nanocellulose. *Polymer Degradation and Stability*, **97**, 2378–2387.

[190] Chaudhuri, S., Chakraborty, R., Bhattacharya, P. (2013). Optimization of biodegradation of natural fiber (*Chorchorus capsularis*): HDPE composite using response surface methodology. *Iranian Polymer Journal*, **22**, 865–875.

[191] Singh, B., Sharma, N. (2008). Mechanistic implications of plastic degradation, *Polymer Degradation and Stability*, **93**, 561–584.

[192] Kaith, B.S., Jindal, R., Jana, A.K., Maiti, M. (2010). Development of corn starch based green composites reinforced with *Saccharum spontaneum* L fiber and graft copolymers—Evaluation of thermal, physico-chemical and mechanical properties. *Bioresource Technology*, **101**, 6843–6851.

[193] Kim, H.S., Kim, H.J., Lee, J.W., Choi, I.G. (2006). Biodegradability of bio-flour filled biodegradable poly(butylene succinate) bio-composites in natural and compost soil. *Polymer Degradation and Stability*, **91**, 1117–1127.

[194] Dong, Y., Ghataura, A., Takagi, H., Haroosh, H.J., Nakagaito, A.N., Lau, K.T. (2010). Polylactic acid (PLA) biocomposites reinforced with coir fibres: Evaluation of mechanical performance and multifunctional properties. *Composites Part A: Applied Science and Manufacturing*, **63**, 76–84.

[195] Kumar, R., Yakubu, M.K., Anandjiwala, R.D. (2010). Biodegradation of flax fiber reinforced poly lactic acid. *Express Polymer Letters*, **4**, 423–430.

[196] Muneer, F., Johansson, E., Hedenqvist, M.S., Gallstedt, M., Newson, W.R. (2014). Preparation, properties, protein crosslinking and biodegradability of plasticizer solvent free hemp fiber reinforced wheat gluten, glutenin, and gliadin composites. *BioResources*, **9**, 5246–5261.

5

NANOCELLULOSE-BASED GREEN NANOCOMPOSITE MATERIALS

QI ZHOU[1,2] AND NÚRIA BUTCHOSA[1]

[1] *School of Biotechnology, Royal Institute of Technology (KTH), AlbaNova University, Stockholm, Sweden*

[2] *Wallenberg Wood Science Center, Stockholm, Sweden*

Biodegradable Green Composites, First Edition. Edited by Susheel Kalia.

5.1 INTRODUCTION

Glass, aramid, and carbon fiber-reinforced plastics have dominated around 90% of all high-performance fiber-reinforced composite materials in the composite industries, and their production continues to grow. With increasing concerns for the climate and environment, in terms of the next generation of materials, it has been recognized globally that paradigm-shifting research is required to improve the performance of materials based on renewable resources. However, attempts to develop very-high-performance natural cellulosic fiber-based composite materials using intact cells from hemp, flax, and cotton have failed mainly due to the inherent imperfections of the secondary cell walls of natural fiber cells. These issues have been recently addressed by replacing with nanocelluloses, which are three orders of magnitude smaller than the intact fiber cells. This is owing to the reason that nanoscale building blocks are individually exceptionally strong since they are close to ideal, defect-free materials [1]. Nanocelluloses are defined as cellulose structures with one dimension in the nanometer range. They are isolated from biological sources that comprise nanosized cellulose fibrils. They exhibit unique properties such as a very large specific surface area, reactive surface of —OH groups, high strength and modulus, and low thermal expansion coefficient and gas permeability. Green nanocomposites from biodegradable polymers and nanocelluloses have demonstrated substantially improved mechanical properties, thermal stability, and barrier properties. Such green biocomposites are expected to have a great potential as alternatives for petroleum-based polymers since they are overall renewable and biodegradable without emitting any toxic compounds to the environment. In this chapter, we review the preparation, structure, and properties of nanocellulose-based green nanocomposites with particular focus on the polymer matrices and the effect of nanocellulose incorporation on thermal and mechanical, barrier, optical, and antimicrobial properties of the composites.

5.2 NANOCELLULOSE

Nanocelluloses are attracting a lot of attention as the new nanoscale building blocks for composite materials. The different terminologies related to cellulose units are summarized in Table 5.1 [2–12]. There are basically three types of nanocellulose being widely utilized in green nanocomposite materials, that is, cellulose nanocrystals (CNCs), cellulose nanofibrils (CNFs), and bacterial cellulose (BC), as simply categorized by their structure and production methods [13, 14]. CNCs and CNFs are generally isolated from biological sources that comprise nanosized cellulose fibrils and fibril aggregates. BC is biosynthesized bottom up from glucose by bacteria. Bacteria from the genus *Acetobacter* extrude 70–150 nm wide cellulose ribbons through their cellular membrane in a similar manner as plant cells, but the secreted BC has the advantage over plant cellulose of being virtually pure and free of any noncellulosic polymer.

TABLE 5.1 Terminologies related to cellulose units[a]

Term	Diameter
Cellulose microfibrils	3 nm thick in plants and more than 20 nm in certain algae and tunicates [2, 3]
Microcrystalline cellulose (MCC)	15–30 nm up to about 5 μm due to aggregation [4]
Cellulose nanocrystals (CNCs)	From 3 to 50 nm depending on the source [5, 6]
Cellulose whiskers	10–20 nm from tunicate mantle [7]
Microfibrillated cellulose (MFC)	25–100 nm from wood pulp fibers [8]
Nanofibrillated cellulose (NFC)	5–30 nm from wood pulp fibers [9–11]

[a]Sehaqui et al. [12], Table 1. Reprinted with permission from Elsevier.

5.2.1 Cellulose Nanocrystals

Cellulose nanocrystals (CNCs), which have also been named as cellulose nanowhiskers, cellulose microcrystals, or nanocrystalline cellulose, are rodlike or conifer-shaped nanoparticles generally prepared by sulfuric acid hydrolysis of native cellulose fibers. CNCs that are able to self-organize into chiral nematic structures in colloidal suspension were first prepared by sulfuric acid hydrolysis of wood pulp fibers in 1992 by Revol et al. [5]. Sulfuric acid hydrolysis results in the formation of sulfate esters on the surface of CNCs. Alternatively, CNCs bearing carboxylic acid groups have also been prepared by hydrochloric acid or hydrobromic acid hydrolysis combined with oxidative carboxylation (2,2,6,6-tetramethylpiperidine 1-oxyl (TEMPO) free radical-mediated oxidation) [15]. Subsequent carbodiimide-mediated coupling of amine-terminated poly(ethylene glycol) to carboxylated cellulose microcrystals improved the stability of the microcrystal suspensions through steric stabilization [16]. CNCs have also been prepared by controlled hydrolysis of cotton with phosphoric acid, and the resulting CNCs exhibit a much higher thermal stability than partially sulfated ones [17]. Endoglucanase has also been used for the controlled hydrolysis of microcrystalline cellulose or recycled pulp to produce CNCs [18]. Recently, we have successfully prepared CNCs bearing high carboxylate content (1.5 mmol/g) by hydrochloric acid hydrolysis of CNFs obtained from TEMPO-mediated oxidation of softwood pulp [19]. The length and width of CNCs are depending on the sources of cellulose [6]. CNCs from cotton are typically 10–20 nm in width and 100–300 nm in length as shown in Figure 5.1a, while the length can be several micrometers for CNCs obtained from tunicate, algal, and bacterial cellulose.

5.2.2 Cellulose Nanofibrils

Cellulose nanofibrils (CNFs), also termed as microfibrillated cellulose (MFC), nanofibrillated cellulose (NFC), or cellulose nanofibers in the literature, are characterized by their high aspect ratio (length to width ratio) and ability to form entangled network structure. The most common source material of CNFs is wood pulp fibers. Pulp fibers are actually chemically and/or mechanically treated wood cells, which consist

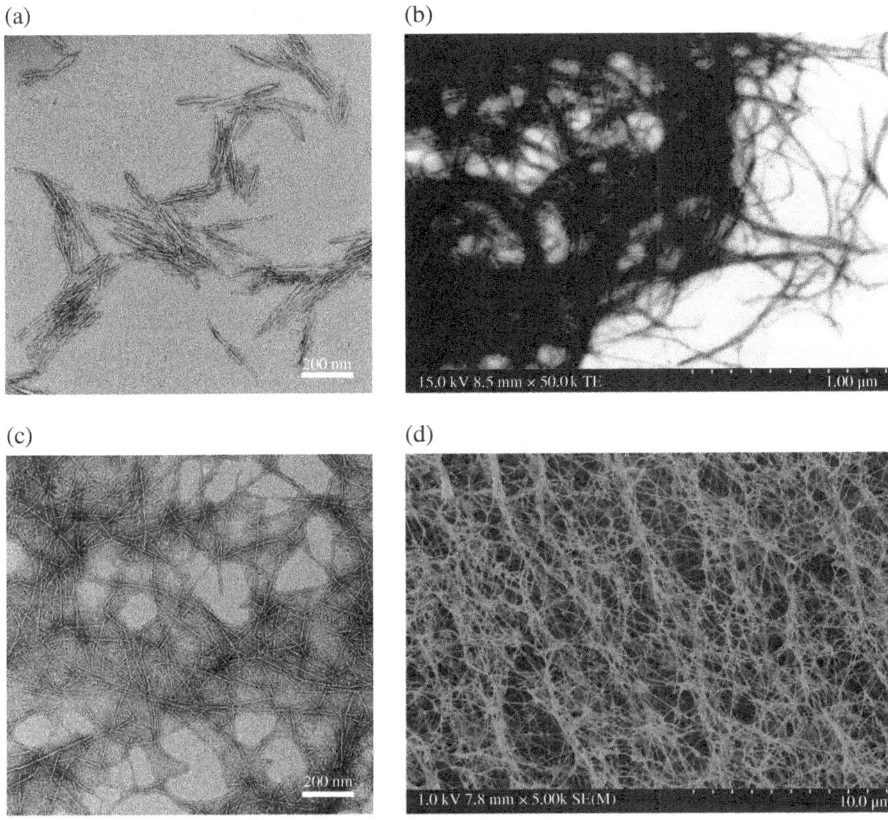

FIGURE 5.1 (a) Transmission electron microscopy (TEM) image of cotton CNCs from dilute water suspension. (b) Micrograph of CNFs from wood pulp obtained by scanning electron microscopy equipped with transmitted electron detector (STEM). (c) TEM image of TEMPO-oxidized CNFs from dilute water suspension. (d) Field emission scanning electron microscopy (FE-SEM) micrograph of the cross section of a fracture surface of a freeze-dried BC pellicle.

of a thin primary cell wall and a thick secondary cell wall that is mostly cellulose. Pulp fibers are long and flexible, typically a few millimeters long and 20–40 μm wide, with a morphology that depends on the tree species. Wood pulp fibers can be treated with different approaches to liberate individualized CNFs, and the extraction process has been extensively investigated in the past decade. Turbak et al. were the first to successfully prepare CNFs by passing a pulp suspension several times through a homogenizer at high pressure [8]. Later on, the high energy consumption of the mechanical disintegration process was reduced by Pääkkö et al. [9] and Henriksson et al. [20] using an enzymatic pretreatment of the wood pulp. Another approach to prepare CNFs is by chemical pretreatment of the wood pulp fibers. The purpose of the chemical modification is to introduce charged groups onto the surface of the nanofibrils that facilitate defibrillation. Using chemical pretreatment methods, well-individualized CNFs have been prepared by either carboxymethylation [21, 22] or TEMPO-mediated oxidation [23, 24] of wood pulp followed by a mild mechanical treatment. CNFs prepared by enzymatic pretreatment combined with mechanical disintegration are typically 5–30 nm in width and a few micrometers in length and tend to form fibril aggregates, as shown in Figure 5.1b [25]. CNFs prepared by TEMPO-mediated oxidation are completely individualized with very uniform nanofibril width distribution (3–5 nm) as shown in Figure 5.1c.

5.2.3 Bacterial Cellulose

Bacteria from the genus *Acetobacter* are aerobic, and the biological function of BC is thought to be as a flotation aid. BC is extracellularly biosynthesized in a virtually pure form, without any other associated polysaccharides that occur in the plant cell walls. Therefore, BC is often used as a model for the study of cellulose. Moreover, BC possesses high crystallinity. The bacterial nanocelluloses (also called nanoribbons) are continuously extruded by the bacteria, creating a three-dimensional network nanostructure as shown in Figure 5.1d. BC pellicles possess outstanding mechanical properties and great water-holding capacity. They can be shaped *in situ* by modifying the container in which the bacteria grow. When a homogeneous water suspension of bacterial CNFs is desired, the BC pellicle has to be disrupted by mechanical means to liberate the nanofibrils. Due to its unique properties, BC has mostly been used for biomedical applications [26], such as artificial blood vessels for microsurgery [27], artificial skin for humans with extensive burns [28], scaffolds for tissue engineering of cartilage [29], and wound dressings [30, 31]. Compared to CNFs from wood, BC ribbons can be modified during biosynthesis by the simple addition of water-soluble polymers in the culture medium of the bacterium. Therefore, cultures of *Acetobacter* represent an ideal biosynthetic system to produce high-strength nanocellulose-based green nanocomposite materials with a biomimetic nanostructure.

5.3 COMPOSITE MATRICES

5.3.1 Cellulose and Cellulose Derivatives

All-cellulose composites (ACCs), in which both the reinforcing fibers and the matrix are cellulose, were first prepared via a two-step process by impregnating aligned ramie fibers with the craft pulp cellulose solution in lithium chloride (LiCl)/*N,N*-dimethylacetamide (DMAc) by Nishino et al. [32]. These materials have overcome the weakness of interfacial adhesion in cellulosic fiber-based biocomposites and possessed excellent mechanical and thermal properties during use. ACCs were also prepared via a one-step process by converting a selective dissolved cellulose fiber surface into a matrix using LiCl/DMAc as a solvent [33]. In these ACCs, the cellulose fibers were either random in plan as in the conventional filter paper [34] or aligned in the longitudinal direction [35]. Cellulose solvents such as *N*-methylmorpholine-*N*-oxide (NMMO), sodium hydroxide (NaOH)/urea, and ionic liquids (ILs) [36, 37] have also been used for cellulose dissolution and the production of ACCs. However, the major issue of this conventional preparation method for ACCs is the limitations and drawbacks of cellulose solvents, which are contradicting the principles of "green" process and manufacturing, as reviewed by Huber et al. [38]. CNFs have ultrahigh specific surface area (SSA) and can be easily processed in water suspensions. All-cellulose biocomposites with fiber network structures at both micro- and nanoscale were successfully prepared by filtering and drying of a fibrous water suspension of wood pulp fibers and CNFs [12, 39]. CNFs showed excellent reinforcing capacity in these wood cellulose composites. When comparing the wood fiber reference and the biocomposite reinforced with 10% CNFs, the tensile strength increased from 98 to 160 MPa and the work to fracture was more than doubled (from 1.7 to 4.4 MJ/m^3). The CNF network improves load transfer between wood fibers and delays the development of damage into large scale at higher stress and strain.

Cellulose derivatives have also been often used as the nanocellulose-based green nanocomposite matrices due to chemical similarity between matrix and reinforcing filler. TEMPO-oxidized CNFs, carboxymethylated CNFs, conventional CNFs prepared by mechanical homogenization, and CNCs have all been used to reinforce hydroxypropyl cellulose (HPC) [40–43]. Comparison of high temperature storage modulus performances suggests stronger fibril–matrix interfaces in the TEMPO-oxidized CNF-filled composite than the others. In our previous work, we have prepared nanostructured biocomposites with ultrahigh strength using BC that was modified during biosynthesis by a simple addition of hydroxyethyl cellulose (HEC) in the culture medium of the bacterium [44]. The biosynthesized BC fibrils were coated with a monolayer of HEC and self-assembled to form loose bundles of fibril aggregates with lager lateral dimensions. Such *in situ* prepared BC/HEC nanocomposites demonstrated dramatically increased tensile strength and work to fracture compared to the neat BC. Furthermore, wood-based CNF/HEC biocomposites have also been prepared [45]. The biocomposites showed remarkable toughness and significant thermomechanical stability due to the CNF network, which also provides favorable resistance to creep.

5.3.2 Hemicelluloses and other Polysaccharides

Hemicelluloses are the noncellulose cell wall polysaccharides of agricultural and forest plants including mainly xyloglucans, xylans, and mannans. Xyloglucan is the major structural polysaccharide in the primary cell wall of plants and can be largely obtained from seeds of tamarind tree (*Tamarindus indica*) owing to its storage function [46]. Xylans and mannans can be found in an enormous amount of by-products produced from pulp and paper processing and cultivation of cereal crops [47]. Recently, considerable interest has been directed to hemicellulose-based functional biopolymers and biomaterials as substitutes for synthetic polymer-based engineered plastic materials. Hemicellulose-based biomaterials are bioactive, biocompatible, and having oxygen barrier properties, which give them potential in numerous applications, such as drug delivery, tissue engineering, and food packaging. In order to improve the film-forming ability of hemicelluloses and the properties of the resulting films, nanocomposite films of xylan, arabinoxylan, arabinoglucuronoxylan, glucomannan, or galactoglucomannans reinforced with either CNFs [48–53] or CNCs [54–58] have been prepared and characterized. Nanocelluloses generally reduce the moisture uptake and improve the mechanical and barrier properties in these composite films.

Hemicelluloses and cellulose derivatives such as carboxymethyl cellulose [59–63], methyl cellulose [59], pectin [64–67], xyloglucan [62, 66, 68, 69], glucomannan [64, 70], arabinoxylan [71], or xylan [61, 64] have also been added in the culture medium of *Acetobacter* and shown to affect the aggregation of the BC fibrils. The structure and formation of the cellulose network are affected by spontaneous interference of the added polymers with cellulose assembly. A xyloglucan/BC composite hydrogel was prepared and used as a model to study the roles of cellulose and xyloglucan in determining the mechanical properties of primary plant cell walls [72] and the effects of plant cell wall enzymes on its mechanical properties [73].

Green nanocomposites based on other polysaccharides, such as alginate [74–79], agar [80], and pullulan [81], have also been prepared using CNCs and CNFs as reinforcement fillers. The addition of nanocelluloses significantly reduced the water sensitivity of the polysaccharide matrices and increased the mechanical properties of the nanocomposite films. These novel green nanocomposites demonstrate good potential in food packaging applications.

5.3.3 Starch

Native starch has a granular structure and contains approximately 25% amylose and 75% amylopectin. Both polysaccharides are composed of α-D-glucose units, but whereas amylopectin is an extensively branched macromolecule, amylose is either linear or only slightly branched. Starch is highly moisture sensitive, and amylopectin-rich starch variants are particularly brittle. Native starch can be converted into thermoplastic starch (TPS) under heat and shear in the presence of plasticizer. Under these conditions, starch granules can be destructurized, melted, and plasticized to form a material that has a thermoplastic behavior. However, TPS is also highly water sensitive with weak mechanical properties affected by the relative humidity. Svagan et al. were the first to introduce wood MFC network into a starch matrix in the form of a 50/50 amylopectin/glycerol blend by using a solution casting method [82]. The composite at 70 wt% of MFC reinforcement showed a layered nanocomposite structure and good MFC dispersion and a unique combination of high modulus (6.2 GPa), tensile strength (160 MPa), and strain to failure (8.1%). Plackett et al. then investigated two types of CNFs as reinforcement in amylopectin films and found that stiffer films can be obtained when using CNFs with a significant content of hemicellulose at relative high loadings [83]. Composite films from amylopectin-rich waxy maize starch and kenaf bast fibers CNFs [84], as well as potato starch and rice straw CNFs [85], have also been prepared using the same film casting and evaporation process. Pregelatinization of starch was performed at elevated temperature in order to ensure good mixing with CNFs and plasticizer. The addition of CNFs increased mechanical properties and decreased the moisture sensitivity of the TPS film significantly. Karimi et al. have also studied the effect of casting techniques on the properties of

CNF/TPS composite films and discovered that simultaneously mixing and processing all starting materials resulted in the most homogeneous material and better performance [86]. CNF/starch nanocomposite from an agricultural waste residue, cassava bagasse, had also been prepared using the same pregelatinization and mixing procedure but finally dried by hot-pressing at 140°C [87]. Furthermore, Svagan et al. have also successfully prepared biomimetic CNF/amylopectin foams, with up to 40 wt% CNF-reinforced cell walls, simply by lyophilization [88]. The effect of CNF/starch ratio on the morphology and mechanical properties of the freeze-dried foam was further studied by Yildirim et al., and the optimum properties were obtained from 6% starch and 1.5% CNF combinations [89]. More interestingly, three-dimensional starch/CNF scaffolds with interconnected pores were fabricated using film casting, salt leaching, and freeze-drying methods [90]. Such scaffolds have uniform porous structure, suitable pore size, good mechanical properties, and biodegradability with potential for tissue engineering matrix. Green biocomposites of TPS and CNFs have also been manufactured by using a twin-screw extrusion process [91, 92]. The results showed that the addition of CNFs improved the mechanical properties and reduced the moisture uptake of the TPS. Nanocomposites from corn TPS and cotton CNCs have also been prepared by conventional extrusion process, and the addition of CNCs improved the mechanical properties of the TPS matrix when the concentration of CNCs was above 2.5 wt% [93]. Cotton CNCs have also been used to reinforce starch from chayote (*Sechium edule* Sw.) tubers that was plasticized with glycerol using solution casting method. The resulting composite films demonstrated better mechanical properties and lower water vapor permeability values [94]. BC/starch biocomposites have also been fabricated by either bottom-up biosynthesis of a coherent network of BC nanofibers in the presence of partially gelatinized starch granules [95] or a solution impregnation method using pure BC sheets [96].

5.3.4 Chitin and Chitosan

Chitin is a linear high molecular weight polysaccharide consisting of (1→4)-β-linked 2-deoxy-2-acetamido-D-glucopyranose monomers. It is the second most abundant carbohydrate in nature, widely available as a by-product of sea food-processing industries. Chitosan is produced commercially by N-deacetylation of chitin and composed primarily of glucosamine or 2-amino-2-dexoy-D-glucose units. However, chitosan is less common in nature. It is biosynthesized through N-deacetylation of chitin and present in the cell wall of filamentous fungi [97]. Chitosan has been widely used in packaging and biomedical applications. Chitosan can form tough and flexible films. Water vapor barrier and mechanical properties of chitosan films can be improved by the addition of CNCs [98]. Methyl adipoyl chloride-functionalized CNCs were reacted with amino groups of the chitosan to produce bionanocomposites, which demonstrated a significant improvement in mechanical performance with a remarkable decrease in water uptake [99]. Similar improvement in the properties was also achieved in the chitosan/CNF biocomposite films [100, 101]. Interestingly, it was found that CNFs have stronger reinforcing effects for the chitosan in the wet state compared to the dry matrix [102]. Biocomposites of CNFs with chitosan functionalized with antibacterial agent methylisothiazolinone [103] or benzalkonium chloride [104] were employed to enhance the antibacterial and mechanical properties of agar or sodium alginate films. Bacterial cellulose/chitosan composites were prepared by immersing the BC pellicles in the chitosan-dissolved acetic acid solution followed by freeze-drying, and the resulting porous material has the potential in tissue engineering or other medical applications [105].

5.3.5 Proteins

Gelatin is a natural protein polymer that is derived from the hydrolysis of collagen. Gelatin has previously found widespread applications as food ingredients, pharmaceutical capsules, tablet coatings, paper additives, cosmetic ingredients, or additives to photographic silver emulsions. Recently, it has been shown that gelatin-based scaffolds have a wide application range within the field of both soft and hard tissue engineering [106]. Gelatin has a relatively low antigenicity compared to collagen and can promote cell adhesion, differentiation, and proliferation. Three-dimensional cellulose fiber/gelatin microscaffolds were found to be able to provide optimal porosity and tensile strength for cell cultures [107]. BC/gelatin composite can be prepared by *in situ* biosynthesis with gelatin dissolved in the culture medium or by postmodification using either impregnation or adsorption method [108, 109]. The resulting composites showed higher water absorption capacity and improved rehydration. Further study using enzymatically modified gelatin revealed that the polar functional groups of gelatin and its hydrolysates as well as BC's porous network contributed to the rehydration ability of composites [110]. In order to improve the mechanical strength, BC/gelatin composites were cross-linked with procyanidin, genipin, transglutaminase, or 1-ethyl-3-(3-dimethylaminopropyl) carbodiimide (EDC) [111, 112].

Collagen is a fibrous structural protein that can be found in animal skin, bone, cornea, tendon, and other tissues. Collagen has also been widely used for applications in tissue engineering such as wound dressings owing to its good adhesion ability to skin. BC/collagen nanocomposite was prepared via surface modification of BC by glycine esterification using Fmoc-based chemistry (9-fluo-renylmethyloxycarbonyl (Fmoc)) followed by impregnation with aqueous collagen solution and cross-linking with EDC [113]. Such BC-based composite was demonstrated as alternative biomaterial for bone tissue engineering/regeneration procedures. Wood CNFs have also been used to reinforce collagen-based implantable scaffolds that were cross-linked with glutaraldehyde [114]. The addition of CNFs and cross-linking of collagen showed positive impact on mechanical properties and moisture stability. Alternatively, CNFs were oxidized into dialdehyde CNFs by $NaIO_4$ and then mixed with collagen in water. The mixtures were freeze-dried to afford CNF/collagen composite aerogels that exhibited good biocompatibility as well as high level of cell activity and proliferation [115].

Apart from gelatin and collagen, poly(L-lysine) (PLL) is promising candidate for a scaffold material. BC nanofibers have been coated with PLL by using procyanidins as cross-linker and used as novel templates for bone-like apatite deposition [116]. Silk fibroin (SF) fibers have been used commercially as biomedical sutures owing to their excellent mechanical properties and biocompatibility. Transparent CNC/SF composite films [117, 118] and electrospun SF nanofibers [119] reinforced with CNCs have been prepared and demonstrated improved mechanical properties. Thermoresponsive hybrid nanocellulose hydrogels have been prepared from a mixture of negatively charged TEMPO-oxidized BC and positively charged elastin-like polypeptide. The hybrid hydrogel was noncytotoxic and demonstrated potential for biomedical applications since the encapsulated cells remained alive and continued to proliferate inside the hydrogel [120].

Protein-based materials have also been widely exploited as potential food packaging materials because of their good barrier properties against oxygen and aromatic compounds. These proteins are low-cost agricultural products such as corn, milk, and soybean. Wheat gluten (*Triticum aestivum*) is a by-product from the wheat starch industry and one of the most frequently studied proteins due to its good film-forming properties. CNCs modified with TEMPO-mediated carboxylation have been used as a reinforcing filler for wheat gluten, and the nanocomposite containing 7.5 wt% CNC has the lowest water vapor permeation value and water solubility, the best tensile strength, and high storage modulus [121]. Caseins, the milk proteins, are amphiphilic containing both hydrophobic and hydrophilic domains. CNCs have also been used to reinforce sodium caseinate [122] and casein-functionalized CNCs have been dispersed in polylactic acid (PLA) owing to the amphiphilic nature of casein [123]. Soybean proteins (SPs), a by-product of the soy oil industry, can be used in adhesives, films, gels, emulsions, as well as medical applications [124]. Aerogels can be prepared by freeze-drying denatured SPs in acidic media, and their mechanical properties can be improved by reinforcing with CNFs [125]. Polyols converted from soybean oil have been used for the synthesis of green polyurethanes similar to commonly used castor oil. The properties of soy polyol-based polyurethanes have also been improved by reinforcing with BC nanofibrils [126] and CNCs [127].

5.3.6 Polylactic Acid and Poly(ε-Caprolactone)

Poly(lactic acid) (PLA) is a biodegradable polyester derived from renewable resources such as corn, wheat, and potato starch or sugarcane. It is one of the most promising thermoplastic biopolymers with biocompatibility and good mechanical properties. We have previously reviewed the preparation, structure, and properties of PLA/nanocellulose composites [128]. In order to obtain well-dispersed hydrophilic reinforcing nanocelluloses in hydrophobic PLA matrix, surface modification of nanocellulose is essential for the preparation of biocomposites. PLA/nanocellulose composites have been fabricated by solvent casting, melt processing, electrospinning, or papermaking process. PLA/nanocellulose biocomposites frequently exhibit significantly improved thermal and mechanical and various other properties compared to those of pure PLA. The tensile modulus and strength of neat PLA are generally improved with an increase of nanocellulose content. Increased storage modulus at temperature above T_g is often observed with the incorporation of nanocellulose in PLA, as well as decreased gas permeability.

Poly(ε-caprolactone) (PCL), produced by ring opening polymerization (ROP) of petroleum-derived ε-caprolactone, is an aliphatic polyester that is biodegradable and biocompatible with a hydrophobic nature. PCL is particularly interesting as a promising matrix component in bionanocomposites. In order to disperse hydrophilic fillers into PCL matrix to obtain high-performance composites, the use of compatibilizers or surface modification of the hydrophilic fillers is required. Thus, CNCs have been either chemically modified with *n*-octadecyl isocyanate [129, 130] or covalently grafted with PCL by a "grafting-from" approach based on *in situ* catalyzed ROP [131, 132]. The PCL/modified CNC composite films were prepared by a casting/evaporation process using either dichloromethane or toluene as the solvent and liquid medium. PCL with different lengths has also been grafted onto CNFs using ROP. It was shown that longer graft lengths significantly improved the stability of the dispersion of CNFs in organic solvent [133]. PCL has also been grafted onto porous CNF nanopaper with high specific surface area (ca. $300\,m^2/g$). The grafted amount of PCL on nanopaper was significantly higher than that on the conventional filter paper. This method allowed the fabrication of PCL/CNF composites with high CNF content with excellent nanostructural control and superior mechanical performance particularly at elevated temperatures [134].

5.3.7 Inorganic Nanoparticles

Polymeric–inorganic hybrid composites mimicking natural materials, such as bone and nacre, have attracted attention owing to the unique combination of light weight, remarkable strength, and toughness. Nature bone is a biocomposite of hydroxyapatite mineral crystals arranged in an organic collagen matrix. Nanocomposites of bacterial cellulose and hydroxyapatite have been extensively investigated [135–139]. Nanocelluloses have been assembled with inorganic platelets to mimicking the brick-and-mort microstructure of nacre. Large, flat clay nanopaper hybrid composites from wood CNFs and montmorillonite (MMT) [140–142], bentonite [143], or talc [144] platelets with have been prepared by using a water-based papermaking procedure [145, 146] instead of the classical layer-by-layer deposition techniques [1]. Such CNF/nanoclay composites demonstrated a combination of functionalities such as fire retardancy, optical transparency, high strength and ductility with low density, and thermal stability. Hybrid composites from TEMPO-oxidized or carboxymethylated CNFs and nanoclay such as MMT [147, 148], synthetic saponite [149], and lithium exchanged vermiculite [150] have also been prepared by a solution casting and evaporation procedure. Such solution-casted transparent composite films showed high strength and flexibility and low oxygen permeability. Ho et al. have prepared a series of composites of cationic CNFs (trimethylammonium-modified) and layered silicates using a process of high-shear homogenization with microfluidizer, pressure filtration, and vacuum hot-press techniques [151, 152]. Thirteen types of layered silicates that belong to smectite, kaolin, talc, vermiculite, and mica groups have been utilized and compared. It was found that montmorillonite flakes covered the cationic CNFs and formed a rather dense structure due to the best interaction level, while composites with 50 wt% mica R120 in a cationic CNF matrix demonstrated best mechanical and barrier properties.

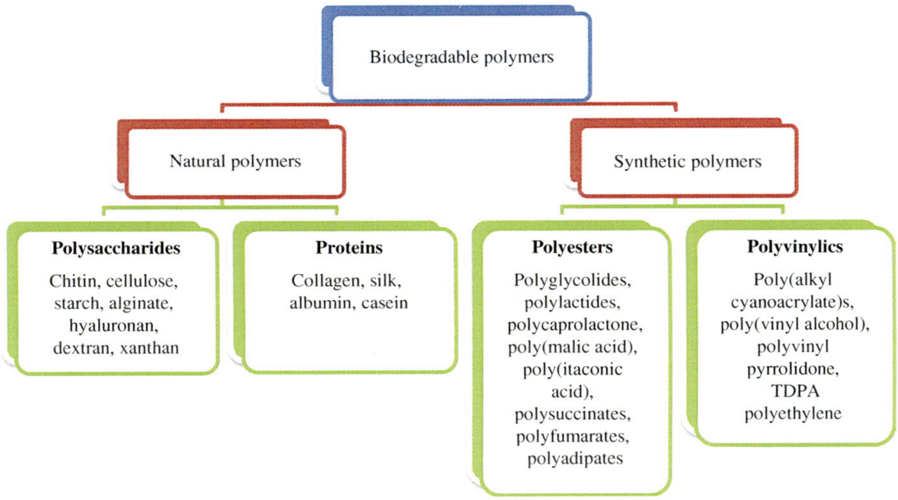

FIGURE 1.1 Classification of biodegradable polymers in four families.

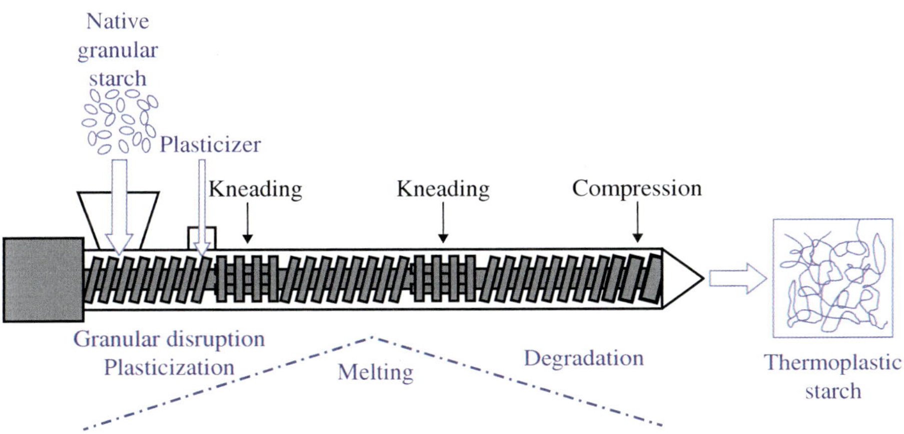

FIGURE 1.2 Thermoplastic starch formation. *Source*: Xie et al. [11], figure 37. Reproduced with permission from Elsevier. Copyright 2015.

Biodegradable Green Composites, First Edition. Edited by Susheel Kalia.

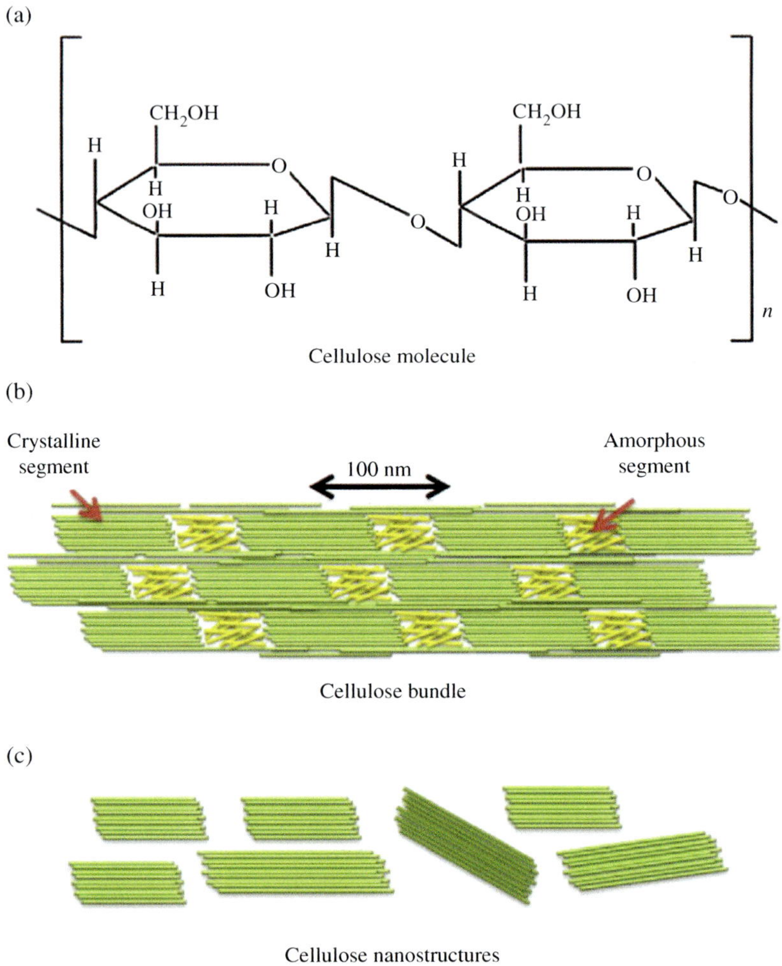

FIGURE 1.3 Schematics of (a) single cellulose chain repeat unit, showing the directionality of the 1–4 linkage and intrachain hydrogen bonding (dotted line), (b) idealized cellulose microfibril showing one of the suggested configurations of the crystalline and amorphous regions, and (c) cellulose nanocrystals after acid hydrolysis dissolved the disordered regions. *Source*: Reddy et al. [24], figure 6. Reproduced with permission from Elsevier. Copyright 2015.

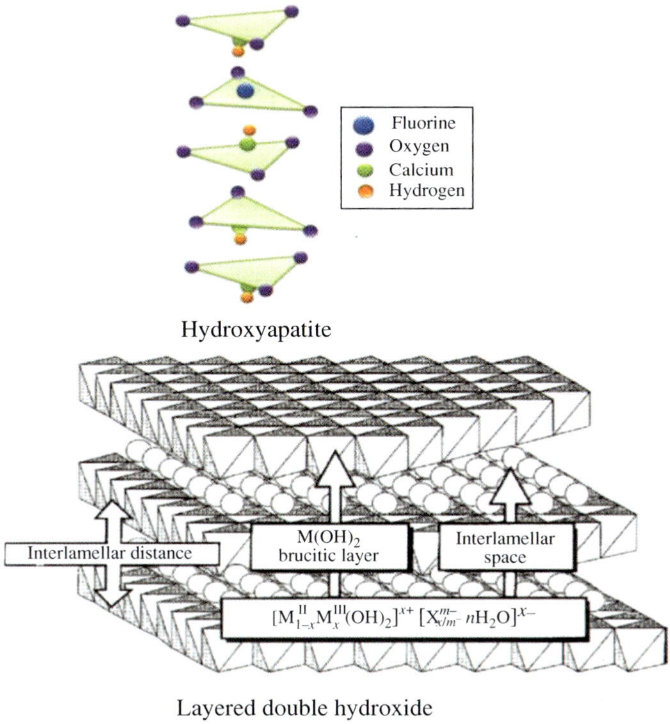

FIGURE 1.4 Schematic representation of hydroxyapatite and layered double hydroxide chemical structure. *Source*: Leroux and Besse [38], figure 13. Reprinted with permission from American Chemical Society. Copyright 2015.

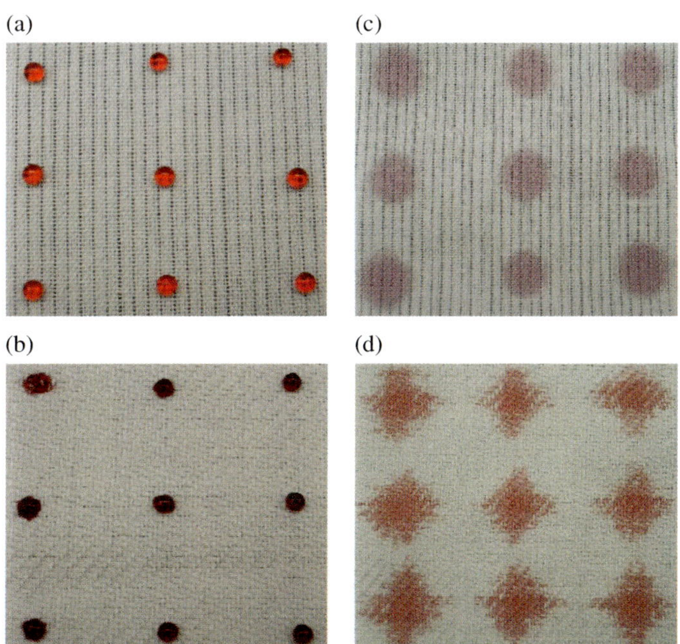

FIGURE 2.7 Comparison of the droplets on the untreated (a: cotton, b: wool) and O_2 plasma-treated (c: cotton, d: wool) fabric surface [1].

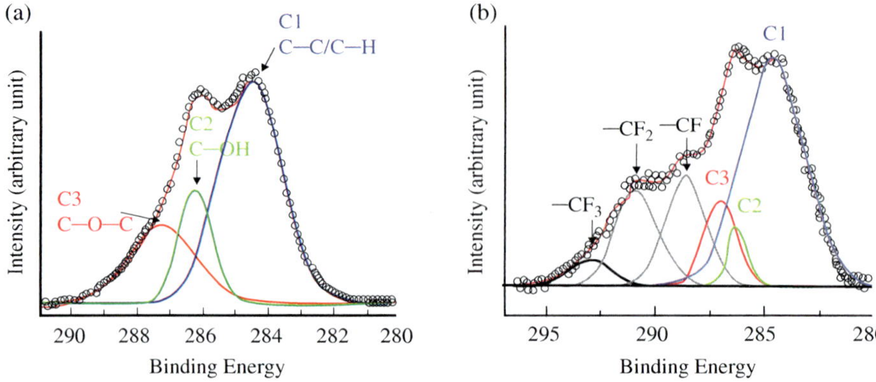

FIGURE 2.10 C1s XPS spectra of fabric surface: (a) untreated cotton and the molecular structure of cellulose; (b) cotton treated at the pressure of 0.05 Torr for 1 min [40]. *Source*: Kamlangkla et al. [40]. Reprinted with permission from Elsevier. Copyright 2014.

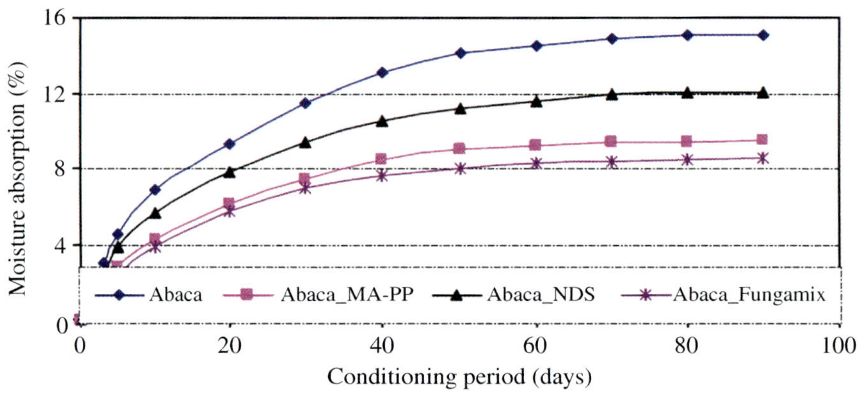

FIGURE 3.16 Moisture absorption of modified and unmodified abaca/PP composites. *Source*: Bledzki et al. [76]. Reprinted with permission from Elsevier. Copyright 2010.

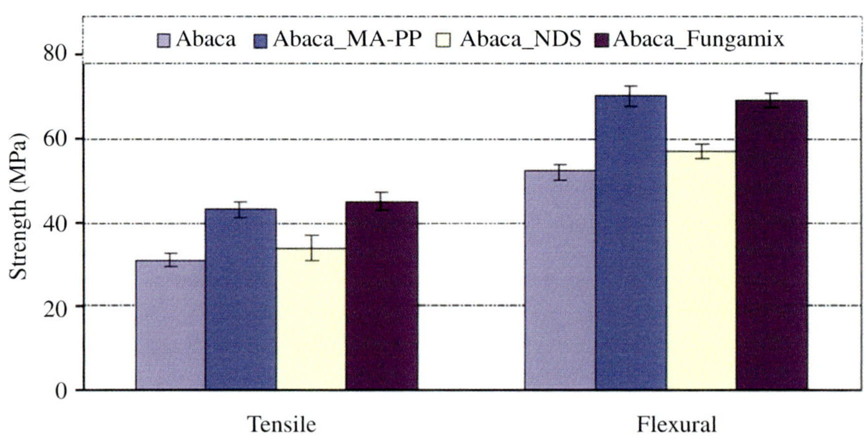

FIGURE 3.17 Tensile and flexural strength of modified and unmodified abaca/PP composites. *Source*: Bledzki et al. [76]. Reprinted with permission from Elsevier. Copyright 2010.

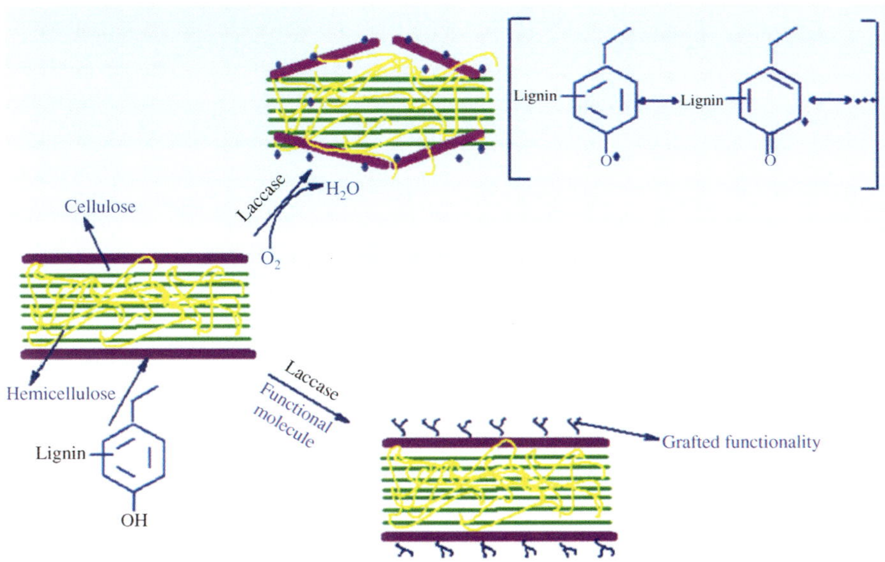

FIGURE 3.18 Laccase-assisted grafting of functional molecule on lignocellulosics. *Source*: Kalia et al. [82], figure 2. Reprinted with permission from Elsevier. Copyright 2014.

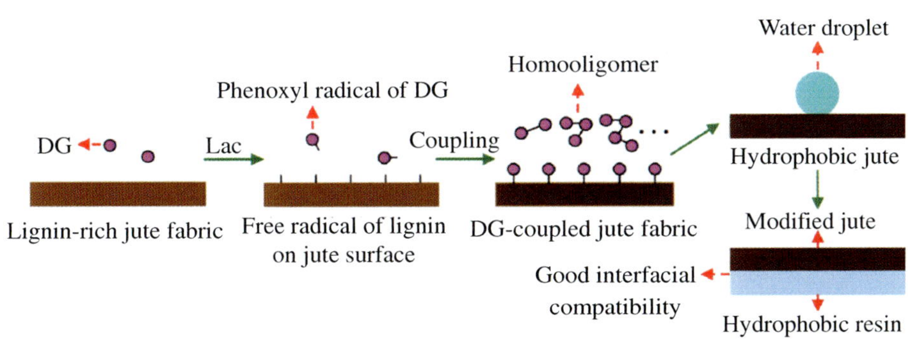

FIGURE 3.19 Schematic illustration of the grafting reaction of dodecyl gallate (DG) on the lignin-rich jute fabric catalyzed by laccase. *Source*: Dong et al. [97]. Reprinted with permission from Elsevier. Copyright 2014.

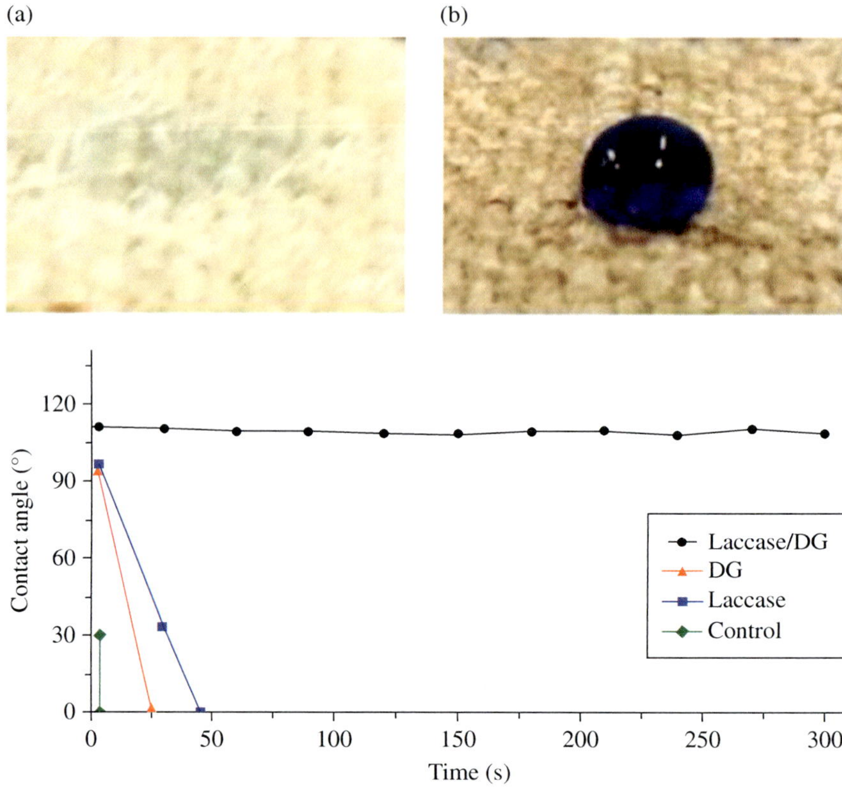

FIGURE 3.20 Contact angles over the time for the control jute (control), the laccase-treated jute (laccase), the DG-alone-treated jute (DG), and the laccase-/DG-treated jute (laccase/DG). Insert: digital photographs of water drop captured on the surface of control jute after 5 s (a) and of treated jute with laccase and DG after 30 min (b). *Source*: Dong et al. [97]. Reprinted with permission from Elsevier. Copyright 2014.

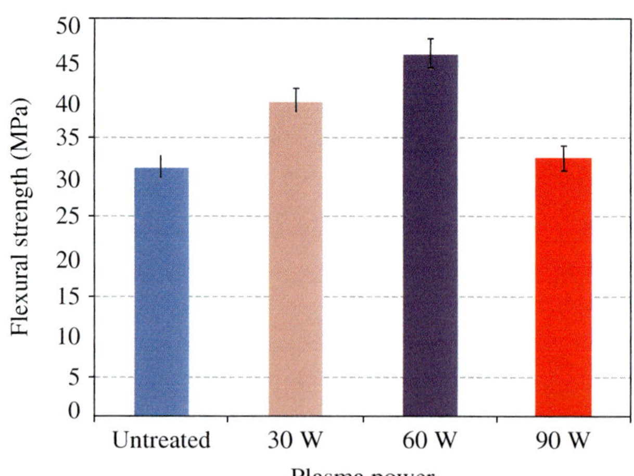

FIGURE 4.8 Effect of plasma power on flexural properties of jute/HDPE composites. *Source*: Seki et al. [106]. Public Domain.

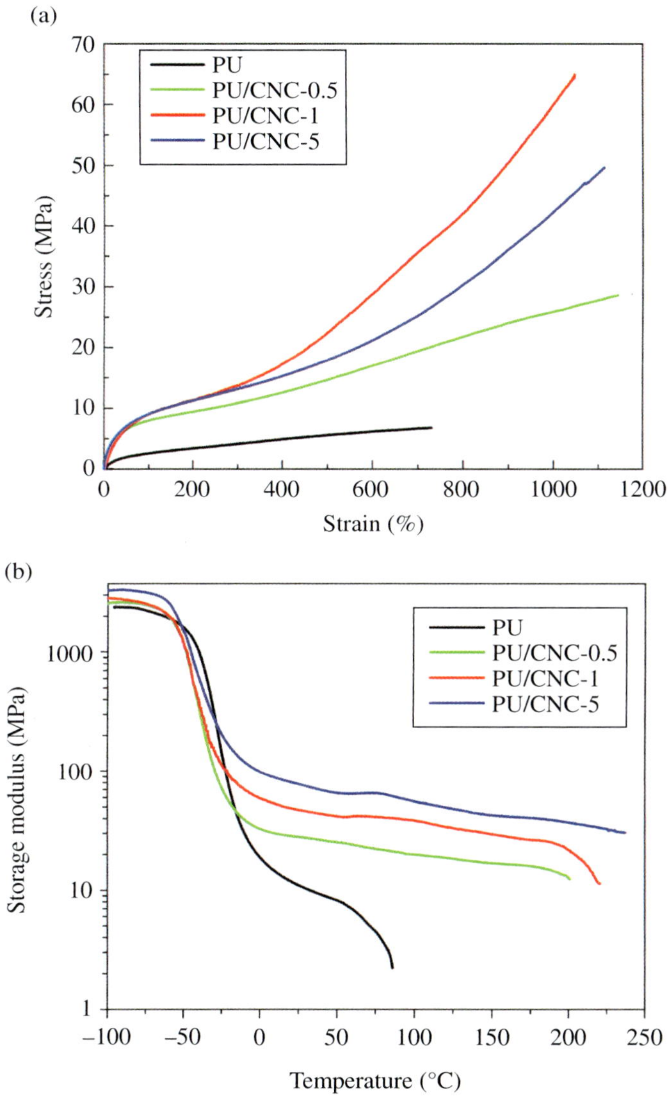

FIGURE 5.3 (a) Stress–strain curves of pure PU and PU/CNC nanocomposites. (b) Storage modulus of the PU and PU/CNC nanocomposites as a function of temperature. *Source*: Pei et al. [153]. Reprinted with permission from the American Chemical Society.

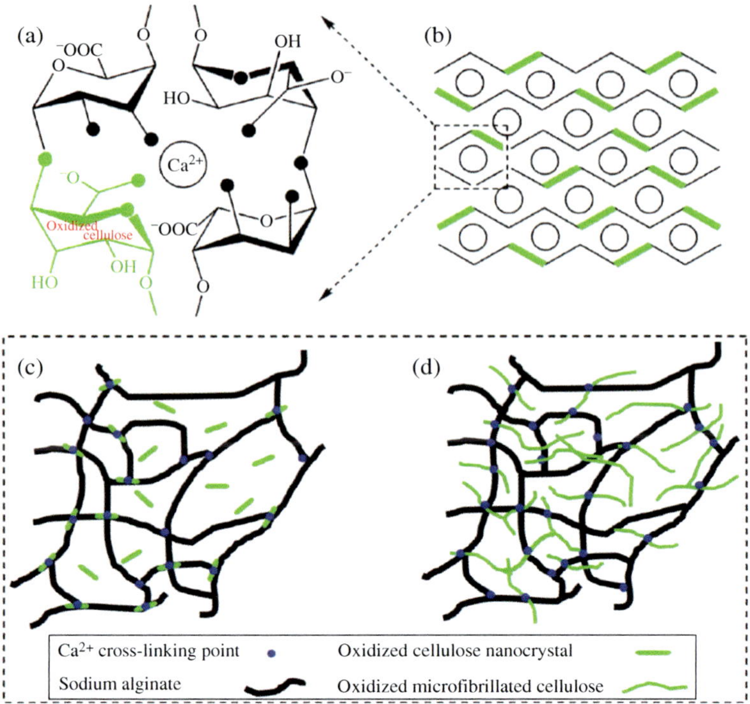

FIGURE 5.4 Schematic representation of the structure of egg-box junction zones in TEMPO-mediated oxidation of cellulose/alginate/calcium system: (a) coordination of Ca^{2+} in a cavity created by a pair of guluronate sequences along alginate chains; (b) laterally associated egg-box multimer in composite sponge; (c) oxidized cellulose nanocrystals (OCN) in the cross-linking of alginate sponge; (d) construction of the semi-interpenetrating polymer network (SIPN) composed of oxidized microfibrillated cellulose (OMFC) and alginate. The black solid circles in panel a represent the oxygen atoms possibly involved in the coordination with Ca^{2+}. *Source*: Lin et al. [77]. Reprinted with permission from the American Chemical Society.

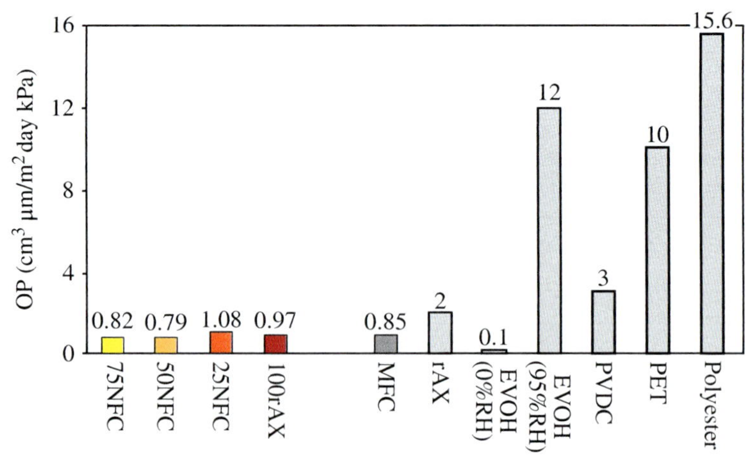

FIGURE 5.5 Oxygen permeability of rye arabinoxylan film and its composite films; literature values for MFC (carboxymethylated), rAX, EVOH (polyethylene vinyl alcohol), PVDC (polyvinylidene chloride), polyester and PET (poly(ethylene terephthalate)); all tests done at 23°C and 50% RH, if not specified differently. *Source*: Stevanic et al. [49], figure 11. Reprinted with permission from Springer Science+Business Media.

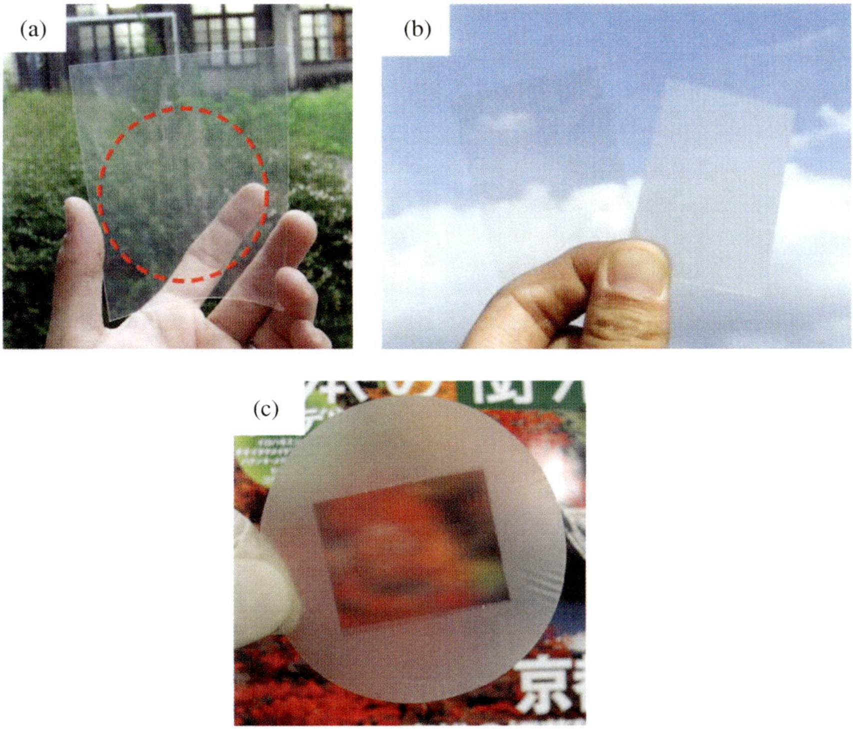

FIGURE 5.7 Various approaches to preparing functional transparent cellulose sheets. (a) Lamination of optically transparent plastics (inside the red circle). (b) Deposition of transparent resins on the surface of cellulose sheets (left, surface-coated sheet; right, uncoated sheet). (c) Drawing of transparent patterns on the sheet using a domestic ink-jet printer. *Source*: Nogi et al. [169], figure 4. Reproduced with permission from John Wiley & Sons.

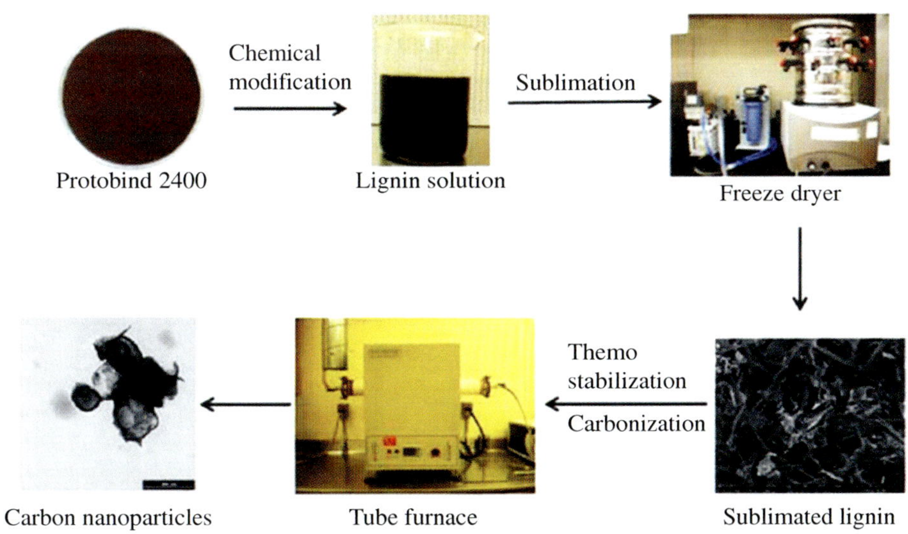

FIGURE 7.4 Schematic of the various steps involved in the synthesis of carbon nanoparticles from lignin. *Source*: Gonugunta et al. [70], Figure 1, p. 149. Reproduced with permission from World Journal of Nano Science Engineering.

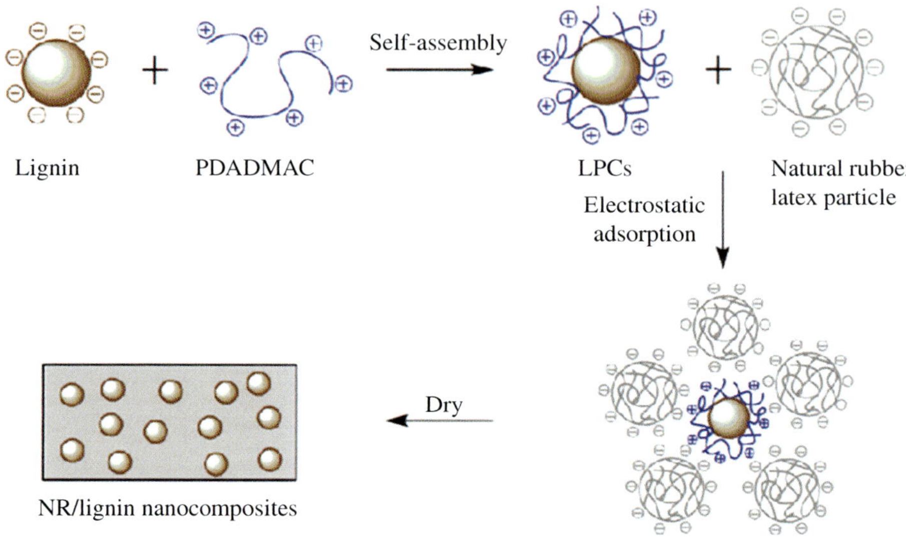

FIGURE 7.5 Schematic illustration of process for NR/lignin nanocomposites. *Source*: Jiang et al. [71], Figure 3. Reproduced with permission from eXPRESS Polymer Letters.

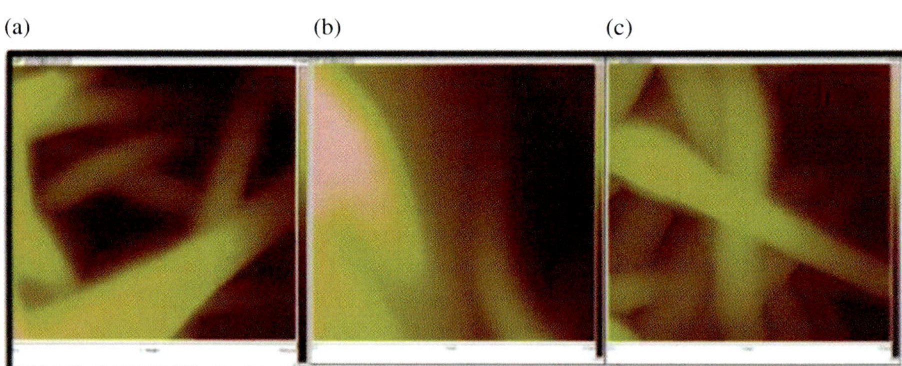

FIGURE 7.9 AFM images of PNIPAM modified lignin nanofibers (Lig-PN-2). (a) Lig-PN-2 in air, (b) Lig-PN-2 in water, and (c) Lig-PN-2 in 0.5 M Na_2SO_4 aqueous solution. *Source*: Gao et al. [69]. Reprinted with permission from American Chemical Society.

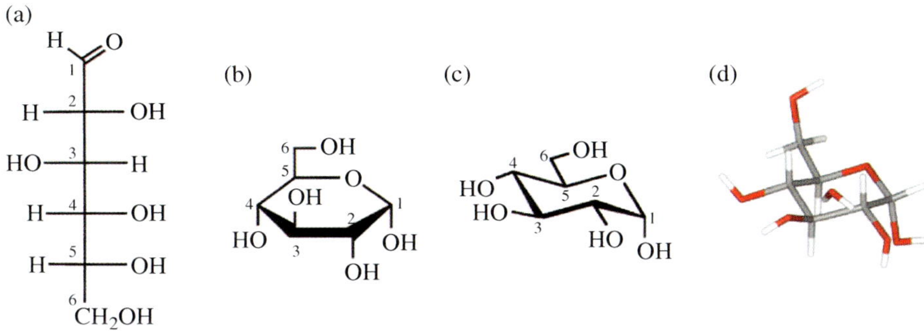

FIGURE 8.5 Structures of α-D-glucopyranose (D-glucose) shown using different projections: (a) Fisher, (b) Haworth, (c) in 4C1 conformation, and (d) a 3D model of glucose in the 4C1 conformation. *Source*: Damager et al. [9]. Reproduced with permission from American Chemical Society; Copyrights: American Chemical Society.

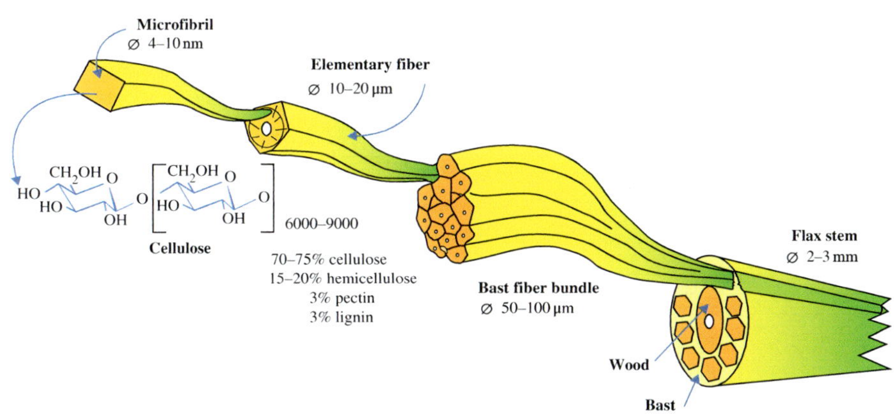

FIGURE 8.6 Schematic of hierarchical structure of cellulose in comparison to starch granule. *Source*: http://www.jeccomposites.com/news/composites-news/industrial-applications-natural fibre-reinforced-composites, 2009. Reproduced with permission from JEC.

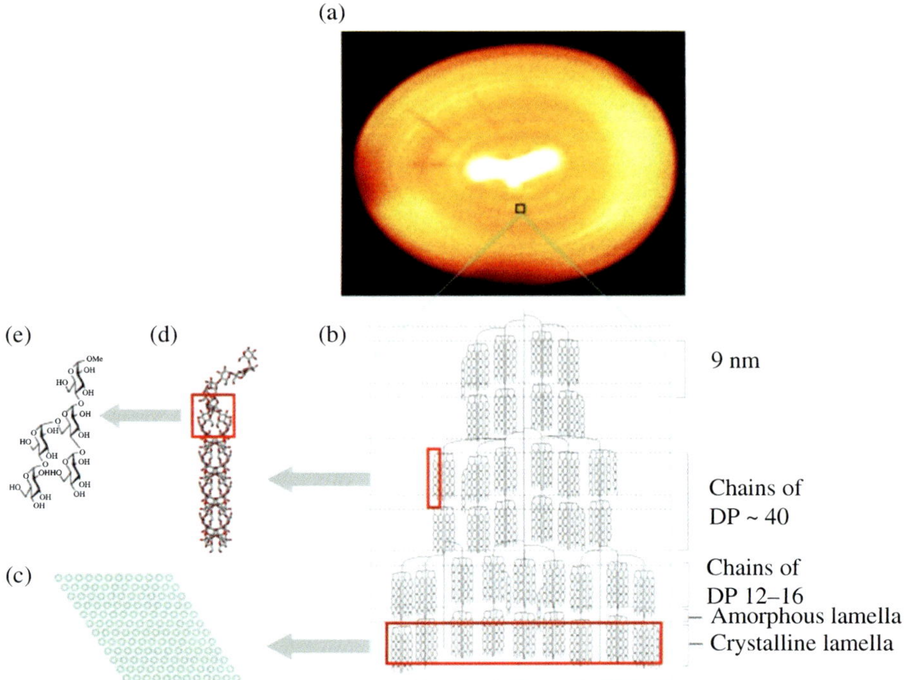

FIGURE 8.8 Principle of top-down strategy of starch structure determination. *Source*: Damager et al. [9]. Reproduced with permission from American Chemical Society; Copyrights: American Chemical Society.

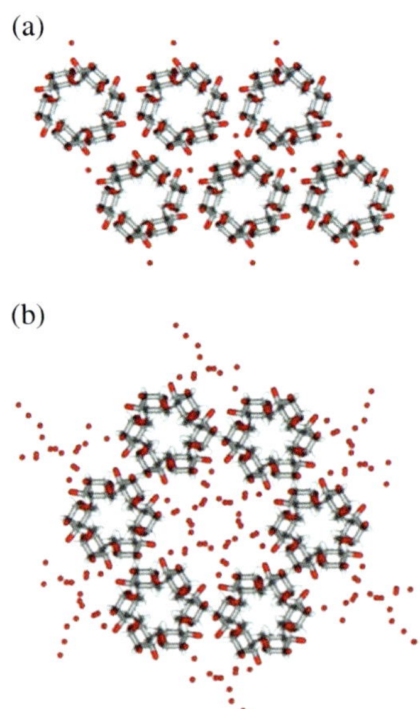

FIGURE 8.10 Hexagonal packing of A-type and B-type starch crystalline polymorphs. *Source*: Damager et al. [9]. Reproduced with the kind permission of American Chemical Society, Copyrights: American Chemical Society. Original is from: Imberty & Pe´rez, S. 1988; John Wiley & Sons, Inc.

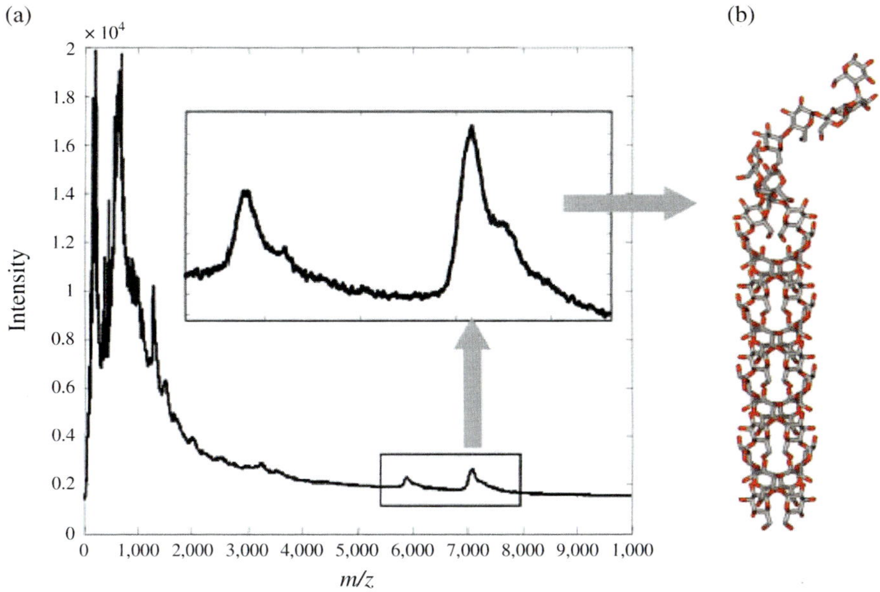

FIGURE 8.12 (a) MALDI-TOF spectrum of large amylopectin fragments. (b) A representative of α-glucan structure giving rise to the distinct larger fragments obtained. *Source*: Engelsen et al. [107]. Reproduced with permission from Elsevier.

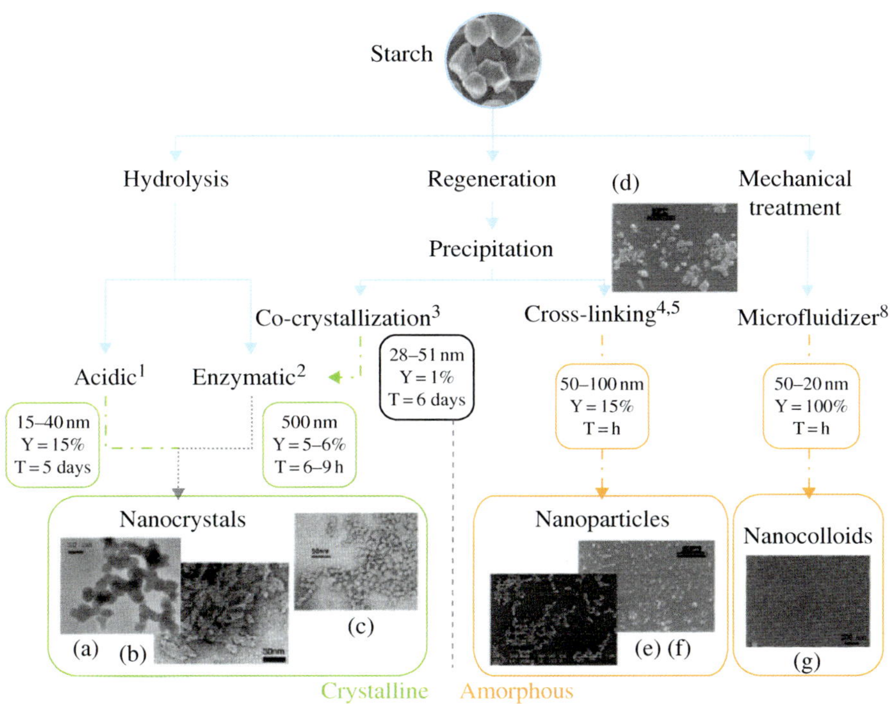

FIGURE 8.13 Different ways of producing crystalline and amorphous starch nanoparticles and TEM photographs of resulting nanoparticles. Le Corre et al. [17]. Reprinted with permission from American Chemical Society; *Source*: (a) Garcia et al. [184]. Reproduced with permission with Wiley-VCH Verlag GmbH & Co. KGaA; (b) Putaux et al. [90]; (c and e) Tan et al. [185]. Reprinted with permission from American Chemical Society; (d and f) Ma et al. [39]. Reprinted with permission from American Chemical Society; (g) Liu et al. (2009).

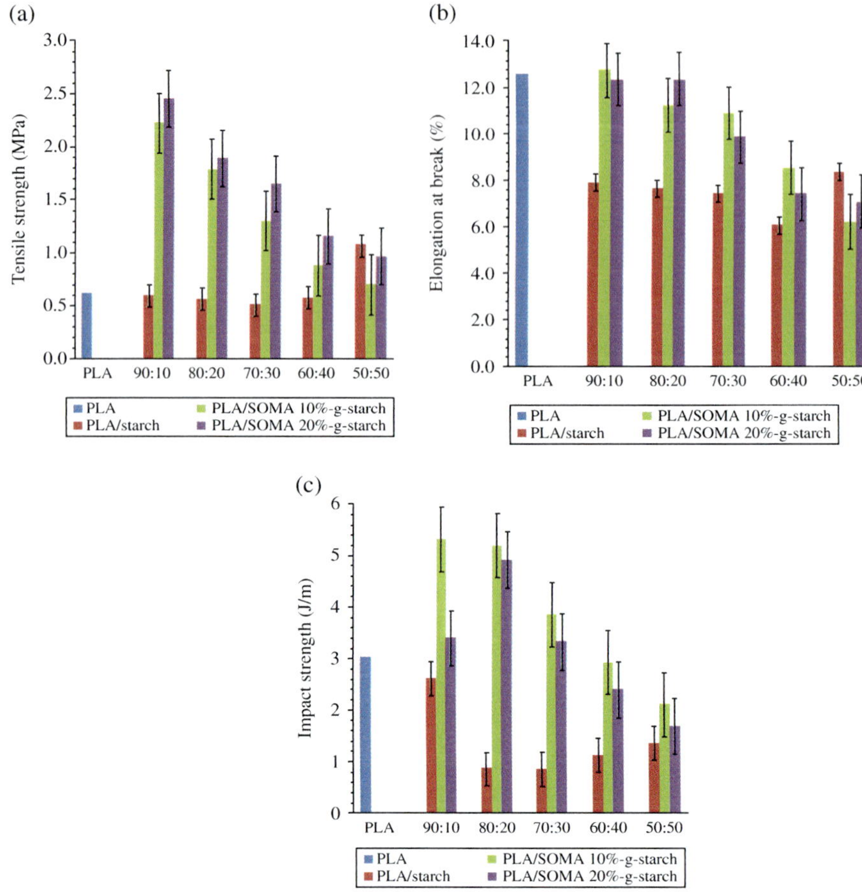

FIGURE 8.15 Plots of mechanical properties of PLA matrix and its composites (PLA/starch, PLA/10 and 20% SOMA-g-STARCH); (a) tensile strength, (b) percent elongation, and (c) impact strength. *Source*: Kiangkitiwan and Srikulkit [212]. CC public domain.

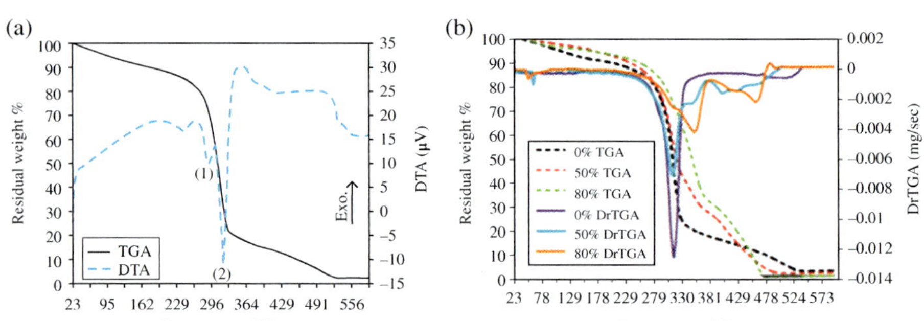

FIGURE 8.17 Plots showing the results of thermal analysis of hot pressed TPS-based matrix and composites with different date palm fiber content: (a) TGA/DTA curves, (b) TGA/ DrTGA curves. *Source*: Ibrahim et al. [51]. Reproduced with permission from Hindawi Publishing Corporation.

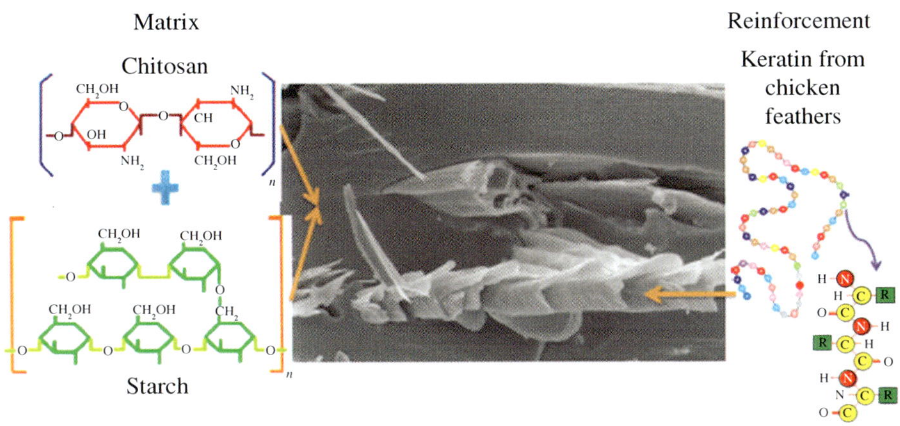

FIGURE 8.18 Schematic view of the constituents (chitosan–tarch–keratin fiber) of the composite obtained. *Source*: Reproduced with the kind permission of the publishers of Cynthia et al. [255].

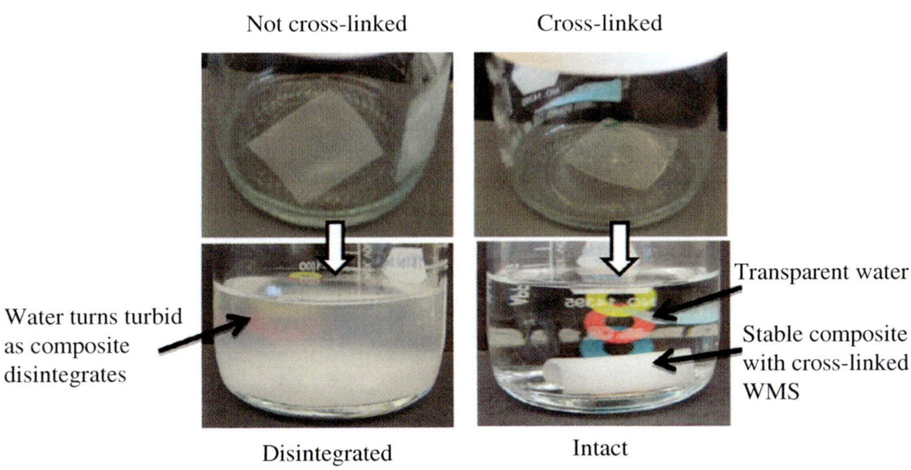

FIGURE 8.26 Various steps involved in obtaining the nanocomposite. *Source*: Dastidar and Netravali [161]. Reprinted with permission from American Chemical Society. Copyright 2013.

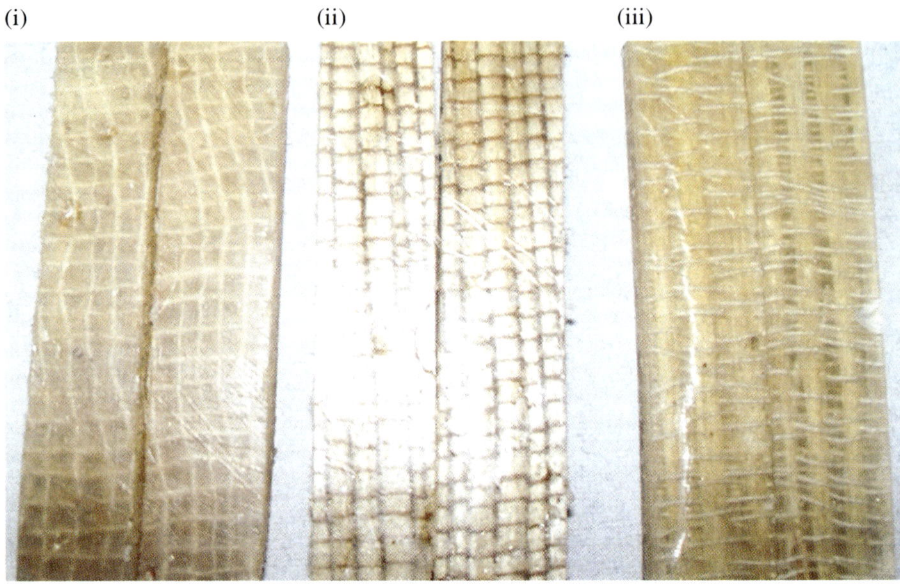

FIGURE 9.4 Green composite laminates processed through film stacking technique. (i) PLA/nettle. (ii) PLA/sisal. (iii) PLA/*Grewia optiva*.

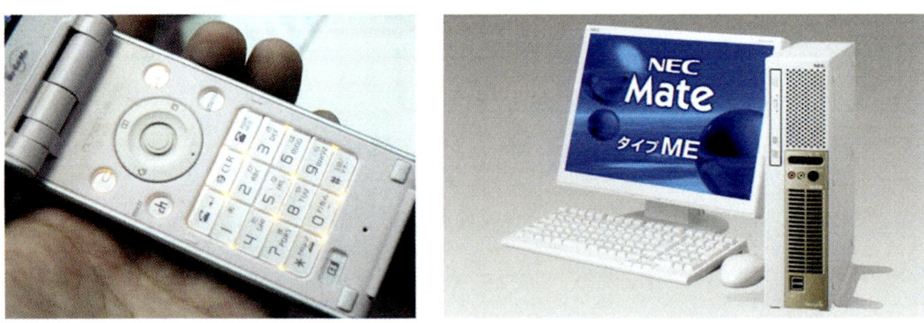

FIGURE 10.1 Mobile phone (FOMA N701iECO) casing made of kenaf fiber-reinforced polylactic acid (left). Mate Type ME desktop business personal computer that in 90% of its plastic housing uses NuCycle™ (right). *Source*: Courtesy of NEC Corporation.

5.4 COMPOSITE PROPERTIES

5.4.1 Thermal and Mechanical Properties

In nanostructured composite materials, the size, surface structure, charge, and functionality of the nanoscale building blocks are essential. Surface properties will control the interactions between nanocelluloses as well as with their matrix, which ultimately dictates the structure and macroscale properties of nanocellulose-based composite material. Svagan et al. [82] revealed that the nanoscale network structure of MFC is important for the unique combination of strength and strain to failure in the all-polysaccharide nanocomposites of starch matrix reinforced with MFC as shown in Figure 5.2. As the MFC content increases, Young's modulus and the strength both increase while the strain to failure decreases, which indicates high compatibility of MFC with starch matrix. The nanostructured characteristics of MFC, in combination with favorable MFC-matrix adhesion, delay material damage during deformation and explain the observed ductility and high toughness.

Regarding CNC-based composites, the reinforcing effect of tunicate CNCs in poly(styrene-co-butyl acrylate) matrix was first demonstrated by Favier et al. [7]. A remarkable enhancement in the storage modulus was observed by dynamic mechanical analysis in the shear mode, and the increase was especially significant above the glass–rubber transition temperature. This was attributed to the formation of a rigid network by the CNCs in the host polymer matrix, which was governed by the percolation mechanism. In this mechanism, the network is expected to form above the critical volume fraction at the percolation threshold, V_{Rc}, which depends on the aspect ratio (L/d) of the filler. However, we found that CNC/polyurethane (PU) nanocomposite with ultrahigh tensile strength and strain to failure was obtained with only 0.5 wt% CNCs as shown in Figure 5.3. This value is much lower than the calculated percolation value (3.3 wt%). Such significant modulus reinforcement effect without formation of a cellulose network is because of the introduction of a discrete reinforcement phase of nanoscale rodlike stiff CNCs, in combination with increased effective rubber cross-link density due to covalent linking of CNCs to the PU hard segments at the interface of filler and matrix.

Lin et al. [77] prepared cross-linked polysaccharides sponges by freeze-drying of amorphous alginate reinforced with TEMPO-oxidized nanocelluloses in the presence of a Ca^{2+} ionic cross-linking agent and compared the role of TEMPO-oxidized CNCs and CNFs in such composite system. They proposed an interesting model as shown in Figure 5.4. TEMPO-oxidized CNFs are involved in both cross-linking with alginate and the network structural construction of the composite, resulting in a semi-interpenetrating polymer network. Thus, the composite sponge derived from oxidized CNFs exhibited ultrahigh porosity, even pore interconnection, good structural stability, and improve water absorption and retention. On the other hand, TEMPO-oxidized CNCs only acted as the coupling points with alginate, resulting in remarkable enhanced mechanical strength of the composite sponges. Inspired by the different effects of the two oxidized nanocelluloses, the incorporation of both nanocelluloses in the alginate-based sponges will be promising.

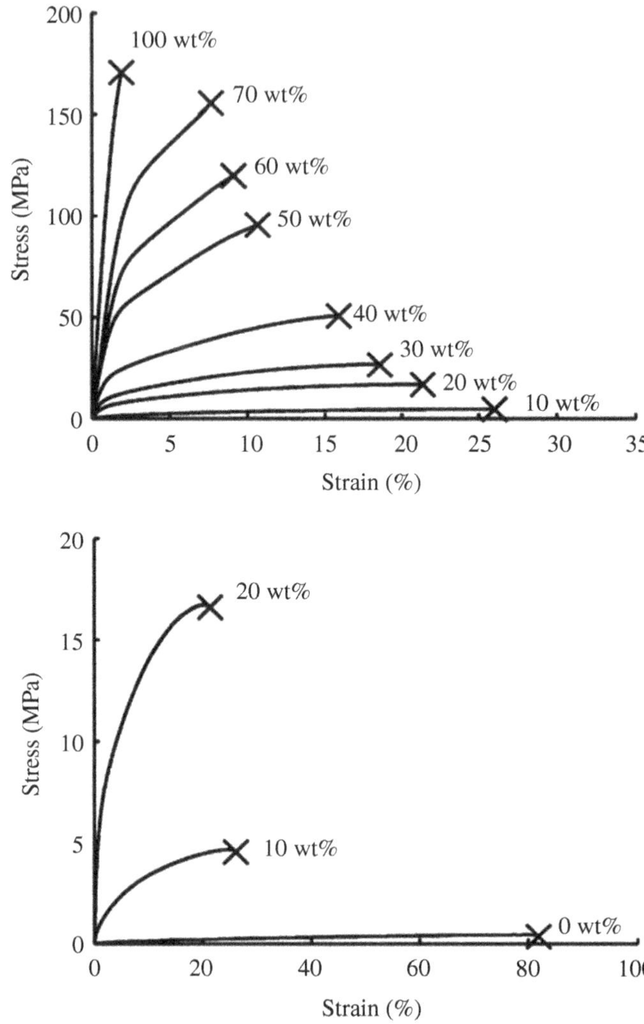

FIGURE 5.2 Typical tensile curves for MFC–amylopectin–glycerol composites with varying MFC content and fixed matrix composition: 50 wt% glycerol and 50 wt% amylopectin. The MFC contents are indicated in the figure. *Source*: Svagan et al. [82]. Reprinted with permission from the American Chemical Society.

5.4.2 Barrier Properties

TEMPO-oxidized CNF-coated PLA film demonstrated excellent oxygen barrier properties similar to synthetic polymer films such as poly(vinylidene chloride). The unmodified PLA film (25 μm thick) had an oxygen permeability of about 746 ml/m^2 day Pa, which was decreased to 1 by a 0.4 μm thin coating layer of CNFs [154]. Films of CNFs that were prepared by enzymatic pretreatment combined

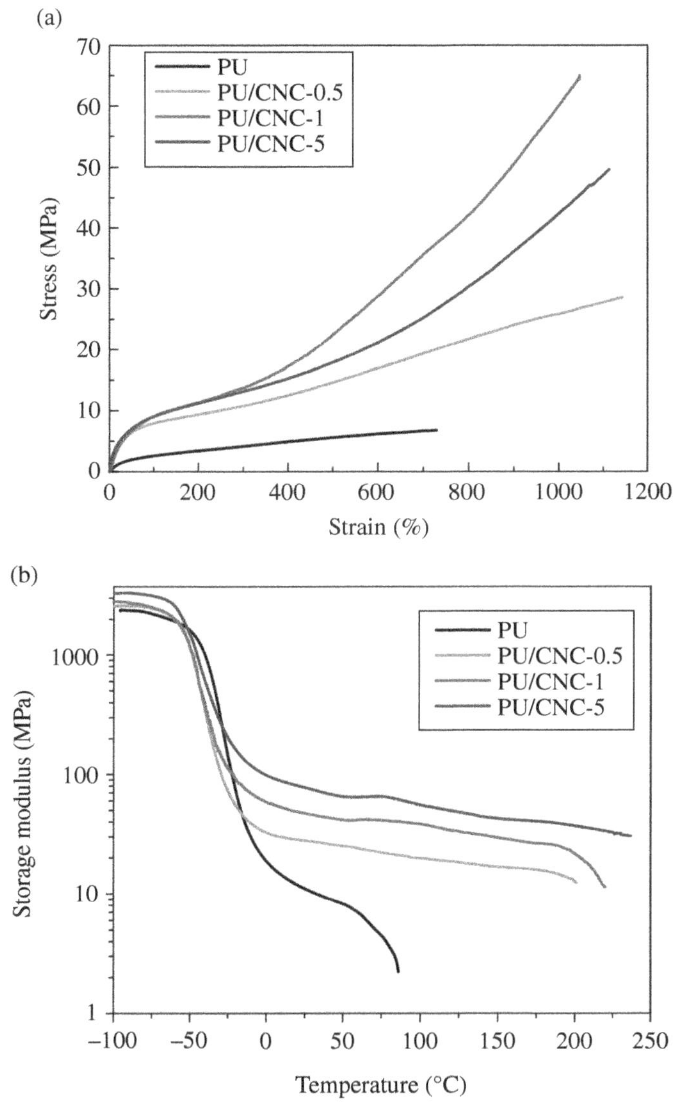

FIGURE 5.3 (a) Stress–strain curves of pure PU and PU/CNC nanocomposites. (b) Storage modulus of the PU and PU/CNC nanocomposites as a function of temperature. *Source*: Pei et al. [153]. Reprinted with permission from the American Chemical Society. (*See insert for color representation of the figure.*)

with mechanical homogenization had an oxygen permeability (OP) value of $2.5 \pm 0.8\,cm^3\,\mu m/m^2$ day kPa as reported by Stevanic et al. [52]. Such CNFs contained 86% cellulose and 14% hemicelluloses from a bleached Norway spruce (*Picea abies*) sulfite pulp. The OP of composite films of spruce galactoglucomannan and arabinoglucuronoxylan reinforced with 85 and 80% CNFs, respectively, showed very low

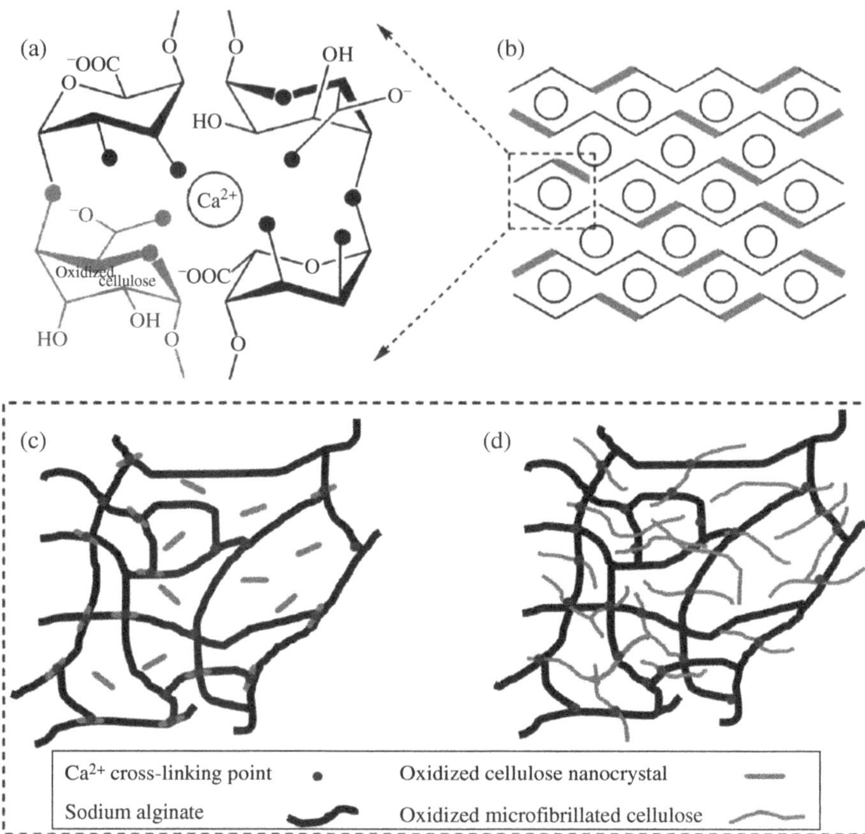

FIGURE 5.4 Schematic representation of the structure of egg-box junction zones in TEMPO-mediated oxidation of cellulose/alginate/calcium system: (a) coordination of Ca^{2+} in a cavity created by a pair of guluronate sequences along alginate chains; (b) laterally associated egg-box multimer in composite sponge; (c) oxidized cellulose nanocrystals (OCN) in the cross-linking of alginate sponge; (d) construction of the semi-interpenetrating polymer network (SIPN) composed of oxidized microfibrillated cellulose (OMFC) and alginate. The black solid circles in panel a represent the oxygen atoms possibly involved in the coordination with Ca^{2+}. *Source*: Lin et al. [77]. Reprinted with permission from the American Chemical Society. (*See insert for color representation of the figure.*)

values from 1.2 to 1.8 $cm^3\,\mu m/m^2$ day kPa at relative humidity (RH) of 50%. This decrease could be explained by the presence of hemicellulose molecules filling the voids in the CNF network. Stevanic et al. [49] further prepared composite films from wood CNFs with a high molecular weight rye arabinoxylan (rAX) and compared the OP values with commercial barrier plastics as shown in Figure 5.5. With the reinforcement, the rAX films with 50–70% CNFs by weight had an OP value of 0.8 $cm^3\,\mu m/m^2$ day kPa at 50% RH, which was in the range of polyethylene vinyl alcohol (EVOH). CNCs have also been incorporated into xylan to prepare high oxygen barrier biodegradable composite films [56]. Oat spelt xylan reinforced with

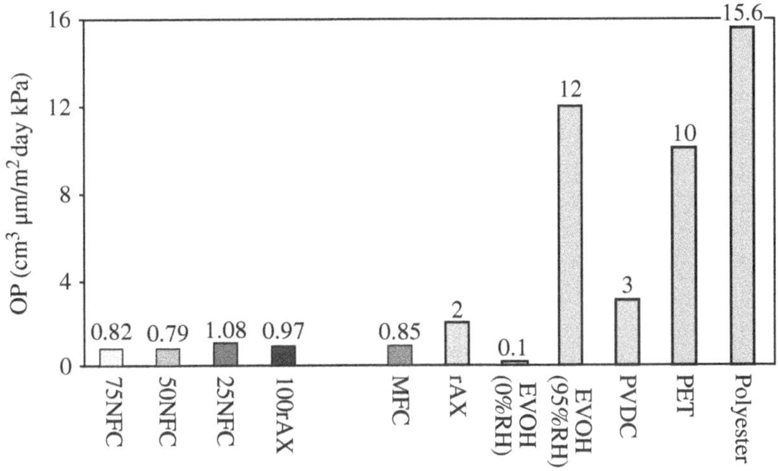

FIGURE 5.5 Oxygen permeability of rye arabinoxylan film and its composite films; literature values for MFC (carboxymethylated), rAX, EVOH (polyethylene vinyl alcohol), PVDC (polyvinylidene chloride), polyester and PET (poly(ethylene terephthalate)); all tests done at 23°C and 50% RH, if not specified differently. *Source*: Stevanic et al. [49], figure 11. Reprinted with permission from Springer Science+Business Media. (*See insert for color representation of the figure.*)

50% sulfonated CNCs showed an OP value of only 0.1799 cm^3 μm/m^2 day kPa because of the high degree of crystallinity of CNCs and the dense structure of the composite. Owing to the same reason, composites of agar [80], starch [94], gluten [121], alginate [79], or chitosan [98] reinforced with CNCs have also presented significantly improved water vapor barrier properties.

5.4.3 Antimicrobial Properties

The development of composites based on nanocelluloses with antimicrobial properties has attracted great interest in recent years. A variety of techniques have been reported, including graft polymerization of monomers carrying antimicrobial groups on cellulose via controlled living polymerization methods [155, 156], grafting of antimicrobial polymers on cellulose via an adsorption-curing process using silane chemistry [157], immobilization of antibacterial peptides or cell lytic enzymes on amino or aldehyde groups of functionalized cellulose [158, 159], incorporation of low molecular weight antibacterial agents such as 1,3,5-triazine derivatives [160, 161], impregnation with silver nanoparticles or nanoclusters [162–164], and incorporation of chitosan that was functionalized with antibacterial agents [103, 104]. However, these approaches require either strong reaction conditions for chemical immobilization of antimicrobial compounds or the use of leachable biocides such as heavy metals.

Recently, we have demonstrated an environmentally friendly approach for the production of BC-based composites with bactericidal activity by the incorporation of chitin nanocrystals [165]. The antibacterial activity of chitin nanocrystals (ChNCs) prepared from α-chitin powders by acid hydrolysis (A-ChNC), TEMPO-mediated oxidation (T-ChNC), and partial deacetylation (D-ChNC) was assessed for the first time although chitosan has been widely used as an antimicrobial agent. The D-ChNC nanocrystals showed the strongest antibacterial activity, with 99 ± 1% inhibition of bacterial growth compared to control sample (Fig. 5.6a), indicating a strong relationship between the growth inhibition of *Escherichia coli* cells and the degree of acetylation value of the ChNCs. BC/D-ChNC nanocomposites were successfully prepared by the incorporation of D-ChNC into BC nanofiber networks via either *in situ* biosynthesis or postmodification. The D-ChNC film sample showed a reduction in bacterial colonies by log 5.0 after 3 h incubation (Fig. 5.6b), while the BC/D-ChNC-p (50/50) nanocomposite showed a log 3.0 (±0.2) reduction of viable *E. coli* cells compared to the control BC sample. These results indicate that the bactericidal activity of the BC/D-ChNC nanocomposites is strongly dependent on the D-ChNC nanocrystal content.

5.4.4 Optical Properties

Nanocellulose-based composite materials show great promise in optical applications. Since the particle size of nanocellulose is smaller than 25 nm, the refractive index difference between the matrix and nanocelluloses can be ignored and optical scattering losses for the composites will be minimized [166]. Yano et al. reported the first example of a transparent composite reinforced with BC nanofibers at a fiber content as high as 70 wt% with a low thermal expansion coefficient and high mechanical strength [167]. They further succeeded in depositing an organic light-emitting diode (OLED) onto the transparent BC nanocomposite with a modulus on the order of 21 GPa. In order to realize roll-to-roll processing on BC nanocomposites in the field of electronic devices, Nogi and Yano prepared transparent, foldable, and high thermal stable nanocomposites by reinforcing a low-modulus transparent resin with 5 wt% of planar BC nanofiber networks with native layered structure [168]. Furthermore, Nogi et al. demonstrated the preparation of transparent nanopaper using CNFs with a diameter of 15 nm from wood fibers. The optical transparency of the nanopaper can be achieved by different surface-smoothing approaches such as the lamination of optically transparent plastics, deposition of transparent resins, and ink-jet printing as shown in Figure 5.7 [169].

Recently, extensive effort has been devoted to improving the optical haze of transparent cellulose nanopapers made of CNFs for application as substrates in optoelectronic devices such as solar cells, outdoor displays, and OLED light systems, with tunable performance [170–173]. Fang et al. prepared a highly transparent (optical transmittance of ca. 96%) paper with high haze (ca. 60%) using TEMPO-oxidized microsized wood fibers by vacuum filtration [173]. The paper was laminated on an organic solar cell and a power conversion efficiency (PCE) enhancement of 10% was observed. Fang et al. further demonstrated that the transmission haze of the paper from TEMPO-oxidized wood cellulose fibers can be tuned the incorporation of TEMPO-oxidized CNFs with different weight ratios [172]. In addition, a MoS_2 field effect transistor was fabricated on the transparent paper of pure CNFs, showing a large current on/off ratio, $I_{on}/I_{off} \sim 1 \times 10^5$.

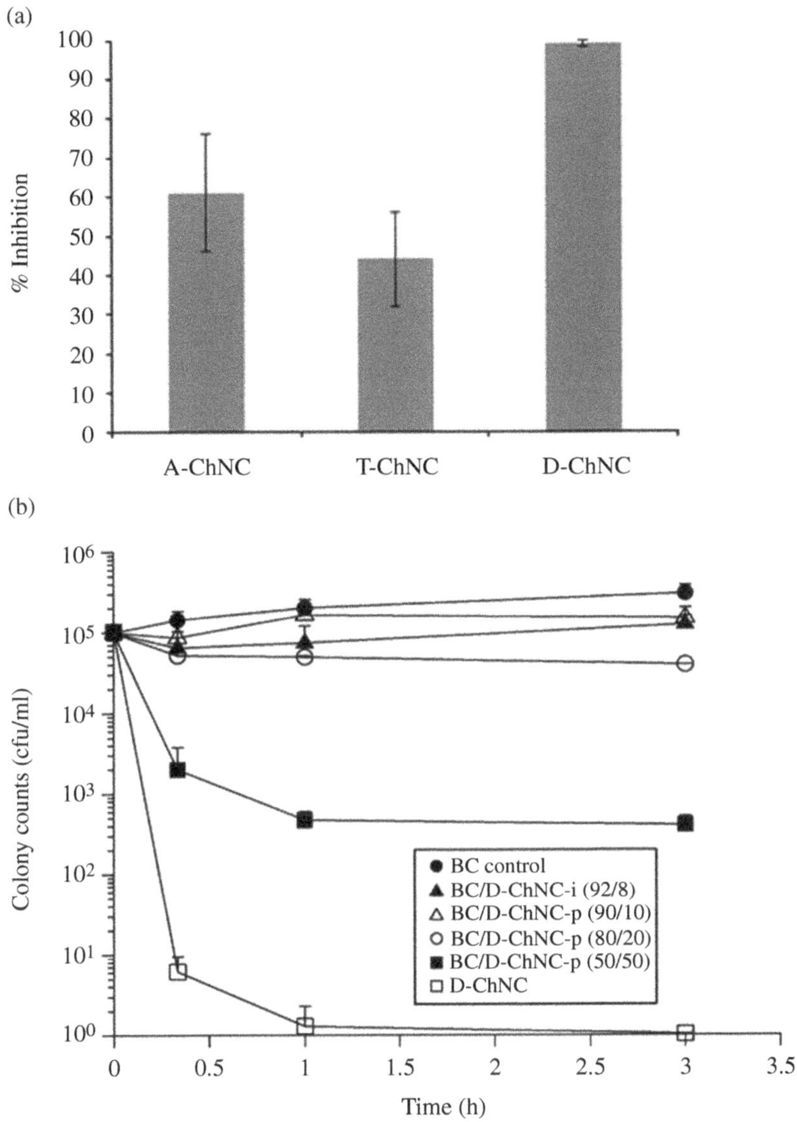

FIGURE 5.6 (a) Growth inhibition of *Escherichia coli* cells in the liquid nutrient-rich medium LB (5 mg chitin nanocrystals were suspended with 10^4 cfu of *E. coli* cells in LB for 3 h). Percentages of inhibition are expressed relative to a control that did not contain any chitin nanocrystals. (b) Time-dependent killing efficacy of BC/ChNC nanocomposites against *E. coli*. Each data point is the average of three independent experiments, and error bars represent standard deviations. The tests were conducted using 7 mg nanocomposites in 3 ml water containing *E. coli* cells at an initial concentration of 10^5 cfu/ml. *Source*: Brown et al. [165]. Reproduced with permission from the Royal Society of Chemistry.

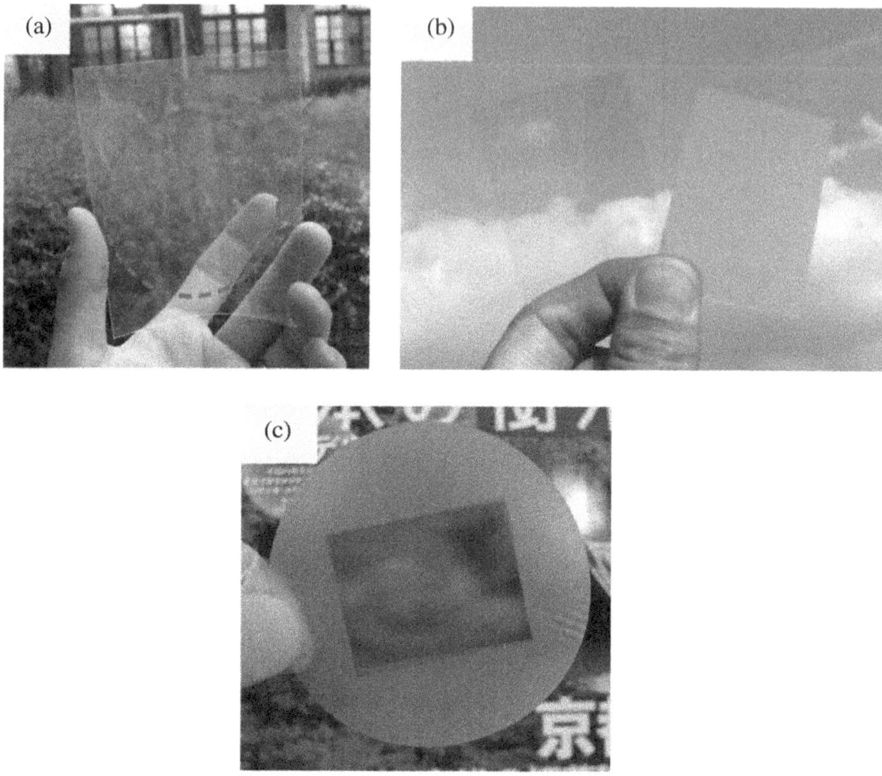

FIGURE 5.7 Various approaches to preparing functional transparent cellulose sheets. (a) Lamination of optically transparent plastics (inside the red circle). (b) Deposition of transparent resins on the surface of cellulose sheets (left, surface-coated sheet; right, uncoated sheet). (c) Drawing of transparent patterns on the sheet using a domestic ink-jet printer. *Source*: Nogi et al. [169], figure 4. Reproduced with permission from John Wiley & Sons. (*See insert for color representation of the figure.*)

5.5 CONCLUSIONS

With the world focus on reducing our dependence on petroleum-based materials, the fabrication of green materials based on renewable resources has become of a high priority. Nanocellulose is expected to become a major contributor to the green economy in the coming decades and should serve as a strategic platform for sustainable development. The research and development in nanocellulose-based green nanocomposites have been critical examined. The size, surface structure, and functionality of nanocelluloses control their interactions with the polymer matrix, which ultimately dictates the structure and properties of the composites. Green nanocomposites based on cellulose nanofibril networks or completely dispersed nanofibrils or nanocrystals in polymer matrix show high strength, high toughness, low moisture adsorption, low thermal expansion, high thermal stability, high optical transparency, and excellent water vapor and oxygen barrier properties. Nanocellulose-based green nanocomposite materials have huge potential as the next generation of materials in the fields of packaging, medical, electronic device, and automotive sectors.

REFERENCES

[1] Podsiadlo, P., Kaushik, A.K., Arruda, E.M., Waas, A.M., Shim, B.S., Xu, J.D., Nandivada, H., Pumplin, B.G., Lahann, J., Ramamoorthy, A., Kotov, N.A. (2007). Ultrastrong and stiff layered polymer nanocomposites. *Science*, **318**, 80–83.

[2] Brown, R.M. (1996). The biosynthesis of cellulose. *Journal of Macromolecular Science Pure and Applied Chemistry*, **A33**, 1345–1373.

[3] Jarvis, M. (2003). Chemistry—cellulose stacks up. *Nature*, **426**, 611–612.

[4] Battista, O.A., Smith, P.A. (1962). Microcrystalline cellulose. *Industrial and Engineering Chemistry*, **54**, 20–29.

[5] Revol, J.F., Bradford, H., Giasson, J., Marchessault, R.H., Gray, D.G. (1992). Helicoidal self-ordering of cellulose microfibrils in aqueous suspension. *International Journal of Biological Macromolecules*, **14**, 170–172.

[6] Beck-Candanedo, S., Roman, M., Gray, D.G. (2005). Effect of reaction conditions on the properties and behavior of wood cellulose nanocrystal suspensions. *Biomacromolecules*, **6**, 1048–1054.

[7] Favier, V., Chanzy, H., Cavaille, J.Y. (1995). Polymer nanocomposites reinforced by cellulose whiskers. *Macromolecules*, **28**, 6365–6367.

[8] Turbak, A.F., Snyder, F.W., Sandberg, K.R. (1983). Microfibrillated cellulose, a new cellulose product: properties, uses, and commercial potential. *Journal of Applied Polymer Science: Applied Polymer Symposium*, **37**, 815–827.

[9] Pääkkö, M., Ankerfors, M., Kosonen, H., Nykänen, A., Ahola, S., Österberg, M., Ruokolainen, J., Laine, J., Larsson, P.T., Ikkala, O., Lindström, T. (2007). Enzymatic hydrolysis combined with mechanical shearing and high-pressure homogenization for nanoscale cellulose fibrils and strong gels. *Biomacromolecules*, **8**, 1934–1941.

[10] Nakagaito, A.N., Yano, H. (2004). The effect of morphological changes from pulp fiber towards nano-scale fibrillated cellulose on the mechanical properties of high-strength plant fiber based composites. *Applied Physics A: Materials Science and Processing*, **78**, 547–552.

[11] Saito, T., Kimura, S., Nishiyama, Y., Isogai, A. (2007). Cellulose nanofibers prepared by TEMPO-mediated oxidation of native cellulose. *Biomacromolecules*, **8**, 2485–2491.

[12] Sehaqui, H., Allais, M., Zhou, Q., Berglund, L.A. (2011). Wood cellulose biocomposites with fibrous structures at micro- and nanoscale. *Composites Science and Technology*, **71**, 382–387.

[13] Klemm, D., Kramer, F., Moritz, S., Lindström, T., Ankerfors, M., Gray, D., Dorris, A. (2011). Nanocelluloses: a new family of nature-based materials. *Angewandte Chemie, International Edition*, **50**, 5438–5466.

[14] Moon, R.J., Martini, A., Nairn, J., Simonsen, J., Youngblood, J. (2011). Cellulose nanomaterials review: structure, properties and nanocomposites. *Chemical Society Reviews*, **40**, 3941–3994.

[15] Filpponen, I., Argyropoulos, D.S. (2010). Regular linking of cellulose nanocrystals via click chemistry: synthesis and formation of cellulose nanoplatelet gels. *Biomacromolecules*, **11**, 1060–1066.

[16] Araki, J., Wada, M., Kuga, S. (2001). Steric stabilization of a cellulose microcrystal suspension by poly(ethylene glycol) grafting. *Langmuir*, **17**, 21–27.

[17] Espinosa, S.C., Kuhnt, T., Foster, E.J., Weder, C. (2013). Isolation of thermally stable cellulose nanocrystals by phosphoric acid hydrolysis. *Biomacromolecules*, **14**, 1223–1230.

[18] Filson, P.B., Dawson-Andoh, B.E., Schwegler-Berry, D. (2009). Enzymatic-mediated production of cellulose nanocrystals from recycled pulp. *Green Chemistry*, **11**, 1808–1814.

[19] Salajkova, M., Berglund, L.A., Zhou, Q. (2012). Hydrophobic cellulose nanocrystals modified with quaternary ammonium salts. *Journal of Materials Chemistry*, **22**, 19798–19805.

[20] Henriksson, M., Henriksson, G., Berglund, L.A., Lindström, T. (2007). An environmentally friendly method for enzyme-assisted preparation of microfibrillated cellulose (MFC) nanofibers. *European Polymer Journal*, **43**, 3434–3441.

[21] Wågberg, L., Decher, G., Norgren, M., Lindström, T., Ankerfors, M., Axnäs, K. (2008). The build-up of polyelectrolyte multilayers of microfibrillated cellulose and cationic polyelectrolytes. *Langmuir*, **24**, 784–795.

[22] Wågberg, L., Winter, L., Ödberg, L., Lindström, T. (1987). On the charge stoichiometry upon adsorption of a cationic polyelectrolyte on cellulosic materials. *Colloids and Surfaces*, **27**, 163–173.

[23] Saito, T., Nishiyama, Y., Putaux, J.L., Vignon, M., Isogai, A. (2006). Homogeneous suspensions of individualized microfibrils from TEMPO-catalyzed oxidation of native cellulose. *Biomacromolecules*, **7**, 1687–1691.

[24] Isogai, A., Saito, T., Fukuzumi, H. (2011). TEMPO-oxidized cellulose nanofibers. *Nanoscale*, **3**, 71–85.

[25] Butchosa, N., Zhou, Q. (2014). Water redispersible cellulose nanofibrils adsorbed with carboxymethyl cellulose. *Cellulose*, **21**, 4349–4358.

[26] Gatenholm, P., Klemm, D. (2010). Bacterial nanocellulose as a renewable material for biomedical applications. *MRS Bulletin*, **35**, 208–213.

[27] Klemm, D., Schumann, D., Udhardt, U., Marsch, S. (2001). Bacterial synthesized cellulose—artificial blood vessels for microsurgery. *Progress in Polymer Science*, **26**, 1561–1603.

[28] Fontana, J.D., Desouza, A.M., Fontana, C.K., Torriani, I.L., Moreschi, J.C., Gallotti, B.J., Desouza, S.J., Narcisco, G.P., Bichara, J.A., Farah, L.F.X. (1990). *Acetobacter* cellulose pellicle as a temporary skin substitute. *Applied Biochemistry and Biotechnology*, **24–5**, 253–264.

[29] Svensson, A., Nicklasson, E., Harrah, T., Panilaitis, B., Kaplan, D.L., Brittberg, M., Gatenholm, P. (2005). Bacterial cellulose as a potential scaffold for tissue engineering of cartilage. *Biomaterials*, **26**, 419–431.

[30] Czaja, W., Krystynowicz, A., Bielecki, S., Brown, R.M. (2006). Microbial cellulose—the natural power to heal wounds. *Biomaterials*, **27**, 145–151.

[31] Alvarez, O.M., Patel, M., Booker, J., Markowitz, L. (2004). Effectiveness of a biocellulose wound dressing for the treatment of chronic venous leg ulcers: results of a single

center randomized study involving 24 patients. *Wounds: A Compendium of Clinical Research and Practice*, **16**, 224–233.

[32] Nishino, T., Matsuda, I., Hirao, K. (2004). All-cellulose composite. *Macromolecules*, **37**, 7683–7687.

[33] Gindl, W., Keckes, J. (2005). All-cellulose nanocomposite. *Polymer*, **46**, 10221–10225.

[34] Nishino, T., Arimoto, N. (2007). All-cellulose composite prepared by selective dissolving of fiber surface. *Biomacromolecules*, **8**, 2712–2716.

[35] Soykeabkaew, N., Arimoto, N., Nishino, T., Peijs, T. (2008). All-cellulose composites by surface selective dissolution of aligned ligno-cellulosic fibres. *Composites Science and Technology*, **68**, 2201–2207.

[36] Shibata, M., Teramoto, N., Nakamura, T., Saitoh, Y. (2013). All-cellulose and all-wood composites by partial dissolution of cotton fabric and wood in ionic liquid. *Carbohydrate Polymers*, **98**, 1532–1539.

[37] Huber, T., Bickerton, S., Mussig, J., Pang, S., Staiger, M.P. (2013). Flexural and impact properties of all-cellulose composite laminates. *Composites Science and Technology*, **88**, 92–98.

[38] Huber, T., Mussig, J., Curnow, O., Pang, S.S., Bickerton, S., Staiger, M.P. (2012). A critical review of all-cellulose composites. *Journal of Materials Science*, **47**, 1171–1186.

[39] Alcalá, M., González, I., Boufi, S., Vilaseca, F., Mutjé, P. (2013). All-cellulose composites from unbleached hardwood kraft pulp reinforced with nanofibrillated cellulose. *Cellulose*, **20**, 2909–2921.

[40] Eyholzer, C., Lopez-Suevos, F., Tingaut, P., Zimmermann, T., Oksman, K. (2010). Reinforcing effect of carboxymethylated nanofibrillated cellulose powder on hydroxypropyl cellulose. *Cellulose*, **17**, 793–802.

[41] Johnson, R.K., Zink-Sharp, A., Renneckar, S.H., Glasser, W.G. (2009). A new bio-based nanocomposite: fibrillated TEMPO-oxidized celluloses in hydroxypropylcellulose matrix. *Cellulose*, **16**, 227–238.

[42] Zimmermann, T., Pohler, E., Geiger, T. (2004). Cellulose fibrils for polymer reinforcement. *Advanced Engineering Materials*, **6**, 754–761.

[43] Zimmermann, T., Pohler, E., Schwaller, P. (2005). Mechanical and morphological properties of cellulose fibril reinforced nanocomposites. *Advanced Engineering Materials*, **7**, 1156–1161.

[44] Zhou, Q., Malm, E., Nilsson, H., Larsson, P.T., Iversen, T., Berglund, L.A., Bulone, V. (2009). Nanostructured biocomposites based on bacterial cellulosic nanofibers compartmentalized by a soft hydroxyethylcellulose matrix coating. *Soft Matter*, **5**, 4124–4130.

[45] Sehaqui, H., Zhou, Q., Berglund, L.A. (2011). Nanostructured biocomposites of high toughness-a wood cellulose nanofiber network in ductile hydroxyethylcellulose matrix. *Soft Matter*, **7**, 7342–7350.

[46] Kochumalayil, J., Sehaqui, H., Zhou, Q., Berglund, L.A. (2010). Tamarind seed xyloglucan—a thermostable high-performance biopolymer from non-food feedstock. *Journal of Materials Chemistry*, **20**, 4321–4327.

[47] Stepan, A.M., Ansari, F., Berglund, L., Gatenholm, P. (2014). Nanofibrillated cellulose reinforced acetylated arabinoxylan films. *Composites Science and Technology*, **98**, 72–78.

[48] Mikkonen, K.S., Pitkanen, L., Liljestrom, V., Bergstrom, E.M., Serimaa, R., Salmen, L., Tenkanen, M. (2012). Arabinoxylan structure affects the reinforcement of films by microfibrillated cellulose. *Cellulose*, **19**, 467–480.

[49] Stevanic, J.S., Bergstrom, E.M., Gatenholm, P., Berglund, L., Salmen, L. (2012). Arabinoxylan/nanofibrillated cellulose composite films. *Journal of Materials Science*, **47**, 6724–6732.

[50] Peng, X.W., Ren, J.L., Zhong, L.X., Sun, R.C. (2011). Nanocomposite films based on xylan-rich hemicelluloses and cellulose nanofibers with enhanced mechanical properties. *Biomacromolecules*, **12**, 3321–3329.

[51] Mikkonen, K.S., Stevanic, J.S., Joly, C., Dole, P., Pirkkalainen, K., Serimaa, R., Salmen, L., Tenkanen, M. (2011). Composite films from spruce galactoglucomannans with microfibrillated spruce wood cellulose. *Cellulose*, **18**, 713–726.

[52] Stevanic, J.S., Mikkonen, K.S., Xu, C.L., Tenkanen, M., Berglund, L., Salmen, L. (2014). Wood cell wall mimicking for composite films of spruce nanofibrillated cellulose with spruce galactoglucomannan and arabinoglucuronoxylan. *Journal of Materials Science*, **49**, 5043–5055.

[53] Hansen, N.M.L., Blomfeldt, T.O.J., Hedenqvist, M.S., Plackett, D.V. (2012). Properties of plasticized composite films prepared from nanofibrillated cellulose and birch wood xylan. *Cellulose*, **19**, 2015–2031.

[54] Mikkonen, K.S., Mathew, A.P., Pirkkalainen, K., Serimaa, R., Xu, C.L., Willfor, S., Oksman, K., Tenkanen, M. (2010). Glucomannan composite films with cellulose nanowhiskers. *Cellulose*, **17**, 69–81.

[55] Saxena, A., Elder, T.J., Ragauskas, A.J. (2011). Moisture barrier properties of xylan composite films. *Carbohydrate Polymers*, **84**, 1371–1377.

[56] Saxena, A., Elder, T.J., Kenvin, J., Ragauskas, A.J. (2010). High oxygen nanocomposite barrier films based on xylan and nanocrystalline cellulose. *Nano-Micro Letters*, **2**, 235–241.

[57] Saxena, A., Elder, T.J., Pan, S., Ragauskas, A.J. (2009). Novel nanocellulosic xylan composite film. *Composites Part B: Engineering*, **40**, 727–730.

[58] Saxena, A., Ragauskas, A.J. (2009). Water transmission barrier properties of biodegradable films based on cellulosic whiskers and xylan. *Carbohydrate Polymers*, **78**, 357–360.

[59] Seifert, M., Hesse, S., Kabrelian, V., Klemm, D. (2004). Controlling the water content of never dried and reswollen bacterial cellulose by the addition of water-soluble polymers to the culture medium. *Journal of Polymer Science Part A: Polymer Chemistry*, **42**, 463–470.

[60] Hirai, A., Tsuji, M., Yamamoto, H., Horii, F. (1998). In situ crystallization of bacterial cellulose—III. Influences of different polymeric additives on the formation of microfibrils as revealed by transmission electron microscopy. *Cellulose*, **5**, 201–213.

[61] Uhlin, K.I., Atalla, R.H., Thompson, N.S. (1995). Influence of hemicelluloses on the aggregation patterns of bacterial cellulose. *Cellulose*, **2**, 129–144.

[62] Yamamoto, H., Horii, F., Hirai, A. (1996). In situ crystallization of bacterial cellulose .2. Influences of different polymeric additives on the formation of celluloses I-alpha and I-beta at the early stage of incubation. *Cellulose*, **3**, 229–242.

[63] Haigler, C.H., White, A.R., Brown, R.M., Cooper, K.M. (1982). Alteration of in vivo cellulose ribbon assembly by carboxymethylcellulose and other cellulose derivatives. *Journal of Cell Biology*, **94**, 64–69.

[64] Tokoh, C., Takabe, K., Sugiyama, J., Fujita, M. (2002). CP/MAS (13)C NMR and electron diffraction study of bacterial cellulose structure affected by cell wall polysaccharides. *Cellulose*, **9**, 351–360.

[65] Kacurakova, M., Smith, A.C., Gidley, M.J., Wilson, R.H. (2002). Molecular interactions in bacterial cellulose composites studied by 1D FT-IR and dynamic 2D FT-IR spectroscopy. *Carbohydrate Research*, **337**, 1145–1153.

[66] Gu, J., Catchmark, J.M. (2014). Roles of xyloglucan and pectin on the mechanical properties of bacterial cellulose composite films. *Cellulose*, **21**, 275–289.

[67] Chanliaud, E., Gidley, M.J. (1999). In vitro synthesis and properties of pectin/*Acetobacter* xylinus cellulose composites. *Plant Journal*, **20**, 25–35.

[68] Whitney, S.E.C., Brigham, J.E., Darke, A.H., Reid, J.S.G., Gidley, M.J. (1995). In-vitro assembly of cellulose/xyloglucan networks—ultrastructural and molecular aspects. *Plant Journal*, **8**, 491–504.

[69] Hayashi, T., Marsden, M.P.F., Delmer, D.P. (1987). Pea xyloglucan and cellulose. 5. Xyloglucan-cellulose interactions in vitro and in vivo. *Plant Physiology*, **83**, 384–389.

[70] Tokoh, C., Takabe, K., Sugiyama, J., Fujita, M. (2002). Cellulose synthesized by *Acetobacter* xylinum in the presence of plant cell wall polysaccharides. *Cellulose*, **9**, 65–74.

[71] Winter, H., Barakat, A., Cathala, B., Saake, B. (2006). Preparation of arabinoxylan and its sorption on bacterial cellulose during cultivation. *Macromolecular Symposia*, **232**, 74–84.

[72] Whitney, S.E.C., Gothard, M.G.E., Mitchell, J.T., Gidley, M.J. (1999). Roles of cellulose and xyloglucan in determining the mechanical properties of primary plant cell walls. *Plant Physiology*, **121**, 657–663.

[73] Chanliaud, E., De Silva, J., Strongitharm, B., Jeronimidis, G., Gidley, M.J. (2004). Mechanical effects of plant cell wall enzymes on cellulose/xyloglucan composites. *Plant Journal*, **38**, 27–37.

[74] Abdollahi, M., Alboofetileh, M., Behrooz, R., Rezaei, M., Miraki, R. (2013). Reducing water sensitivity of alginate bio-nanocomposite film using cellulose nanoparticles. *International Journal of Biological Macromolecules*, **54**, 166–173.

[75] Abdollahi, M., Alboofetileh, M., Rezaei, M., Behrooz, R. (2013). Comparing physico-mechanical and thermal properties of alginate nanocomposite films reinforced with organic and/or inorganic nanofillers. *Food Hydrocolloids*, **32**, 416–424.

[76] Sirvio, J.A., Kolehmainen, A., Liimatainen, H., Niinimaki, J., Hormi, O.E.O. (2014). Biocomposite cellulose-alginate films: promising packaging materials. *Food Chemistry*, **151**, 343–351.

[77] Lin, N., Bruzzese, C., Dufresne, A. (2012). TEMPO-oxidized nanocellulose participating as crosslinking aid for alginate-based sponges. *ACS Applied Materials & Interfaces*, **4**, 4948–4959.

[78] Huq, T., Salmieri, S., Khan, A., Khan, R.A., Le Tien, C., Riedl, B., Fraschini, C., Bouchard, J., Uribe-Calderon, J., Kamal, M.R., Lacroix, M. (2012). Nanocrystalline cellulose (NCC) reinforced alginate based biodegradable nanocomposite film. *Carbohydrate Polymers*, **90**, 1757–1763.

[79] Azeredo, H.M.C., Miranda, K.W.E., Rosa, M.F., Nascimento, D.M., de Moura, M.R. (2012). Edible films from alginate-acerola puree reinforced with cellulose whiskers. *LWT—Food Science and Technology*, **46**, 294–297.

[80] Reddy, J.P., Rhim, J.W. (2014). Characterization of bionanocomposite films prepared with agar and paper-mulberry pulp nanocellulose. *Carbohydrate Polymers*, **110**, 480–488.

[81] Trovatti, E., Fernandes, S.C.M., Rubatat, L., Perez, D.D., Freire, C.S.R., Silvestre, A.J.D., Neto, C.P. (2012). Pullulan-nanofibrillated cellulose composite films with improved thermal and mechanical properties. *Composites Science and Technology*, **72**, 1556–1561.

[82] Svagan, A.J., Samir, M.A.S.A., Berglund, L.A. (2007). Biomimetic polysaccharide nanocomposites of high cellulose content and high toughness. *Biomacromolecules*, **8**, 2556–2563.

[83] Plackett, D., Anturi, H., Hedenqvist, M., Ankerfors, M., Gallstedt, M., Lindstrom, T., Siro, I. (2010). Physical properties and morphology of films prepared from microfibrillated cellulose and microfibrillated cellulose in combination with amylopectin. *Journal of Applied Polymer Science*, **117**, 3601–3609.

[84] Karimi, S., Dufresne, A., Tahir, P.M., Karimi, A., Abdulkhani, A. (2014). Biodegradable starch-based composites: effect of micro and nanoreinforcements on composite properties. *Journal of Materials Science*, **49**, 4513–4521.

[85] Nasri-Nasrabadi, B., Behzad, T., Bagheri, R. (2014). Preparation and characterization of cellulose nanofiber reinforced thermoplastic starch composites. *Fibers and Polymers*, **15**, 347–354.

[86] Karimi, S., Tahir, P.M., Dufresne, A., Karimi, A., Abdulkhani, A. (2014). A comparative study on characteristics of nanocellulose reinforced thermoplastic starch biofilms prepared with different techniques. *Nordic Pulp & Paper Research Journal*, **29**, 41–45.

[87] Teixeira, E.D., Pasquini, D., Curvelo, A.A.S., Corradini, E., Belgacem, M.N., Dufresne, A. (2009). Cassava bagasse cellulose nanofibrils reinforced thermoplastic cassava starch. *Carbohydrate Polymers*, **78**, 422–431.

[88] Svagan, A.J., Samir, M.A.S.A., Berglund, L.A. (2008). Biomimetic foams of high mechanical performance based on nanostructured cell walls reinforced by native cellulose nanofibrils. *Advanced Materials*, **20**, 1263–1269.

[89] Yildirim, N., Shaler, S.M., Gardner, D.J., Rice, R., Bousfield, D.W. (2014). Cellulose nanofibril (CNF) reinforced starch insulating foams. *Cellulose*, **21**, 4337–4347.

[90] Nasri-Nasrabadi, B., Mehrasa, M., Rafienia, M., Bonakdar, S., Behzad, T., Gavanji, S. (2014). Porous starch/cellulose nanofibers composite prepared by salt leaching technique for tissue engineering. *Carbohydrate Polymers*, **108**, 232–238.

[91] Hietala, M., Rollo, P., Kekalainen, K., Oksman, K. (2014). Extrusion processing of green biocomposites: compounding, fibrillation efficiency, and fiber dispersion. *Journal of Applied Polymer Science*, **131**, 39981.

[92] Hietala, M., Mathew, A.P., Oksman, K. (2013). Bionanocomposites of thermoplastic starch and cellulose nanofibers manufactured using twin-screw extrusion. *European Polymer Journal*, **49**, 950–956.

[93] Teixeira, E.D., Lotti, C., Correa, A.C., Teodoro, K.B.R., Marconcini, J.M., Mattoso, L.H.C. (2011). Thermoplastic corn starch reinforced with cotton cellulose nanofibers. *Journal of Applied Polymer Science*, **120**, 2428–2433.

[94] Aila-Suarez, S., Palma-Rodriguez, H.M., Rodriguez-Hernandez, A.I., Hernandez-Uribe, J.P., Bello-Perez, L.A., Vargas-Torres, A. (2013). Characterization of films made with chayote tuber and potato starches blending with cellulose nanoparticles. *Carbohydrate Polymers*, **98**, 102–107.

[95] Grande, C.J., Torres, F.G., Gomez, C.M., Troncoso, O.P., Canet-Ferrer, J., Martinez-Pastor, J. (2009). Development of self-assembled bacterial cellulose-starch nanocomposites. *Materials Science and Engineering C: Biomimetic and Supramolecular Systems*, **29**, 1098–1104.

[96] Wan, Y.Z., Luo, H.L., He, F., Liang, H., Huang, Y., Li, X.L. (2009). Mechanical, moisture absorption, and biodegradation behaviours of bacterial cellulose fibre-reinforced starch biocomposites. *Composites Science and Technology*, **69**, 1212–1217.

[97] Muzzarelli, R.A.A., Boudrant, J., Meyer, D., Manno, N., DeMarchis, M., Paoletti, M.G. (2012). Current views on fungal chitin/chitosan, human chitinases, food preservation, glucans, pectins and inulin: a tribute to Henri Braconnot, precursor of the carbohydrate polymers science, on the chitin bicentennial. *Carbohydrate Polymers*, **87**, 995–1012.

[98] Azeredo, H.M.C., Mattoso, L.H.C., Avena-Bustillos, R.J., Ceotto, G., Munford, M.L., Wood, D., McHugh, T.H. (2010). Nanocellulose reinforced chitosan composite films as affected by nanofiller loading and plasticizer content. *Journal of Food Science*, **75**, N1–N7.

[99] de Mesquita, J.P., Donnici, C.L., Teixeira, I.F., Pereira, F.V. (2012). Bio-based nanocomposites obtained through covalent linkage between chitosan and cellulose nanocrystals. *Carbohydrate Polymers*, **90**, 210–217.

[100] Dehnad, D., Emarn-Djomeh, Z., Mirzaei, H., Jafari, S.M., Dadashi, S. (2014). Optimization of physical and mechanical properties for chitosan-nanocellulose biocomposites. *Carbohydrate Polymers*, **105**, 222–228.

[101] Fernandes, S.C.M., Freire, C.S.R., Silvestre, A.J.D., Neto, C.P., Gandini, A., Berglund, L.A., Salmen, L. (2010). Transparent chitosan films reinforced with a high content of nanofibrillated cellulose. *Carbohydrate Polymers*, **81**, 394–401.

[102] Nordqvist, D., Idermark, J., Hedenqvist, M.S. (2007). Enhancement of the wet properties of transparent chitosan-acetic-acid-salt films using microfibrillated cellulose. *Biomacromolecules*, **8**, 2398–2403.

[103] Liu, K., Lin, X.X., Chen, L.H., Huang, L.L., Cao, S.L. (2014). Dual-functional chitosan-methylisothiazolinone/microfibrillated cellulose biocomposites for enhancing antibacterial and mechanical properties of agar films. *Cellulose*, **21**, 519–528.

[104] Liu, K., Lin, X.X., Chen, L.H., Huang, L.L., Cao, S.L., Wang, H.W. (2013). Preparation of microfibrillated cellulose/chitosan-benzalkonium chloride biocomposite for enhancing antibacterium and strength of sodium alginate films. *Journal of Agricultural and Food Chemistry*, **61**, 6562–6567.

[105] Cai, Z., Chen, P., Jin, H.J., Kim, J. (2009). The effect of chitosan content on the crystallinity, thermal stability, and mechanical properties of bacterial cellulose-chitosan composites. *Proceedings of the Institution of Mechanical Engineers—Part C: Journal of Mechanical Engineering Science*, **223**, 2225–2230.

[106] Dubruel, P., Unger, R., Van Vlierberghe, S., Cnudde, V., Jacobs, P.J.S., Schacht, E., Kirkpatrick, C.J. (2007). Porous gelatin hydrogels: 2. In vitro cell interaction study. *Biomacromolecules*, **8**, 338–344.

[107] Xing, Q., Zhao, F., Chen, S., McNamara, J., DeCoster, M.A., Lvov, Y.M. (2010). Porous biocompatible three-dimensional scaffolds of cellulose microfiber/gelatin composites for cell culture. *Acta Biomaterialia*, **6**, 2132–2139.

[108] Taokaew, S., Seetabhawang, S., Siripong, P., Phisalaphong, M. (2013). Biosynthesis and characterization of nanocellulose-gelatin films. *Materials*, **6**, 782–794.

[109] Chen, H.H., Lin, S.B., Hsu, C.P., Chen, L.C. (2013). Modifying bacterial cellulose with gelatin peptides for improved rehydration. *Cellulose*, **20**, 1967–1977.

[110] Lin, S.B., Hsu, C.P., Chen, L.C., Chen, H.H. (2009). Adding enzymatically modified gelatin to enhance the rehydration abilities and mechanical properties of bacterial cellulose. *Food Hydrocolloids*, **23**, 2195–2203.

[111] Wang, J., Wan, Y.Z., Luo, H.L., Gao, C., Huang, Y. (2012). Immobilization of gelatin on bacterial cellulose nanofibers surface via crosslinking technique. *Materials Science & Engineering, C: Materials for Biological Applications*, **32**, 536–541.

[112] Chang, S.T., Chen, L.C., Lin, S.B., Chen, H.H. (2012). Nano-biomaterials application: morphology and physical properties of bacterial cellulose/gelatin composites via crosslinking. *Food Hydrocolloids*, **27**, 137–144.

[113] Saska, S., Teixeira, L.N., de Oliveira, P.T., Gaspar, A.M.M., Ribeiro, S.J.L., Messaddeq, Y., Marchetto, R. (2012). Bacterial cellulose-collagen nanocomposite for bone tissue engineering. *Journal of Materials Chemistry*, **22**, 22102–22112.

[114] Mathew, A.P., Oksman, K., Pierron, D., Harmad, M.F. (2012). Crosslinked fibrous composites based on cellulose nanofibers and collagen with in situ pH induced fibrillation. *Cellulose*, **19**, 139–150.

[115] Lu, T.H., Li, Q., Chen, W.S., Yu, H.P. (2014). Composite aerogels based on dialdehyde nanocellulose and collagen for potential applications as wound dressing and tissue engineering scaffold. *Composites Science and Technology*, **94**, 132–138.

[116] Gao, C., Wan, Y.Z., Lei, X.W., Qu, J., Yan, T., Dai, K.R. (2011). Polylysine coated bacterial cellulose nanofibers as novel templates for bone-like apatite deposition. *Cellulose*, **18**, 1555–1561.

[117] Noishiki, Y., Nishiyama, Y., Wada, M., Kuga, S., Magoshi, J. (2002). Mechanical properties of silk fibroin-microcrystalline cellulose composite films. *Journal of Applied Polymer Science*, **86**, 3425–3429.

[118] Cho, S.Y., Lee, M.E., Choi, Y., Jin, H.J. (2014). Cellulose nanofiber-reinforced silk fibroin composite film with high transparency. *Fibers and Polymers*, **15**, 215–219.

[119] Huang, J., Liu, L., Yao, J.M. (2011). Electrospinning of Bombyx mori silk fibroin nanofiber mats reinforced by cellulose nanowhiskers. *Fibers and Polymers*, **12**, 1002–1006.

[120] Cheng, J., Park, M., Hyun, J. (2014). Thermoresponsive hybrid hydrogel of oxidized nanocellulose using a polypeptide crosslinker. *Cellulose*, **21**, 1699–1708.

[121] Rafieian, F., Shahedi, M., Keramat, J., Simonsen, J. (2014). Mechanical, thermal and barrier properties of nano-biocomposite based on gluten and carboxylated cellulose nanocrystals. *Industrial Crops and Products*, **53**, 282–288.

[122] Pereda, M., Arnica, G., Racz, I., Marcovich, N.E. (2011). Structure and properties of nanocomposite films based on sodium caseinate and nanocellulose fibers. *Journal of Food Engineering*, **103**, 76–83.

[123] Gu, J., Catchmark, J.M. (2013). Polylactic acid composites incorporating casein functionalized cellulose nanowhiskers. *Journal of Biological Engineering*, **7**, 31.

[124] Reddy, N., Yang, Y.Q. (2011). Potential of plant proteins for medical applications. *Trends in Biotechnology*, **29**, 490–498.

[125] Arboleda, J.C., Hughes, M., Lucia, L.A., Laine, J., Ekman, K., Rojas, O.J. (2013). Soy protein-nanocellulose composite aerogels. *Cellulose*, **20**, 2417–2426.

[126] Seydibeyoglu, M.O., Misra, M., Mohanty, A., Blaker, J.J., Lee, K.Y., Bismarck, A., Kazemizadeh, M. (2013). Green polyurethane nanocomposites from soy polyol and bacterial cellulose. *Journal of Materials Science*, **48**, 2167–2175.

[127] Luo, X.G., Mohanty, A., Misra, M. (2013). Green nanocomposites from soy-based biopolyurethane with microcrystalline cellulose. *Macromolecular Materials and Engineering*, **298**, 412–418.

[128] Zhou, Q., Berglund, L.A. (2015) PLA-nanocellulose Biocomposites. In A.Jiménez, M.Peltzer, R.Ruseckaite (Eds.), *Poly(Lactic Acid) Science and Technology: Processing, Properties, Additives and Applications*, pp. 225–242, The Royal Society of Chemistry, Cambridge.

[129] Hassan, M.L., Bras, J., Hassan, E.A., Fadel, S.M., Dufresne, A. (2012). Polycaprolactone/modified bagasse whisker nanocomposites with improved moisture-barrier and biodegradability properties. *Journal of Applied Polymer Science*, **125**, E10–E19.

[130] Siqueira, G., Bras, J., Follain, N., Belbekhouche, S., Marais, S., Dufresne, A. (2013). Thermal and mechanical properties of bio-nanocomposites reinforced by Luffa cylindrica cellulose nanocrystals. *Carbohydrate Polymers*, **91**, 711–717.

[131] Habibi, Y., Dufresne, A. (2008). Highly filled bionanocomposites from functionalized polysaccharide nanocrystals. *Biomacromolecules*, **9**, 1974–1980.

[132] Habibi, Y., Goffin, A.L., Schiltz, N., Duquesne, E., Dubois, P., Dufresne, A. (2008). Bionanocomposites based on poly(epsilon-caprolactone)-grafted cellulose nanocrystals by ring-opening polymerization. *Journal of Materials Chemistry*, **18**, 5002–5010.

[133] Lonnberg, H., Larsson, K., Lindstrom, T., Hult, A., Malmstrom, E. (2011). Synthesis of polycaprolactone-grafted microfibrillated cellulose for use in novel bionanocomposites-influence of the graft length on the mechanical properties. *ACS Applied Materials & Interfaces*, **3**, 1426–1433.

[134] Boujemaoui, A., Carlsson, L., Malmstrom, E., Lahcini, M., Berglund, L., Sehaqui, H., Carlmark, A. (2012). Facile preparation route for nanostructured composites: surface-initiated ring-opening polymerization of epsilon-caprolactone from high-surface-area nanopaper. *ACS Applied Materials & Interfaces*, **4**, 3191–3198.

[135] Wan, Y.Z., Hong, L., Jia, S.R., Huang, Y., Zhu, Y., Wang, Y.L., Jiang, H.J. (2006). Synthesis and characterization of hydroxyapatite-bacterial cellulose nanocomposites. *Composites Science and Technology*, **66**, 1825–1832.

[136] Zimmermann, K.A., LeBlanc, J.M., Sheets, K.T., Fox, R.W., Gatenholm, P. (2011). Biomimetic design of a bacterial cellulose/hydroxyapatite nanocomposite for bone healing applications. *Materials Science & Engineering, C: Materials for Biological Applications*, **31**, 43–49.

[137] Hong, L., Wang, Y.L., Jia, S.R., Huang, Y., Gao, C., Wan, Y.Z. (2006). Hydroxyapatite/bacterial cellulose composites synthesized via a biomimetic route. *Materials Letters*, **60**, 1710–1713.

[138] Hutchens, S.A., Benson, R.S., Evans, B.R., O'Neill, H.M., Rawn, C.J. (2006). Biomimetic synthesis of calcium-deficient hydroxyapatite in a natural hydrogel. *Biomaterials*, **27**, 4661–4670.

[139] Grande, C.J., Torres, F.G., Gomez, C.M., Bano, M.C. (2009). Nanocomposites of bacterial cellulose/hydroxyapatite for biomedical applications. *Acta Biomaterialia*, **5**, 1605–1615.

[140] Liu, A.D., Walther, A., Ikkala, O., Belova, L., Berglund, L.A. (2011). Clay nanopaper with tough cellulose nanofiber matrix for fire retardancy and gas barrier functions. *Biomacromolecules*, **12**, 633–641.

[141] Liu, A.D., Berglund, L.A. (2013). Fire-retardant and ductile clay nanopaper biocomposites based on montmorrilonite in matrix of cellulose nanofibers and carboxymethyl cellulose. *European Polymer Journal*, **49**, 940–949.

[142] Liu, A.D., Berglund, L.A. (2012). Clay nanopaper composites of nacre-like structure based on montmorrilonite and cellulose nanofibers-improvements due to chitosan addition. *Carbohydrate Polymers*, **87**, 53–60.

[143] Gabr, M.H., Phong, N.T., Abdelkareem, M.A., Okubo, K., Uzawa, K., Kimpara, I., Fujii, T. (2013). Mechanical, thermal, and moisture absorption properties of nano-clay reinforced nano-cellulose biocomposites. *Cellulose*, **20**, 819–826.

[144] Liimatainen, H., Ezekiel, N., Sliz, R., Ohenoja, K., Sirvio, J.A., Berglund, L., Hormi, O., Niinimaki, J. (2013). High-strength nanocellulose-talc hybrid barrier films. *ACS Applied Materials & Interfaces*, **5**, 13412–13418.

[145] Walther, A., Bjurhager, I., Malho, J.M., Ruokolainen, J., Berglund, L., Ikkala, O. (2010). Supramolecular control of stiffness and strength in lightweight high-performance nacre-mimetic paper with fire-shielding properties. *Angewandte Chemie, International Edition*, **49**, 6448–6453.

[146] Walther, A., Bjurhager, I., Malho, J.M., Pere, J., Ruokolainen, J., Berglund, L.A., Ikkala, O. (2010). Large-area, lightweight and thick biomimetic composites with superior material properties via fast, economic, and green pathways. *Nano Letters*, **10**, 2742–2748.

[147] Wang, J.F., Cheng, Q.F., Lin, L., Jiang, L. (2014). Synergistic toughening of bioinspired poly(vinyl alcohol)-clay-nanofibrillar cellulose artificial nacre. *ACS Nano*, **8**, 2739–2745.

[148] Wu, C.N., Saito, T., Fujisawa, S., Fukuzumi, H., Isogai, A. (2012). Ultrastrong and high gas-barrier nanocellulose/clay-layered composites. *Biomacromolecules*, **13**, 1927–1932.

[149] Wu, C.N., Yang, Q.L., Takeuchi, M., Saito, T., Isogai, A. (2014). Highly tough and transparent layered composites of nanocellulose and synthetic silicate. *Nanoscale*, **6**, 392–399.

[150] Aulin, C., Salazar-Alvarez, G., Lindstrom, T. (2012). High strength, flexible and transparent nanofibrillated cellulose-nanoclay biohybrid films with tunable oxygen and water vapor permeability. *Nanoscale*, **4**, 6622–6628.

[151] Ho, T.T.T., Zimmermann, T., Ohr, S., Caseri, W.R. (2012). Composites of cationic nanofibrillated cellulose and layered silicates: water vapor barrier and mechanical properties. *ACS Applied Materials & Interfaces*, **4**, 4832–4840.

[152] Ho, T.T.T., Ko, Y.S., Zimmermann, T., Geiger, T., Caseri, W. (2012). Processing and characterization of nanofibrillated cellulose/layered silicate systems. *Journal of Materials Science*, **47**, 4370–4382.

[153] Pei, A.H., Malho, J.M., Ruokolainen, J., Zhou, Q., Berglund, L.A. (2011). Strong nanocomposite reinforcement effects in polyurethane elastomer with low volume fraction of cellulose nanocrystals. *Macromolecules*, **44**, 4422–4427.

[154] Fukuzumi, H., Saito, T., Wata, T., Kumamoto, Y., Isogai, A. (2009). Transparent and high gas barrier films of cellulose nanofibers prepared by TEMPO-mediated oxidation. *Biomacromolecules*, **10**, 162–165.

[155] Lee, S.B., Koepsel, R.R., Morley, S.W., Matyjaszewski, K., Sun, Y.J., Russell, A.J. (2004). Permanent, nonleaching antibacterial surfaces. 1. Synthesis by atom transfer radical polymerization. *Biomacromolecules*, **5**, 877–882.

[156] Roy, D., Knapp, J.S., Guthrie, J.T., Perrier, S. (2008). Antibacterial cellulose fiber via RAFT surface graft polymerization. *Biomacromolecules*, **9**, 91–99.

[157] Andresen, M., Stenstad, P., Moretro, T., Langsrud, S., Syverud, K., Johansson, L.S., Stenius, P. (2007). Nonleaching antimicrobial films prepared from surface-modified microfibrillated cellulose. *Biomacromolecules*, **8**, 2149–2155.

[158] Miao, J.J., Pangule, R.C., Paskaleva, E.E., Hwang, E.E., Kane, R.S., Linhardt, R.J., Dordick, J.S. (2011). Lysostaphin-functionalized cellulose fibers with antistaphylococcal activity for wound healing applications. *Biomaterials*, **32**, 9557–9567.

[159] Nakamura, M., Iwasaki, T., Tokino, S., Asaoka, A., Yamakawa, M., Ishibashi, J. (2011). Development of a bioactive fiber with immobilized synthetic peptides designed from the active site of a beetle defensin. *Biomacromolecules*, **12**, 1540–1545.

[160] Chen, Z.B., Luo, J., Sun, Y.Y. (2007). Biocidal efficacy, biofilm-controlling function, and controlled release effect of chloromelamine-based bioresponsive fibrous materials. *Biomaterials*, **28**, 1597–1609.

[161] Ringot, C., Sol, V., Barriere, M., Saad, N., Bressollier, P., Granet, R., Couleaud, P., Frochot, C., Krausz, P. (2011). Triazinyl porphyrin-based photoactive cotton fabrics: preparation, characterization, and antibacterial activity. *Biomacromolecules*, **12**, 1716–1723.

[162] Pinto, R.J.B., Marques, P.A.A.P., Neto, C.P., Trindade, T., Daina, S., Sadocco, P. (2009). Antibacterial activity of nanocomposites of silver and bacterial or vegetable cellulosic fibers. *Acta Biomaterialia*, **5**, 2279–2289.

[163] Dankovich, T.A., Gray, D.G. (2011). Bactericidal paper impregnated with silver nanoparticles for point-of-use water treatment. *Environmental Science & Technology*, **45**, 1992–1998.

[164] Diez, I., Eronen, P., Osterberg, M., Linder, M.B., Ikkala, O., Ras, R.H.A. (2011). Functionalization of nanofibrillated cellulose with silver nanoclusters: fluorescence and antibacterial activity. *Macromolecular Bioscience*, **11**, 1185–1191.

[165] Butchosa, N., Brown, C., Larsson, P.T., Berglund, L.A., Bulone, V., Zhou, Q. (2013). Nanocomposites of bacterial cellulose nanofibers and chitin nanocrystals: fabrication, characterization and bactericidal activity. *Green Chemistry*, **15**, 3404–3413.

[166] Beecroft, L.L., Ober, C.K. (1997). Nanocomposite materials for optical applications. *Chemistry of Materials*, **9**, 1302–1317.

[167] Yano, H., Sugiyama, J., Nakagaito, A.N., Nogi, M., Matsuura, T., Hikita, M., Handa, K. (2005). Optically transparent composites reinforced with networks of bacterial nanofibers. *Advanced Materials*, **17**, 153–155.

[168] Nogi, M., Yano, H. (2008). Transparent nanocomposites based on cellulose produced by bacteria offer potential innovation in the electronics device industry. *Advanced Materials*, **20**, 1849–1852.

[169] Nogi, M., Iwamoto, S., Nakagaito, A.N., Yano, H. (2009). Optically transparent nanofiber paper. *Advanced Materials*, **21**, 1595–1598.

[170] Fang, Z.Q., Zhu, H.L., Li, Y.Y., Liu, Z., Dai, J.Q., Preston, C., Garner, S., Cimo, P., Chai, X.S., Chen, G., Hu, L.B. (2014). Light management in flexible glass by wood cellulose coating. *Scientific Reports*, **4**, 5842.

[171] Zhu, H.L., Parvinian, S., Preston, C., Vaaland, O., Ruan, Z.C., Hu, L.B. (2013). Transparent nanopaper with tailored optical properties. *Nanoscale*, **5**, 3787–3792.

[172] Fang, Z.Q., Zhu, H.L., Bao, W.Z., Preston, C., Liu, Z., Dai, J.Q., Li, Y.Y., Hu, L.B. (2014). Highly transparent paper with tunable haze for green electronics. *Energy & Environmental Science*, **7**, 3313–3319.

[173] Fang, Z.Q., Zhu, H.L., Yuan, Y.B., Ha, D., Zhu, S.Z., Preston, C., Chen, Q.X., Li, Y.Y., Han, X.G., Lee, S., Chen, G., Li, T., Munday, J., Huang, J.S., Hu, L.B. (2014). Novel nanostructured paper with ultrahigh transparency and ultrahigh haze for solar cells. *Nano Letters*, **14**, 765–773.

6

POLY(LACTIC ACID) HYBRID GREEN COMPOSITES

MAHBUB HASAN[1], AZMAN HASSAN[2], AND ZAINOHA ZAKARIA[3]

[1] *Department of Materials and Metallurgical Engineering, Bangladesh University of Engineering and Technology, Dhaka, Bangladesh*

[2] *Department of Polymer Engineering, Faculty of Chemical Engineering, Universiti Teknologi Malaysia, Johor, Malaysia*

[3] *Department of Chemistry, Faculty of Science, Universiti Teknologi Malaysia, Johor, Malaysia*

Biodegradable Green Composites, First Edition. Edited by Susheel Kalia.

6.1 INTRODUCTION

The use of bio-based polymers as sustainable materials is of increasing interest due to their potential to substitute certain petroleum-derived composites. Currently there have been increasing concerns regarding shortages of landfills, persistence of non-biodegradable petroleum-based polymers in the environment, and the dependence on finite and depleting nonrenewable petroleum resources. Green composites combining bioplastics with natural/synthetic fibers/fillers have emerged as promising alternatives to conventional polyolefin/glass fiber composites because they offer a wide variety of advantages, such as low density, renewability, biodegradability, and better cost versus a certain required performance [1]. These composites are attracting attention both in industry and in academia, because they could allow complete degradation in soil or by composting process and do not emit any toxic components [2].

A green composite can be turned into a hybrid green composite by incorporating two or more different types of natural/synthetic fibers/fillers in a single matrix. Recently hybrid composites have received considerable attention to the scientific community because of their capability of providing designers new freedom of tailoring composites and thus achieving properties that cannot be attained in binary systems containing one type of filler/fiber dispersed in a matrix. It has also provided a more cost-effective utilization of expensive fillers/fibers by replacing them partially with less expensive fibers. Although several fillers/fibers can be incorporated into the hybrid system, a combination of only two types of fillers/fibers would be more beneficial. By careful selection of the reinforcing fillers/fibers, the performance properties of the resulting composite can be significantly improved, while the material cost can be substantially reduced [3]. The properties of hybrid composites depend on the individual components in which there is a more favorable balance between the inherent advantages and disadvantages. The disadvantages of one filler/fiber could be overcome by the advantages of the other filler/fiber. Besides, the advantages of one type of filler/fiber could also complement by the advantages of the other filler/fiber in hybrid composite that contains two or more types of fibers/fillers. Through proper material design, a hybrid composite with balance in cost and performance could be obtained [4].

Among all bio-based polymers, **poly(lactic acid)** (PLA) has been the front-runner due to its attractive mechanical properties, renewability, biodegradability, and relatively low cost [5]. The first attempt to prepare PLA was ascribed to Pelouze in 1845 by condensing L-lactic acid (LA) and removing water continuously, leading to low molecular weight PLA [6]. Recently, Cargill has set up a continuous manufacture of renewable LA, followed by its direct ring-opening polymerization. The process starts from the fermentation of corn glucose into L-LA, the condensation of L-LA into LA oligomers, and the catalytic depolymerization under reduced pressure to give pure LA containing minimum 95% of L-LA. The resulting LA can be subsequently polymerized into PLA using $Sn(Oct)_2$ as catalyst to give high molecular weight PLA [7, 8]. The final properties of PLA do not depend on the molecular weight, but more on the content in optical impurities between both LA enantiomers within PLA chains. PLA homopolymers prepared from optically pure L-LA or D-LA are semicrystalline polyesters with a melting temperature of about 175°C and a glass transition temperature of 55–60°C, whereas for a content in D-lactide below 93%, the resulting PLA materials are amorphous with glass transition temperature of 55–60°C [8, 9]. In addition to its biodegradability and renewability, PLA exhibits a Young's modulus of around 3 GPa, a tensile strength between 50 and 70 MPa with an elongation at break of ca. 4%, and an impact strength close to 2.5 kJ/m^2 [10]. By comparison with commodity polymers such as polyethylene, polypropylene, polystyrene, and *polyethylene terephthalate* [11], the mechanical properties of semi-crystalline P(L-LA) are attractive, particularly its Young's modulus, making it as an excellent substitute for commodity polymers in short-time packaging.

The present chapter describes hybrid green composites manufactured from PLA. The manufacturing techniques of various PLA hybrid green composites are first elaborated. The properties and applications of those composites are then detailed. The conclusions are mentioned in the last paragraph of the chapter.

6.2 MANUFACTURING TECHNIQUES OF PLA HYBRID GREEN COMPOSITES

Various techniques are adopted so far in order to manufacture PLA hybrid green composites. The most common ones are described in the following.

6.2.1 Melt Mixing/Blending

Melt mixing or blending is the most common technique applied so far for PLA hybrid green composite manufacturing. Antar et al. manufactured carbon PLA/nanotube/graphite conductive hybrid green composites using blending. PLA was doped with graphite and carbon nanotube (CNT) in order to improve thermoelectrical properties. PLA was blended with various amounts of graphite and CNT. A Brabender internal mixer, equipped with contrarotating blades, was used to disperse conductive fillers into the matrix. Before blending, all the materials were dried overnight in an oven at 100°C to remove eventual traces of moisture. Blending was carried out at 60 rpm for 10 min at 20°C above the matrix melting temperature that is 180°C. The ratio of CNT to graphite used was 1 : 3 [12]. The graphite was useful in improving both thermal conductivity and Seebeck coefficient, whereas CNT was employed to improve the dimensionless figure of merit (ZT). The ZT values could be enhanced by increasing confinement of conducting fillers and decreasing the thermal conductivity of the matrix with voids.

Shi et al. prepared PLA/bamboo/talc hybrid green biocomposites using melt mixing and blending. Bamboo fiber and talc were initially mixed in Haake Rheomix 600p (Haake Technik GmbH, Coesfeld, Germany) and melted at 180°C for 10 min using a rotor speed of 60 rpm. Bamboo and talc were thus premixed at different ratio, added to PLA, and then blended together under strong shearing. These hybrid green composites were compression molded at 180°C into sheets for subsequent measurements [13]. The heat distortion temperature (HDT) was greatly improved with simultaneous addition of 20% bamboo fiber and 20% talc, which was otherwise impossible to obtain using only a single fiber with PLA. The improvement in HDT was believed to be due to transcrystallization that occurred in the hybrid green composite.

Chang et al. manufactured PLA/montmorillonite (MMT)/synthetic mica hybrid green composites using solution casting technique. Dispersions of Na^+-MMT and Na^+-mica were added to solutions of ammonium salts of hexadecylamine (C16). Organophilic clays were obtained through a multistep process and were called C16-MMT and C16-mica, respectively. In a 500 ml beaker, 4.0 g of hexadecylamine, 2.5 ml of concentrated HCl, and 200 ml of water were placed. This solution was heated at 80°C. In 400 ml of water, 10 g of clay was dispersed at 80°C. This dispersion of clay was added to the ammonium salt of the hexadecylamine solution, and the mixture was stirred vigorously for 1 h. The precipitate was isolated by filtration, placed in a 250 ml beaker with 300 ml of hot water, and stirred for 1 h. The product was then filtrated and freeze-dried. C16-mica was made using the same procedure of that used for C16-MMT. Since the synthesis procedures for polymer/organoclay hybrid green nanocomposites with different weight percent (wt%) of organoclays are very similar, only a representative example for the preparation of PLA/C16-MMT (2 wt%) is mentioned. A mixture of 50.0 g of a DMA dispersion containing 0.08 g of C_{16}-MMT, 4.0 g of a PLA solution, and excess DMA was stirred vigorously at room temperature for 4 h. The solution was spread onto a $10 \times 10\,cm^2$ glass plate and the solvent was evaporated in a vacuum at 70°C for 1 day. The films were then cleaned in an ultrasonic cleaner three times for 5 min each. These films, with the solvent removed, were dried again in a vacuum oven at 70°C for 1 day. The thickness of the films was 10–15 mm. C16-mica was more effective in improving the physical strength of PLA films compared to C16-MMT as C16-mica was believed to enhance the interfacial interaction between the clay and PLA. However only 4 wt% C16-MMT was enough to obtain the highest initial modulus, whereas 2–8 wt% of C16-mica was needed for the same purpose [14].

Lee et al. prepared wood flour/talc-filled PLA hybrid green composites using melt compounding technique. The talc content in the hybrid green composite was fixed at 10 wt%, whereas the wood flour content was varied from 10 to 40 wt%. The wood flour provided low cost, low density, high specific strength, and biodegradability. On the other hand, talc had positive influence of modulus, strength, creep, and elastic recovery performance of the PLA hybrid green composites. A 19 mm corotating twin-screw extruder with 40 L/D ratio (Bautek Co., Uijeongbu City, Korea) and a pelletizer (Bautek Co., Uijeongbu City, Korea) were used to compound the blends. Silane was simply integral blended with other components without any additional treatment. The compounding temperature was 180°C with screw speeds in the range of 100–150 rpm. The extruded strands were air-cooled and pelletized with a pelletizer (Bautek Co., Uijeongbu City, Korea). The hybrid pellets were dried at 90°C for 24 h in a vacuum oven to remove the absorbed moisture and cooled to room temperature and then injection molded at 190°C to form test samples [15].

PLA/cellulose nanowhisker/nanoclay hybrid green composites were manufactured by Hong and Kim using melt mixing and compression molding technique.

Cellulose is abundant in nature and biodegradable. It provided good mechanical reinforcement in PLA. On the other hand, nanoclay provided improved mechanical, thermal, physical, and fire ratardancy properties to pure PLA. Melt mixing was conducted in Haake PolyDrive Rheomix R 600 mixer equipped with a roller blades rotating at 60 rpm. Melt mixing temperatures were set to 180, 190, and 200°C, respectively. PLA was melted for 5 min first. The fillers and additives were then loaded as fast as possible, and melt mixing of all the ingredients was followed. The content of cellulose fillers was fixed at 100 phr, while the nanoclay content was varied in between 0 and 5 phr. This was followed by compression molding in Carver hydraulic hot press at 180°C under 1000 psi pressure for 6 min [16].

Santangelo et al. prepared PLA/CNT/smectite clay hybrid green composites using melt mixing and hot compression. CNT provided higher strength and better thermal and electrical conductivity, while chemically modified smectite clay was compatible with organic polymer and provided low cost. In order to prepare PLA composites with varying (1–5 wt%) filler content, PLA and appropriate amount of filler powders were initially milled in the solid state in a Retsch centrifugal ball mill. Samples were milled for 1 h in a 50 cm^3 cylindrical steel jar with five steel balls of 10 mm diameter with rotating speed of 650 rpm. PLA sample milled in the absence of fillers was taken as a reference. In order to clarify the role of each filler, composites were prepared with fixed contents of either sole filler [17].

Eng et al. prepared PLA/nanoclay/polycaprolactone (PCL)/oil palm mesocarp fiber hybrid green composites by melt blending technique. The composition of PLA and PCL was kept constant at 85 and 15 wt%, respectively, in blend, while the clay content was kept constant at 1 wt%. Only the content of oil palm mesocarp fiber was varied from 0 to 30 wt%. Oil palm mesocarp fiber provided low cost, low density, and biodegradability, while the nanoclay addition resulted in improved mechanical and dynamic mechanical properties of the PLA hybrid green composites. PLA, PCL, nanoclay, and oil palm mesocarp fiber were manually premixed in a container and fed into Brabender Plastograph EC at 170°C with rotor speed of 50 rpm for 10 min. The products were then compression molded into sheets of 1 mm thickness by an electrically heated hydraulic press with a force of 1500 kN at 160°C for 10 min. The sample sheets were then used for further characterization [18].

6.2.2 Extrusion/Injection Molding

Extrusion and injection molding are the second most common technique applied until now for manufacturing of PLA hybrid green composites. Huda et al. manufactured PLA/talc/recycled newspaper hybrid green composites using extrusion and injection molding techniques. The talc and recycled newspaper filler content were set at 10 and 30 wt%, respectively. The newspaper cellulose filler provided low cost, low weight, and biodegradability, while the talc filler provided improved flexural, impact, and dynamic mechanical properties and better dispersion of fillers inside PLA. Prior to processing, the recycled newspaper cellulose fiber and PLA were dried under vacuum at 80°C for 24 h. The PLA matrix, however, was not dried. The polymer and the cellulose fibers were extruded at 100 rpm with a Micro 15 cc compounding system (DSM Research, Geleen, the Netherlands) at 183°C for 10 min. The extruder had a screw length 150 mm, a L/D of 18, and a net capacity of 15 cm^3. In order to obtain the desired specimen samples for various measurements and analysis, the molten composite samples were transferred after extrusion through a preheated cylinder to a mini-injection molder, which was preset with the desired temperature and cooling system. Injection-molded samples were placed in sealed polyethylene bags in order to prevent moisture absorption [19, 20].

Gorrasi et al. prepared PLA/clay mineral/CNT hybrid green composites using an extruder. Initially polymer and fillers were dried under vacuum at 40°C for 24 h before processing. The hybrid composites were prepared using a microextruder (Thermo Haake MiniLab Compounder) equipped with corotating screws and a closed loop for recirculation. The temperature was set at 180°C. The filler contents in the green hybrid composites were varied at 1, 3, 4, and 5 wt%. The clay mineral resulted in better mechanical and thermal properties, while CNT reduced photodegradation rate of the hybrid green nanocomposites. All the ingredients were mixed for 15 min at 150 rpm screw speed. The time was set minimum to that allowed proper filler distribution inside the matrix. Several hybrid composites were prepared by varying filler concentration. Finally all the extruded samples were molded in a hot press at 180°C for 3 min before cooling to room temperature [21].

Microfibrillated cellulose (MFC) and bamboo fiber-reinforced PLA hybrid green composites were manufactured by Okubo et al. using extrusion and injection molding techniques. Bamboo fiber bundles were used as primary load-carrying reinforcement, whereas MFC created an interphase in the polymer matrix around the bamboo fiber that prevented sudden crack growth. In order to investigate the influence of MFC dispersion on the properties of PLA, two different weight fractions (1 and 2 wt%) of MFC/PLA hybrid green composites were prepared. The MFC was first mixed by hand into the PLA and then processed using a laboratory-scale three-roll mill (EXAKT 80E, EXAKT Technologies, Inc.), consisting of three chrome-plated hardened steel rollers (80 mm in diameter) rotating at varying angular velocities where gap and speed settings on the mill are controlled electronically and the shearing force imparted by the rollers contributes to the separation of cellulose fibrils from the swollen and agglomerated state. The MFC/PLA mixture was processed in the three-roll mill at progressively decreasing gap settings of 70, 50, 35, 25, 15, 10, and 5 μm with the speed of the third roll fixed at 100 rpm. At each gap setting, the entire MFC/PLA mixture was passed through 10 times. During the milling process, deionized water was added in order to prevent the solution from drying on the roller surfaces. To assess the influence of processing on the MFC/PLA composite properties, two minimum gap settings of 35 and 5 μm were utilized to give varying levels of dispersion. For comparison, MFC/PLA composites were also produced without subsequent processing in the three-roll mill. After dispersion, the MFC/PLA mixtures were dried by casting the mixture in shallow pans and allowing the water to evaporate for several days. To produce test specimens made of MFC-reinforced PLA hybrid composites, a microscale injection molder was utilized in combination with a 15 cm^3 twin-screw extruder (DSM X- Plore). The PLA was mixed in the extruder at 140°C with a screw speed of 150 rpm for 5 min, and the molten material was then extruded into the preheated barrel on the injection molder [22].

Nyambo et al. prepared PLA/wheat straw/corn stover/soy stalk hybrid green composites with the help of an extruder and an injection molding machine. Wheat straw, corn stover, and soy stalk agricultural residues were cheap, which helped maintain the carbon dioxide balance and gave an incentive to the farmers. The hybrid agricultural residues also provided better mechanical properties to green composites as compared to single fiber/filler composite. Dry wheat straw, corn stover, and soy stalk fibers were ground in a Fritsch Universal Cutting Mill Pulverisette 19 and sieved through a mesh of size 0.25 mm. The PLA and the fibers were dried in an oven at 80°C for 4 h. A total of 30 wt% of the dried fibers and 70 wt% PLA were extruded in a DSM Micro 15 cc twin-screw compounder having a speed of 100 rpm for 3 min at 180°C for all three processing zones. A reference sample of unmodified PLA without any fiber and a hybrid composite consisting of PLA combined with 10 wt% of soy stalks, corn stover, and wheat straw was prepared using the same procedure. Test specimens were prepared by injection molding in a DSM Micro 12 cc molder. The molten composites obtained were transferred from the extruder to the injection molder using a cylinder preheated at 180°C [1].

6.2.3 Other Techniques

Xie et al. produced tricobalt tetraoxide and graphene-reinforced PLA hybrid green composites using *in situ* chemical reaction and melt mixing techniques. Initially tricobalt tetraoxide/grapheme nanofillers were synthesized with the help of chemical reaction process and then redispersed into the PLA matrix by melt mixing. Tricobalt tetraoxide/graphene hybrid nanofillers reduced the fire hazard of PLA when dispersed into it [23].

PLA/SiO_2/TiO_2 hybrid green composites were prepared by Yeh et al. using sol–gel technique. The hybrid inorganic SiO_2/TiO_2 provided higher hardness, durability, and thermal stability to flexible and tougher organic PLA polymer. Sodium metasilicate hydrate aqueous solution (60 g of SMS in 150 ml of water) was added dropwise into 150 ml of hydrochloric acid (3.6 mol=l) in a few minutes with stirring at 80°C, followed by further stirring for 15 min. Thereafter, 150 ml of THF and then 90 g of sodium chloride were added to this solution with stirring. After reaction for 30 min and then standing for 10 min, the solution was into converted two immiscible phases. The organic layer, silicic acid–THF solution designated as Sol A, was separated and dried with 30 g of anhydrous sodium sulfate. The silicic acid–THF solution extracted contained about 78.1 g/l of SiO_2. The mixture Sol B was prepared by dissolving a stoichiometric amount of TTIP, H_2O, HCl (as the catalyst), and acetic acid in isopropanol and then stirring at room temperature for 30 min to obtain a homogeneous solution. The two solutions (A and B) were mixed at a SiO_2/TiO_2 weight ratio of 1 : 1. The total content of SiO_2/TiO_2 was varied at 3, 7, 10, 13, and 20 wt%. The final mixture was stirred for 25 h at room temperature. For the preparation of hybrids with different amounts of SiO_2 and TiO_2, a determined amount of PLA was melted in a Brabender Plastograph 200 nm Mixer W50EHT instrument with a blade-type rotor (50 rpm) and a temperature of 190–200°C. When the PLA had melted completely, the SiO_2/TiO_2 solution was added to the mixer to continue the *in situ* sol–gel process for another 15 min. Prior to characterization, each sample was dried at 105°C in a vacuum oven for 3 days to remove residual solvents. The hybrid products were pressed into thin plates using a hot press at 190°C, and they were put into a dryer for cooling. Next the cooled plate was made into standard specimens for characterization [24].

6.3 PROPERTIES OF PLA HYBRID GREEN COMPOSITES

6.3.1 Mechanical Properties

Various researchers in the past reported improvement in mechanical properties of PLA hybrid green composites when fibers/fillers were incorporated into the matrix. Some of mechanical properties PLA hybrid green composites are described in the following.

6.3.1.1 Tensile Properties Yeh et al. prepared PLA/SiO_2/TiO_2 hybrid green composites by varying SiO_2/TiO_2 in between 0 and 20 wt% and found that the tensile strength increased up to 10 wt% SiO_2/TiO_2. The decrease in tensile strength at higher filler content was due to the separation between the organic and inorganic phases in the hybrid system [24].

Gorrasi et al. prepared PLA/clay mineral/CNT hybrid green composites and showed the effect of filler content and UV radiation time on the elastic moduli. The elastic modulus of the PLA hybrid green composites was higher than unfilled PLA due to the reinforcing effect of the fillers. The elastic modulus of pure PLA decreased with radiation time due to UV-induced chain cleavage. The introduction of hybrid fillers prevented the loss of mechanical consistency of PLA. The moduli of PLA hybrid green composites remained constant up to 300 h photodegradation. This might be explained by the formation of entangled and interconnected network of polymeric chains and hybrid fillers, surviving the chain cleavage and preserving the mechanical consistency of the samples up to very high irradiation time [21].

PLA/MMT/synthetic mica hybrid green composites were prepared by Chang et al. using solution casting technique. The tensile properties were subsequently evaluated. The ultimate tensile strengths of the PLA hybrid green composite increased remarkably with the filler content and possessed a maximum value for clay content of 4 wt%. Above the critical wt% filler loadings corresponding to the ultimate strengths, the strength values of all the hybrids started to decrease mainly due to the agglomeration of clay particles. The initial modulus increased linearly with increasing filler content up to a critical value of 4 wt%. When the filler content was greater than this critical value, the modulus started to decrease. The big enhancement of the modulus is ascribed to the resistance exerted by the filler itself, as well as to the orientation and the higher aspect ratio of the mica layers. Additionally, the stretching resistance of the oriented backbone of the polymer chain in the gallery also contributed to enhancing the modulus. The elongation at break of PLA clearly increased with the introduction of fillers. However, a certain wt% of filler loading corresponds to a maximum value of the elongation at break. Thus in a certain filler content range, the introduction of filler led to both strengthening and toughening of the PLA matrix [14].

Lee et al. prepared wood flour/talc-filled PLA hybrid green composites and evaluated their tensile properties. The tensile strength and Young's modulus increased with increase in both filler loading. 1% silane treatment also increased the tensile properties of treated filler-reinforced PLA hybrid green composites compared to the raw filler-reinforced PLA hybrid green composites [15].

MFC and bamboo fiber-reinforced PLA hybrid green composites were manufactured, and tensile properties were reported by Okubo et al. The tensile strength and stiffness increased slightly with filler content; however the fracture behavior of the PLA hybrid green composites became drastically different with the addition of filler. The unreinforced PLA and the PLA hybrid green composites where the cellulose was not milled show brittle fracture behavior with little or no plastic deformation. For both set of PLA hybrid green composites prepared using the calendaring approach, there was a significant increase in specimen ductility [22].

Nyambo et al. prepared PLA/wheat straw/corn stover/soy stalk hybrid green composites and found variation in tensile properties with fiber loading. The tensile strength of PLA hybrid green composites decreased with fiber loading from 10 to 40 wt% due to poor adhesion between the fiber and matrix. On the other hand, the tensile modulus increased with fiber loading due to stiffness effect of fiber reinforcement in PLA [1].

PLA/cellulose nanowhisker/nanoclay hybrid green composites were manufactured by Hong and Kim. Their report indicated that cellulose nanowhisker-reinforced hybrid green composites had higher tensile strength compared to the microcrystalline cellulose-reinforced PLA hybrid green composites [16].

Eng et al. prepared PLA/nanoclay/PCL/oil palm mesocarp fiber hybrid green composites by melt blending technique. The tensile strength and elongation at break were found to be decreased with increase in palm mesocarp fiber content in their work [18].

6.3.1.2 Flexural Properties Flexural properties are some of the important parameters influencing the performance of hybrid green composites in actual applications. Huda et al. manufactured PLA/talc/recycled newspaper hybrid green composites and reported flexural properties of the same. The flexural strength and flexural modulus increased with filler loading. This was due to better stress transfer between the filler and matrix and better dispersion of filler materials inside PLA [19, 20, 25]. Nyambo et al. prepared reported flexural properties of PLA/wheat straw/corn stover/soy stalk hybrid green composites in their work. The flexural strength decreased, while the flexural modulus increased with fiber loading [1]. PLA/PCL/oil palm mesocarp fiber hybrid green composites were prepared by Eng et al. They reported flexural properties from their work. Incorporation of 1% clay in PLA hybrid green composites resulted in better flexural strength. This might be due to due to better interfacial adhesion between the matrix and filler, which improved stress transfer from the matrix to the filler. When higher amount of palm fiber was incorporated into the hybrid green composites, the flexural strength decreased. This might be due to the increment of population of fiber defects and fiber ends with increased fiber content. The fiber defects might have also acted as source of stress concentration in hybrid composites [26]. The flexural modulus on the other hand increased with fiber loading due to higher stiffness of the palm fiber compared to the PLA matrix [18].

6.3.1.3 Impact Strength The impact strength is the ability of a material to withstand fracture or the amount of energy required to propagate a crack. Nyambo et al. prepared PLA/wheat straw/corn stover/soy stalk hybrid green composites and reported decrease in impact energy with addition of fiber. This might be due to decrease in toughness and elongation at break with fiber loading [1]. PLA/PCL/oil palm mesocarp fiber hybrid green composites were prepared by Eng et al. They reported a decrease in impact strength with fiber loading. When fiber was incorporated in the matrix, poor energy absorbing capability of the fiber inhibits deformation and ductile mobility of polymer molecules. This decreased the ability of composite to absorb energy during crack propagation. Besides, less energy was required to initiate a crack as fiber created high-stress concentration region [18, 27]. On the other hand, Huda et al. manufactured PLA/talc/recycled newspaper hybrid green composites using extrusion and injection molding techniques. They reported improved impact strength with filler loading due to better interfacial adhesion between the filler and PLA matrix [20, 28]. Table 6.1 shows the variation of mechanical properties of PLA-based green hybrid composites.

6.3.2 Dynamic Mechanical Properties

Dynamic mechanical thermal analyzer (DMTA) is a frequently used characterization machine for evaluating dynamic mechanical properties of hybrid green composites. Nyambo et al. prepared PLA/wheat straw/corn stover/soy stalk hybrid green composites and reported dynamic mechanical properties in their publication. The storage modulus increased with filler addition due to stiffening effect. However increase in temperature decreased the modulus due to softening of the PLA hybrid green composite. The glass transition temperature was not affected by filler addition. The tan delta peaks decreased in intensity with fiber addition. This fiber hindered the mobility of PLA chains, reducing the tan delta peak intensity [1, 29].

Shi et al. evaluated the dynamic mechanical properties of PLA/bamboo/talc hybrid green biocomposites manufactured using melt mixing and blending techniques. The storage modulus values found in their work increased with bamboo fiber and talc. The heat treatment of the green composites also had the same effect on storage modulus. On the other hand, glass transition temperature was not affected by fiber and talc addition, although the heat treatment increased that temperature of PLA hybrid green composites [13].

PLA/cellulose nanowhisker/nanoclay hybrid green composites were manufactured by Hong and Kim. The dynamic mechanical properties were evaluated later using DMTA. The storage modulus increased with the addition of nanowhisker and nanoclay as they improved the stiffness of PLA matrix. Like storage modulus, the

TABLE 6.1 Variation of mechanical properties of PLA-based green hybrid composites[a]

	Tensile properties			Flexural properties		
Type of composite	Tensile strength (MPa)	Young's modulus (GPa)	Elongation at break (%)	Flexural strength (MPa)	Flexural modulus (MPa)	Impact strength (J/m)
PLA/SiO_2/TiO_2	47–77	—	—	—	—	—
PLA/clay mineral/CNT	—	1.35–2.41	—	—	—	—
PLA/MMT/synthetic mica	17–44	0.17–0.64	550–1270	—	—	—
PLA/wood flour/talc	36–71	1.3–2.1	—	—	—	—
PLA/MFC/bamboo	45–54	3.5–4.8	—	—	—	—
PLA/wheat straw/corn stover/soy stalk	45–65	2.8–5.2	—	81–100	—	23–28
PLA/cellulose nanowhisker/nanoclay	26–46	—	—	—	—	—
PLA/nanoclay/ PCL/oil palm mesocarp fiber	22–46	0.58–0.90	—	12–42	1700–2700	30–180
PLA/talc/recycled newspaper	—	—	—	36–113	40–94	13.1–38.7

[a] From Refs. [1, 14–16, 18–22, 24].

glass transition temperature also increased with nanowhisker and nanoclay loading. This might be due to considerable restriction of mobility of PLA chains with the inclusion of nanomaterials inside PLA [16].

Eng et al. prepared nanoclay PLA/PCL/oil palm mesocarp fiber hybrid green composites and reported dynamic mechanical properties. There was increase in storage modulus with fiber and nanoclay addition due to the stiffening effect, whereas increase in temperature decreased the storage modulus. This was due to loss of stiffness at higher temperatures. The addition of fiber, on the other hand, decreased the tan delta values in the glass transition region as fiber decreased the molecular mobility of the PLA hybrid green composites. The mechanical loss to overcome interfriction between molecular chains was also reduced [18].

Huda et al. manufactured PLA/talc/recycled newspaper hybrid green composites and evaluated storage modulus and tan delta values with the help of DMTA. The storage modulus of PLA hybrid green composites was higher than that of the unfilled PLA matrix, which indicates the stress transfer from the matrix to the cellulose fiber [30]. The storage modulus increased with addition of talc because of the greater stiffness of the talc that generated a stiffer interface in the PLA matrix [31]. On the other hand, the height of the tan delta peak decreased with the presence of cellulose content. One possible explanation of this phenomenon is that neat PLA had a sharp and intense peak because there is no restriction to the chain motion, whereas the presence of the cellulose surfaces hindered the chain mobility, resulting in the reduction of sharpness and height of the tan delta peak [19, 20]. Table 6.2 shows the variation of dynamic mechanical and thermal properties of PLA-based green hybrid composites.

6.3.3 Thermal Properties

6.3.3.1 Thermogravimetric Analysis Thermogravimetric analysis (TGA) is a common technique of examining thermal properties of composites. PLA/MMT/synthetic mica hybrid green composites were prepared by Chang et al. using melt mixing technique. They mentioned that the thermal degradation temperature decreased with increase in clay content in PLA hybrid green composites [14]. Lee et al. reported that thermal decomposition temperature of wood flour/talc-filled PLA hybrid composites decreased with wood flour loading. This may be due to decomposition of wood components in the range of 200–380°C [15]. Eng et al. prepared nanoclay PLA/PCL/oil palm mesocarp fiber hybrid green composites. They mention that decomposition onset temperature increased with increase in clay content in the PLA hybrid green composites. Clay hindered the permeability of volatile degradation products, which increased thermal stability [18]. Huda et al. manufactured PLA/talc/recycled newspaper hybrid green composites and evaluated thermal properties using TGA. The thermal decomposition started in the range of 310–320°C, while thermal stability increased with fiber addition in PLA [19]. Gorrasi et al. prepared PLA/clay mineral/CNT hybrid green composites using an extruder. They reported that thermal degradation temperature decreased with increase in radiation time. The higher the radiation time, the faster the decomposition rate [32]. This was probably due to the formation

TABLE 6.2 Variation of dynamic mechanical and thermal properties of PLA-based green hybrid composites[a]

	Dynamic mechanical properties			Thermal properties	
Type of composite	Maximum storage modulus (MPa)	Maximum loss modulus (GPa)	Maximum tan delta	Maximum glass transition temperature (°C)	Maximum thermal degradation temperature (°C)
PLA/wheat straw/corn stover/soy stalk	4,500	860	1.90	61	—
PLA/cellulose nanowhisker/nanoclay	2,300	—	—	—	—
PLA/nanoclay/ PCL/oil palm mesocarp fiber	—	8200	2.80	—	221
PLA/talc/recycled newspaper	12,000	—	2.2	—	320
PLA/MMT/synthetic mica	—	—	—	—	370
PLA/wood flour/talc	—	—	—	63	380
PLA/SiO_2/TiO_2	—	—	—	68	370
PLA/clay mineral/CNT	—	—	—	59	—

[a] From Refs. [1, 14–16, 18–21, 24].

of low molecular weight species during photoaging [21]. PLA/SiO_2/TiO_2 hybrid green composites were prepared by Yeh et al. using sol–gel technique. They reported that initial decomposition temperature increased with SiO_2/TiO_2 loading [24].

6.3.3.2 Differential Scanning Calorimetry Thermal properties of organic/inorganic hybrids depend on the interaction of polymer chain and inorganic network and consequently uniform distribution of reinforced fillers [33]. Nyambo et al. prepared PLA/wheat straw/corn stover/soy stalk hybrid green composites in their work. They reported that addition of fiber in PLA matrix did not affect the melting and glass transition temperatures; however there was reduction in enthalpy of melting, enthalpy of crystallization, and crystallization temperature with the addition of same [1].

Lee et al. prepared wood flour/talc-filled PLA hybrid composites and reported thermal properties against wood flour loading. The glass transition temperature slightly decreased with wood flour, whereas the alkali treatment increased that temperature due to better adhesion between PLA and wood flour. Lee at al. also mentioned that the melting temperature also decreased with wood flour loading in the composites compared to PLA. This was due to the formation of voids in the PLA hybrid green composites [15].

Huda et al. manufactured PLA/talc/recycled newspaper hybrid green composites and evaluated thermal properties using differential scanning calorimetry (DSC). Addition of newspaper fiber did not change the melting and glass transition temperatures much. However talc-filled hybrid green composites had higher glass transition temperature compared to PLA. On the other hand, there was an increase in crystallization temperature with the addition of talc and newspaper in PLA [20].

Gorrasi et al. prepared PLA/clay mineral/CNT hybrid green composites and described their thermal properties. Compared to PLA, cold crystallization occurred at lower-temperature PLA hybrid green composites. This was most probably induced by the clay mineral/CNT that acted as nucleation agents in the system. The melting temperatures of hybrid green composites were also lower compared to pure PLA. However, the crystallinity was same for PLA and hybrid green composites. On the other hand, the glass transition temperature of PLA hybrid green composites increased after UV radiation treatment [21].

MFC and bamboo fiber-reinforced PLA hybrid green composites were manufactured by Okubo et al. The reported that there was no change in meting point, melt enthalpy, and degree of crystallinity when cellulose was added to PLA [22].

Yeh et al. prepared PLA/SiO_2/TiO_2 hybrid green composites by varying SiO_2/TiO_2 and showed the variation of thermal properties against SiO_2/TiO_2 content. SiO_2/TiO_2 hindered the motion of long-chain polymer segments, which in turn increased the glass transition temperature of PLA/SiO_2/TiO_2 hybrid green composites [24].

6.3.4 Surface Morphology

Surface morphology study of hybrid composites is very important as it indicates interaction and adhesion between the fiber/filler and the matrix. The degree of dispersion of the fiber/filler in the matrix can also be known from morphology. Nyambo et al. prepared PLA/wheat straw/corn stover/soy stalk hybrid green composites and evaluated surface morphology using scanning electron microscope (SEM). The SEM images show moderate dispersion of the fiber at high magnification and platelike structures at low magnification. The images showed evidence of voids and fiber pullout that might have contributed to the reduction of mechanical properties. Besides, there was clear evidence of fiber and fiber fracture in the wheat straw composites. This is reasonable because no compatibilizer was added and no surface modification of the fibers was performed to improve interfacial adhesion [1].

Lee et al. prepared silane-treated wood flour/talc-filled PLA hybrid green composites by melt compounding and studied surface morphology of the composites. The manufacturing method led to a uniform distribution of wood flour in the PLA matrix. The system without silane treatment showed a poor compatibility between the PLA matrix and wood flour. The surface of wood flour particle was believed to be delaminated from the PLA matrix, and microsized voids were formed during tensile testing. The silane treatment significantly improved the compatibility, leading to less filler/matrix debonding [15].

Huda et al. manufactured PLA/talc/recycled newspaper hybrid green composites and reported that the fibers were well dispersed in the PLA matrix. This indicates that the fibers were separated during the extrusion process. The findings also indicate better adhesion between the fiber and matrix [19, 20].

Okubo et al. manufactured MFC and bamboo fiber-reinforced PLA hybrid green composites and reported the morphological properties of manufactured composites. When the bamboo fiber specimen was produced with the unmilled MFC/PLA, the bamboo fiber was cut sharply on the fracture surface near the crack path. In the region of the fiber, the fracture surface was very smooth, indicating brittle crack propagation. In contrast, considerable fiber pullout was observed on the specimen where the MFC had been milled and had a rough fracture surface, suggesting that large plastic deformation occurred [22].

PLA/SiO_2/TiO_2 hybrid green composites were manufactured by Yeh et al. They reported that SiO_2/TiO_2 particles were agglomerated as a large cluster indicating poor dispersion of the particles and poor adhesion and compatibility between the particles and PLA. This was because the interfacial force between the phases was facilitated only by hydrogen bonds [24].

6.3.5 Electrical Properties

Evaluation of electrical properties of hybrid green composites is important when those are applied to electrical applications. Santangelo et al. prepared PLA/CNT/smectite clay hybrid green composites using melt mixing and hot compression. They reported that incorporation of hybrid clay-based fillers improved the electrical properties of the insulating pristine polymer, rendering PLA semiconductive [17]. Gorrasi et al. prepared PLA/clay mineral/CNT hybrid green composites and determined their electrical properties. The electrical conductivity of PLA was found constant at different irradiation times. However the electrical conductivity of the composites with hybrid fillers showed an increasing trend irradiation time. This was due to the formation of percolation pathway by the CNT [21].

6.4 APPLICATIONS OF PLA HYBRID GREEN COMPOSITES

Natural fiber/filler-reinforced biopolymers are nowadays replacing synthetic fiber/filler-reinforced polymer due to their advantages such as low density, renewability, biodegradability, and low cost [1]. These composites are attracting attention both in industry and in academia, because they allow complete degradation in soil and do not emit any toxic or noxious components [2]. Present filed of applications of PLA-based green hybrid composites include automotive industry, thermoelectrical industries, agriculture, packaging, biomedical engineering, plastic industry, and so on [1, 12, 17, 19]. Capability of various manufacturing techniques of producing these PLA hybrid green composites with tailor-made properties will propel new field of application in the near future.

6.5 CONCLUSIONS

Using biodegradable polymers as matrices, natural/synthetic hybrid fiber/filler-reinforced plastics become the most environmental friendly materials as they can be composed at the end of their life cycle. These hybrid green composites provide a good combination of properties and processability at a moderate cost. Among all biopolymers, PLA has the greatest potential to replace petroleum-based synthetic polymers. Some of the environmental benefits of PLA are its low production energy and reduction of greenhouse gas production. Because of its ability to be thermally crystallized, impact modified, stress crystallized, filled, copolymerized, and processed in most polymer processing equipment, it may be the best polymer with the broadest range of applications. The mechanical, dynamic mechanical, thermal, morphological, electrical, and other properties of PLA hybrid green composites described in the present chapter have already been evaluated. Thus the PLA hybrid green composites might find applications in a variety of research and industrial field in the near future.

REFERENCES

[1] Nyambo, C., Mohanty, A.K., Misra, M. (2010) Polylactide-based renewable green composites from agricultural residues and their hybrids. *Biomacromolecules* **11**, 1654–1660.

[2] Lee, S.H., Wang, S., Teramoto, Y. (2008) Isothermal crystallization behavior of hybrid biocomposite consisting of regenerated cellulose fiber, clay, and poly(lactic acid). *Journal of Applied Polymer Science* **108**, 870–875.

[3] Majeed, K., Jawaid, M., Hassan, A. et al. (2013) Potential materials for food packaging from nanoclay/natural fibres filled hybrid composites. *Materials & Design* **46**, 391–410.

[4] John, M.J., Thomas, S. (2008) Biofibres and biocomposites. *Carbohydrate Polymers* **71** (3), 343–364.

[5] Auras, R., Harte, B., Selke, S. (2004) An overview of polylactides as packaging materials. *Macromolecular Bioscience* **4**, 835–864.

[6] Carothers, W.H., Dorough, G.L., Van, N.F.J. (1932) Studies of polymerization and ring formation X. the reversible polymerization of six-membered cyclic esters. *Journal of the American Chemical Society* **54**, 761–772.

[7] Södergård, A., Stolt, M. (2002) Properties of lactic acid based polymers and their correlation with composition. *Progress in Polymer Science* **27**, 1123–1163.

[8] Stridsberg, K., Ryner, M., Albertsson, A.C. (2002) Controlled ring-opening polymerization: polymers with designed macromolecular architecture. *Advances in Polymer Science* **157**, 41–65.

[9] Jamshidi, K., Hyon, S.H., Ikada, Y. (1998) Thermal characterization of polylactides. *Polymer* **29**, 2229–2234.

[10] Anderson, K.S., Schreck, K.M., Hillmyer, M.A. (2008) Toughening polylactide. *Polymer Reviews* **48**, 85–108.

[11] Mark, J.F. (2009) *Polymer Data Handbook*, Oxford University Press, London.

[12] Antar, Z., Feller, J.F., Noel, H. et al. (2012) Thermoelectric behavior of melt processed carbon nanotube/graphite/poly(lactic acid) conductive biopolymer nanocomposites (CPC). *Materials Letters* **67**, 210–214.

[13] Shi, Q.F., Mou, H.Y., Li, Q.Y. et al. (2012) Influence of heat treatment on the heat distortion temperature of poly(lactic acid)/bamboo fiber/talc hybrid biocomposites. *Journal of Applied Polymer Science* **123**, 2828–2836.

[14] Chang, J.H., An, Y.U., Cho, D. et al. (2003) Poly(lactic acid) nanocomposites: comparison of their properties with montmorillonite and synthetic mica (II). *Polymer* **44**, 3715–3720.

[15] Lee, S.Y., Kang, I.A., Yoon, G.H.D.H. et al. (2008) Thermal and mechanical properties of wood flour/talc-filled polylactic acid composites: effect of filler content and coupling treatment. *Journal of Thermoplastic Composite Materials* **21**, 209–223.

[16] Hong, J., Kim, D.S. (2013) Preparation and physical properties of polylactide/cellulose nanowhisker/nanoclay composites. *Polymer Composites* **34** (2), 293–298.

[17] Santangelo, S., Gorrasi, G., Lieto, R.D. et al. (2011) Polylactide and carbon nanotube/smectite-clay nanocomposites: preparation, characterization, sorptive and electrical properties. *Applied Clay Science* **53**, 183–194.

[18] Eng, C.C., Ibrahim, N.A., Zainuddin, N. et al. (2014) Enhancement of mechanical and dynamic mechanical properties of hydrophilic nanoclay reinforced polylactic acid/polycaprolactone/oil palm mesocarp fiber hybrid composites. *International Journal of Polymer Science* Article ID 715801, DOI org/10.1155/2014/715801.

[19] Huda, M.S., Mohanty, A.K., Drzal, L.T. et al. (2005) Mechanical and thermo-mechanical studies of poly (lactic acid) PLA/talc recycled newspaper fiber hybrid composites, Global Environmental Conference, USA, Paper ID 33, pp. 1–12.

[20] Huda, M.S., Drzal, L.T., Mohanty, A.K. (2007) The effect of silane treated and untreated talc on the mechanical and physico-mechanical properties of poly(lactic acid)/newspaper fibers/talc hybrid composites. *Composites Part B* **38**, 367–379.

[21] Gorrasi, G., Miloni, C., Piperopoulos, E. et al. (2013) Hybrid clay mineral-carbon nanotube-PLA nanocomposite films. Preparation and photodegradation effect on their mechanical, thermal and electrical properties. *Applied Clay Science* **71**, 49–54.

[22] Okubo, K., Fujii, T., Thostenson, E. (2009) Multi-scale hybrid biocomposite: processing and mechanical characterization of bamboo fiber reinforced PLA with microfibrillated cellulose. *Composites Part A* **40**, 469–475.

[23] Xie, X., Li, Y., Liu, Z.Q. et al. (2009) Low-temperature oxidation of CO catalysed by Co_3O_4 nanorods. *Nature* **458**, 746–749.

[24] Yeh, J.T., Chai, W.L., Wu, C.S. (2008) Study on the preparation and characterization of biodegradable polylactide/SiO_2–TiO_2 hybrids. *Polymer-Plastics Technology and Engineering* **47**, 887–894.

[25] Nishino, T., Hirao, K., Kotera, M. et al. (2003). Kenaf reinforced biodegradable composite. *Composites Science and Technology* **63**, 1281–1286.

[26] Sawpan, M.A., Pickering, K.L., Fernyhough, A. (2012). Flexural properties of hemp fibre reinforced polylactide and unsaturated polyester composites. *Composites Part A* **43** (3), 519–526.

[27] Park, B.D., Balatinecz, J.J. (1997) Mechanical properties of wood fiber/toughened isotactic polypropylene composites. *Polymer Composites* **18** (1), 79–89.

[28] Riccieri, J.E., Carvalho, L.H.D., V'azquez, A. (1999) Interfacial properties and initial step of the water sorption in unidirectional unsaturated polyester vegetable fiber composites. *Polymer Composites* **20** (1), 29–37.

[29] Pothana, L.A., Oommenn, Z., Thomas, S. (2003) Dynamic mechanical analysis of banana fiber reinforced polyester composites. *Composites Science and Technology* **63**, 283–293.

[30] Rana, A.K., Mitra, B.C., Banerjee, A.N. (1999) Short jute fiber-reinforced polypropylene composites: dynamic mechanical study. *Journal of Applied Polymer Science* **71**, 531–539.

[31] Martinez, J.G., Laguna, M., Areso, O.S. et al. (2002) A dynamic–mechanical study of the role of succinyl-fluorescein grafted atactic polypropylene as interfacial modifier in polypropylene/talc composites: effect of grafting degree. *European Polymer Journal* **38**, 1583–1589.

[32] Islam, M.S., Pickering, K.L., Foreman, N.J. (2010) Influence of accelerated ageing on the physico-mechanical properties of alkali-treated industrial hemp fiber reinforced poly(lactic acid) (PLA) composites. *Polymer Degradation and Stability* **95**, 59–65.

[33] Wei, Y., Yang, D. (1992) Thermal stability and hardness of new polyacrylate-SiO_2 hybrid sol-gel materials. *Materials Letters* **13**, 261–266.

7

LIGNIN/NANOLIGNIN AND THEIR BIODEGRADABLE COMPOSITES

Anupama Rangan[1], M.V. Manjula[2], K.G. Satyanarayana[3] and Reghu Menon[2]

[1] *Department of Pharmaceutical Chemistry, Vivekananda College of Pharmacy, Bengaluru, Karnataka, India*
[2] *Department of Physics, Indian Institute of Science, Bengaluru, Karnataka, India*
[3] *Poornaprajna Institute of Scientific Research (PPISR), Bengaluru, Karnataka, India*

Biodegradable Green Composites, First Edition. Edited by Susheel Kalia.

7.1 INTRODUCTION

7.1.1 Renewable Bioresources-Sustainability and Biodegradability Issues

Lignocellulose is the most promising biomass that is abundant, renewable, and inexpensive and that has come into the limelight in the past decade. Indeed, in the recent years, the significance of this biodegradable resource for the enhancement and maintenance of industrial societies has gained immense importance. Developing a sustainable global economy by maximizing the utilization of the lignocellulosic mass is a challenge in the forefront of industrial research [1]. Many methods involving mechanical, physical, physicochemical, chemical, and enzymatic technologies are being explored to utilize lignocellulose for varied applications across industries. An increased effort in the field of bioresources has led to a considerable improvement in the use of green biotechnology related to lignocellulose biomass. The ever increasing costs of fossil fuels and their greenhouse effects are a major concern about global warming. Hence, there is a core demand to explore alternative cheaper and eco-friendly energy resources [2]. One such effort is the renewable energy Directive of 2009/28/EC of the European Parliament and of the Council of 23 April 2009, according to which a common framework for the production and promotion of energy from renewable sources was to be established and has set mandatory renewable energy targets for each member state (http://europa.eu/legislation_summaries/energy/renewable_energy/en0009_en.htm). Further, each member state is required to meet 20% of the European Union's overall energy consumption from renewable energy resources by 2020. In 2007, the biomass energy corresponded to around 67% of the European

TABLE 7.1 Percent composition of lignocellulose components in various lignocellulosic materials[a]

Lignocellulosic material	Lignin (%)	Hemicelluloses (%)	Cellulose (%)	Reference
Sugarcane bagasse	20	25	42	[6]
Sweet sorghum	21	27	45	[6]
Hardwood	18–25	24–40	40–55	[5]
Softwood	25–35	25–35	45–50	[5]
Corn cobs	15	35	45	[7]
Corn stover	19	26	38	[8]
Rice straw	18	24	32.1	[7]
Nut shells	30–40	25–30	25–30	[9]
Newspaper	18–30	25–40	40–55	[10]
Grasses	10–30	25–50	25–40	[5]
Wheat straw	16–21	26–32	29–35	[11]
Banana waste	14	14.8	13.2	[12]
Bagasse	23.33	16.52	54.87	[13]
Sponge gourd fibers	15.46	17.44	66.59	[13]

The above data has been taken from the references mentioned in the last column. Details of these references can be seen in the original paper referred above.

[a] From Iqbal et al. [14], Table 1, p. 3159. Reproduced with permission from BioResources Journal.

Union gross renewable energy consumption and this was expected to grow significantly in response to the ambitious newly adopted directive [3].

Lignocellulose chiefly consists of cellulose, hemicellulose, lignin, and pectin as the major constituents with resins, waxes, proteins, and extractives as minor constituents [4, 5]. The percentage composition of lignocellulose components in various lignocellulosic materials is shown in Table 7.1 [14].

Cellulose, being most abundant in nature, has been extensively studied. Cellulose in various forms has been in the limelight for applications across many industries. After cellulose, lignin is the second most abundant natural renewable biopolymer [15]. Most of the studies carried out with respect to lignin suggest that the amount and the composition of lignin vary predominantly with the source of origin; but several factors such as growing, harvesting, and drying of the lignocellulosic biomass and the pretreatment methods can have a significant effect on the final composition of extracted lignin [16]. In the recent years, lignin is regarded as a raw material with a high recovery potential, accessible from renewable sources, and with low costs and a negligible environmental pollution hazard [17, 18].

One of the application areas of the earlier discussed biomass product, that is, lignin, is in polymer composites. These composites reinforced with biopolymer fillers such as lignin having dimensions in the nanometer range are known as "nanocomposites." These have attracted greater interest from researchers than the microcomposites of similar materials due to unexpected synergistic properties derived from the two components. Most of the industrial applications of plant products such as in papermaking and ethanol production from biomass produce lignin as a major by-product. Development of lignin-based composites has been extensively studied as the impressive properties of lignin, such as its high abundance, low weight, environmental friendliness, and its antioxidant, antimicrobial, and biodegradable nature, along with its CO_2 neutrality and reinforcing capability, make it an ideal candidate for the development of novel polymer composite materials [19]. More details on the lignin-based composites are discussed later (see Section 7.5).

7.1.2 Nanotechnology and Application of Nanotechnology (Specifically for Cellulose and Lignin)

In the last few decades, as nanotechnology is revolutionizing the use of materials, application of lignocellulosic biomass holds a great promise to provide the key materials platform for the sustainable production of renewable, recyclable, and environmentally preferable goods and products. As proposed by Wegner and Jones [20], nanotechnology can be used to tap the enormous undeveloped potential that trees possess—as photochemical "factories" that produce rich sources of raw materials using sunlight and water. It has become critically important to not only produce products and goods for better quality of life but also produce these materials by concepts of sustainability. In this context, the merging of nanotechnology and lignocellulosic biomass utilization is vital.

Using lignocellulosic products based on nanotechnology can impart greater functionality whereby there will be reduction in the use of materials in production. This is not only applicable to wood-based products but also to any product for which current methods produce hazardous waste by-products. Lignin, an undervalued natural polymer, is rapidly gaining importance, especially with the application of nanotechnological innovations. There are many patents that are filed covering mostly the production of nanolignin, which helps in the realization of the vast potential of nanolignin [21, 22].

Considering the previously mentioned information on ecological/environmental necessities, which in turn lead to the need for the use of renewable bioresources to take care of sustainability of the resource and biodegradability issues and the role played by nascent nanotechnology for the better use of lignin, this chapter presents the following: (i) details on lignin and its types, structure, and properties, (ii) method of synthesis of nanolignin and its characterization, and (iii) applications of lignin as polymer composites and current applications in the development of green composites and potential applications of nanolignin in various sectors, particularly in medical sector as multi-functional nanoparticles. Finally, perspectives of lignin-based materials and concluding remarks are included.

7.2 LIGNIN

7.2.1 Structure, Chemical Nature, Complexity, and Linkage Heterogeneity

Chemically, lignin is a highly branched polyphenolic polyether. It consists of a complex structure that results from the condensation of the phenyl propane units obtained from the main precursors like *p*-hydroxy phenyl alcohol, guaiacyl alcohol, and

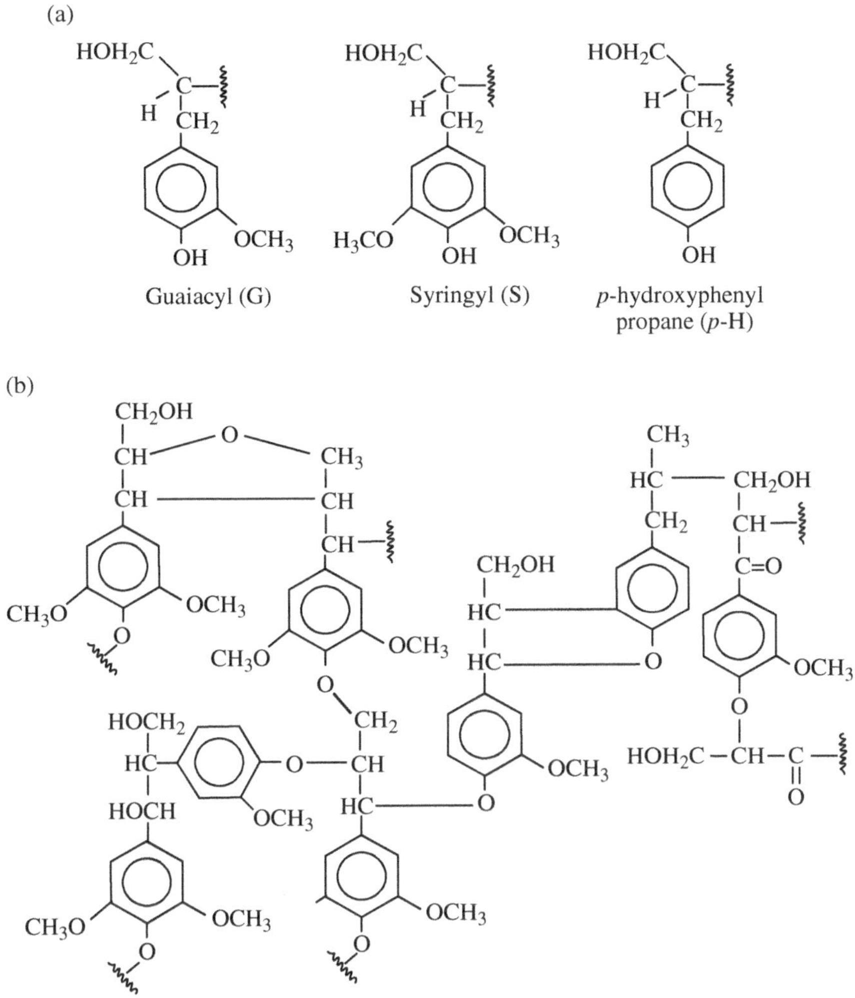

FIGURE 7.1 (a) Schematic representation of the structural units of lignin and (b) structure of lignin proposed by Alder. *Source*: Tejado et al. [23]. Reproduced with permission from Elsevier.

syringyl alcohol. The differential amounts of these three hydroxyl cinnamyl alcohols along with different degrees of methoxylation/ethoxylation lead to structural diversity in lignins. Both schematic representation of the structural units of lignin and the structure of lignin as proposed by Adler are shown in Figure 7.1 [23].

As can be seen from the figure, there are both carbon–carbon and carbon–oxygen bonds that occur between the monomer units in lignin. The carbon–oxygen links between a *p*-hydroxy moiety and the β-end of the propenyl group (β–O–4) are the most common bonds between monomer units in lignin. Depending upon the type of lignin (whether it is procured from soft wood or hard wood), varied degree of cross-linking is possible in lignin, and depending upon the degree of substitution, the rigidity of the structure varies [24].

An interesting aspect of lignin is that it carries both phenolic and aliphatic hydroxyl groups that can be used for further modification or polymerization. Unlike cellulose, lignin is aromatic, which can be an advantage for various applications. The presence of these hydrophobic and hydrophilic functional moieties makes it an attractive polymer to work with for any suitable application.

According to the International Lignin Institute, about 40–50 million tonnes of lignin per annum is produced worldwide (http://www.ili-lignin.com/aboutlignin.php). In spite of the huge amounts of lignin, which are readily available, lignin is normally considered as a nonpreferred by-product [25]. Lignin's very limited solubility and complex structure having very broad distribution of molecular weight and microstructure have been the potential reasons for considering lignin as a waste product from biomass. In view of these, only 2% of this is used in agricultural or other sectors as binders (pellets for animal feeds, bricks, ceramics, etc.) [26, 27]. As one sees, a growing demand for traditional applications (paper manufacture, biomass fuels, composting, animal feed, etc.), newer applications, and potential markets for lignin has been identified in the recent years. The anticipation of energy and environmental crises is forcing to reevaluate the efficient utilization of natural resources. It is also compelling researchers to find alternative uses for natural, renewable resources using clean technologies. Thus, lignin holds considerable potential and can meet the current energy demand of the modern world. In a world where there is excessive dependence on petroleum, it becomes essential to discover alternative sources for liquid fuels. In addition, advances in the area of biotechnology could help in such an endeavor with discovery and characterization of new enzymes and production in homologous or heterologous systems. The final outcome should be a low-cost conversion of lignin into biofuels and biochemicals. Application of nanotechnology, green biotechnology, and novel technological advances in methods and processing of lignin can be the main focus of the future research that can help in the earlier mentioned endeavors.

Currently lignin is processed and used for many applications; majority of it is as low-value products, which are dependent on the source and type of process [28, 29]. Listed in the succeeding text are a few types of commercially available lignin and their applications.

7.2.2 Types, Structure, Properties, and Uses of Modified/Processed Lignin

Structural heterogeneity and the chemical complexity of lignin have been a challenge for the characterization of its naturally occurring form. Accordingly, various general characterization techniques have been used to understand lignin in its totality. Since chemical composition is dependent on various factors, obtaining a consistent reference standard for quantification has been found to be a difficult task. Indeed many labs use Indulin AT (earlier marketed by Sigma-Aldrich Corporation, USA, or Westvaco, USA) or Protobind (from ALM India Pvt. Ltd.) for this purpose. Qualitatively, lignin-based materials have been evaluated by chemical methods or ultraviolet (UV) spectroscopy [30]. Although most of the properties of lignin can be determined, the difficulty arises in determining accurately its molecular weight due to varying functionalities and high cross-linking. However, the methods that have been usually used for the characterization of lignin nanoparticles will be discussed later (Section 7.4).

Different types of lignins are produced during processes that involve various treatments of lignocellulosic materials. These are called "technical lignins," which are readily available in large amounts. These include kraft lignin, lignosulfonates, and soda lignin. In contrast, hydrolysis, organosolv, and ionic liquid lignins constitute only a small part of potentially valuable lignins. They are produced in rather small amounts, but have a potential to be developed into industrial-scale products [31]. A summary of the chemical composition of various technical lignins mentioned earlier is shown in Table 7.2 [32], while some details of these are given in the succeeding text:

7.2.2.1 ***Kraft Lignin*** This is produced in the sulfate (kraft) cooking process [23]. This lignin contains higher amounts of phenolic hydroxyl groups due to extensive cleavage of β-aryl bonds during processing. Kraft lignin is applied for the production of low-tonnage or pilot-scale products.

Examples of kraft lignin-based products are carriers for fertilizers and pesticides [33, 34], carbon fibers [35, 36], blends with thermoplastic polymers [31, 37], binders and resins [23, 38, 39], ion-exchange resins [40, 41], and activated carbons [42, 43]. In addition to these applications, kraft lignin can be used for the production of low-molecular-weight compounds such as vanillin, hydroxylated aromatics, quinines, aldehydes, and aliphatic acids.

7.2.2.2 ***Soda Lignin*** This is obtained from soda or soda–anthraquinone pulping processes. Soda lignin is sulfur-free. Thus the chemical composition of soda lignin is closer to that of natural lignin in comparison with kraft lignin and lignosulfonates [32, 44, 45].

Soda lignin is used for the production of phenolic resins [23], animal nutrition [32, 46], and dispersants [44, 47].

7.2.2.3 ***Lignosulfonates*** These are water-soluble anionic polyelectrolytes that contain a large number of charged groups. Lignosulfonates have a variety of functional groups: phenolic hydroxyl groups, carboxylic groups, and sulfur-containing groups [48]. Unlike kraft lignin and soda lignin, lignosulfonates are soluble in water, have a relatively high molecular weight with a broad distribution of molecular weights, and quite high ash content [32, 49]. Lignosulfonates have unique colloidal properties. This is due to the presence of a variety of functional groups and other structural features [48, 50]. Due to these properties, lignosulfonates are used as stabilizers in colloidal suspensions, dispersing agents [32, 51], binders for drilling agents [32, 52], detergents glues and animal feeds [53], particleboards [32, 54], surfactants [32, 55], adhesives [38, 49], and cement additives [56, 57].

TABLE 7.2 Chemical composition of "technical" lignins[a]

Parameter	SL	KL	HL	OSL	LS	ILL
Ash (%)	0.7–2.3	0.5–3.0	1.0–3.0	1.7	4.0–8.0	0.6–2.0
Moisture content (%)	2.5–5.0	3.0–6.0	4.0–9.0	7.5	5.8	—
Carbohydrates (%)	1.5–3.0	1.0–2.3	10.0–22.4	1–3	—	0.1
Acid-soluble lignin (%)	1.0–11	1–4.9	2.9	1.9	—	—
Nitrogen (%)	0.2–1.0	0.05	0.5–1.4	0–0.3	0.02	—
Sulfur (%)	0	1.0–3.0	0–1.0	0	3.5–8.0	1.5
Molecular weight (M_W)	1,000–3,000 (up to 15,000)	1,500–5,000 (up to 25,000)	5,000–10,000	500–5,000	1,000–50,000 (up to 150,000)	≈2,000
Polydispersity	2.5–3.5	2.5–3.5	4.0–11.0	1.5	4.2–7.0	—

The above data has been taken from many references, which can be seen in the original paper referred above.

HL, hydrolysis lignin; ILL, ionic liquid lignin; KL, kraft lignin; LS, lignosulfonates; OSL, organosolv lignin; SL, soda lignin.

[a]From Vishtal and Kraslawski [32], Table 1, p. 3552. Reproduced with permission from BioResources Journal.

7.2.2.4 Organosolv Lignin These are obtained from organosolv pulping process. In this process, a mixture of organic solvents such as acetic acid, formic acid, ethanol, and peroxy-organic acids and water is used as a cooking medium [58]. Although ionic liquids such as alkylbenzene sulfonate have also been used to dissolve lignins [59], it has not been applied on commercial scale yet.

Organosolv lignin is used as a filler in the formulations of inks, varnishes, and paints [60].

7.2.2.5 Hydrolysis Lignin These are obtained from either dilute acid hydrolysis [61] or more commonly by pretreatments combined with subsequent enzymatic hydrolysis and fermentation of wood [62]. They are highly porous, thus making them suitable for application as sorbents.

Table 7.3 lists an outline of commercial lignin-based products and their applications with specific details on market volumes and values.

7.3 NANOLIGNIN AND METHODS OF PREPARATION OF NANOLIGNIN

As research in the field of nanolignin is still in infancy, successful lignin utilization may be possible through the application of various advances in the field of nanotechnology. There is a considerable effort in this endeavor and the various methods for the preparation of nanolignin or nanolignin-based products have been published in recent times [67]. These are discussed in the succeeding text:

7.3.1 Precipitation Method

Frangville et al. [67] have developed a method for the fabrication of novel biodegradable nanoparticles from lignin, which are apparently nontoxic for microalgae and yeast. The results of this study indicate that the stability of the lignin nanoparticles is dependent on the pH of the solution before precipitation. Two alternative methods have been compared for the synthesis of lignin nanoparticles in this study. These methods are schematically shown in Figure 7.2. In the first method (a), lignin nanoparticles are precipitated from a solution of Indulin AT in ethylene glycol by the slow addition of aqueous hydrochloric acid. These precipitated nanoparticles can cross-link to form pH-stable lignin nanoparticles after dialysis. In the second method (b), lignin is dissolved in aqueous sodium hydroxide and the nanoparticles of lignin are precipitated by the addition of nitric acid. It was observed that the lignin nanoparticles produced by route (b) dissolve above pH 5, while the ones produced by route (a) are stable up to pH 10, even without the cross-linking step. These lignin nanoparticles produced by route (a) were found to have densely packed lignin domains. This explains their stability at higher pH. The lignin nanoparticles obtained by route (b) were highly porous, consisting of smaller lignin domains.

TABLE 7.3 List of commercial lignin-based products and their applications with specific details on market volumes and values[a]

Function	Application	Typical addition level of lignosulfonates	Potential/current volume	Estimated/current volume
Binder and adhesive	Animal feed pellets (also as lubricant)	1–4%		Current $0.05/lb [63]
	Ores			
	Molding applications			
	Printing inks			
	Foundry cores			
	Minerals			
Additive	Concrete admixture		Current: $4.00–5.00/gal as an aqueous solution [63]	
	Oil well drilling muds		Current $1.00–2.00/lb [63]	
	Tanning agents			
	Rubbers			
	Industrial cleaners	0.05–2.0% in water [64]		
	Cement manufacture (dispersant and grinding aid)	0.02–0.2% on dry cement [64]		
	Animal feed molasses	Up to 11% approved by the US FDA (1992) [64]		
	Papermaking additives (bleaching, deinking, scale inhibition, hydrogen peroxide transport stabilization)		Potential: 40 million lb/year [63]	
Dispersant	Carbon black	3–8% [64]		
	Pesticides	2–10%	Current: 18,000–20,000 tonnes/year [65]	
	Gypsum wallboard	1–3 lb per 100 ft of ½ in. wallboard [63]		
	Inks and pigments			
	Dyestuffs	Up to 50% [65]	Current: 25,000 tonnes/year (90% of the total market) [65]	
	Bricks	0.5–2.0% on green clay [64]		
	Ceramics			
	Refractories	4–10% on refractory solids [64]		

Emulsifier, stabilizer, extender	Road surfaces (dust control)		
	Wax/water and oil/water emulsion	2.5–3.5% [64]	
	Asphalt emulsion	1–2% by weight [65]	
	Soils		
Metal sequestering agent	Industrial water, water treatment (e.g., in boilers and cooling towers)	1–1000 ppm [64]	Potential: over 20 million lb/year [63] Estimates: $0.50/0.60/lb for liquids (sales price for aminopolycarboxylates) [63]
	Agricultural micronutrients		
	Water treatment		
Coreactant	Phenol–formaldehyde and urea–formaldehyde resins	10–15% resin solids replaced [64]	Current: small Potential: significant
Others	Negative plate expander in lead acid batteries		
	Vanillin manufacture		
	Kitty litter		
	Fertilizers		
	Protein coagulants		
	Protective colloids in steam boilers		

[a] From Toumi et al. [66]. http://www.star-colibri.eu/files/files/Deliverables/D2.1-Report-19-04-2010.pdf and Garkulak and Lebo [28]. Reproduced with permission from Star-COLIBRI.

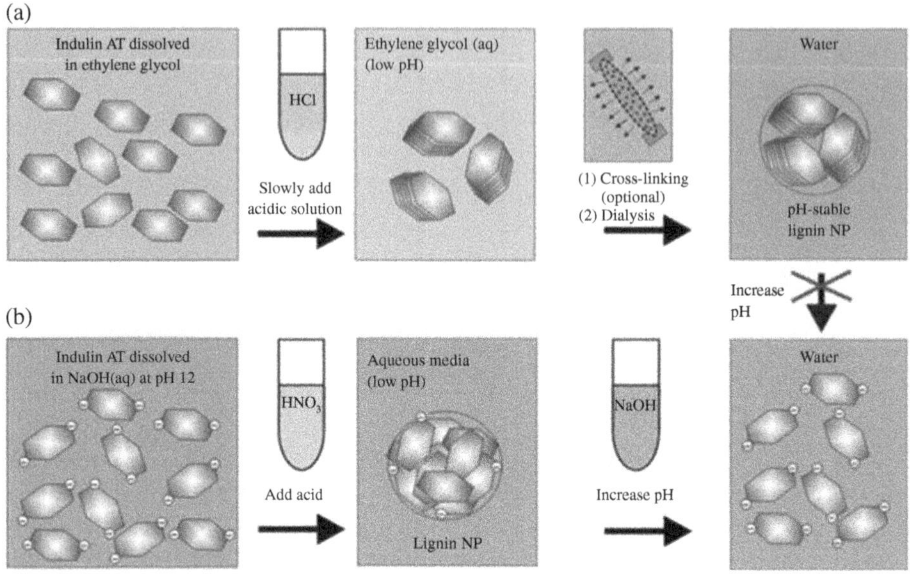

FIGURE 7.2 Schematic representation of preparation routes for lignin nanoparticles: (a) precipitation of lignin nanoparticles from a solution of lignin in ethylene glycol and (b) precipitation of lignin nanoparticles from an alkaline solution of lignin. *Source*: Frangville et al. [67]. Reproduced with permission from Wiley-VCH Verlag GmbH & Co. KGaA.

7.3.2 Chemical Modification Method

Lignin nanoparticles have been obtained by hydroxymethylation of lignin under special conditions and by the amendment of commercial products through epoxidation [18]. These hydroxymethylated lignin-based nanoparticles were characterized by dimensional distribution. The nanoparticles obtained through epoxidation of the Protobind commercial-type lignin were identified in the separated supernatants. Lignin-based nanoparticles modified in the presence of copper ions showed high biological wood stability, as assessed by mass loss and contact angle determinations [68]. The veneer samples analyzed after burial in soil for 6 months showed low mass loss values and high contact angle values. Chemical treatment leads to high hydrophobicity resulting in long-term stability.

7.3.3 Electrospinning Followed by Surface Modification

A series of ionic responsive poly(*N*-isopropylacrylamide) (PNIPAM) surface-modified lignin nanofiber mats were prepared by aqueous surface-initiated atom transfer radical polymerization (SI-ATRP) [69]. By adjusting the initial monomer concentration and surface initiator density, PNIPAM brushes with various molecular weights, thicknesses, and grafting densities were immobilized on electrospun lignin nanofiber mats. ATR-FTIR, scanning electron microscopy (SEM), thermogravimetric analysis (TGA), XPS, and water contact angle measurements were carried out that confirmed

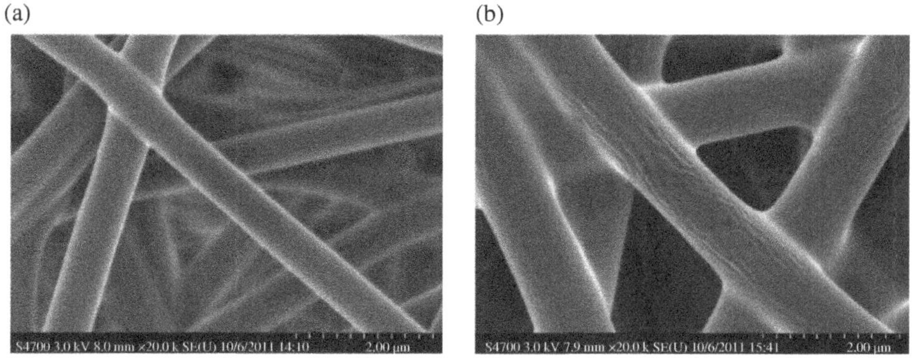

FIGURE 7.3 SEM images of lignin nanofiber mat surface before (a) and after (b) PNIPAM surface modification. *Source*: Gao et al. [69]. Reprinted with permission from American Chemical Society.

successful surface modification. This study showed that the PNIPAM-modified lignin nanofiber mats (Lig-PN) had lower critical solution temperature (LCST) and was similar to that of PNIPAM. Surface modification also led to ionic responsive characteristics. As the ionic concentration was increased, the water contact angles of the Lig-PN increased correspondingly. The change in the surface morphology of the lignin nanofibers is shown in Figure 7.3. As seen by these SEM figures, an increase in lignin nanofiber mat diameter is clearly evident after SI-ATRP. The authors of this study proposed that these nanofibrous functionalized materials may be used for the development of thermoresponsive separation and purification devices.

7.3.4 Freeze Drying Followed by Thermal Stabilization and Carbonization

Carbon nanoparticles have been synthesized using lignin as a renewable feedstock by employing a freeze-drying process followed by thermal stabilization and carbonization [70]. The whole process is shown schematically in Figure 7.4. As shown in the figure, aqueous solution of lignin was treated with various concentrations of potassium hydroxide, sonicated and solidified using liquid nitrogen. The solidified lignin samples were freeze-dried in order to achieve porous lignin. A gradual heating up to 250°C led to thermal stabilization of the freeze-dried lignins. It was observed that the gradual heating helped retain the porous microstructure during the subsequent carbonization process. These thermostabilized lignin samples were carbonized in a tubular furnace at 700°C in a nitrogen atmosphere whereby carbon nanoparticles were obtained.

7.3.5 Supercritical Antisolvent Technology

This is a patented method that involves multiple steps involving dissolution and precipitation. The black liquor from paper industry is refined with dissolution and precipitation techniques and finally dried at a low temperature to obtain refined industrial

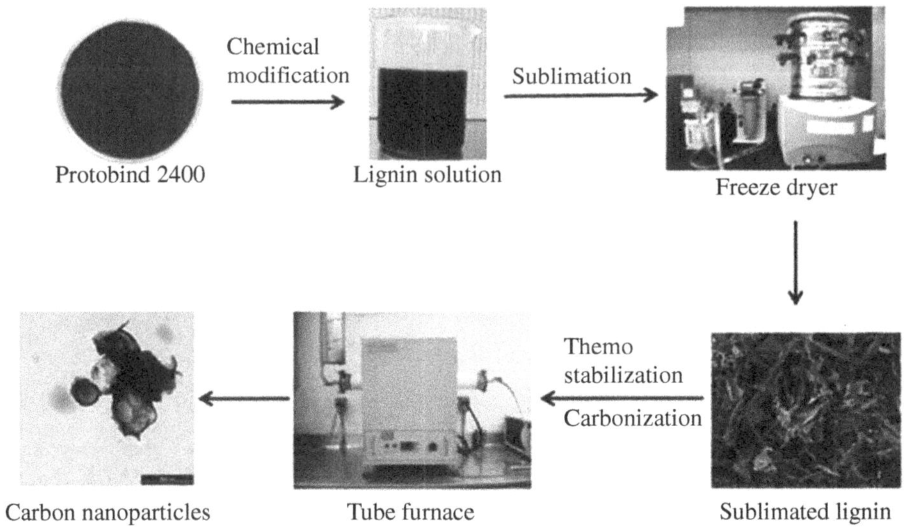

FIGURE 7.4 Schematic of the various steps involved in the synthesis of carbon nanoparticles from lignin. *Source*: Gonugunta et al. [70], Figure 1, p. 149. Reproduced with permission from World Journal of Nano Science Engineering. (*See insert for color representation of the figure.*)

lignin [22]. The dried lignin is then dissolved in dioxane, followed by centrifugal separation of the solution to obtain insoluble lignin. This insoluble lignin can be reused. Nanolignin particles have been prepared from the lignin–dioxane solution by using a carbon dioxide supercritical device and adopting a supercritical antisolvent method. It is reported that the operation process has the following main characteristics: (i) the raw material is papermaking industrial waste pollutant whereby the pollution problem of the papermaking industry could be solved; (ii) high-performance nanolignin is obtained; (iii) there will not be any pollution in this method since the supercritical antisolvent technology is adopted and the solvent can be recycled; and finally (iv) the product obtained would not have any solvent residue. All these make this method to open up a broad prospect for wider applications of the lignin.

7.3.6 Chemomechanical Methods

Another patent has been obtained for the preparation of nanolignin with a controllable particle size, which involves the following steps: (i) alkali lignin is pretreated, (ii) this is followed by grinding and preparation of an alkali lignin suspension liquid, (iii) the alkali lignin is converted into lignin sol, and (iv) then this lignin sol is dried to obtain nanolignin [21]. This method has the following advantages: (i) controlled particle size between 10 and 70 nm with an average particle size of about 30 nm is obtained, (ii) the highest particle content would have a particle size of about 26 nm, and finally (iii) it is a simple processing method with an easy operation that is, most importantly, free of environmental pollution.

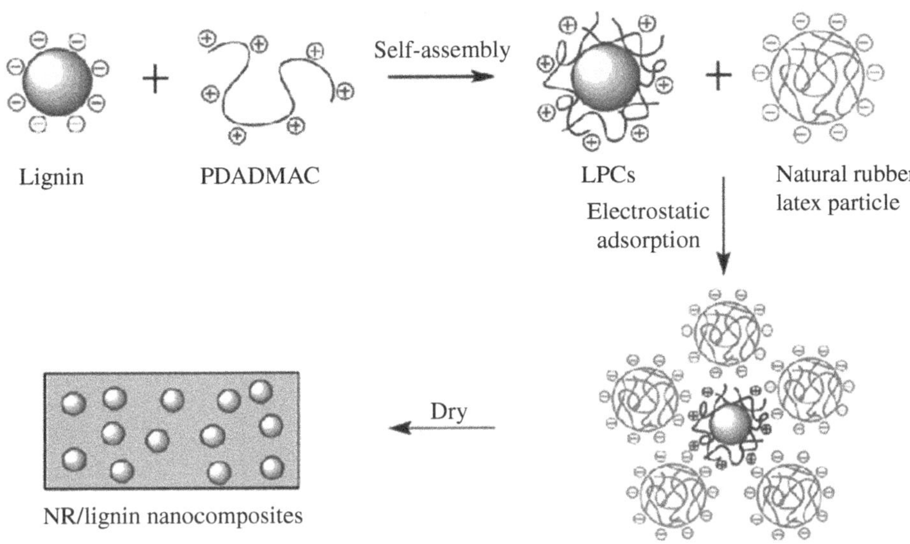

FIGURE 7.5 Schematic illustration of process for NR/lignin nanocomposites. *Source*: Jiang et al. [71], Figure 3. Reproduced with permission from eXPRESS Polymer Letters. (*See insert for color representation of the figure.*)

7.3.7 Nanolignin by Self-assembly

A novel strategy has been proposed to prepare nanolignin and its composites with natural rubber [71]. Schematically the process is shown in Figure 7.5. The purified industrial sulfate lignin in an alkaline solution of pH 12 was slowly added to a 2% solution of an alkaline solution of poly(diallyldimethylammonium chloride). The resulting solution was coprecipitated by adjusting the pH to 2 with sulfuric acid. Finally, the obtained mixtures were filtered, washed with water, and dried in vacuum oven at 50°C. In this process, it has been reported that the nanolignin could be obtained by fabricating colloidal lignin–poly(diallyldimethylammonium chloride) complexes (LPCs) via self-assembly technology. Additional analysis of the LPCs indicated intensive interaction of the polymer with lignin by cation–π and π–π interactions. These nanolignin particles (average particle size <100 nm) were found to be stable in aqueous solution. The nanoscale dispersions of LPCs significantly enhanced thermal stability and mechanical properties of natural rubber and LPC-based nanocomposites.

7.3.8 Template-Mediated Synthesis of Lignin-based Nanotubes and Nanowires

Vermerris and coworkers, using a method to obtain lignin-based nanotubes and nanowires, covalently linked isolated lignin to the inner walls of activated alumina membranes and then added layers of dehydrogenation polymer onto

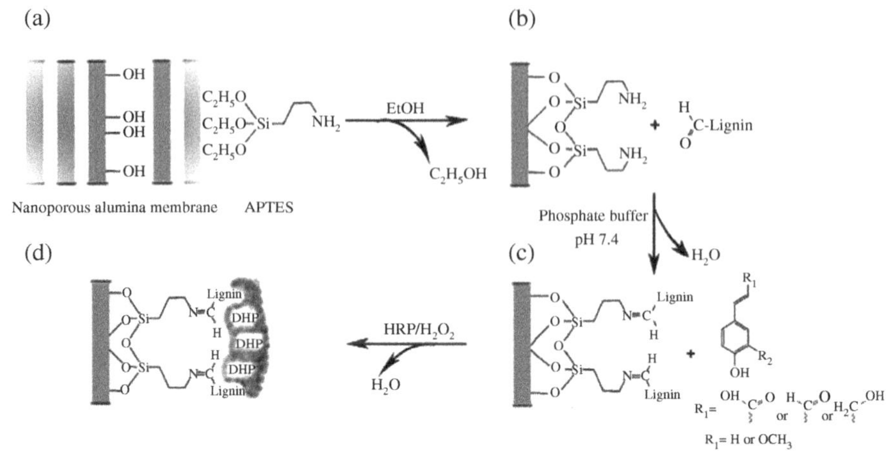

FIGURE 7.6 Schematic illustration of the synthesis of lignin nanotubes. (a) Activation of alumina membrane with APTES; four vertically oriented pores are shown in this image, but for the sake of clarity the reactive sites are only shown on one of the pore walls. The inactivated (sputter-coated) upper and lower surfaces of the membrane are indicated by the dark-gray caps. (b) Schiff's base reaction between the amino groups on the APTES-activated membrane and aldehyde moieties in the thioglycolate lignin forms a base layer; only one pore wall is displayed. (c) Hydroxycinnamic acids, hydroxycinnamaldehydes or hydroxycinnamyl alcohols are added and (d) polymerize onto the lignin base layer through a reaction catalyzed by horseradish peroxidase (HRP)/H^2O^2. DHP = dehydrogenation polymer. *Source*: Caicedo et al. [72]. Reproduced with permission from IOP.

this base layer via a peroxidase-catalyzed reaction [72]. This is schematically shown in Figure 7.6. The method involves activation of alumina membrane with (3- aminopropyl)triethoxysilane (APTES). This is followed by a Schiff's base reaction between the amino groups on the APTES-activated membrane and aldehyde moieties in the thioglycolate lignin. In the next step, hydroxycinnamic acids, hydroxycinnamaldehydes, or hydroxycinnamyl alcohols are added and polymerized onto the lignin base layer through a reaction catalyzed by horseradish peroxidase and hydrogen peroxide. With this method, it has been possible to change the thickness of the polymer layer deposited within the pores by using phenolic monomers that had different reactivities. It has been reported that the synthesized nanotubes had a wall thickness of approximately 15 nm and the nanowires had a nominal diameter of 200 nm. These studies have shown that the novel nanotubes were flexible and could be biofunctionalized easily and specifically, as shown by in vitro assays with biotin and concanavalin A. In addition, it was observed that their intrinsic optical properties could also be varied as a function of their chemical composition. Thus it has been proposed that these lignin-based nanotubes could be applied for a variety of new applications including drug delivery.

7.4 CHARACTERIZATION OF LIGNIN NANOPARTICLES

It is generally known that the qualitative identification and isolation of lignin has been the focus of most studies involving lignin. As mentioned earlier (see Section 7.2.2), many of the characterization methods used for lignin have also been used in the case of nanolignin despite the structural heterogeneity and chemical complexity these present. It must be noted that although qualitative characterization is usually carried out in most lignin studies, quantitative analysis has been a challenge due to structural heterogeneity and the chemical complexity of lignin, besides the absence of a global reference standard for lignin. Therefore, some of the methods that have been usually used for their characterization are given in the succeeding text:

7.4.1 Microscopy

Physical methods of characterization include surface morphology studies by microscopy. SEM, transmission electron microscopy (TEM), and atomic force microscopy (AFM) have been commonly used for morphological characterization. With standard procedures for the preparation of samples, these methods are very fundamental in assessing the formation of nanoparticles and for ascertaining their size and shape.

Considering the difficulty in the use of optical microscope to study the images in the micrometer range, SEM seems to be the better choice since it allows for clear perception of the fine structure details. SEM image of lignin nanotubes and nanowires formed by template-mediated biosynthesis provides insight into the surface characteristics of these particles. Figure 7.7 shows SEM images of lignin nanotubes showing the porous nature and the surface morphology of the nanotubes [72].

As the size of the particles falls into the nanometer category, further magnification of the particles is necessary to ascertain their size and shape. This is usually carried out with TEM and AFM. TEM images provide topographical, morphological, compositional, and crystalline information. The images allow researchers to view samples at a molecular level, making it possible to analyze structure and texture. Figure 7.8 shows the TEM images of lignin nanoparticles obtained by the acid precipitation of 0.56 wt.% Indulin AT in ethylene glycol and after dialysis. Figure 7.8a shows the high-resolution image of lignin nanoparticles after being negatively stained with uranyl acetate. The image indicates that lignin molecules might be clustered in stack-like structures. On the other hand, Figure 7.8b shows the low-magnification TEM image of the same, indicating that the domains of small lignin molecule clusters form highly porous nanoparticles [67].

AFM is a widely used tool for imaging, measuring, and manipulating matter at the nanoscale. AFM images provide insight into the surface morphology and texture of the lignin nanofibers. Moreover, surface modifications at nanoscale can be visualized by AFM techniques. AFM techniques were also used to study the effect of ion concentration on the surface-modified lignin nanofiber mats [69]. This is shown in Figure 7.9. AFM studies were conducted in air, water, and 0.5 M Na_2SO_4 aqueous solution. Dry PNIPAM-modified lignin nanofibers were clearly seen in air. However,

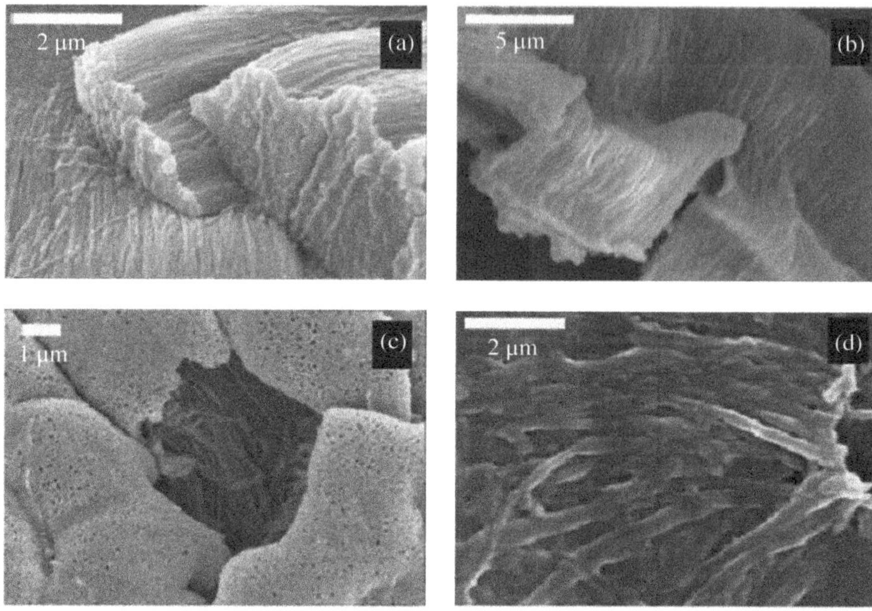

FIGURE 7.7 SEM images of lignin nanotubes illustrating (a) polymer growth both in the pores and on the top and bottom surfaces of the template (bar = 2 μm), (b) pliable bundles of lignin nanotubes reflect the material's flexibility (bar = 5 μm), (c) Au/Pd sputter-coating of the top and bottom of the membrane suppresses surface synthesis, as shown by the clean membrane surface observed in a partially dissolved membrane fragment that enables visualization of the lignin nanotubes (bar = 1μm) and (d) free stabilized lignin nanotubes after the membrane has been dissolved (bar = 2 μm). *Source*: Caicedo et al. [72]. Reproduced with permission from IOP.

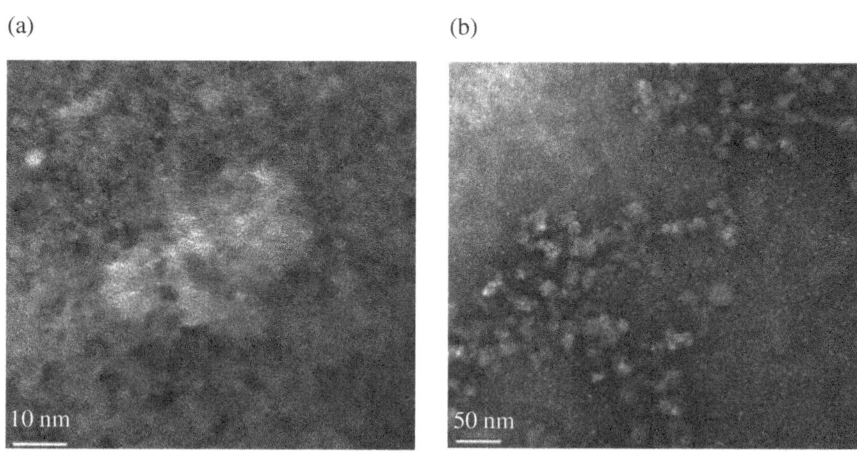

FIGURE 7.8 TEM images of lignin nanoparticles obtained by acid precipitation: (a) high-resolution image and (b) low-magnification image. *Source*: Frangville et al. [67]. Reproduced with permission from Wiley-VCH Verlag GmbH & Co. KGaA.

FIGURE 7.9 AFM images of PNIPAM modified lignin nanofibers (Lig-PN-2). (a) Lig-PN-2 in air, (b) Lig-PN-2 in water, and (c) Lig-PN-2 in 0.5 M Na_2SO_4 aqueous solution. *Source*: Gao et al. [69]. Reprinted with permission from American Chemical Society. (*See insert for color representation of the figure.*)

in water, the nanofibers were not easily recognized. This suggests that the extended PNIPAM brushes covered the profile of lignin nanofibers and obscured the images. The nanofibers were clearly visible when these fibers were exposed to an aqueous solution of 0.5 M sodium sulfate. This study indicates that the ionic concentration induces expansion and contraction in the lignin nanofiber surface grafted PNIPAM brushes.

7.4.2 Thermal Analysis

Thermal analytical methods are emerging to be indispensable in polymer nanomaterial research. Differential scanning calorimetry (DSC), TGA, dynamic mechanical thermal analysis (DMTA), and thermal mechanical analysis (TMA) are some of the methods commonly used for the characterization of nanopolymers and nanocomposite materials [73]. As dispersion is minimized to a nanoscale, properties such as melting, crystallization, cure kinetics, and glass transition present a peculiar change. To study the phenomena occurring during a thermal scan of nanofillers and polymer nanocomposites, DSC has been widely applied [74]. Vyazovkin had used DSC to study the influence of nanoparticles on the morphology of polymer matrices. DMTA allows for the measurement of stiffness and energy losses as a function of temperature, and hence it is frequently used in nanocomposite characterization [75]. The variations in the degree and the scale of dispersion of nanofillers can be easily studied using DMTA techniques. TMA has been mainly used to measure the coefficients of thermal expansion (CTE) of nanocomposite materials to compare them with those of the matrix. On the other hand, TGA has been used to analyze the effect of the introduction of nanofillers into a polymer matrix on the thermal stability of the polymer [76].

7.4.3 X-Ray Diffraction

X-ray diffraction offers a number of different dedicated methods to investigate nanostructures. These are X-ray reflectometry (XRR), high-resolution X-ray diffraction (HRXRD), X-ray diffuse scattering (XDS), in-plane grazing incidence diffraction (IP-GID), small angle X-ray scattering (SAXS) in transmission, and grazing incidence SAXS in reflection (GISAXSR).

XRR determines layer thickness, roughness, and density, while HRXRD helps to verify layer thickness, roughness, chemical composition, lattice spacing and mismatches, relaxation, and so on. On the other hand, XDS is used to determine lateral and transversal correlations, distortions, density, and porosity; while the IP-GID is used to study lateral correlations of the thinnest organic and inorganic layers, and depth profiling. SAXS in transmission or GISAXSR is used to determine the size, shape, distribution, orientation, and correlation of nanoparticles present in solids or solutions [77].

7.4.4 Other Methods

Some of the other methods used by various researchers to characterize nanolignin include particle size analyzers/particle shape analyzers, which are particularly useful in applications that include colloids, emulsions, nanoparticles, and submicron suspensions. These analyzers determine the size distribution of solid or liquid particulates, and some of these may also measure zeta potential, molecular weight, and second viral coefficient [78, 79].

7.5 LIGNIN COMPOSITES/NANOLIGNIN-BASED "GREEN" COMPOSITES

Considerable efforts are now being made to effectively utilize waste lignin as one of the components in polymer matrices for high-performance composite applications. Different types of lignin-reinforced polymer composites starting from synthetic to biodegradable polymer matrices have been prepared [80, 81]. Details of some of these efforts are given in the succeeding text:

7.5.1 Lignin-based Thermoplastic Polymer Composites

Thermoplastic composites consisting of polystyrene (PS) polymers reinforced with 0–80 wt.% lignin have been prepared by blending them through compounding in an internal batch mixer with and without the addition of a linear triblock copolymer based on styrene–ethylene/butylene–styrene (SEBS) [82]. These composites have been characterized by mechanical and rheological properties as well as morphologies. SEM studies of the composites have helped determine the state of dispersion and the interfacial adhesion between the lignin particles and the PS matrix. In the case of mechanical properties, with the increase in lignin content in the blend, increase in both the flexural and torsion moduli and decrease in the tensile properties have been observed. Further, the addition of a compatibilizer has been reported to improve the tensile properties of the blend. Analysis of the shear rheological behavior of the lignin–PS blends has revealed that the viscosity, dynamic moduli, and activation energy were very sensitive to contents of both lignin and compatibilizer.

7.5.2 Rubber-based Lignin Composites

One of the main applications of lignin that have been reported is in the development of rubber-reinforced lignin composites. In fact, there are some patents filed particularly in the United States in this regard [83–86]. Most of these patents are related to improvement of properties of rubber such as tack. More details of these can be seen in the references mentioned earlier. Other recent studies on these composites are briefly mentioned in the succeeding text:

A number of polymer composite systems have been prepared using styrene–butadiene rubber (SBR) as the matrix and lignin as reinforcement along with layered double hydroxide (LDH) by melt compounding method [87]. In these studies it was observed that the incorporation of lignin in the rubber composites significantly improved the mechanical properties such as tensile strength, elongation at break, modulus, and hardness of the composites. Briefly, the method involves the synthesis of lignin–LDH complex by *in situ* method. This was followed by preparation of SBR/lignin–LDH composites by melt compounding method. It was observed by X- ray diffraction analysis that crystal lignin–LDH was successfully obtained, while good dispersion of lignin–LDH in SBR matrix was observed in TEM studies. Further improved properties such as tensile strength, elongation at break, 300% modulus, and hardness of R have also been reported for lignin–LDH/SBR composites compared to those of LDH/SBR composites. On the other hand, thermal studies have indicated reduction of the degradation temperature of the lignin–LDH/SBR by 10% weight loss, whereas 50% weight loss (T_{50}) was much higher than that of pristine LDH/SBR. These results have been attributed to the barrier property of the well-dispersed lignin–LDH in SBR matrix.

Another study has reported the development of natural rubber latex-based composites incorporating lignin derived from rubber wood as filler with varying loading and its influence on the oil resistance, tensile strength, and morphology of the resulting films [88]. It has been reported that the addition of lignin above 2 wt.% enhanced oil resistance of lignin-filled rubber latex films, while the tensile strength values also increased by 60 and 73% at lignin loading of 10 and 12 wt.%, respectively. Observed increase in tensile strength values has been explained as due to the chain entanglement and better dispersion of lignin in the rubber matrix observed in the microscopy studies.

7.5.3 Lignin-reinforced Biodegradable Composites

Biodegradable polymers such as polybutylene succinate (PBS)-based composites have been prepared using lignin as reinforcement by melt mixing method using varying amounts of lignin and polymeric methylene diphenyl diisocyanate (PMDI) compatibilizer [89]. Improved tensile and flexural properties of the composites have been reported for the 65% lignin-incorporated PBS composites. Further simultaneous enhancement of tensile, flexural, and impact strength values has been reported with the incorporation of 1% PMDI in 50% lignin-filled PBS composites. In addition, increase in heat deflection temperature (HDT) of the matrix has been observed with the incorporation of both lignin and PMDI. These observed results have been explained based on the SEM studies. Figure 7.10 shows fractographs of the lignin–PBS composites obtained by SEM for PBS–50% lignin composite (Fig. 7.10a and c) and PBS–65% lignin composite (Fig. 7.10b and d) at two different magnifications, while Figure 7.10e is that for the PBS–50% lignin composite containing 1% PMDI at one magnification.

7.5.4 Lignin-reinforced Foam-based Composites

Studies on the preparation and characterization of the mechanical properties of water-blown rigid biofoam-based composites prepared from soy oil-based polyurethane and different amounts of lignin (0–15 wt.%) as reinforcement have been reported [90]. First, a novel soy-based polyurethane biofoam (BioPU) from two polyols (soybean oil-derived polyol SOPEP and petrochemical polyol Jeffol A-630) and poly(diphenylmethane diisocyanate) (pMDI) was prepared. Then, using lignin (from bioethanol production) as a reactive reinforcing filler, the composites were prepared with and without lignin and also using water as a blowing agent. The composites thus produced were characterized by density and thermal and mechanical properties. The density of the lignin-reinforced BioPU samples increased with increasing lignin content, with composites with 10% lignin exhibiting the best mechanical properties. In fact, it was observed that compressive and flexural strength values increased from 0.39 ± 0.05 MPa to 0.46 ± 0.02 MPa and from 0.54 ± 0.12 to 0.87 ± 0.17, respectively, when the lignin content was increased from 0 to 10%, respectively. Similarly, the glass transition temperature (T_g) and storage modulus have been found to increase with increasing lignin content compared to the neat foam composites without lignin. These observations were confirmed by dynamic mechanical analysis (DMA) results also. Thus, it can be concluded that simple method used in this study helps one in exploring the potential of using lignin in polyurethane applications.

7.5.5 Lignin-based Composite Coatings

Production of composite coatings with organosolv lignin (Lig) has been reported using silver-doped hydroxyapatite (Ag/HAP) and HAP coatings on titanium by an electrophoretic deposition method [91]. The authors have opined that higher lignin concentrations have protected the HAP lattice during sintering, thus improving the stability of coating. Indeed, lignin has been shown to have a significant effect on microstructure, morphology, and thermal behavior of prepared biocomposite coatings. Additionally, these composite coatings were analyzed for corrosion stability, cytotoxicity, and antimicrobial efficiency, which suggested that Ag/HAP/Lig could

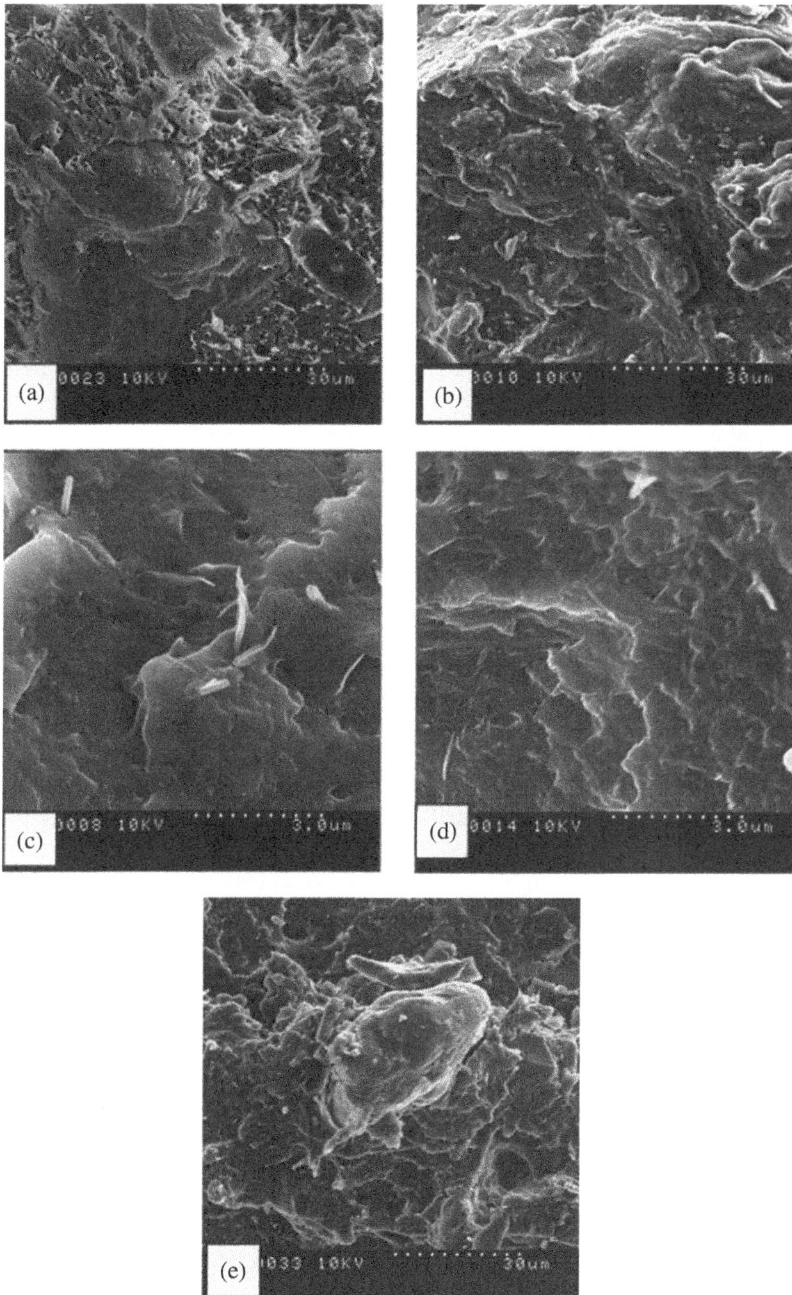

FIGURE 7.10 SEM fractographs of PBS–lignin composites: (a) 50% lignin content (lower magnification) and (c) the same at higher magnification; (b) 65% lignin content (lower magnification) and (d) the same at higher magnification; (e) PBS–50% lignin + 1% PMDI (lower magnification). *Source*: Sahoo et al. [89]. Reprinted with permission from Elsevier.

be classified as nontoxic against healthy immunocompetent peripheral blood mononuclear cells (PBMCs). The authors have proposed that the biopolymer lignin would allow the formation of a compact, well-adherent, and homogeneous coating due to the establishment of different hydrogen bonds between the functional groups of hydroxyapatite and lignin. The toxicity studies highlighted that lignin, a natural polymer, could provide a significant advantage over the brittle hydroxyapatite for applications in bone implantation.

7.5.6 Synthesis of Lignin–PLA Copolymer Composites

Considering the fact that very often difficulties are faced to blend lignin with other biopolymers besides improving the miscibility of lignin with other bioplastics, Chung and coworkers have developed a catalytic and solvent-free method for the synthesis of a lignin-based polylactic acid (PLA) copolymer [92]. Graft polymerization of lactide onto lignin through catalysis using triazabicyclodecene has been attempted in this study. By varying the lignin/lactide ratio and reacetylation treatment, to get a lignin-*g*-poly(lactic acid) copolymer with control of chain length of the PLA has been found to be possible. An end group analysis has revealed high grafting efficiency and preferential grafting on lignin aliphatic hydroxyls over phenolic hydroxyls. These lignin-*g*-PLA copolymers have been used as dispersion modifiers in PLA-based materials to enhance UV absorption and reduce brittleness without sacrificing the value of Young's modulus.

7.5.7 Nanolignin-based "Green" Composites

Although the previously mentioned examples of lignin-based composites are used for applications in industry, preparation of such bio-based composites using nanolignin would put lignin utilization in the high-value product category. Recent studies have highlighted this perspective. Del Saz-Orozco and coworkers have reported the preparation and characterization of nanolignin-reinforced foam-based composites preparing different formulations of the phenolic foams [93]. Further, optimization of lignin nanoparticle-reinforced foams was done using analysis of variance approach. It was observed that the incorporation of nanolignin significantly improved the mechanical properties of the foams. For example, the compressive modulus and compressive strength were found to increase up to 128 and 174%, respectively, compared to the unreinforced foams. Because the area of nanolignin-based green composites is a nascent area of research, it provides great opportunities to be explored.

7.6 POTENTIAL APPLICATIONS OF LIGNIN/NANOLIGNIN

A successful and efficient utilization of lignin remains a daunting task regardless of the intended applications. Meeting these challenges will be the key to the successful application of lignin in technological endeavors. Accordingly some current applications such as in composites and "green" composites have already been mentioned in previous sections of this chapter; some future potential applications of lignin such as nanomedicine and multifunctional nanoparticles are briefly discussed in the succeeding text:

The application of nanotechnology to modify drug delivery appears to be a promising approach. The absorption and bioavailability of drug molecules play a vital role in their efficacy. The reduction in particle size to nanoscale can indeed lead to increased dissolution of drugs and further lead to increased bioavailability. Nanoparticles can be used for both controlled and targeted drug delivery. Their smaller size, controlled drug release potential, targeting ability, enhancement of therapeutic efficacy, and reduction of toxicity are some of their important attributes. Nanoparticle drug delivery in cancer chemotherapy enhances therapeutic effectiveness and reduces side effects of the drug payloads by improving their pharmacokinetics. Currently, several nanoparticle-based chemotherapeutics have emerged on the market, while many are undergoing various stages of clinical or preclinical development [94]. Advances in biomaterials research have led to the emergence of biocompatible and biodegradable polymeric nanoparticles for drug delivery applications. Several synthetic polymers approved by the US FDA such as poly(lactic-*co*-glycolic acid) (PLGA) and polycaprolactone (PCL) and several natural polymers such as chitosan and polysaccharides have been investigated extensively for nanoparticle synthesis [95, 96]. Lignin, being a biopolymer possessing inherent antioxidant properties and versatile functional moieties on the surface, is an ideal candidate to explore the option of using lignin nanoparticles as drug delivery carriers. In cancer chemotherapy and nanoparticle-based combination therapy, which includes coadministration of multiple drugs, the versatility and hydrophilic and hydrophobic nature of lignin molecules can be of great advantage. Compared with liposome, polymeric nanoparticles generally have higher stability, sharper size distribution, more tunable physicochemical properties, sustained and more controllable drug release profiles, and higher loading capacity for poor water-soluble drugs. The lignin platform also offers higher synthetic freedom that allows particles to be tailored for specific needs.

7.7 PERSPECTIVES AND CONCLUDING REMARKS

The global market for nanotechnology was valued at about US$ 20 billion in 2011, while the total sales are expected to reach about US$ 49 billion by 2017. The global market for nanofiber-based products increased from $128.3 million in 2011 to $151.7 million in 2012 (http://www.bccresearch.com/market-research). Thus, the criteria for cleaner and safer environment have directed scientific research efforts towards biopolymer-based materials that can be easily degraded and bio-assimilated towards the end of their life cycle. Lignin as the second most abundant natural polymer can and should be developed as a rich resource. The present limitations and challenges in the utilization of lignins can be side-stepped by modifying the lignin particles with respect to their size and chemical nature. Lignin-based nanocomposites are a fast-growing area of research. The properties of lignin, its surface morphology, and interfacial characteristics play a crucial role in the synthesis and application of lignin nanocomposites. Apart from the composites, lignin nanoparticles have been used as natural UV barriers and antibacterial agents in the finishing process of protective textiles [97].

The general perception is that lignin is amorphous and the structural heterogeneity that arises due to polymerization of the various alcohols and their methoxylations limits the application of lignin in the high-value product category. These disadvantages may be overcome by the conversion of macro- or microscopic lignins into nanostructures. As it is evident in many cases, there is a significant change in the physicochemical properties of the materials as they are converted into nanoforms. Thus, coupling their distinct biodegradability, naturally less toxicity, and extensive functionalization capacity for surface modifications with nanostructure, the potential of lignins can be unleashed in the future. Lignin, being chemically distinct from polysaccharides and triglycerides, are not easily attacked by fungal or bacterial enzymes. This feature can be of added value for a wide range of lignin applications.

ACKNOWLEDGMENTS

The authors are grateful to Prof. Susheel Kalia, editor of *Biodegradable Green Composites*, for inviting them to contribute this chapter and to the publisher John Wiley & Sons, Inc. They express their sincere thanks to all the publishers (American Chemical Society, Elsevier, eXPRESS Polymer Letters, IOP Publishing, Wiley-VCH Verlag GmbH & Co., World Journal of Nano Science and Engineering) and their contact persons, Rights Links, who have helped in getting the necessary permissions, and thank all the authors who have given their kind permission to reproduce the figures and tables used from their published work free of cost. Prof. Anupama Rangan would like to thank Janatha Educational Society, Rajajinagar, Bengaluru 560055 (India), for their motivation and encouragement for research. Dr. K. G. Satyanarayana would like to express sincere thanks to Poornaprajna Institute for Scientific Research (PPISR), Bengaluru, with whom he has been associated, for their encouragement.

REFERENCES

[1] Kumar, P., Barrett, D.M., Delwiche, M.J., and Stroeve, P. (2009). Methods for pretreatment of lignocellulosic biomass for efficient hydrolysis and biofuel production. *Ind. Eng. Chem.*, **48**, 3713–3729.

[2] Shahzadi, T., Mehmood, S., Irshad, M., Anwar, Z., Afroz, A., Zeeshan, N., Rashid, U., and Sughra, K. (2014). Advances in lignocellulosic biotechnology: A brief review on lignocellulosic biomass and cellulases. *Adv. Biosci. Biotechnol.*, **5** (3), 246–251.

[3] Uslu, A. Gomez, N.C., and Belda, M.S. (2010). Demand for lignocellulosic biomass in Europe-Policies, supply and future demand for lignocellulosic biomass. Published by **Elobio**- Energy research Centre of the Netherlands (ECN). From: http://www.elobio.eu/fileadmin/elobio/user/docs/Additional_D_to_WP3-Linocellulosic_biomass.pdf. Accessed on October 20, 2014.

[4] Deobald, L.A. and Crawford, D.L. (1997). Lignocellulose biodegradation. In: Hurst C.J., Knudsen G.R., Stetzenbach L.D., and Walter M.V. (Eds) *Manual of Environmental Microbiology* (pp. 730–737). Washington, DC: ASM Press.

[5] Matherbe, S. and Cloete, T.E. (2002). Lignocellulose biodegradation: Fundamentals and applications. *Rev. Environ. Sci. Biotechnol.*, **1**(2), 105–114.

[6] Kim, M. and Day, D.F. (2011). Composition of sugarcane, energy cane, and sweet sorghum suitable for ethanol production at Louisiana sugar mills. *J. Ind. Microbiol. Biotechnol.*, **38**(7), 803–807.

[7] Prasad, S., Singh, A., and Joshi, H.C. (2007). Ethanol as an alternative fuel from agricultural, industrial and urban residues. *Resour. Conserv. Recycl.*, **50**, 1–39.

[8] Zhu, Y., Lee, Y.Y., Elander, R.T. (2005). Optimization of dilute acid pretreatment of corn stover using a high solids percolation reactor. *Appl. Biochem. Biotechnol.*, **121–124**, 1045–1054.

[9] Abbasi, T. and Abbasi, S.A. (2010). Biomass energy and environmental impacts associated with its production and utilization. *Renew. Sustain. Energy Rev.*, **144**, 919–937.

[10] Howard, R.L., Abotsi, E., Jansen van Rensburg, E.L., and Howard, S. (2003). Lignocellulose biotechnology: Issues of bioconversion and enzyme production. *Afr. J. Biotechnol.*, **2**, 603–619.

[11] McKendary, P. (2002). Energy production from biomass: Overview of biomass. *Bioresour. Technol.*, **83**, 37–43.

[12] John, F., Monsalve, G., Medina, P.L.V., and Ruiz, C.A.A. (2006). Ethanol production of banana shell and cassava starch. *Dyna. Univ. Nacl. Colomb.*, **73**, 21–27.

[13] Guimaraes, J.L., Frollini, E., Da Silva, C.G., Wypych, F., and Satyanarayana, K.G. (2009). Characterization of banana, sugarcane bagasse and sponge gourd fibers of Brazil. *Ind. Crop Prod.*, **30**(3), 407–415.

[14] Iqbal, H.M.N., Kyazze, G, and Keshavarz, T. (2013). Advances in the valorization of lignocellulosic materials by biotechnology: An overview. *BioResources*, **8**(2), 3157–3176.

[15] Boerjan, W., Ralph, J., and Baucher, M. (2003). Lignin biosynthesis. *Annu. Rev. Plant Biol.*, **54**, 519–546.

[16] Ross, K. and Mazza, G. (2010). Characteristics of lignin from flax shives as affected by extraction conditions. *Int. J. Mol. Sci.*, **11**(10), 4035–4050.

[17] Matsushita, Y., Wada, S., Fukushima, K., and Yasuda, S. (2006). Surface characteristics of phenol–formaldehyde–lignin resin determined by contact angle measurement and inverse gas chromatography. *Ind. Crop. Prod.*, **23**(2), 115–121.

[18] Popa, V.I., Căpraru, A.-M., Grama, S., and Mălutan, T. (2011). Nanoparticles based on modified lignins with biocide properties. *Cellul. Chem. Technol.*, **45**(3–4), 221–226.

[19] Nagarajan, V., Mohanty, A.K., and Misra, M. (2013). Sustainable green composites: Value addition to agricultural residues and perennial grasses. *ACS Sustain. Chem. Eng.*, **1**, 325–333.

[20] Wegner, T.H. and Jones, E.P. (2009). A fundamental review of the relationships between nanotechnology and lignocellulosic biomass. In: L.A. Lucia and O.J. Rojas (Eds) *The Nanoscience and Technology of Renewable Biomaterials*. (pp. 1–41), Chichester- UK.; doi: 10.1002/9781444307474.ch1: John Wiley & Sons, Ltd.

[21] Zhiming, L., Chao, L., and Haiying, W. (2013). Preparation method of nanolignin with controllable particle size. Chinese Patent CN. 103145999A. Dated June 12, 2013.

[22] Zhiqiang, S., Wenfeng, S., Jun, Z., Baoyou, Z., Bin, Z.Y., Lei, Y., Baishi, Z., Xiuhua, Z., Jian, Z., Luming, H. (2011). Method for preparing nano lignin by using supercritical anti-solvent technology. Chinese Patent CN. 102002165A, Dated April 6, 2011.

[23] Tejado, A., Peña, C., Labidi, J., Echeverria, J.M., and Mondragon, I. (2007). Physico-chemical characterization of lignins from different sources for use in phenol-formaldehyde resin synthesis. *Bioresour. Technol.*, **98**(8), 1655–63.

[24] Doherty, W.O.S., Mousavioun, P., and Fellows, C.M. (2011). Value-adding to cellulosic ethanol: Lignin polymers. *Ind. Crop. Prod.*, **33**, 259–276.

[25] Hatakeyama, H. and Hatakeyama, T. (2010). Lignin structure, properties, and applications. In: A. Abe, K. Dusek, and S. Kobayashi (Eds) *Biopolymers* (Vol. **232**, pp. 1–63). Berlin: Springer Verlag.

[26] Calvo-Flores, F.G. and Dobado, J.A. (2010). Lignin as renewable raw material. *ChemSusChem*, **3**, 1227–1235.

[27] Lora, J.H. and Glasser, W.G. (2002). Recent industrial applications of lignin: A sustainable alternative to nonrenewable materials. *J. Polym. Environ.*, **10**(1–2), 39–48.

[28] Garkulak, J.D. and Lebo, S.E. (2000). *Commercial use of lignin based materials. In: Lignin: Historical, Biological and Materials Perspective*. American Chemical Society Symposium Series, **742**, Washington, DC. p. 304–320.

[29] Stewart, D. (2008). Lignin as a base material for materials applications: Chemistry, application and economics. *Ind. Crop. Prod.*, **27**(2), 202–207.

[30] Parasuraman, P., Singh, R., Bolton, T.S., Omori, S., and Francis, R.C. (2007). Estimation of hardwood lignin concentrations by UV spectroscopy and chlorine demethylation. *BioResources*, **2**(3), 459–471.

[31] Gosselink, R.J.A. (2011). Lignin as a renewable aromatic resource for the chemical industry. Lignin as a renewable aromatic resource for the chemical industry. Ph.D. Thesis (p. 191). Wageningen Universiteit (Wageningen University), Wageningen, December 6, 2011.

[32] Vishtal, A. and Kraslawski, A. (2011). Challenges in industrial applications of technical lignins. *BioResources*, **6**(3), 3547–3568.

[33] Ramírez, F., González, V., Crespo, M., Meier, D., Faix, O., & Zúñiga, V. (1997). Ammoxidized kraft lignin as a slow-release fertilizer tested on *Sorghum vulgare*. *Bioresour. Technol.*, **61**(1), 43–46.

[34] Zhang, Y.-H.P. (2008). Reviving the carbohydrate economy via multi-product lignocellulose biorefineries. *J. Ind. Microbiol. Biotechnol.*, **35**(5), 367–75.

[35] Kadla, J.F., Kubo, S., Gilbert, R.D., Venditti, R.A., Compere, A.L., and Griffith, W.L. (2002). Lignin-based carbon fibers for composite fiber applications. *Carbon*, **40**(15), 2913–2920.

[36] Kubo, S., Uraki, Y., and Sano, Y. (1998). Preparation of carbon fibers from softwood lignin by atmospheric acetic acid pulping. *Carbon*, **7**, 1119–1124.

[37] Cazacu, G., Pascu, M.C., Profire, L., Kowarski, A.I., Mihaes, M., and Vasile, C. (2004). Lignin role in a complex polyolefin blend. *Ind. Crop. Prod.*, **20**(2), 261–273.

[38] Alanso, M.V., Oliet, M., Pérez, J.M., Rodriguez, F. Astarloa, G., and Echeverria, J. (2004). Use of a methylolated softwood ammonium lignosulfonate as partial substitute of phenol in resol resins manufacture. *J. Appl. Polym. Sci.*, **84**(2), 643–650.

[39] Turunen, M., Alvila, L., Pakkanen, T., and Rainio, J. (2003). Modification of phenol-formaldehyde resol resin by lignin, starch, and urea. *J. Appl. Polym. Sci.*, **88**, 582–588.

[40] Dizhbite, T., Zakis, G., Kizima, A., Lazareva, E., Rossinskaya, G., Jurkjane, V., Telysheva, G., and Vieturs, U. (1999). Lignin: A useful bioresource for the production of sorbtion-active materials. *Bioresour. Technol.*, **67**(3), 221–228.

[41] Zoumpoulakis, L. and Simitzis, J. (2001). Ion exchange resins from phenol/formaldehyde resin-modified lignin. *Polym. Int.*, **50**, 277–283.

[42] Carrott, P.J. and Ribeiro Carrott, M.M. (2007). Lignin from natural adsorbent to activated carbon: Review. *Bioresour. Technol.*, **98**(12), 2301–2312.

[43] Fierro, V., Muniz, G., Basta, A.H., El-Saied, H., and Celzard, A. (2010). Rice straw as precursor of activated carbons: Activation with ortho-phosphoric acid, *J. Hazad. Mater.*, **181**, 27–34.

[44] Nadif, A., Hunkeler, D., and Käuper, P. (2002). Sulfur-free lignins from alkaline pulping tested in mortar for use as mortar additives. *Bioresour. Technol.*, **84**(1), 49–55.

[45] Wormeyer, K., Ingram, T., Saake, B., Brunner, G., and Smirnova, I. (2011). Comparison of different pretreatment methods for lignocellulosic materials. Part II: Influence of pretreatment on the properties of rye straw lignin, *Bioresour. Technol.*, **102**(8), 4157–4164.

[46] Baurhoo, B., Ruiz-Feria, C.A., and Zhao, X. (2008). Purified lignin: Nutritional and health impact on farm animals—A review. *Anim. Feed Sci. Technol.*, **144**(3), 175–184.

[47] Gopalakrishnan, K., Ceylan, H., and Kim, S. (2013). Renewable biomass-derived lignin in transportation infrastructure strengthening applications. *Int. J. Sustain. Eng.*, **6**(4), 316–325.

[48] Areskogh, D., Li, J., Gellerstedt, G., and Henriksson, G. (2010). Structural modification of commercial lignosulphonates through laccase catalysis and ozonolysis. *Ind. Crop. Prod.*, **32**(3), 458–466.

[49] Shulga, G., Soloddovniks, P., Shakels, V., and Anisckevicha, O. (2005). New semi-interpenetrating polymer networks incorporating soluble lignin. Proceedings of the 7th ILI Forum, Barcelona, International Lignin Institute, Lausanne, April 28–29, pp. 159–162.

[50] Areskogh, D. and Henriksson, G. (2011). Immobilisation of laccase for polymerisation of commercial lignosulphonates. *Process Biochem.*, **46**(5), 1071–1075.

[51] Fagerholm-Byman, H., Mikkola, P., Rosenholm, J.B., Lindén, E., and Carlsson, R. (1999). The influence of lignosulphonate on the properties of single and mixed Si_3N_4 and ZrO_2 suspensions. *J. Eur. Ceram. Soc.*, **19**(1), 41–48.

[52] Zhang, L.M. and Yin, D.-Y. (2002). Preparation of a new lignosulphonate-based thinner: Introduction of ferrous ions. *Colloids. Surf. A.*, **210**(1), 13–21.

[53] Lora, J. (2008). Industrial commercial lignins: Sources, properties and applications. In: Belgacem, M., and Gandini, A. (Eds) *Monomers, Polymers and Composites from Renewable Resources*. (pp. 225–242), Amsterdam: Elsevier.

[54] Jin, C.D., Li, S., Liu, L., Ye, J.W., and Du, C.D. (2011). Research of process on making bamboo particle board by laccase-treated calcium lignosulfonate. *Adv. Mater. Res.*, **179–180**, 1031–1036.

[55] Palmqvist, L., Lyckfeldta, O., Carlsröm, E., Davous, P., Kauppi, A., and Holmberg, K. (2006). Dispersion mechanism in aqueous alumina suspensions at high solids loadings. *Colloids Surf. A*, **274**(1), 100–109.

[56] Ansari, A. and Pawlik, M. (2007). Floatability of chalcopyrite and molybdenite in the presence of lignosulfonates. Part I. Adsorption studies. *Miner. Eng.*, **20**(6), 600–608.

[57] Grierson, L.H., Knight, J.C., and Maharaj, R (2005). The role of calcium ions and lignosulphonate plasticizer in the hydration of cement. *Cem. Concr. Res.*, **35**(4), 631–636.

[58] Xu, F., Sun, J.-X., Sun, R., Fowler, P., and Baird, M.S. (2006). Comparative study of organosolv lignins from wheat straw. *Ind. Crop. Prod.*, **23**(2), 180–193.

[59] Tan, S.S.Y., MacFarlane, D.R., Upfal, J., Edye, L.A., Doherty, W.O.S., Patti, A.F., Pringle, J.M., and Scott, J.L. (2009). Extraction of lignin from lignocellulose at atmospheric pressure using alkylbenzenesulfonate ionic liquid. *Green Chem.*, **11**(3), 339–345.

[60] Belgacem, M.N., Blayo, A., and Gandini, A. (2003). Organosolv lignin as a filler in inks, varnishes, and paints. *Ind. Crop. Prod.*, **18**(2), 145–153.

[61] Zhang, J., Deng, H., Lin, L., Sun, Y., Pan, C., and Liu, S. (2010). Isolation and characterization of wheat straw lignin with a formic acid process. *Bioresour. Technol.*, **101**(7), 2311–2316.

[62] Chandra, R.P., Bura, R., Mabee, W.E., Berlin, A., Pan, X., and Saddler, J.N. (2007). Substrate pretreatment: The key to effective enzymatic hydrolysis of lignocellulosics? *Adv. Biochem. Eng. Biotechnol.*, **108**, 67–93.

[63] Holladay, J.E., Bozell, J.J., White, J.F., and Johnson, D. (2007). Top value-added chemicals from biomass volume ii: Results of screening for potential candidates from biorefinery lignin. From: http://www1.eere.energy.gov/bioenergy/pdfs/pnnl-16983.pdf, Accessed on October 21, 2014.

[64] Northey, R.A. (1992). Low-cost uses of lignin. In: *Materials and Chemicals from Biomass*. Symposium Series, (Vol. **476**, pp. 146–175). Washington, DC: American Chemical Society.

[65] Pye, E.K. (2006). Industrial lignin production and applications. In: B. Kamm, P.R. Gruber, and M. Kamm (Eds) *Biorefineries-Industrial Processes and Product* (Vol. **2**, pp. 165–200). Weinheim, Germany; Wiley-VCH Verlag GmbH & Co. KGaA.

[66] Toumi, S., Mäkinen, T., Niemela, K. In: Report of Project 'Star-COLIBRI', No. 241535-FP7 entitled "Strategic Targets for 2010-Collaboration Iniative on Biorefineries". Available at http://www.star-colibri.eu/files/files/%20Deliverables/D2.1-Report-19-04-2010.pdf. Chapter 2: Market and Consmer Data, Section: A-Wood (Lignin)-Products from Lignin, p. 43–55.

[67] Frangville, C., Rutkevičius, M., Richter, A.P., Velev, O.D., Stoyanov, S.D., and Paunov, V.N. (2012). Fabrication of environmentally biodegradable lignin nanoparticles. *ChemPhysChem*, **2012**, 13, 4235–4243.

[68] Qian, Y., Zhang, Q., Qiu, X., and Zhu, S. (2014). CO_2-responsive diethylaminoethyl-modified lignin nanoparticles and their application as surfactants for CO_2/N_2-switchable Pickering emulsions. *Green Chem.*, **16**, 4963–4968.

[69] Gao, G., Dallmeyer, J.I., and Kadla, J.F. (2012). Synthesis of lignin nanofibers with ionic-responsive shells: Water-expandable lignin-based nanofibrous mats. *Biomacromolecules*, **13**(11), 3602–3610.

[70] Gonugunta, P., Vivekanandhan, S., Mohanty, A.K., and Misra, M. (2012). A study on synthesis and characterization of biobased carbon nanoparticles from lignin. *World J. Nano Sci. Eng.*, **2**(3), 148–153.

[71] Jiang, C., He, H., Jiang, H., Ma, L., and Jia, D.M. (2013). Nano-lignin filled natural rubber composites: Preparation and characterization. *eXPRESS Polym. Lett.*, **7**(5), 480–493.

[72] Caicedo, H.K., Dempere, L.A., and Vermerris, W. (2012). Template-mediated synthesis and bio-functionalization of flexible lignin-based nanotubes and nanowires. *Nanotechnology*, **23**, 105605–105617.

[73] Corcione, C.E. and Frigione, M. (2012). Characterization of nanocomposites by thermal analysis. *Materials*, **5**, 2960–2980.

[74] Vyazovkin, S. (2008). Thermal analysis. *Anal. Chem.*, **80**, 4301–4316.

[75] Meneghetti, P. and Qutubuddin, S. (2006). Synthesis, thermal properties and applications of polymer-clay nanocomposites. *Thermochim. Acta*, **442**(1–2), 74–77.

[76] Shen, L., Lin, Y., Du, Q., and Zhong, W. (2006). Studies on structure–property relationship of polyamide-6/attapulgite nanocomposites. *Compos. Sci. Technol.*, **66**(13), 2242–2248.

[77] Ito, Y., Omote, K., and Harada, J. (2003). A new small angle X-ray scattering technique for determining nano-scale pore/particle size distributions in thin film. *Adv. X-ray Anal.*, **46**, 56–60.

[78] Intra, P. and Tippayawong, N. (2008). An overview of differential mobility analyzers for size classification of nanometer-sized aerosol particles. *Songklanakarin J. Sci. Technol.*, **30**(2), 243–256.

[79] Kulkarni, N. and Muddapur, U. (2014). Biosynthesis of metal nanoparticles: A review. *J. Nanotechnol.*, **2014**, 1–8.

[80] Misra, M., Khan, N., Satyanarayana, N., and Mohanty, A.K. (2010). Electrospun nanocomposite fibres: Processing, properties and applications. In: A.D. Ngo, J. Denault, S.V. Hoa, A. Nakai, T. Morii, and H. Hamada (Eds) *Design, Manufacturing and Applications of Composites: Proceedings of the Eighth Joint Canada–Japan Workshop on Composites: École de Technologie Supérieure, Montréal, Québec; Industrial Materials Institute, Boucherville, Québec: July 26–29, 2010* (pp. 137–146). Lancaster, PA: DEStech Publications.

[81] Mohanty, A.K., Vivekanandhan, S., and Misra, M. (2010). Advanced green composites and nanocomposites: New industrial biomaterials from refinery. In: A.D. Ngo, J. Denault, S.V. Hoa, A. Nakai, T. Morii, and H. Hamada (Eds) *Design, Manufacturing and Applications of Composites: Proceedings of the Eighth Joint Canada–Japan Workshop on Composites: École de Technologie Supérieure, Montréal, Québec, Canada; Industrial Materials Institute, Boucherville, Québec, Canada: July 26–29, 2010* (pp. 3–7). Lancaster, PA: DEStech Publications.

[82] Reza Barzegari, M., Alemdar, A., Zhang, Y., and Rodrigue, D. (2012). Mechanical and rheological behavior of highly filled polystyrene with lignin. *Polym. Compos.*, **33**(3), 353–361.

[83] Benko, D.A., Hahn, B.A., Cohen, M.P., Dirk, S.M., and Cicotte, K.N. (2014). Functionalized lignin, rubber containing functionalized lignin and products containing such rubber composition. US Patent No. US 8,664,305 B2, Dated March 4, 2014.

[84] Boutsicaris, S.P. (1984). Lignin reinforced synthetic rubber. US Patent No. 4,477,612, October 16, 1984.

[85] Dimitri, M.S. and Force, C.G. (1981). Tackifying composition for rubber. US Patent No. 4,287,104 dated: September 1, 1981.

[86] Doughty, J.B. and Charleston, S.C. (1966). Lignin reinforced rubber and method of preparation thereof. US Patent No. 3,247,135. April 19, 1966.

[87] Xiao, S., Feng, J., Zhu, J., Wang, X., Yi, C., and Su, S. (2013). Preparation and characterization of lignin-layered double hydroxide/styrene-butadiene rubber composites. *J. Appl. Polym. Sci.*, **130**, 1308–1312.

[88] Asrul, M., Othman, M., Zakaria, M., and Fauzi, M.S. (2013). Lignin filled unvulcanised natural rubber latex: Effects of lignin on oil resistance, tensile strength and morphology of rubber films. *Inter. J. Eng. Sci. Invent.*, **2**(5), 38–43.

[89] Sahoo, S., Misra, M., and Mohanty, A.K. (2011). Enhanced properties of lignin-based biodegradable polymer composites using injection moulding process. *Compos. A: Appl. Sci. Manuf.*, **42**, 1710–1718.

[90] Luo, X., Mohanty, A., and Misra, M. (2013). Lignin as a reactive reinforcing filler for water-blown rigid biofoam composites from soy oil-based polyurethane. *Ind. Crop. Prod.*, **2013**(47), 13–19.

[91] Erakovic, S., Jankovic, A., Tsui, G.C.P., Tang, C.-Y., Miskovic-Stankovic, V., and Stevanovic, T. (2014). Novel bioactive antimicrobial lignin containing coatings on titanium obtained by electrophoretic deposition. *Int. J. Mol. Sci.*, 2014, **15**(7), 12294–12322.

[92] Chung, Y.-L., Olsson, J.V., Li, R.J., Frank, C.W., Waymouth, R.M., Billington, S.L., and Sattely, E.S. (2013). A renewable lignin–lactide copolymer and application in biobased composites. *ACS Sustain. Chem. Eng.*, **1**, 1231–1238.

[93] Del Saz-Orozco, B., Oliet, M., Alonso, M.V., Rojo, E., and Rodríguez, F. (2012). Formulation optimization of unreinforced and lignin nanoparticle-reinforced phenolic foams using an analysis of variance approach. *Compos. Sci. Technol.*, **72**(6), 667–674.

[94] Brannon-Peppas, L. and Blanchette, J.O. (2004). Nanoparticle and targeted systems for cancer therapy. *Adv. Drug Deliv. Rev.*, **56**(11), 1649–1659.

[95] Aryal, S., Hu, C.M., and Zhang, L. (2010). Polymer–cisplatin conjugate nanoparticles for acid-responsive drug delivery. *ACS Nano* **4**(1), 251–258.

[96] Zhang, L., Radovic-Moreno, A.F., Alexis, F., Gu, F.X., Basto, P.A., Bagalkot, V., Jon, S., Langer, R.S., and Farokhzad, O.C. (2007). Co-delivery of hydrophobic and hydrophilic drugs from nanoparticle-aptamer bioconjugates. *ChemMedChem*, **2**(9), 1268–1271.

[97] Zimniewska, M., Batog, J., Bogacz, E., and Romanowska, B. (2012). Functionalization of natural fibres textiles by improvement of nanoparticles fixation on their surface. *J. Fiber Bioeng. Inform.*, **5**(3), 321–339.

WEB SITE REFERENCES

[1] http://europa.eu/legislation_summaries/energy/renewable_energy/en0009_en.htm. Accessed on October 20, 2014.

[2] http://www.ili-lignin.com/aboutlignin.php. Accessed on September 27, 2014.

[3] http://www.star-colibri.eu/files/files/Deliverables/D2.1-Report-19-04-2010.pdf. Accessed on October 16, 2014.

[4] http://www.bccresearch.com/market-research. Accessed on October 20, 2014.

8

STARCH-BASED "GREEN" COMPOSITES

K.G. Satyanarayana[1] and V.S. Prasad[2]

[1] *Poornaprajna Institute of Scientific Research (PPISR), Bengaluru, Karnataka, India*
[2] *Chemical Sciences & Technology Division, National Institute for Interdisciplinary Science & Technology (NIIST-CSIR), Thiruvananthapuram, Kerala, India*

Biodegradable Green Composites, First Edition. Edited by Susheel Kalia.

8.1 INTRODUCTION

Growing environmental awareness and societal needs have become catalytic agents for the development of "eco-friendly" materials. One such new material is "green" composites. This consists of natural matrix such as starch incorporated with natural fibers such as lignocellulosic fibers (banana, coir, sisal, sugarcane bagasse, wood, etc.) [1–4]. Such materials are completely "biodegradable" due to the action of microorganisms by their conversion in to CO_2 and water, which are absorbed by the plants. Thus, they are easily composted after their life whereby completing nature's carbon cycle. Despite their limitations such as low processability and reduction of the ductility, low impact strength, high moisture absorption, etc., which can be overcome, in recent times, these "green" composites have been attracting attention to replace the traditional composites based on nonrenewable petroleum-based feedstocks such as synthetic polymers and fibers (glass, carbon, etc.) in various sectors due to their competitive properties. This is because of various attributes such as abundant availability, renewability, low cost of the constituents, and reduced weight along with their manufacturing methods, which are similar to the conventional polymer composites. Application areas include buildings and constructions (doors, furniture, false ceilings, decorative purposes, etc.), automotive (interiors), etc. Besides nonrenewability, synthetic polymer based materials are toxic to the environment, nonbiodegradable, and expensive. In fact, manufacturing processes of "green" composites do not cause any skin irritation during their preparation, unlike the synthetic materials based on conventional composites. In view of all the positive attributes of "green" composites, publications including specific journals devoted to this subject, textbooks, and even some industries named after these materials have been established in recent times.

Starch is one of various biopolymers used as matrix as well as reinforcement/filler in the development of "green" composites. This chapter presents various types of natural starches, their modifications and blends, structure and properties, preparation of various types of starch-based "green" composites including nanocomposites, various approaches to the synthesis of these composites, and their structure and properties. In addition, current and future applications of starch-based "green" composites and some perspectives of these composites are also discussed.

8.1.1 Starch

Since there have been many reviews as well as chapters particularly during the last two decades or so having very detailed discussions about their sources, availability, structure, and properties along with some applications of starch, these are briefly mentioned here [3, 5–35]. Starch, being the main reserve carbohydrate in plants and accounting for more than 50% of the humankind's food in terms of calories, is the most abundant plant-based material in nature after cellulose [20, 32]. It is found in plant roots (cassava), stalks (sago), and crop seeds (rice, maize, wheat, legumes; peas, beans, lentils), most of which form the staple crops [3, 17, 36, 37]. Of course, geographic and climatic factors (tempered cold, tropical, subtropical zones, and inundated areas) and the desired functional properties of the corresponding starch determine the choice of these plants [13]. Thus, it represents one of the renewable resources. Of the crop seeds, maize with about 82% forms the main source of starches, followed by wheat with 8%, potatoes with 5%, and cassava the balance 5% [7]. The total amount of starch produced annually in these crops is over a billion tons, majority of which is consumed directly as food or fed to domestic animals. It is reported that the global annual starch production by man approximates 3000 million tons, and the industrial production of pure, refined starch now exceeds 60 million tons [38]. More than a decade ago, the world starch market including native and modified starches was estimated at about 48 million tons with the value of starch output worth €15 billion per year owing to the industrialists and researchers seeking new properties or potential high-value application [31, 38]. In fact, starch being cheap and easy to produce and therefore for centuries starch has also been extensively used in numerous nonfood applications, such as an adhesive for wood and as gum in the textile industry [10], in addition to expected important role in the future in providing resources for the increasing demand for CO_2 neutral energy [10, 13]. As mentioned earlier, starch is also used both as a matrix and reinforcing filler in biodegradable or "green" plastic materials [9, 11, 14, 18–20, 29, 32, 39–42]. Other emerging area of application of starch getting increased attention is the "green" nanocomposites, which show unique properties compared to conventional composite because of the nanometric size effect even at low filler content [9, 10, 12, 14, 16, 18–21, 23–25, 28, 29, 31, 32, 34, 43–48]. Nanostarch as nanofillers seemed to exhibit strong reinforcing effects, and studies have shown their positive impact in barrier packaging as well [23, 24, 49].

Many of these applications require costly modifications of the extracted starch. These nonfood applications of the starch along with possibility of its conversion into chemicals (ethanol, acetone, and organic acids), or in the production of synthetic polymers as well as to produce biopolymer through fermentation or be hydrolyzed and employed as a monomer or oligomer and also in the production of new polymeric materials (as fillers for other polymers) by grafting with a variety of reagents, make it one of the most versatile materials for potential use in polymer technology [8, 10]. However, direct use of starch as a renewable resource material would be more economical, and therefore over the last few decades, this has been a major research area in material science [8]. In addition, continuous efforts to develop new materials based on starch to meet the increasing demand for such materials particularly based on renewable resources along with new uses for starch would lead to increase the demand for starch production, which then looks for commercial sources of starch.

Natural starch normally occurs as discrete and partially crystalline microscopic granules, which are held together by an extended micellar network of associated molecules [17]. The starch industry extracts and refines starches by wet grinding, sieving, and drying. In whole grain starch extraction, the powder is treated with caustic soda solution to separate the proteins. It is either used as extracted from the plant as native starch or modified to get specific properties in the form of modified starch. Chemical modification of starch by grafting to improve the properties of the composites have been reported recently [50–53]. Some available reviews on starch discuss not only about such modifications of starch, mostly by chemical methods leading to tailor-made materials [26], but also its chemistry, structure, and technology [17, 18, 32].

8.1.1.1 Thermoplastic Starch Some of the attractive properties of starch include: natural and abundant availability, recyclability and renewability, biodegradability, low cost, ease of assimilation (important in food industry), and ease of subjecting it to chemical/thermal changes whereby starches show thermoplastic behavior, which helps in its suitability for various nonfood applications [54]. One such application is its use in composite technology to prepare "biodegradable" or "green" composites when used with either other natural polymers or with plant fibers or other type of natural materials. This is possible provided some of the limitations of starch such as high water solubility, poor melt processability, and brittleness due to anarchical growth of one of their constituents [14, 55] are overcome, which are inherent in the granular structure of starch. It is reported that such a structure can be disrupted by heating (action of high temperature and shear stress) or by adding plasticizers [33, 56, 57]. These processes are reported to break the structure of starches (semicrystalline form) in the case of plasticizers, they interact with the polymer chains reducing the interaction between adjacent chains and enhancing film flexibility [58] and cause destructuring of the starch chains, which would lead to intermolecular rearrangement [59] and thus make the starch to flow and thus rendering it suitable for thermoplastic processing [55]. Resulting material thus obtained is termed "*thermoplastic starch*" (TPS), which can be obtained by dry mixing and heating/shearing. It can also be obtained by a casting process consisting of gelatinization of an aqueous dispersion of starch and plasticizer in the dry form to get a thin film [60, 61].

Therefore, attempts have been made to make them flexible by using a suitable plasticizer, such as glycerol, sorbitol, other polyols, and even water [22, 33, 37, 62, 63] to make them suitable for engineering applications including in the preparation of the above mentioned composites through extrusion, injection, or compression molding [33, 56] or use of a batch mixer connected to a torque rheometer [62]. Necessary characteristic of a plasticizer is its chemical affinity with the starch used. Thus, polyols like glycerol or sorbitol and water, whose hydroxyls establish hydrogen bonds with starch hydroxyls, meet this condition [58]. The plasticizing effect by these materials has been explained by the free-volume theory according to which the plasticizer generates free-volume regions that allow higher mobility of polymeric chain segments [64].

Besides, change in properties of such TPS takes place by these plasticizers, which depend on the amount of plasticizer used. It is observed that glycerol is the mostly used plasticizer in the TPS production due to its availability, biodegradability, and affinity with starch [65–70]. In addition, glycerol provides stability and compatibility with hydrophilic packaging chains in starches [71], while its content affects the TPS properties. For example, in case water is used as a plasticizer of starch, due to impediment caused by its high mobility in the matrix (starch), it should be in very small amounts [68, 69]. In this case, the mechanism of plasticization of starch is based on reducing the hydrogen bonds between macromolecules along with the reduction of dipole–dipole interactions due to alterations in the dielectric constant [72].

Thermal or other treatment techniques have also been attempted to overcome the limitations of starch-based materials as well as to improve their properties [73–75]. The first attempt to find the effect of annealing of wheat starch to find the effect on gelatinization revealed that gelatinization would take place only when the starch was heated above its glass transition but not high enough to cause irreversible melting or gelatinization [74]. On the other hand, annealing of starch below the gelatinization temperature has been found to lead to discrete molecular organization with the formation of a more organized structure with less free energy, which in turn affected the mechanical and other properties [73].

8.1.1.2 Starch Nanocrystals Various researchers have used different nomenclatures for the crystalline part of starch obtained by hydrolysis but to a different extent (from the highest or most to the least). These include "starch crystallite," "starch nanocrystal (SNC)," "microcrystalline starch," and "hydrolyzed starch." Figure 8.1 shows plots of percent hydrolysis versus days for different starch sources. These starch nanoparticles can be amorphous as well [17].

The aforementioned plots have used data as follows: from References: Waxy maize (plain line) from [76]; waxy maize, normal maize, amylomaize V, amylomaize VII, and oat and rice starch (dash line) from [77]; high amylose (65%) (dotted line) from [78]; wrinkled pea, mung bean, lentil, and corn (dot-filled line) from [79].

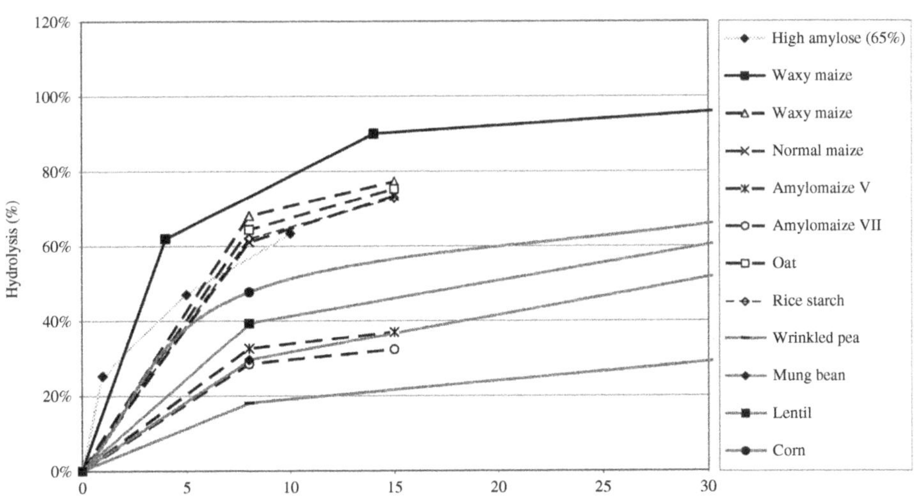

FIGURE 8.1 Hydrolysis kinetics of different starches. *Source*: Le Corre et al. [17]. Reprinted with permission from American Chemical Society. Copyright 2010.

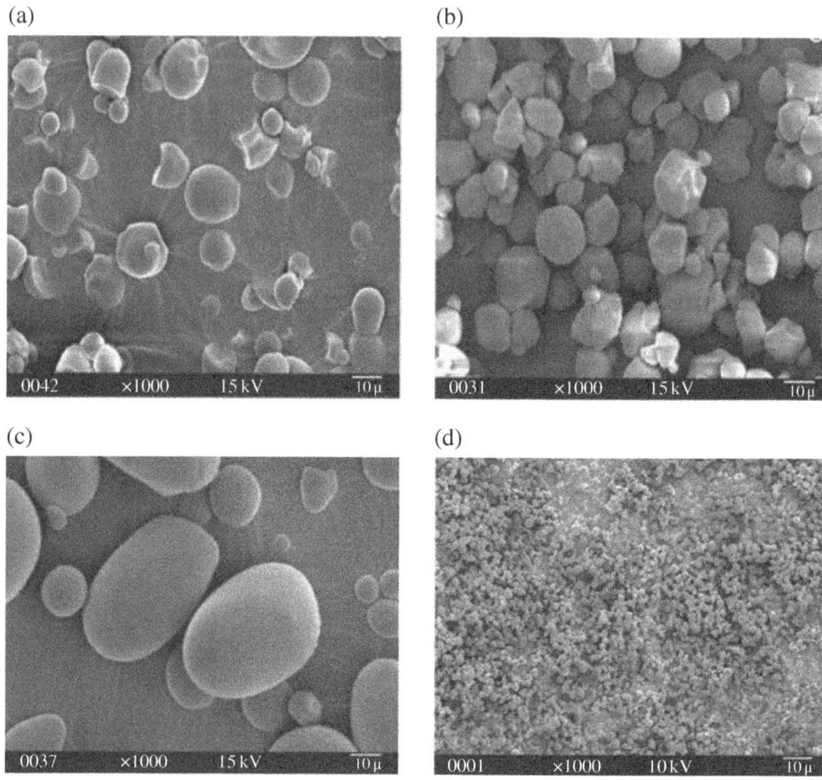

FIGURE 8.2 SEM photographs of starch and starch nanocrystals: (a) cassava starch (CaS), (b) maize starch, (c) potato starch (PoS), and (d) sugarcane starch (×1000). KGaA. *Source*: Figueira et al. [85]. Reproduced with permission from Food Sci. Technol (Campinas) Sociedade Brasileira de Ciencia e Tecnologia de Alimentos - (SBCTA).

The first reported method of obtaining SNCs by acid hydrolysis of starch is by Dufresne et al. [80]. There are also reports of obtaining starch nanoparticles that are hydrophobically modified in the aqueous media using long-chain fatty acids under mild conditions [53, 81, 82]. SNCs have been prepared by acid hydrolysis using different cellulose sources (both plant and non-plant based) such as ramie, cotton, sugar beet, tunicin, MCC, and bacterial cellulose (BC) and characterized using various techniques such as TEM, AFM, E-SEM, small-angle neutron scattering, and polarized and depolarized dynamic light scattering (DLS, DDLS) [83, 84]. Similarly, various researchers have observed the SNCs derived from various plants through TEM or SEM. Some of these are shown in Figure 8.2a–d, wherein one can see them to be of various sizes, with varying lengths, widths, and thickness [85].

Apart from the conventional methods of acid hydrolysis mentioned earlier [86, 87], Bhattacharya et al. have developed a novel method for preparing SNCs using pea starch (PS) [88]. The method in brief is addition of plasticizer to the starch, followed by two steps, namely, homogenization and ethanol wash. Finally, freeze drying

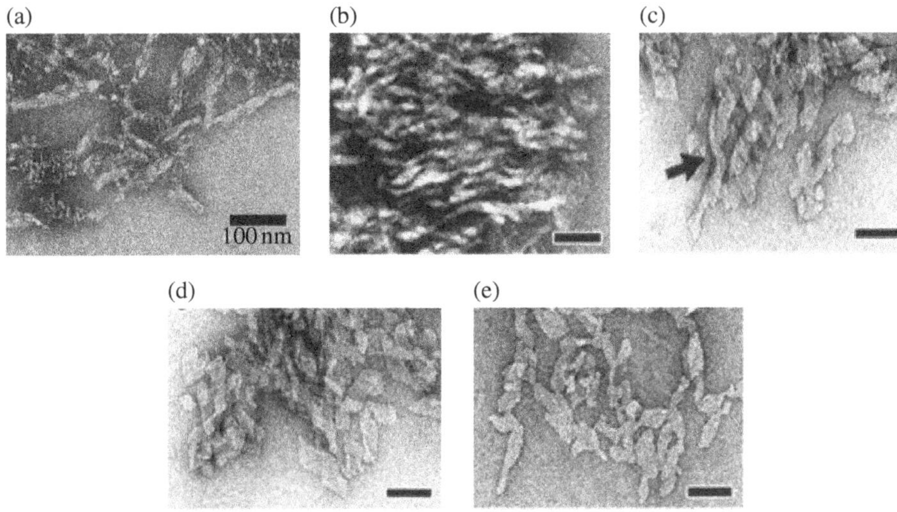

FIGURE 8.3 TEM photographs of nanocrystals obtained from (a) pea starch (PS) and (b–e) waxy maize starch with different time of acid hydrolysis. Pyramidal stack of crystals are shown by arrow in (c). Reproduced with the kind permissions from John Wiley & Sons and American Chemical Society. Publishers of Refs. *Source*: (a): Zheng (H) et al. [89, figure 1, p. 476] and (b–e): Putaux et al. [90], respectively. Copyright of John Wiley 2009 & Sons and American Chemical Society 2003.

of the final sample was done. They have reported size of nanocrystals of PS as 50–100 nm. On the other hand, protein nanocrystals prepared by repetitive solubilization and crystallization by the same authors were reported to be 350 nm in the optimized final stage consisting of solubilization and crystallization, while particle size of 870 nm was observed in the first stage of protein isolate.

One can also see the photographs of sugarcane starch granules (Fig. 8.2d), which have been recently extracted from sugarcane juice. It is also reported that these granules are globular, oval, and semispherical shapes. They are the smallest sized ones (1–3 μm) compared to those of 20–50 μm size of potato, 10–15 μm size of maize, and 8–20 μm size of cassava starch (CaS) granules. Figueira and coworkers [85] have stated that the size of sugarcane starch particles reported by them was in the range (1–10 μm) reported earlier by others. In fact, it is reported that the dimensions of starch nanoparticles range from 20 to 200 nm, width from 10 to 30 nm, and thickness from 5 to 10 nm [18]. Individual and isolated platelets of SNCs can also be seen despite the SNCs tend to self-aggregate.

Figure 8.3a–e shows TEM photographs of nanocrystals obtained from PS hydrolyzed by H_2SO_4 [89] (Fig. 8.3a) and (Fig. 8.3b–e) waxy maize starch (WMS) with different time of acid hydrolysis using 2.2N HCl at 36°C [90]. While the micrographs shown in (Fig. 8.3b–d) are for hydrolysis time of after 2 weeks, the one shown in (Fig. 8.3d) reveal nearly individual waxy maize SNCs obtained after 6 weeks of hydrolysis. A pyramidal stack of crystals is shown by arrow in (Fig. 8.3c).

The heterogeneity in particle size and the shape of the nanocrystals were reported due to the differences in starch types [91, 92]. For example, bigger nanocrystals

(40–70 nm for potato) 30–80 nm for pea and different shapes such as grape like as well as round crystals have also been reported [17, 53, 81, 93].

Further, contrary to cellulose nanocrystals, SNCs are not 100% crystalline, rather they are only 45% crystalline with variations depending on the botanical origin [94].

SNCs are also layered particles that can be utilized in composite preparation as in layered silicates [94, 95]. These nanocrystals have been reported to possess a reactive surface covered with hydroxyl groups [18]. This is reported to provide a possibility for its modification via a chemical reaction strategy. Hence they can have strong reinforcing effects, which are reflected in the positive effects of some characteristics, such as barrier properties in packaging applications [23, 24, 49]. Further, the reinforcing effect of SNCs is generally ascribed to the formation of a hydrogen-bonded percolating filler network above a given matrix. In 2008 onward grafted SNCs were used in biodegradable matrices in particular polycaprolactone (PCL)-grafted SNC through microwave assisted in presence of polymerizations [96]. SNCs, the nanoscale biofiller derived from native starch granules, have been compounded with many different kinds of polymer matrices, which will be discussed later. The intrinsic rigidity of SNCs, special platelet-like morphology, strong interfacial interactions, and the percolation network organized by nanocrystals, all contribute to optimized mechanical performance, thermal properties, solvent absorption, and barrier properties. To improve interfacial adhesion between SNCs and polymer matrices, a series of chemical modifications based on "chemical reaction," "graft to," and "graft from" strategies have been carried out and, in some cases, markedly enhanced the performance of the nano materials [97].

It has also been mentioned that the purpose of chemical modification of SNCs is to contribute to specific functions as well as to expand the applications of SNCs. An estimate of the surface hydroxyl group content of approximately 14% of the total amount available has been reported [98]. For example, reactive hydroxyl groups of only 0.0025 μmol have been reported to exist in 1.0 μg of freeze-dried SNCs.

Similarly, amylopectin-rich WMS (trade name Waxylis 200)-based nanoscale monocrystalline starch particles have been successfully modified with two different agents, namely, stearic acid chloride, and poly(ethylene glycol) methyl ether (PEGME), using 2,4-TDI to graft the PEGME to the starch surface [99]. While FTIR, XPS spectroscopy, DSC, and contact angle measurements confirmed successful surface modification evidenced by the complete surface coverage, X-ray diffraction and DSC studies revealed that the starch crystalline structure was not affected. Further, extensive crystallization of the stearate moieties grafted to the starch nanoparticle surface was revealed by the XRD studies, which was found to form a crystalline hydrophobic shell around the hydrophilic SNC. On the other hand, individualization of nanoparticles was revealed by TEM studies, which has been attributed to the reduction of polar and hydrogen-bonding forces. Figure 8.4 shows the effect of surface modification by two different agents used.

The authors have observed crystallization of the grafts on the starch surface. Thus changes (nanocrystal surface with plasticizing chains) have been observed in the character of the nanocrystal surfaces and their activity due to surface modifications. The authors have indicated testing of these materials as toxic agent-trapping devices

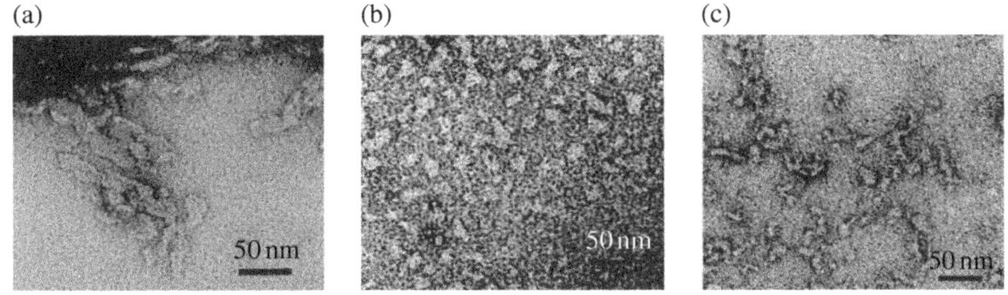

FIGURE 8.4 Transmission electron micrographs of (a) unmodified starch nanocrystals and starch nanocrystals surface modified with (b) stearates and (c) polyethylene glycol methyl ether. *Source*: Thielemans et al. [99]. Reprinted with permission from American Chemical Society.

and in co-continuous composites for possible applications as compatibilized polymer additives, surface-active particles, and co-continuous nanocomposite precursors.

Considering that the reinforcing effect of these SNCs being generally attributed to the formation of hydrogen-bonded percolating filler network above a given matrix, nanocomposites with these as fillers have been prepared. Details of nanocomposites systems studied are given later under Section 8.2.2.3.

8.1.1.3 Structure and Properties of Starch/TPS Starch and cellulose, as natural polysaccharides, are fascinating macromolecules possessing structural diversity and functional versatility. Although starch is a familiar substance in our everyday life, its importance to mankind and its structural complexity are rarely appreciated. Starch is made of polymers of the simple sugar, glucose, and is synthesized inside the plastid compartment of plant cells as insoluble granules. It is mainly composed of two polysaccharides (glucosidic macromolecules): the linear and helical *amylose* and the branched *amylopectin*. Each amylopectin molecules in starch contains 10,000–100,000 glucosyl units, which are linked together to form a branched cluster structure. The molecular architecture of amylopectin is critical in allowing the polymer to adopt secondary and tertiary structures that result in the formation of a semicrystalline, insoluble granule [17].

Pure starch is a white powder, tasteless and odorless, insoluble in cold water or alcohol, consisting of microscopic granules with diameters ranging from 2 to 100 μm depending on the botanical origin and with a specific density of 1.5 [13, 29]. The basic formula of starch is $(C_6H_{10}O_5)_n$ and the glucose monomer of which is called alpha D-glucopyranose or D-glucose. Depending on their botanical origin, starch raw materials have different conversion factors, size, shape, and chemical content [17]. The size, amylose content, and crystallinity of starch from different sources are shown in Table 8.1.

Interesting and detailed reviews and articles on starch structure have been published [9, 17, 18, 100]. Starch composition was first determined by studying the residue of its total acid hydrolysis. In most common types of starches, the weight percentage of amylose ranges between 72 and 82%, and the amylopectin ranges from

TABLE 8.1 Characteristics of starches from different botanic origins[a]

Botanic source	Maize	Wheat	Rice	Cassava	Potato	Pea
Av. granular size (μm)	30	30	2–7	3–30	40–100	2–40
Amylose content (%)	25–28	25–29	20–25	28	20–25	33–48
Crystallinity (%)	39	36	38–51	—	25	18–36

[a] Le Corre et al. [17]. Reprinted with permission from American Chemical Society.

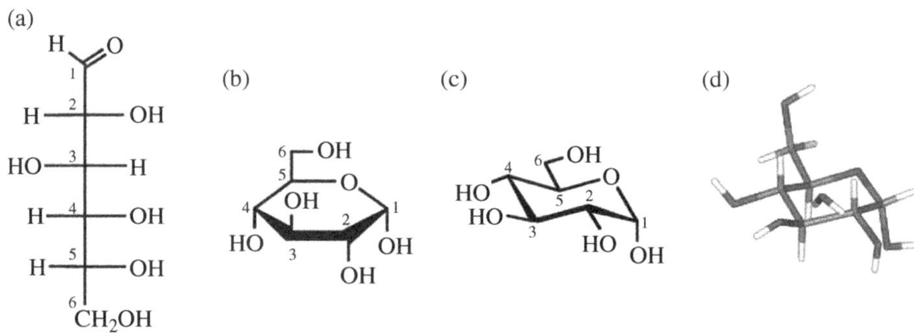

FIGURE 8.5 Structures of α-D-glucopyranose (D-glucose) shown using different projections: (a) Fisher, (b) Haworth, (c) in 4C1 conformation, and (d) a 3D model of glucose in the 4C1 conformation. *Source*: Damager et al. [9]. Reproduced with permission from American Chemical Society; Copyrights: American Chemical Society. (*See insert for color representation of the figure.*)

18 to 28%. In starch amylose and amylopectin occur in various ratios depending on the plant species [18, 32].

In the case of polysaccharides, having typical polyglucan structure, containing only glucose repeat units, a broad variety of structures appear, resulting from the stereochemistry of the anomeric C-atom, from the regiochemistry of the glycoside linkage and from the pattern of branching as in starch and cellulose. The second largest polyglucan, starch, consists of two primary polymers containing D-glucose, the linear α (1→4) linked amylose and the amylopectin composing additional α (1→6) linked branches compared to cellulose, the most important polysaccharide in nature other than starch, which is having a very uniform molecular structure of α (1→4) linked polyglucan and used as a raw material for the commercial production of fibers and in paper industry. Structure of D-glucose monomer unit of starch polymer is shown in Figure 8.5a–d using different projections [9]. A schematic of hierarchical structure of cellulose in comparison to starch granule is shown in Figure 8.6. The structures of amylose, amylopectin, in comparison to animal starch glycogen are shown in Figures 8.7, 8.8, and 8.9 [9, 25]. In Figure 8.8, which shows the principle of top-down strategy of starch structure determination, the cross section of wheat starch granule is shown in (a) taken using a confocal microscope (Mikkel A. Glaring). A schematic drawing of layered structure of amylopectin with alternating amorphous and crystalline lamellae repeating with 9 nm spacing is shown

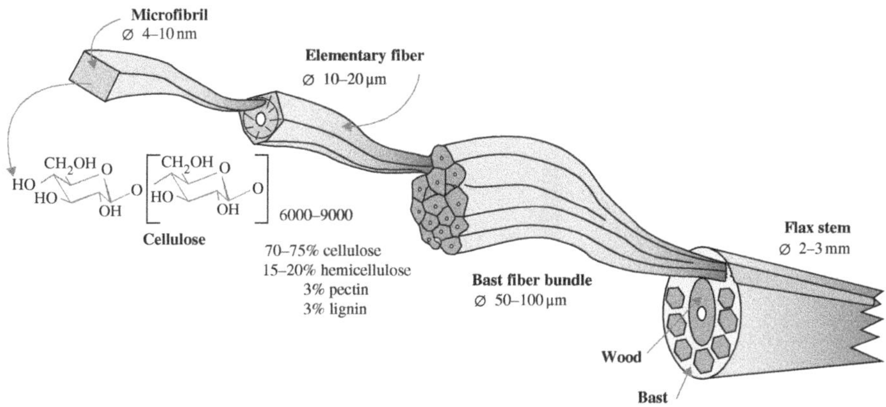

FIGURE 8.6 Schematic of hierarchical structure of cellulose in comparison to starch granule. *Source*: http://www.jeccomposites.com/news/composites-news/industrial-applications-natural fibre-reinforced-composites, 2009. Reproduced with permission from JEC. (*See insert for color representation of the figure.*)

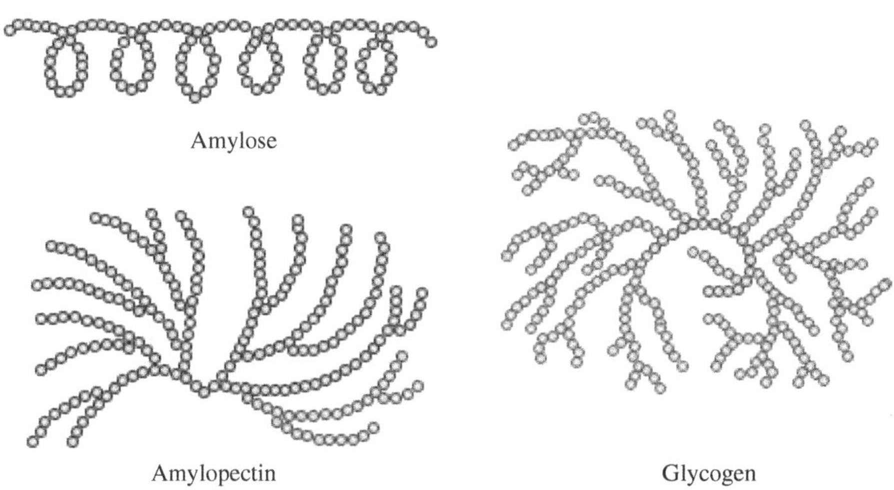

FIGURE 8.7 Structure of amylose, amylopectin, and glycogen. *Source*: Reproduced from Open Access, 2014, http://biology.tutorvista. com/biomolecules/carbohydrates.html.

in b. A cross section of the parallel helices pack to form either hexagonal (A-type or B-type or pseudohexagonal packing) is shown in c. The state-of-the art modeling concerning tiny double helical structure and nanocrystalline packing are shown in d, while e shows the structure of chemically synthesized branched pentasaccharide mimicking the branch point in starch and glycogen.

Packing of A-type and B-type starch crystalline polymorphs is shown in Figure 8.10 [9, 25, 44, 101]. It is reported that while the dense structure of the A-type only allows few structured water molecules (red dots), the more open B-type makes space for more structured water molecules.

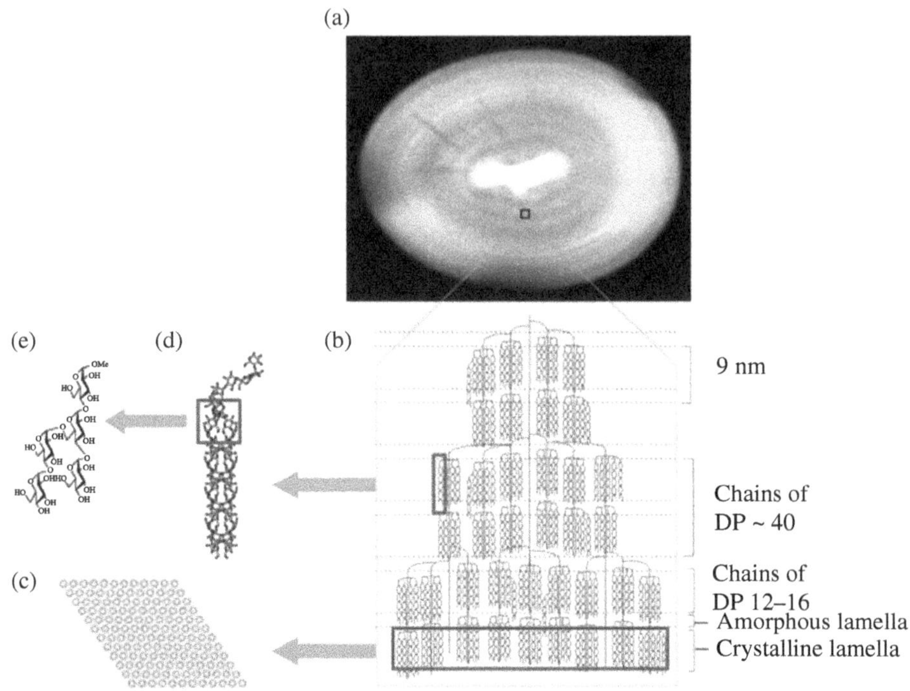

FIGURE 8.8 Principle of top-down strategy of starch structure determination. *Source*: Damager et al. [9]. Reproduced with permission from American Chemical Society; Copyrights: American Chemical Society. (*See insert for color representation of the figure.*)

In addition to these polysaccharides mainly isolated from different carbohydrate plant sources, various fungi and bacteria synthesize polyglucans such as curdlan [α (1→3) linked], scleroglucan, schizophyllan [α (1→3)] linked main chain and [α (1→6) linked branches], and pullulan (α (1→4) and α (1→6) linked). The most important polysaccharide for medical and industrial applications produced by bacterial strains is dextran, a family of neutral polysaccharides consisting of α (1→6) linked D-glucose main chain with varying proportions of linkages and branches [49].

The simple and compact structure of starch and its human analogue glycogen has proven to be very successful for providing energy to living organisms and as energy storage reservoirs in biological systems [15, 21, 102, 103]. The success is based on the hyperefficient molecular packing of the glucose. The metabolism and architecture of these two polymers are highly dependent on the presence of water. The extent of crystallinity, aggregation, and hydration is of fundamental importance for starch. In the case of starch, crystallinity has been attributed mainly to the amylopectin and not to linear amylase, which presents a conformation that hinders its regular association with other chains. Starches exhibit poor melt processability and are highly water soluble, are difficult to process and brittle due to anarchical growth of their

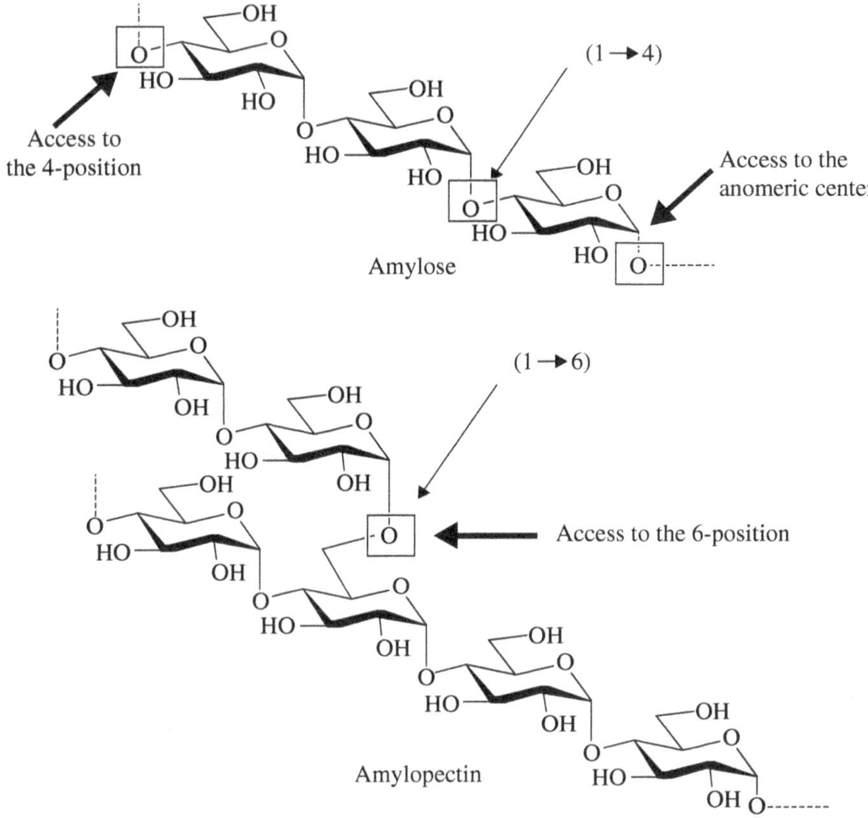

FIGURE 8.9 Representative partial structures of amylose and amylopectin the two polymers of starch. *Source*: Damager et al. [9]. Reproduced with permission from American Chemical Society; Copyrights: American Chemical Society.

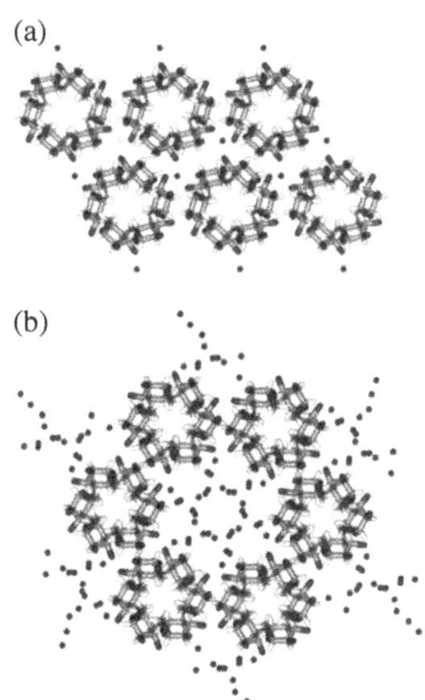

FIGURE 8.10 Hexagonal packing of A-type and B-type starch crystalline polymorphs. *Source*: Damager et al. [9]. Reproduced with the kind permission of American Chemical Society, Copyrights: American Chemical Society. Original is from: Imberty & Pérez, [101]; Reproduced with the kind permission from John Wiley & Sons, Inc. (*See insert for color representation of the figure.*)

constituents, and hence need a plasticizer [27, 55] to make them suitable for engineering applications.

Recently starch has been synthesized [5, 6, 104, 105] by sets of specific enzyme activities, which can lead to the design of advanced and healthy foods to combat obesity and other lifestyle-related diseases [25] (Fig. 8.11). Large and more complex α-glucans are gradually being chemically synthesized.

Furthermore, the structure of hydrolyzed potato starch (PoS) with respect to its molecular weight and degree of polymerization (DP) were studied recently using advanced analytical techniques such as MALDI-TOF mass spectroscopy. Figure 8.11 shows the spectrum obtained indicating two fragments with molecular masses of 5898 and 7078 Da corresponding to 33 and 40 glucose units, respectively, both having a secondary smaller peak (shoulder) corresponding to one additional glucose unit [106, 107]. In the figure showing the information about amylopectin fragments, A indicates MALDI-TOF spectrum of large amylopectin fragments, while B indicates a representative of the ά-glucan structures giving rise to the distinct larger fragments obtained in the case of potato hydrolyzed starch. This result was similar to the one proposed earlier [108] with dimensions of double helical amylopectin motifs in native granular PoS with single stranded fragments of DP typical for PoS [5, 6, 104–107, 109] (Fig. 8.12).

In the case of TPSs, due to the use of plasticizers change in morphology and properties of starches occur, this change being a function of the amount of plasticizer used. Other factors that influence such changes include processing conditions, amylose–amylopectin ratio, type and amount of plasticizer, and moisture content. One such property change can be illustrated by the study of Bergo et al. [110], who observed glass transition temperature (T_g) of CaS (131.9°C) decreasing with increasing addition of glycerol, reaching 62.2°C for 30% glycerol addition. Similarly, they also observed changes in tensile properties of TPS of CaS. Even the melting points of TPS have also been found to decrease [60, 61].

Other observation has been low degree of crystallinity of amylopectin and amylose, when the TPS films were produced under high temperature and low relative humidity, although in general, the degree of crystallinity of the films increased with amylose content [69]. It is reported that amylose films have better ductility and barrier properties than amylopectin films [68]. It may be noted that TPS absorbs moisture when exposed to humidity and therefore exhibits inferior or relatively lower, yet useful mechanical properties compared to conventional synthetic polymeric materials despite its inherent lower strength and stiffness [111]. For example, Young's modulus (YM), ultimate tensile strength (UTS), and strain at break for TPS range from 75–160 MPa, 5.5–7.5 MPa, and 75–180%, respectively, in comparison to 210–2700 MPa, 55–11,000 MPa, and 120–380% for some of the synthetic polymers such as low-density polyethylene (LDPE), high-density polyethylene (HDPE), PP, and PET.

Another TPS studied processing of PoS with water as plasticizer using injection molding technique and characterizing it for rheological, thermal, and molecular changes [112]. It was observed from the rheological studies that this starch could be processed like polyethylene (PE), while YM value (1.5 GPa) of this thermoplastic

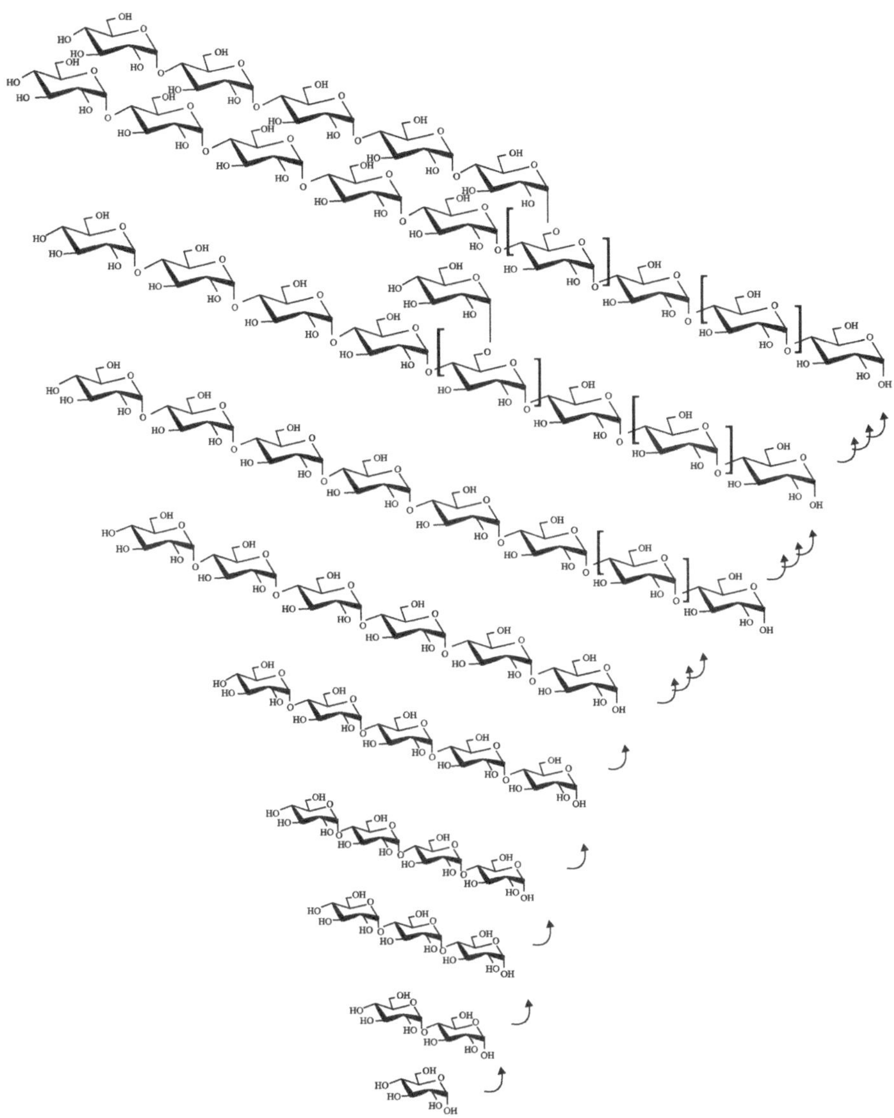

FIGURE 8.11 Principle of the bottom-up strategy of starch structure analysis. *Source*: Damager et al. [9]. Reproduced with permission from American Chemical Society; Copyrights: American Chemical Society.

was similar to that of conventional synthetic polymers [polypropylene (PP) and HDPE]. Stress–strain curves of this TPS were found to be of a typical glassy thermoplastic. Thermal studies carried out by differential scanning calorimetry (DSC) indicated that the starch exhibited gelatinization characteristics similar to those of conventional starch processing at lower temperatures. This was attributed to swelling

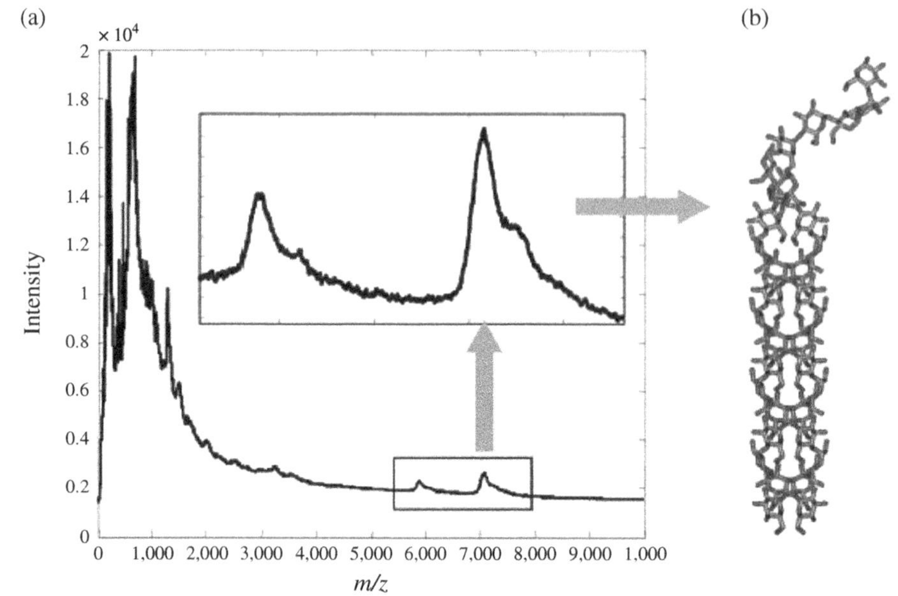

FIGURE 8.12 (a) MALDI-TOF spectrum of large amylopectin fragments. (b) A representative of α-glucan structure giving rise to the distinct larger fragments obtained. *Source*: Engelsen et al. [107]. Reproduced with permission from Elsevier. (*See insert for color representation of the figure.*)

and destruction of starch granules, while observed formation of melt at higher temperatures to destruction of crystallites by thermal and aqueous conditions along with the molecular order.

Knowing that strong inter- and intramolecular interactions among proteins such as wheat proteins can be reduced by plasticizing them with suitable plasticizers such as glycol, wheat starch has been chemically modified using glycol [113]. It was observed that processing of this material got improved, which resulted in TPS exhibiting more flexibility and extensibility with doubling of mechanical properties. For example, at 55% relative humidity, tensile strength increased from 8 to 15 MPa and percent elongation from approximately 70 to 130–140. These have been attributed to the development of cross-linked networks. On the other hand, glass transition temperature (T_g) decreased from 308 to 288 K. However, due to the limited work carried out on wheat starch and other types of starches have hindered expansion of their utilization in composites.

Hence, attempts have been made, as will be discussed later, to overcome the previously mentioned problems along with improving the strength properties of TPS in general, through blending or incorporation of abundant and relatively inexpensive plant materials, such as hard and soft wood fibers or other plant fibers, both micro and nano (jute, coconut, banana, cane bagasse, etc.), to obtain "biocomposites" or "green" composites [55, 114–117], which are totally biodegradable [111, 116].

8.2 STARCH-BASED COMPOSITES

Multicomponent polymeric materials based on starch may be grouped under three groups: (i) combination of starch with plastics, (ii) mechanical mixtures with both fibrous (e.g., micro- and nanosize natural fibers) and nonfibrous (e.g., calcium carbonate, talc, kaolin, mica, silica) materials, and (iii) incorporation of starch at a hyperfine structural level. These include starch-graft biopolymers, miscible or compatible blends of starch/biopolymers, and starch nanocrystal incorporated biopolymer matrices. All the aforementioned possibilities would help in tailoring the mechanical properties of such thermoplastics for possible structural applications depending on the source and processing conditions, or by using fillers, including cellulose micro/nanofibrils and natural "macro" fibers [55, 111, 114]. Accordingly, fabrication methods of these various types of composites also differ, even though most of the conventional methods of polymer-based materials are followed with appropriate modifications. In the following, processing methods to prepare various types of starch-based composites are first discussed (Section 2.1) followed by different composites systems studied including nano and bionano systems (Section 2.1) and then their structure and properties (Section 2.2) and finally their applications (Section 2.3).

Both synthetic and natural polymers used with starch include polyvinyl alcohol (PVA), poly (3-hydroxybutyrate *co*-3-hydroxyvalerate) (PHBV), polyhydroxybutyrate (PHB), etc., while the plant fibers include micro or nanofibers (NFs) of banana, coir, jute, pineapple leaf, sisal, ramie fibers, and even wood fibers. In the case of micro fibers, its content can reach up to 70 wt.%. Other natural materials used as reinforcements in starch include calcium carbonate, graphene, etc. Although there are number of published reports on the aforementioned types of "green" composites based on starch and other natural polymers, only few published mostly with plant fibers as reinforcements since 2000 are given here for brevity [105, 118–153]. One of the main focuses of the aforementioned studies has been the improvement of the mechanical properties of these TPS composites [33].

8.2.1 Processing Techniques/Methods

8.2.1.1 Processing of Starch-based Microcomposites As in the case of conventional microcomposites, intrinsic properties such as solubility, dispersibility, degradation, etc. of both matrix and the reinforcements dictate the type of processing technique used in the preparation of "green" composites including bionanocomposites. From the published literature, it becomes very clear that all the processing/fabrication methods being used for the polymer-based conventional composites can also be used in the case of "green" composites, of course with some modifications depending on the nature (particles, microfibers, NFs, whiskers, etc.) and dimensions of the reinforcements. While doing so, the following factors are important for the processing of "green" composites based on starch as also other natural polymers: proper mixing of filler materials (other polymers/plant fibers (micro or nano) or other fillers such as calcium carbonate, graphene, etc.); proper surface treatments, if required; and finally a high-speed fabrication technique, ofcourse of low cost [154, 155].

As mentioned earlier of this chapter, there are many attractions in the processing of "green" composites and other merits of using plant fibers in composite technology, which can be seen in Refs. [22, 156]. One of the important factors is the ready availability of reinforcements in a suitable form. Hence, researchers have attempted to get the reinforcements in the form required depending on the processing method such as development of three preforming techniques. While Table 8.2 lists briefly a summary of matrix–reinforcement and processing methods used by some of the researchers, more details and other references dealing with the processing aspects of "green" composites including those based on starch matrices on this can be seen in Refs. [22, 156].

Conventional processing methods to prepare polymer-based composites include layup methods, molding (compression, thermo-molding/forming, injection, extrusion, resin transfer) methods, vacuum bonding, and autoclave bonding. Other methods employed to prepare "green" composites particularly to prepare nanocomposites,are solution intercalation, in situ intercalative polymerization, melt intercalation, solution casting, mixing of reinforcements with surfactants followed by freeze drying, and then disperse in some organic solvent [168, 169]. This is followed by the preparation of gel and then by extrusion of the mixture. All these can be classified under (i) dry processing, where composite sheets for compression molding could be obtained; (ii) wet processing, where a sheet-molding product could be obtained; and finally (iii) powder impregnation. However, it may be noted that in the latter method there is no need to use organic solvents.

Also, the process provides ease of mixing and dispersion of multiphase components along with consumption of low energy with the possibility of recycling of powders. In general, on the basis of advantages offered by many of the above techniques, the most preferred technique for the production of "green" composites using plant fibers seems to be press molding or thermo-molding.

On the other hand, solution casting method is the simplest, which involves incorporation of these reinforcements into selected polymeric matrices (hydrophilic/aqueous systems) using a processing medium such as water or solvent. Then evaporation of the liquid can be hastened by freeze drying. In case the reinforcements are magnetic in nature, magnetic field can also be used to achieve the alignment of the reinforcements. In fact, starch-based composites containing bamboo nanocrystals have been prepared by this method [158], which showed good and comparable reinforcing effect even at very low-fiber content (8 wt. %) in comparison to microcomposites containing 30–50 wt. % cellulosic fibers [170, 171].

The casting method of preparing "green" composites has a problem due to the involved complexity particularly within the preparation of nanocomposites such as starch–NFs or whiskers as it involves four constituents (starch, NFs or whiskers, plasticizer and water) [172]. There will also be interfacial problem involved between the constituents particularly the plasticizer water and amylopectin/cellulose

TABLE 8.2 Matrix–reinforcement–processing of starch-based "green" composites

Matrix	Reinforcement	Processing	Process aids/plasticizers/ cross-linkers used	Reference
Corn starch+	Ramie	Ball milling, compression molding	Glycerol	Sreekala et al. [157]
Thermoplastic corn starch+	Bamboo	Compression molding	—	Liu et al. [158]
Cassava starch+	Wheat bran fibers	Solution casting	Glycerol	Fama et al. [159]
Cassava starch+	Coconut fiber	Compression molding		Lomeli et al. [133]
Cassava starch	Bagasse	Baking		Mello and Mali [160]
Starch	Date palm/flax fiber			Ibrahim et al. [51]
Maze starch	Microfibrillated cellulose		BTCA[a]	Dastidar and Netravali [161]
Starch	Jute/hemp	Injection molding	—	Vilaseca et al. [27, 162]
Starch	Sisal short fiber	Injection molding	—	Alvarez et al. [163]
Starch				
Nonextracted cassava starch+	Kenaf fiber	Compression molding	PVA, PEG, glycerol	Song and Kim [164]
				Chen et al. [165]
Sago starch	Clay		Glycerol	Zuraida et al. [166], or Li et al. [167]

[a] BTCA, butane tetracarboxylic acid cross-linker.

interferes with hydrogen bonding of inter whiskers, which would lead to low mechanical properties.

This film casting method can be applied in the preparation of bionanocomposites using BC either by incorporating BC fibrils into a polymer matrix such as starch [61, 173] or by adding the polymer to the culture medium [174]. In fact, this method has been successfully used recently in the preparation of bionanocomposites of CaS containing with and without surface-treated mats of BC (*Acetobacter xylinum*, product of Membracel Produtos Tecnologicos Ltda, Almirante Tamandare, PR, Brazil) [61]. The procedure followed was homogenization of dispersion of BC in sodium azide solution in a blender kept under refrigeration. This mixture was subjected to enzymatic hydrolysis before immersion into boiling water in order to inactivate the enzymes and cooled in an ice bath. Then, thus prepared BC was mixed with CaS plasticized with 30% glycerol. The composite mixture was then heated under reflux, and for 30 min this was stirred at 90–95°C to ensure complete gelatinization of starch and then cast into films. High temperatures and/or long drying periods without vacuum produced inhomogeneous materials with air bubbles visibly trapped in the films. Although the process was optimized, the dispersion of nano BC was not good [175] as evident from the opacity of the obtained composite films. Thus prepared films were then maintained for at 43% RH in a desiccator at room temperature before subjecting them for various tests.

On the other hand, the aforementioned problem can be overcome by resorting to "melt compounding" method, wherein both compounding of the composite constituents and then hot pressing or extrusion of this mixture are carried out. For example, nanocomposite of CaS with cellulose whiskers (CW) was prepared in two steps in this method using sorbitol and glycerol mixture as plasticizer [172]. Using a Rheomixer at fixed speed for fixed time, homogenous mixture of starch and the plasticizer was produced by subjecting it to ultrasonic treatment for about 5 min and then required amount of reinforcement (in the form of aqueous suspension) was added to this. Complete mixture was then hot pressed in suitable steel mold using hot press maintaining the temperature at which neither the volatilization of glycerol nor thermal degradation of reinforcement took place. Finally, obtained composite plates (laminates) were dried at 60–70°C till constant weight was obtained. In order to enhance the wetting between the matrix and the reinforcements, sometimes, coating of the reinforcements with a suitable surfactant or by surface modification whereby could be resorted to, which would improve wetting and also the fiber distribution [126, 168, 176, 177].

Besides, there is also review on the thermal processing of starch-based polymers [178], while some researchers have used the thermo-molding method when both the constituents (matrix and the reinforcement) are in solid form [128, 132, 133, 136]. Used matrices are starches of castor oil cake, cassava, natural corn starch, and hydrolyzed starch, and reinforcements were micro fibers of green coconut, banana, sugarcane bagasse, and nanocellulose fibers of balsa. In all these cases, the mixture of matrix and plasticizer was powdered and then cut fibers were added and homogenously mixed before subjecting the composite mixture to thermo-molding process at suitable temperature and pressure for a fixed time. Composites thus obtained showed improved properties.

In all the aforementioned processes, care should be taken with regard to preservation of mechanical properties of the plant fibers by minimizing attrition and degradation due to mixing along with ensuring good wettability to enable incorporation of a higher volume fraction of fibers with control of fiber orientation and to achieve a high degree of fiber dispersion.

Yet another method of preparing starch-based polymer composites is by the combination of starch with plastics, also called as "modifications of starches." This method includes blending them with other polymers and use of plasticizers such as glycerol and water, and these have been the subject matter of many studies. Details of some examples of this are given later under different blend systems.

One of the examples reported was the blending of a mixture of wheat starch granules with polylactic acid (PLA) or LDPE using both glycerol and water in different proportions, which were used as plasticizers processed by twin-screw extruders [144, 179]. It was observed that biodegradability in the PLA/starch blend was retained, while increase in modulus and decrease in ductility in the LDPE–starch blend due to the starch granules acting as filler. These results indicate that depending on the plasticizer content (glycerol and water) and its dispersion in the blends, plasticization can lead to a wide range of properties. Here the starch granules acted as filler resulting in an increased modulus but loss of ductility of the composites with increasing starch content. This process was found to be very economical and also does not affect their biodegradability of the resulting composite. This result underlines the importance of plasticized starch (PIS) leading to a wide range of properties such as variable biodegradation, depending on the plasticizer content (glycerol and water), and its dispersion in the PLA leading to a wide range of properties. More details can be found in the original references cited previously.

Blends of different amounts (65, 75, 85, and 95 g) of PVA with CaS with water or citric acid or adipic acid or glycerol as plasticizers has been studied using a single-screw extruder [144].

Reactive extrusion, in presence of radical initiators such as 2,2′-azobis(2-methylpropionamidine) dihydrogen chloride or 2,2′-azobis (isobutyronitrile), has been used to prepare PoS grafted with *N*-vinylformamide (NVF) copolymers [180]. The parameters such as screw speed, temperature profile, and initiator type were changed, while ratio of the constituents (NVF/starch/water) was maintained at 40:100:34. Solubility tests in alkali aqueous solution at pH 9.0 carried out to find the raw copolymerization products determined on the basis of extraction with water/ethanol mixture in the ratio of 70:30 revealed that (i) raw reactive products changes between 50 and 75 wt. % and (ii) the dissolution in the range of 18–43 wt. %, while the products obtained at lower extrusion temperature profile (six heating sections at 60°C) exhibited higher solubilities. Prepared starch-g-NVF copolymer showed branches having two–three times shorter molar mass than that of polyacrylamide branches of starch-g-acrylamide copolymers on enzymatic degradation with α-amylase/β-amylase enzyme system.

A new method to produce *in situ* composite developed by Chen et al. [181] was recently reported, wherein grafting reaction of the hydroxyl groups on the surface modified starch (SM-St) with L-lactic acid was made. In this case, a novel biodegradable starch-grafted copolymer, starch-g-poly(L-lactide) (St-g-PLLA), was synthesized *in situ* by the ring-opening graft polymerization of a L-lactide (LLA) monomer onto the surface of the SM-St granules in the presence of $Sn(Oct)_2$ as a catalyst. Grafting efficiency of the PLLA could reach as high as 64 wt. %. Characterization with respect to structure of St-g-PLLA was carried out by IR, DSC, and WAXD, while SEM studies revealed good adhesion between the two components. The authors have also observed improved medium resistance of St-g-PLLA and the mechanical properties of the PLLA composite blending with St-g-PLLA over those of the PLLA/starch blending composite.

There are other methods of preparing "green" composites with matrices other than starch and wood particles or flakes and even fibers of banana, sugarcane bagasse, etc. as reinforcements particularly for preparing particle boards such as three-layer particle boards and graded-density particle boards [3]. As these may not be of much use for starch-based "green" composites, no other details are given here.

Properties of some of these composites have been found to be comparable and even competitive to E-glass/epoxy composites, suggesting that these composites could be used even in various structural applications provided some of the limitations are overcome. These include high hydrophilic nature of composites based on starches due to high solubility of starches in water [115, 120] and high percent elongation at break (30–100%) coupled with low tensile strength at 40–75% relative humidity (RH) [111, 141, 152], when known plasticizers such as glycerol are used. Besides, dependence of properties particularly strength properties due to change in humidity is another factor to be considered as evident from high values of percent elongation at break of 83 ± 17 and 101 ± 15 at 58 and 75% RH values, respectively, in a starch with 30% glycerol matrix [141].

8.2.1.2 Processing of Starch-based Nanocomposites Considering the compatibility between the biopolymers and lignocellulosic NFs/whiskers, study of bionanocomposites of these materials is mostly limited to hydrophilic matrices. However, with surface modifications including chemical grafting [182] of reinforcements, this could be extended to hydrophobic matrices also. Several examples of this can be seen in Ref. [183].

In the case of processing of bionanocomposites, as in the case of microcomposites, intrinsic properties such as solubility, dispersibility, degradation, etc. of both constituents dictate their processing techniques. It has been reported that the complexity involved in using nanocellulose in the preparation of bionanocomposites by casting method compared to conventional melt processing method is one of the major problems encountered [172]. This is evident from the four constituents of starch-based nanocomposites (starch, natural fibers/cellulose, main plasticizer, and water). Along with this, the other problem is the competitive interactions involved between these constituents. Interference of collection of plasticizers (glycerol and water) near the interface zone of amylopectin/cellulose with hydrogen bonding

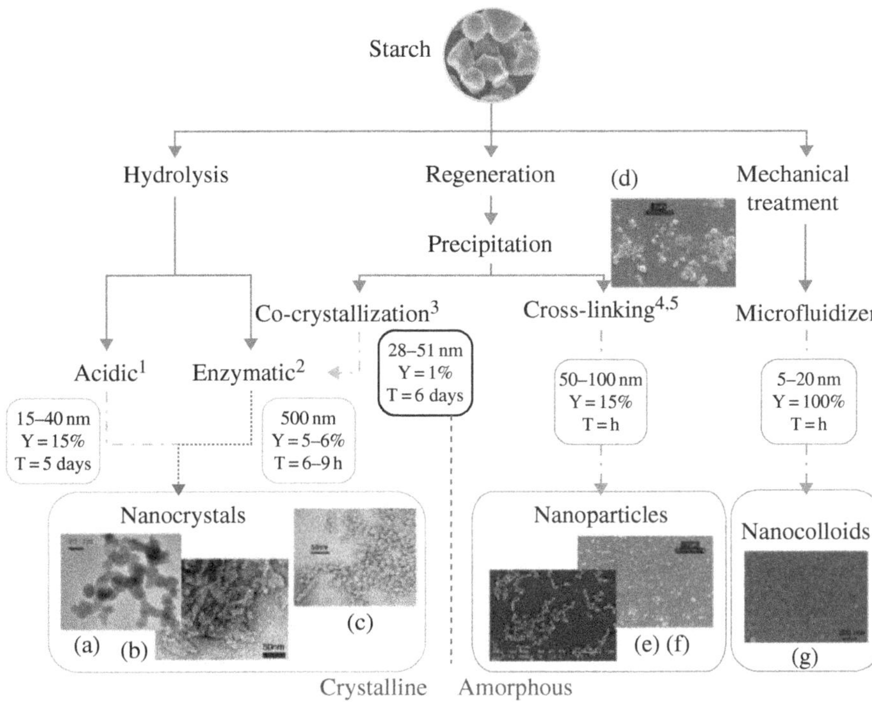

FIGURE 8.13 Different ways of producing crystalline and amorphous starch nanoparticles and TEM photographs of resulting nanoparticles. Le Corre et al. [17]. Reprinted with permission from American Chemical Society; *Source*: (a) Garcia et al. [184]. Reproduced with permission with Wiley-VCH Verlag GmbH & Co. KGaA; (b) Putaux et al. [90]; (c and e) Tan et al. [185]. Reprinted with permission from American Chemical Society; (d and f) Ma et al. [39]. Reprinted with permission from American Chemical Society; (g) Liu et al. [186]. (*See insert for color representation of the figure.*)

of inter whiskers is one such example, which results in lowering of the mechanical properties of nanocomposites.

Accordingly, there are several techniques for preparation of nanocomposites including mixing in solution or melt and processing, which include casting, extrusion/reactive extrusion, injection and compression molding, melt intercalation and solution intercalation, *in situ* intercalative polymerization, mixing of CW with surfactants followed by freeze drying and then disperse in some organic solvent [168, 169], melt intercalation and preparation of gel followed by extrusion. Some of these methods used to synthesize starch nano systems are shown in Figure 8.13, while a combination of 2–3 of these methods have been used by some researchers. For example, *in situ* intercalative polymerization was used recently in the preparation of PS–clay nanocomposites [187], while reactive extrusion, injection molding, and mixing techniques have been used for the processing of starch-based nanocomposites [17, 187].

Figure 8.13 depicts three methods of producing crystalline and amorphous starch nanoparticles, namely, hydrolysis, regeneration, and mechanical treatments. The first method leads to nanocrystals, whereas the latter two methods lead to both amorphous and crystalline particles in the final batch.

The figure also shows transmission electron micrographs obtained in these processes by different researchers (a, [184]; b, [90]; c and e, [185]; d and f, [39]; g, [186]). Micrograph (c) is of SNCs after cocrystallization and enzymatic hydrolysis. Numbers in the figure indicate different aspects of nanocrystals processing. Thus, one and two indicate acidic and enzymatic hydrolysis methods [188] and three indicates preparation of nanocrystals by complex formation between amylose and *n*-butanol [189]. It may be noted that the enzymatic hydrolysis is used to selectively keep crystalline particles, and this process has been believed to lead to blocklets rather than nanoparticles. Further, numbers four and five indicate production of starch nanoparticles by precipitation of gelatinized starch in nonsolvent followed by a cross-linking reaction [39, 185], and finally, six indicates production of starch nanoparticles by fluidization similar to that of microfibrillated cellulose (MFC) [178, 186] intended. It leads to crystalline microparticles turning into amorphous nanoparticles with increasing run numbers.

It may be noted that solution intercalation technique used in the case of preparation of bionanocomposites is based on a solvent system in which starch (biopolymer) is soluble in solution. The nanofiller is then swollen in another solvent such as water, chloroform, or toluene. Then, the biopolymer and solution of swollen nanoparticle are mixed to get the bionanocomposite; the polymer chains intercalate and may replace the solvent molecules within the interlayers of the aggregated nanofiller. Self-assembly in bionanosystems based on xyloglucan/oleylamine-modified TEMPO-oxidized nanocellulose/starch is another method reported recently [190].

8.2.2 Structure and Properties of Starch-Polymer Systems (Blends/Composites)

Tables 8.3, 8.4, and 8.5 summarizes various composites based on starches studied, their processing methods, and some of the characteristics of these systems. More details of these can be seen from the references mentioned against each system.

As mentioned earlier in this chapter, advanced analytical techniques have also been used recently to study the structure of starch with respect to the molecular weight and DP.

8.2.2.1 Starch-Polymer Systems Normally, starches are modified either to improve some of its properties or to make useful materials for structural applications by blending with other polymers. Some detail of studies carried out by many researchers over the years is briefly given in the following pages.

PLA/Starch Systems (Blends/Composites) PLA is one of the most promising biodegradable materials with excellent mechanical properties but has some disadvantages

TABLE 8.3 Matrix–reinforcement–processing and characteristics of biodegradable nanocomposites: properties of starch–polymer composites

System	Properties	Reference
PLA + starch composites	Increased crystallization, mechanical properties increased from 35 to 50 MPa	Tserki et al. [150]
Starch/polycaprolactone composites	TS and percent elongation increased by 20–30%	Drzal et al. [154]
Starch/polypropylene carbonate composites	TS 19.4 MPa; elongation 88.5%	Liu et al. [158]
Polyvinyl acetate–starch composites	Swelling reduction in water	Niu et al. [191]
Starch/polyester composites	No deterioration in TS With 70% starch	Ohkita and Lee [192]
Starch/PVA composites, ammonium polyphosphate	Flame retardant composite with V-0 rating (UL-94)	Wu et al. [193]
Starch/LDPE composites	Degradable composite with 40–70% starch	Inceoglu and Menceloglu [194]
Potato starch, LDPE, LDPE-g-MA 0.5%	TS reduced above 15% starch elongation decreased above 5% starch	Gupta et al. [195]
PHB/starch composites	TS increases from 20 to 90 MPa by 10% starch, above 10% starch TS decreases	Ismail and Gamal [196]

TABLE 8.4 Matrix–reinforcement–processing and characteristics of biodegradable nanocomposites: properties of starch–natural fiber composites

System	Fiber	Mechanical properties	Reference
Cassava starch, glycerol, potassium sorbate	40% Wheat bran fiber	Improved properties	Fama et al. [159]
Cassava starch, glycerol	30% Coir	3.24 MPa (TS) 59.8 MPA (modulus)	Lomelí et al. [132, 133]
Castor oil cake starch, glycerol	Bagasse	Improved TS	Vanderleia [151]; Guimarães et al. [128]
Corn starch, glycerol	25% Banana	Improved TS	Guimarães et al. [128]
Potato starch (PoS)	15% Bagasse	24.5 MPa (TS)	Chen et al. [121, 165]
Starch-based biopolymer (Ribawood)	Jute	Improved TS and flexural strength	Vilaseca et al. [27]
Corn starch/glycerol	Sisal	Modulus 560% higher, TS 162% higher	Corradini et al. [197]
Starch, PVA, PEG, glycerol	Kenaf	TS 9.72 MPa	Song and Kim [164]
Starch	5–20% Sisal	Improved TS	Alvarez et al. [163]
Starch	Hemp	Improved TS	Vilaseca et al. [162]
Cassava starch, glycerol	Bagasse	Suitable for automobile parts	Kaewtatip and Thongmee [198]
Starch	Alpha grass	Improved properties	Espinach et al. [199]

TS, tensile strength.

TABLE 8.5 Matrix–reinforcement–processing and characteristics of biodegradable nanocomposites: properties of starch based other composites including nanocomposites

System	Properties/comments	Reference
Maize starch/glycerol/ kenaf nanofiber	21% decrease in moisture content with 10% nanofiber	Karimi et al. [200]
Corn starch/newspaper (cellulose) nanofiber	The composites containing 8 wt. % of fibers presented the best improvement of mechanical properties and thermal resistance and the highest decrease of percentage water absorption when compared to the others. Can be used as commodity plastics/biodegradable artifacts, for example, organic waste bags and seeding grow bags, being strong, cheap, abundant, and recyclable	Wattanakornsiri et al. [201]
Waxy maize starch (WMS)/1,2,3,4-butane tetracarboxylic acid/ microfibrillated cellulose (MFC)	15% MFC-cross-linked WMS composite films exhibited Young's modulus of over 2.3 GPa, fracture strain of 3.1%, and fracture stress of 39 MPa. The toughness was significantly higher, even without the use of plasticizers such as sorbitol. Good candidate materials for replacing epoxies	Dastidar and Netravali [161]
Potato starch (PS)/ graphene (GN)	1.774 wt. % GN–starch, the tensile strength reached 25.4 MPa, while the tensile strength of PS was only 8.2 MPa. WVP values decreased from 1.5×10^{-10} to 1.1×10^{-10} g/(m s Pa) when the loading of GN–starch increased from 0 to 0.248 wt. %. PS/GN–starch composites could shield most of the UV light. Also, the conductivity of the composite was 9.7×10^{-4} S/cm at a GN–starch content of 1.774 wt. %	Zheng et al. [202]
Corn starch / montmorillonite clay/ urea/ethanolamine	The tensile stress, tensile strain, Young's modulus, and breaking energy, respectively, achieved 29.81 MPa, 152.14%, 734.58 MPa, and 3.05 Nm. Thermal stability and water resistance of composites also improved	Huang and Yu [203]
Pea starch nanocrystal-g-polycaprolactone (PCL)/polylactic acid (PLA) nanocomposite	Elongation and strength were enhanced with 5 wt. % StN-g-PCL	Yu et al. [204]
Oxidized pea starch (OPS)/chitosan (CS)	Mechanical properties testing showed that the carboxyl groups introduced into the starch matrix improved the properties of the OPS/CS composites. The tensile strength of oxidized starch was higher than that of regular starch, water barrier property of OPS blend films improved. Potential applications in agriculture, in drug release and medicine, and in packaging.	Wu et al. [205]

such as stiffness, poor thermal stability particularly occurring during heat processing, and high costs. These limit its applications in general use of plastics such as industrial packaging and agriculture films. Therefore, an improvement of PLA properties has been one of the challenging research topics. Accordingly, attempts have been made to melt blend them with starch (10–30 wt. %) and also incorporate some lignocellulosic fibers such as wood sawdust, hazelnut shells, flax fibers, and corncob. Accordingly, many researchers have studied PLA/starch composites by melt blending [206], twin-screw extruder and compression molding [207], extrusion and uniaxial orientation processing with a pretreated mixing [181], and grafting reaction of the hydroxyl groups on the starch with L-lactic acid [208]. It was observed that although the thermal stability as well as the storage modulus of PLA/starch composites has increased with increasing starch content, oxygen permeability was found to decrease [206].

A study of PLA/starch composites containing 30 and 36 wt. % of glycerol as plasticizer processed by foaming revealed that both tensile strength and YM of these materials increased with increasing PLA content (TS: 16–22 MPa, YM: 1.8–2.4 GPa for 40 wt. % PLA to TS: 52–56 MPa, and YM: 3.6–4 GPa for 100 wt. % PLA) [179]. On the other hand, the viscosity of starch was found to be affected by the plasticizer (glycerol) and residual water content; with increasing shear rates from 1 to 75 s^{-1}, the viscosity decreased from 0.2×10^5 to 0.2×10^3 Pa s and approximately 10^4 to 0.5×10^2 Pa s, for materials with 30 and 36 wt. % glycerol, respectively. However, they observed constant (~0.6×10^3 Pa s) viscosity of PLA over a wide range of shear rates (10^{-2} to <101 s^{-1}) before it decreased to 0.2×10^3 Pa s at shear rates above 10 s^{-1}. In the case of strength values, they were found to remain constant even after 10 weeks of aging at room temperature suggesting stability of this material during that period. The authors opined that PIS was protected by PLA from potential changes in glycerol and water (plasticizer) content of the starch.

On the other hand, surface modification of starch for better compatibility with PLA using epoxidized itaconic acid (EIA) and epoxidized cardanol (Epicard) exhibited increase in tensile strength from about 35 MPa to over 50 MPa and enhanced crystallization ability of PLA [209]. While the former has been attributed to strong interaction formed at the interface between modified starch and PLA resulting in improved interfacial adhesion as observed by SEM studies, the latter was confirmed by DSC and DMA. Surface modification of starch carried out to produce hydrophobicity was confirmed by the FT-IR and ^{1}H NMR studies, as observed by increased contact angle from 44 to 100°.

Improvement of interfacial adhesion between PLA and starch when Epicard was used was evident from the microstructural study of the composite using scanning electron microscopy [209]. Figure 8.14 shows fractographs of (i) 70 PLA-30 starch blend, (ii) 68 PLA-2- cardanol- 30 starch blend, (iii) 68 PLA-2 Epicard- 30 starch, (iv) 45 PLA- 5 Epicard -50 starch blend, and (v) 45 PLA- 5 Epicard -50 starch blend.

It can be seen from Figure 8.14a that the fracture surface shows many holes with sharp edges caused by the removal of starch from the PLA matrix due to the poor interfacial adhesion between the hydrophobic PLA and typically hydrophilic starch [210, 211]. This suggests incompatibility between the matrix and filler. Improvement

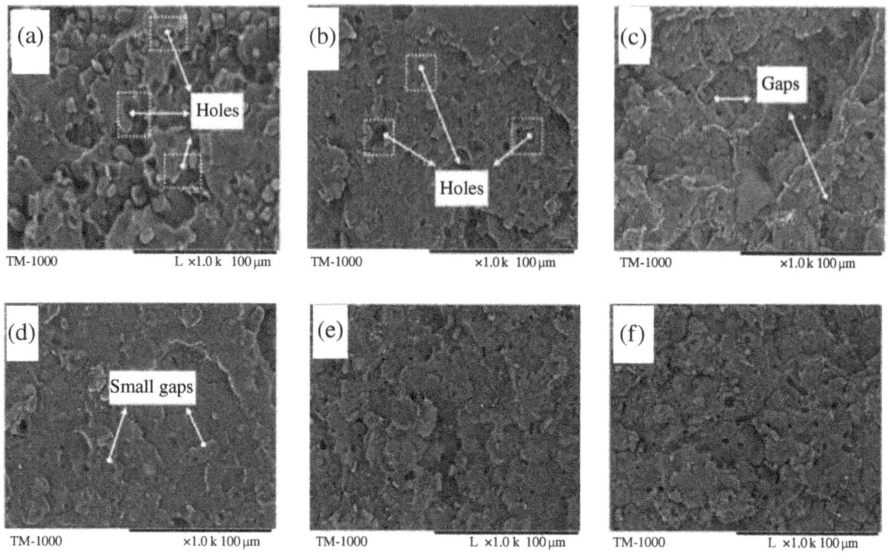

FIGURE 8.14 Scanning electron micrographs of (a) 70 PLA-30 starch blend, (b) 68 PLA-2 cardanol-30 starch blend, (c) 68 PLA-2 Epicard-30 starch blend, (d) 45 PLA-5 Epicard-50 starch blend, (e) 45 PLA-5 Epicard-50 starch blend, and (f) 45 PLA-5 Epicard-60 starch blend. *Source*: Xiong et al. [209]. Reproduced with permission from American Chemical Society. Copyright 2014.

of compatibility and thereby the adhesion between the matrix and the filler starch can be seen by the addition of cardanol or Epicard as depicted in Figures 8.14b, c.

It can also be seen from these two figures that the 2% of cardanol marginally decreases the holes, while the same amount of Epicard almost eliminates the holes. These observations suggest absence of any chemical reaction between starch and cardanol, which could not change the starch, while Epicard does (confirmed by the— ^{1}H NMR spectra of extracted starch particles from PLA/starch blends with and without cardanol and Epicard). In fact, presence of number of pores is also evident in Figure 8.14b, which has been attributed to the occurrence of coalescence of cardanol in the PLA matrix of the ternary blend.

On the other hand, in the other ternary blend of PLA–Epicard–starch, reaction between the latter two during processing (melt blending) could take place (Fig. 8.14b) as confirmed by ^{1}H NMR spectra as mentioned earlier. Marginal increase in adhesion between the matrix and starch could be achieved with increasing amount of only Epicard (5%) as evident from Figure 8.14d but still showing a small amount of unreacted Epicard, while increasing the amount of starch from 30 to 60 wt. %, it can be seen that the interfacial adhesion between PLA and starch via the reaction between starch and Epicard improves due to more reaction between Epicard and increasing amount of starch. This is evident from well-embedded starch particles in the PLA matrix (Fig. 8.14e, f).

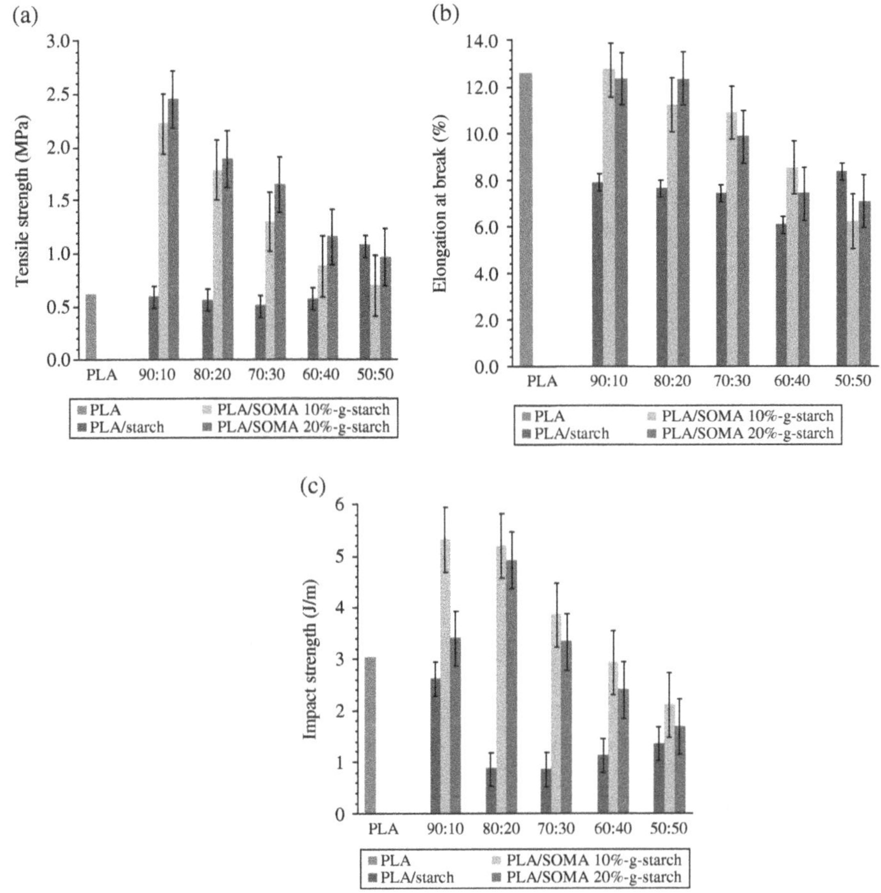

FIGURE 8.15 Plots of mechanical properties of PLA matrix and its composites (PLA/starch, PLA/10 and 20% SOMA-g-STARCH); (a) tensile strength, (b) percent elongation, and (c) impact strength. *Source*: Kiangkitiwan and Srikulkit [212]. CC public domain. (*See insert for color representation of the figure.*)

Considering the limitations of the cost and very poor mechanical properties of PLA compared to synthetic plastics, an attempt has been reported to improve the interfacial adhesion between the matrix (PLA) and reinforcement (CaS powder) [212]. For this purpose, the latter was prepared first by mixing CaS granules and soybean oil maleate grafted with maleic anhydride (MA) using dicumyl peroxide (DCP) as an initiator in a twin-screw extruder to get SOMA-g-STARCH. Then, the mixture of PLA and different amounts of SOMA-g-STARCH was subjected to compression molding to obtain the composite of PLA/SOMA-g-STARCH with compositions of 90 : 10, 80 : 20, 70 : 30, 60 : 40, and 50 : 50. The composite was characterized by mechanical and thermal properties, morphology, and compatibility. Figure 8.15a–c shows the plots of tensile strength, percent elongation at break, and impact strength values, respectively, of PLA, PLA/starch composites, and PLA/SOMA-g-STARCH composites. It can be seen from the figure (Fig. 8.15a) that the tensile strength of melt extruded PLA is only 0.6 MPa, which is unusually low when compared to the literature reported value of 40–50 MPa. This has been attributed to the fact that the PLA used as matrix was twice melt processed and therefore there could be possibility of substantial depolymerization of PLA, which resulted in a significant reduction of its molecular weight. Further reduction was observed in tensile strength for PLA-unmodified starch composites, which has been attributed probably due to phase separation caused by the poor interfacial adhesion between PLA and unmodified starch. However, as can be seen, the tensile strength improved and is higher than those of PLA/starch composites with surface modification of starch with soybean oil maleate (PLA/SOMA-g-STARCH composites) due to improvement of interfacial adhesion between the matrix and PLA and filler starch powder.

On the other hand, from Figure 8.15b, it can be seen that the percent elongations of pure PLA matrix are higher than those of all the PLA/starch composites with and without surface modification, even though slight increase in its value for composite containing 50% starch over the other compositions and almost the same value as that of the matrix for PLA-10% SOMA-g-STARCH composite. These observations have been attributed to imperfect mixing of PLA and starch in addition to formation of boundary cracks in the PLA-unmodified starch composites, while possible occurrence of phase separation in PLA-20% SOMA-g-STARCH composites.

In the case of impact strength, it is found to be very low (3 J/m) for the PLA matrix similar to that observed in the case of tensile strength (Fig. 8.15c), which has been attributed to the rigid nature of PLA. Similarly, it gets further reduced for PLA-unmodified starch composites, which could be attributed to incompatibility between the matrix and the filler unmodified starch. However, impact strength values for modified starch-filled composites could be seen to be higher than those of PLA and PLA/starch composites, which have been attributed to good distribution of filler and improved interfacial adhesion probably due to soybean oil having good plasticity on the particle surface. Based on the previous text, it has also been reported that the best properties (compatibility, surface appearance, and mechanical properties) were exhibited by composites with 90 PLA-10 SOMA-g-STARCH and 80 PLA- and 20 SOMA-g-STARCH.

Another study of extruded PLA and its blends with different types of starch (rice, corn, wheat, and potato) and methylene diphenyl diisocyanate (MDI) using a twin-screw extruder and then compression molded showed type of starch used, and their granular size had no significant effects on mechanical properties (YM, tensile strength, and percent elongation) of the resulting composite in the case of all the starches except for the PoS, while crystallinity of the composites decreased with the increase in the granule size of the starch [207]. In the case of composites PoS with PLA/MDI (45/55/0.5), both tensile strength and elongation showed slight decrease when the granule size of PoS increased from 19 to 63 μm.

Extrusion and uniaxially orientated processing of pretreated mixing of PLA and etherified starch (EST) under a programmed temperature showed that the tensile properties (tensile strength, elongation at break, and modulus) of the composites with a pretreatment procedure were significantly improved compared to those of samples without the pretreatment [208]. This has been attributed to the strong interaction of a small amount of PLA molecules with EST, which might be ascribed to the esterification between the terminal carboxyl groups of PLA and hydroxyl groups in EST. This was confirmed by the SEM studies of morphology of the composites, which underlined that a pretreatment procedure resulted in better dispersion and homogeneity of starch in the PLA matrix than the composites without the pretreatment procedure. This technique has been identified as a reasonable method to produce commercial PLA/starch composites with economic feasibility.

It was found necessary to have surface modification of starch to achieve better compatibility between the starch and PLA. Fully biobased PLA/starch composites with starch having surface modification by EIA and epoxidized cardanol (Epicard) [209] showed superior mechanical and thermal properties.

Yet another study reported preparation of a blend of starch and poly (L-lactic acid), [PLLA] by melt-based processing technique for using it in the cartilage side to overcome inadequate natural rebuilding capacity of a conventional articular cartilage [213]. Based on the potential of tissue engineering to provide an effective alternative to engineer the damaged cartilage, an integrated porous bilayered scaffold was developed with a view to mimic the requirements of cartilage and underlying subchondral bone. Accordingly, a common polymeric component was included in both cartilage and bone components. This maximized the integration at the interface. In order to induce the bioactivity in the bone region, PLLA was reinforced with hydroxyapatite (HA) and bioactive glass (BaG). The surfaces of this studied by SEM and FTIR techniques as a function of soaking time in a simulated body fluid (SBF) revealed formation of a bone-like apatite and apatite layer covering the surface. This was found to increase with increasing soaking time, while the cartilage layer did not exhibit the formation of any apatite-like layer. The blend of starch and PLLA was found to possess an adequate water uptake capability, indicating this as an interesting application of PLA/starch composites.

Development of an *in situ* composite of PLLA–starch (starch-g-poly (L-lactide) (St-g-PLLA)) has been reported based on grafting reaction of the hydroxyl groups on the starch with L-lactic acid [181]. This process involved ring-opening graft polymerization of a LLA monomer onto the surface of the SM-St granules in the presence of $Sn(Oct)_2$ as a catalyst. It was observed that the PLLA grafting efficiency could reach as high as 64 wt. %. Various characterization techniques (IR, DSC, and WAXD) revealed the structure of St-g-PLLA. Mechanical properties of the developed PLLA St-g-PLLA blended composite and the medium resistance of St-g-PLLA were found to be higher than that of the PLLA/starch blending composite. SEM studies of the composite revealed good adhesion between the two components supporting the observed improved mechanical properties of the composite.

Usually, low tensile and yield strength, stiffness and elongation at break, as well as poor moisture stability have been observed in most starch-based composites, warranting attempts to improve these properties. Accordingly, a study has reported the use of starch as fillers in starch-filled polymer blends, TPS with biodegradable synthetic polymer like PLA to produce foamed starch [214]. This study has helped in realizing renewable biodegradable substitutes for the conventional plastics.

TPS/PLA composites have been studied to determine the nonisothermal crystallization kinetics of pure PLA and TPS/PLA composites using a DSC method with three different cooling rates. This study was aimed at finding an appropriate use for the thermoplastic corn starch [215] The authors observed different crystallizations in these TPS/PLA composites due to the different amounts of the contents and different cooling rates. It was also observed that the crystallinity of the PLA–TPS was improved and the mobility of chains was constrained in view of PLA acting as a nucleating agent. To describe the process of nonisothermal crystallization, three theoretical models, namely, the Avrami, Ozawa, and Mo models, have been used. It was observed that the Avrami analysis modified by Jeziorny and the Mo method successfully described the nonisothermal crystallization process of the pure PLA and the TPS/PLA composites. On the other hand, the Ozawa analysis could not give an adequate description. Kinetic parameters such as the Avrami exponent, kinetic crystallization rate constant, relative degree of crystallinity, and crystallization enthalpy, etc. have also been determined at various scanning rates.

One of the studies has reported preparation and characterization of blends of wheat starch with PLA or LDPE, with starch granules using both glycerol and water in different proportions were used as plasticizers [19, 207]. It was observed that biodegradability in the PLA/starch blend was retained, while increase in modulus and decrease in ductility occurred in the LDPE–starch blend. This indicated that depending on the plasticizer content (glycerol and water) and its dispersion in the blends, plasticization can lead to a wide range of properties.

Starch/Polycaprolactone Systems (Blends/Composites) Biodegradable polymers such as starch are often combined with osteoconductive materials, such as BaG for bone regeneration and repair. Toward this end, starch-poly-ε-caprolactone (SPCL) composites with and without BaG fibers have been prepared and characterized. For this purpose, sheets of 30 wt. % starch- 70 wt. % poly-ε-caprolactone (SPCL) were produced using single-screw extrusion and then compression molded with 10–20% layers of BaG fibers to obtain their composites [216]. It was observed that initial mechanical properties of the reinforced composites were at least 50% better than the properties of the matrix. However, on hydrolysis these properties were found to decrease for 2 weeks when the values approached the value same as those of the matrix material. On the other hand, even after 6 weeks of hydrolysis, the mass of the composites was found to decrease only by about 5% indicating that the amount of glass in the composites remained almost constant up to a 6-week period of hydrolysis. The former also suggests that it is possible to enhance initial mechanical properties of SPCL by reinforcing it with BaG fibers. Although observed mechanical properties of the composites are typical for bone fillers, the authors have opined that the strength properties need to be further improved for allowing more demanding bone applications.

With the aim of understanding the effect of different amounts of silane coupling agent on the properties of starch polymer blends and their composites, PCL–starch (starch having sodium octenyl succinate groups) blends and their composites with the incorporation of pine-leaf fibers were prepared by melt mixing at 160°C and characterized for physical properties such as water absorption, mechanical and thermal properties, and morphology studies [217]. It was observed that both the physical properties and morphology of the PCL/starch and PCL/starch /pine-leaf composites were largely improved by the composition of the composite, particularly by the content of the silane coupling agent. In fact, interfacial compatibility between the PCL matrix and the fillers improved considerably as evidenced by the morphological studies of the composites wherein good interfacial adhesion between the PCL matrix and the fillers was observed. These materials have the potential application as biodegradable food packaging materials with natural pine flavor. These composites with the silane coupling agent exhibited lower absorption of water than that of the PCL/starch composite with the silane coupling agent. This has been reported to suggest that pine-leaf powder could inhibit the access of water into the composite, and pine-leaf powder is less hydrophilic compared to starch. Observed comparatively lower water absorption capacity of the PCL/starch and PCL/starch/pine-leaf composites both with the silane coupling agent has been attributed to good distribution of starch and pine-leaf powder in the PCL matrix and successful compatibilization with PCL at the interfaces resulting good interfacial adhesion in the presence of the silane coupling agent.

Starch/PP Carbonate System (Blend/Composites) Polypropylene carbonate (PPC)–corn starch (CS)–poly(methyl acrylate) [PMA] composites were prepared using starch-g-poly(methyl acrylate) (S-g-PMA) copolymer prepared by ceric ammonium nitrate-initiated grafting reaction of PMA onto the corn starch in a plastic order batch mixer at 150°C and characterized for mechanical and thermal properties [218]. Significantly improved tensile properties (stiffness, strength, and elongation) and thermal stability were observed in the prepared material with the incorporation of S-g-PMA into PPC. These have been attributed to better phase adhesion between PPC matrix and S-g-PMA reinforcements than matrix (composite) of PPC/CS.

Considering that the moisture in native starch easily induces the degradation of PPC during the processing, another study has reported the development of a biodegradable PPC/glycerol-plasticized thermoplastic dried starch (GPTPS) composites [75PPC/25GPTPS/1SA] using a screw extruder [39]. Succinic anhydride was found to connect both the glycerol-modified starch and PPC as revealed by the improved interaction between PPC and GPTPS (as determined by FTIR). In view of this, various properties as well as the morphology of the composites have been observed to be improved. For example, tensile strength and elongation at break of the composite were found to be as high as 19.4 MPa and 88.5%, respectively; similarly improved dynamic mechanical thermal properties and significant improvement in thermal stability of composites (determined by TGA) have also been observed. Improved interfacial adhesion (as observed by SEM studies) between PPC and GPTPS has been attributed to these improvements in observed properties.

Poly (Vinyl Acetate)–Starch System (Blend/Composites) Recently, polyvinyl acetate (PVAc)—starch composite (St-g-PVAc) with urea (individual encapsulation of urea within the starch matrix modified by vinyl acetate through *in situ* graft copolymerization) was prepared in the form of membrane by solution casting and characterized with the aim of its possible use in agricultural applications, an interesting controlled release applications of starch composites with polymers [191]. Reduction in swelling of the starch matrix was observed with the introduction of hydrophobic PVAc, whereby urea-nitrogen release time got extended up to 28 h in water. These composite films could biodegrade also in the soil environment.

Starch/PVA System (Blend/Composites) This is another starch polymer composite system studied extensively [129, 193, 219–225]. One study reports the preparation and characterization of starch/PVA containing cellulosic fibers and citric acid as plasticizer and glutaraldehyde (GA) as the cross-linker [221]. These blended composite films exhibited better mechanical properties than the matrix along with antimicrobial activities suggesting prospects for these composites in food packaging applications.

Another study has dealt with the preparation and characterization of PVA, epoxidized natural rubber (ENR), and starch with GA as a cross-linking reagent [223]. The authors have observed that the swelling ratio of the polymer hydrogel decreased with increasing amounts of GA, ENR, PVA, and glycerol, while the swelling behavior of the hydrogel increased with increasing starch content only. It was also observed that the tensile strength of the polymer hydrogel could also be improved by adding quantum-sized silver nanoparticles of 10 nm of diameter.

Using a blend of starch and PVA as matrix and citric acid as plasticizer along with cross-linked with GA and raw and methyl methacrylate (MMA)-grafted *silk fibroin* particles as reinforcements, composite films were prepared by solution casting method [225]. These films showed improved mechanical and water-resistance properties compared to those of PVA (St/PVA) blends. Besides, tensile strength of the composite films increased, while elongation at break decreased.

Another study with solution casting method was used to prepare PVA/starch blends with and without cross-linking, and these are characterized for moisture content, solubility resistance in water, 5% acetic acid, 50% ethanol, and sunflower oil, swelling characteristics in the latter two solutions, and finally tensile properties [222]. It was found that tensile strength, tensile modulus, tear and burst strengths, as well as solubility resistance of cross-linked PVA/starch composite films were significantly improved over the composite films without cross-linking. However, higher tensile elongation, moisture content, moisture absorption, and swelling were observed in the case of composite films without cross-linking compared to those of the cross-linked composite films.

An interesting and useful biodegradable PVA/glycerol-plasticized TPS and its intumescent flame retardant composites have been prepared and both were characterized for flame retardancy and thermal stability [205]. In order to utilize the charring capacity of the polyhydric compounds and also to restrain the reaction between ammonium polyphosphate (APP) and starch during processing, microencapsulated ammonium polyphosphate (MCAPP) was used. While the flame retardancy of TPS and TPS/MCAPP were evaluated by LOI, UL 94, and microscale combustion calorimeter (MCC), thermal stability of both the materials was determined by TG with the thermal degradation and gas products of both the materials (matrix TPS and TPS/MCAPP composite) were monitored by TG-FTIR and dynamic FTIR analysis. It was found that both the matrix and its composites with only 2 wt. % MCAPP passed V-0 in UL 94 test, while the neat TPS did not pass any rating. This passing of the test was attributed to the presence of MCAPP, which reduced the total heat release of TPS sharply in MCC test. The thermal degradation and gas products of TPS and TPS/MCAPP were monitored by TG-FTIR and dynamic FTIR analysis. XPS and SEM measurements were utilized to investigate the chemical structure, as well as the surface morphology of the residual char. PVA/starch composites with 2% APP showed flame retardancy.

PVA–CaS composites with 65–95 g PVA content in 300 g of starch (using water or citric acid or adipic acid or glycerol as plasticizers) prepared using a single-screw extruder were characterized by thermo-gravimetric analysis, mechanical properties, biodegradability, and morphology with a view to study the effect of PVA content on these properties [129]. It was observed that homogeneous surface morphology of the composite was possible only when the PVA content was 65 g in 300 g of starch.

Three decomposition stages were observed for the blend, the first one between 273 and 373 K due to the loss of water and glycerol, the second between 573 and 673 K, and the third between 723 and 773 K due to mainly the oxidative decomposition, respectively, of both PVA and starch. Finally, the residual part of these (a blend containing PVA and starch chains) took place. The authors have opined that these would indicate an improvement in the apparent thermal stability of PVA and starch blends over that of original PVA and starch due to the formation of crystalline form of PVA and mixed "macromolecule" as a result of good interaction between the PVA and starch.

In the case of mechanical properties, based on the stress–strain curves of the blends, the authors have reported an increase in breaking strength with increasing PVA content up to 85 g, which then decreased around 95 g of PVA, while elongation at break was 320%.

On the other hand, consumption of plasticizers followed by faster degradation of starch and PVA was observed in the biodegradability studies of the blend as determined by weight loss (85–90%) on burying in soil for 4 weeks, which was confirmed by the microstructural studies. Based on the aforementioned results, it may be concluded that such starch polymer blends hold promise for a variety of applications.

Continuing the studies on the aforementioned system to find the effect of plasticizer and plasma treatment on the physical and mechanical properties of these blends prepared, good compatibility between the constituents of the blend was observed, despite poor interface between PVA and starch revealed by the platelike starch formation [219]. In view of this good compatibility, both the ultimate stress and strain marginally improved on plasma treatment, while marked improvement (10–45%) of strain was observed by treatment of the blends with water and citric acid, indicating the effectiveness of both treatments.

In the case of corn starch, a study has been reported about the efforts made to overcome its hydrophilicity by resorting to a new coating of amorphous carbon by chemical vapor deposition (CVD) using plasma of methane and subsequently with sulfur hexafluoride (SF6) [224]. It was found that methane/SF6 plasma treatment yielded very high surface hydrophobicity based on the changes observed in contact angle measurements, which was related to the observed rearrangement of surface polymer chains and to the chemical modifications of the surface mentioned earlier. Atomic force microscopy (AFM) images revealed the coating was homogeneous.

Starch–Polyester System (Blend/Composites) Considering that the addition of starch filler improves the rate of degradation of polyesters, studies have been carried out to prepare and characterize starch–aliphatic polyester composites. This is because there would be complete biodegradability and compostability of both the constituents individually and in their blends [226]. These composites exhibited excellent properties when small amounts of compatibilizers such as anhydride functionalized polyesters are added. For example, tensile strength of this composite was found to be comparable with those of synthetic polyester even at 70 wt. % starch [103, 192, 227, 228]. Furthermore, it is well known that the main factors influence the degree of degradation for both these components depending on the type of microorganisms and their populations.

Starch/Polyethylene System (Blend/Composites) It is well known that good attributes such as low price, useful properties, broad supply chains, and mature processing facilities have made the polyolefins very popular. However, they suffer from difficulty to degrade easily in the natural environment. This has attracted increasing interest of researchers to find ways of making polyolefins degradable. Oxidative degradation and blending with natural polymers are two ways of imparting this property to polyolefins.

Blending of corn starch with polyethylene to form the composites revealed about 84% of the degradable materials became biodegradable. In addition, an increase in CO_2 content was also observed after about 300 days of biodegradability studies. This was evaluated by weight loss, CO_2 production, and mechanical properties [220]. As mentioned earlier, these results of corn starch blends also suggest better and wider utilization of these in composite technology [224].

A review by Liu et al. [229] has reported presenting some of the research and developments in the acceleration of the degradation of polyolefin blends and composites through additives. Both LDPE and HDPE have been used as the polymer matrices with 40–70% of various starches such as corn starch, sago, tapioca, and rice starch.

It was observed that modification of LDPE by glucose/starch improved the biodegradability of PE to make LDPE completely degradable [230]. Effects of using compatibilizer, clay, and plasticizers on physicomechanical properties of LDPE are recently studied [194, 195]. In these studies, LDPE/PoS nanocomposite films were prepared by melt extrusion process by first preparing starch–clay nanocomposite by solution intercalation method followed by melt mixing the resultant material with LDPE in presence of maleic anhydride-grafted polyethylene (MAgPE) produced by reactive extrusion as a compatibilizer [195]. A uniform distribution of both starch

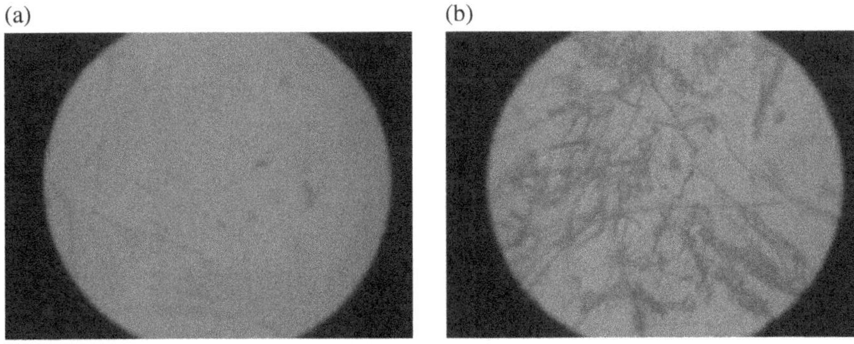

FIGURE 8.16 Macrographs showing the evidence of fungi growth (*Aspergillus Niger*) on surface of LDPE/TS biocomposites. (a) Matrix LDPE and (b) LDPE/TS composite with CaS. (×200). *Source*: Ali et al. [232]. Reproduced with permission from Chemical Engineering Transactions.

particles and clay layers was observed due to the initial intercalation reaction of clay layers with starch molecules, the conversion of starch into TPS by plasticizers, and the use of MAgPE as a compatibilizer, without any need of alkyl ammonium treatment to improve functionalization or specific surface treatment to improve the bonding, in LDPE matrix [195].

Considering the growing interest in polysaccharide-filled plastics and the compatibility issue between PoS and polyethylene, a biodegradable composite of PoS incorporated with 50% LDPE 0.5% grafted maleic anhydride (LDPE-g-MA) and LDPE were prepared in a single-step twin-screw extruder [195, 231]. This was characterized for the compatibility behavior of polar group PS and nonpolar LDPE, tensile properties, and thermal stability. No significant change was observed in the thermal stability and melting temperature of these composites with various combinations of PS, while tensile strength could be maintained at up to 15% mixing of PS, but percentage elongation was intact only up to 5% mixing of PS.

With a view to improve the degradability of injection-molded conventional plastic packaging materials, CaS–LDPE blend was prepared by melt compounded a twin-screw extruder [232]. The CaS used was of food-grade CaS having an average particle size of 32.97 μm and 11.5% average moisture content, while the LDPE having density of 0.922 g/cm^3, melt flow index of 7 g/10 min, and melting temperature (T_m) of 110°C. Processing conditions used were 70–80 rpm screw speed and range of barrel temperatures: 150, 130, 130, and 140°C. After this the material was injection molded to produce sample for evaluation of mechanical and biodegradation properties. It was observed that the blend became very ductile with the addition of starch while both YM and TS decreased. On the other hand, biodegradation carried out by exposing the blend samples to fungi and enzyme environment revealed promotion of growth (on the surface of LDPE/TS biocomposites) as well as increased fungi colony in samples having starch content up to 20% by the presence of starch attack of granular starch present on the surface of the samples by microorganisms till most of the starch was assimilated as a carbon source. This is shown in Figure 8.16, which shows the macrographs of the matrix LDPE (i) and its composite with CaS (ii) showing the evidence of fungi growth (*Aspergillus Niger*), on surface of LDPE/TS biocomposites with and without starch, under microscope. It was concluded that the biocomposites studied were degradable and therefore they could be used as an alternative to the conventional synthetic nondegradable polymers in food packaging and other applications.

PHB–Starch System (Blend/Composites) A biodegradable composite with PHB and commercial corn was prepared using stearic acid as a compatibilizer and glycerol as a plasticizer (to improve adhesion) by solution casting technique [196]. This was characterized for tensile properties, biodegradability, and degradation as well as water diffusion characteristics. Increase of water diffusion and penetration coefficients of water with increasing starch content in the composites was observed. Similarly, with the increase of starch content up to 10 wt. %, the tensile properties (elastic modulus, tensile strength, and strain at rupture) of the PHB–starch composites were found to increase, with tensile strength increasing from 21.2 to 93.9 MPa. However, these properties were found to decrease when starch content was higher than 10 wt. % in the composite. In the case of biodegradation determined using actinomycetes, rate of biodegradation increased proportionally with the starch content in the composite, while it was found to accelerate in a culture medium of pH approximately 7.0 at 30°C. On the other hand, lipase produced by *Streptomyces albidoflavus* did not degrade PHB/starch composites in enzymatic degradation experiments.

Another composite of PoS with and without glycerin (plasticizer) and blended with PHB prepared by solvent casting method showed no change in the melting point of PHB by blending it with starch, while slight increase in T_g was observed, particularly when TPS was used [233]. Besides, higher thermal stability was observed for the blend.

A study with composite of polybutylene succinate (PBS) with corn starch (CS) as filler and lysine diisocyanate (LDI) as a coupling agent showed that crystallization kinetics of the composite and morphological structure of PBS spherulites were affected as observed in DSC study and microscopy observations [192].

Corn starch–PP Blend Another study on corn starch (2–6%) blends with PP prepared by compounding in a Haake torque rheometer at constant temperature (180°C) for different times in a closed system has reported a fine dispersion of starch particles in the PP matrix, exhibiting good distribution and interaction of the starch particles with increasing processing time up to 15 min [234]. It is mentioned that starch stabilized the melt as evident from the reduction in melt flow index and increase in apparent viscosity of the processed material with complete inhibition of the PP melt degradation up to 20 min processing at the starch content of 6 wt. %. Furthermore, it has also been reported that while UTS of the PP film-based composites remained almost constant, an increase in elongation at break for the composites having 6 wt. % of starch was observed compared with those of the controls.

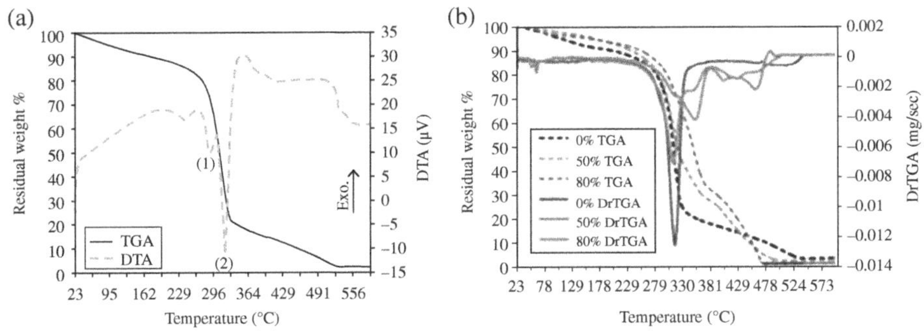

FIGURE 8.17 Plots showing the results of thermal analysis of hot pressed TPS-based matrix and composites with different date palm fiber content: (a) TGA/DTA curves, (b) TGA/ DrTGA curves. *Source*: Ibrahim et al. [51]. Reproduced with permission from Hindawi Publishing Corporation. (*See insert for color representation of the figure.*)

One recent study has reported the thermal analysis of hot pressed corn starch plasticized by glycerin matrix obtained by emulsion technique and its composites containing 20–80 wt. % date palm fibers [51]. Figure 8.17a shows the TGA/DTA curves of the matrix, while Figure 8.17b shows the TGA/DTG curves of its composites with different amounts of fiber content.

It has been reported that obtained DTA curves are similar to those reported earlier [128]. On the other hand, two weight loss phases can be seen in this figure with the first phase being attributed to about 9% weight loss up to 174°C (indicating presence of hydrated/adsorbed water and glycerin) while the second weight loss phase of about 88% between 175 and 530°C being attributed to the thermal decomposition of starch (burning of starch organic matter) with about 60% was between 280 and 340°C. These results are reported to be similar to those observed earlier by other researchers working with other starches [178, 186, 235].

In the case of TGA and DTG results of TPS matrix and its composites containing 50 and 80 wt. % date palm fibers, thermal stability of composites has been reported to increase with increasing amount of date palm fibers content suggesting good adhesion between TPS matrix and reinforcing fibers as reported earlier [235]. In fact, higher onset of degradation temperature (174°C) of these biocomposites compared to those reported for synthetic polymers such as PE- and PP-based composites suggesting better thermal stability of these biocomposites. This has been attributed to increasing amount of the cellulosic matter. As in the case of the matrix, studied composites also showed two major peak regions, the first set occurring at 316°C (due to the decomposition of starch), with these peaks becoming smaller with increasing fiber content, while the second at 356°C was due to the decomposition of cellulosic matter (date palm fibers). Another interesting observation reported has been the broad decomposition of lignin at all temperatures with a significant peak at higher temperature around 468°C as reported by Brebu and Vasile [236]. Evaporation of most of the water added in the preparation of composites in their preparation by emulsion technique has reported to have taken place during preheating and hot-pressing processes resulting in weight loss of about 3–5% at 100°C in the matrix and its composites. Furthermore, decreasing residual weight with increasing fiber content was observed in the case of all the composites.

Other Types of Blends/Composites Developed for Property Modifications Composites of PoS with semi-interpenetrating polymer network (semi-IPN) cryogels based on polyacrylamide (PAAm) gels were synthesized for possible removal of heavy metal ions such as Cu^{2+}, Cd^{2+}, Ni^{2+}, and Zn^{2+} from aqueous solutions by batch mode under noncompetitive conditions [237]. On the other hand, maize and CaSs were modified through microwave esterification using recovered coconut oil and microbial lipase [50]. It was observed that this type of esterification of both starches was advantageous in that it gave degree of substitution (DS) values of 1.55 and 1.1 for maize starch and CaS, respectively. On the other hand, esterification of CaS in solution state and semisolid state for 36 h at 60°C gave DS values of 0.08 and 0.43, respectively. Higher DS values have been attributed to the thermal stability as determined by the TGA and DSC studies because the onset of decomposition was at a higher temperature (492°C) for the modified starches than those of unmodified ones (330°C), and these starches were found to be stable above 600°C. On the other hand, α-amylase digestibility and viscosity were found to be reduced for modified starches.

One of the biggest obstacles in preparing quality materials with biomass has been found to be low interfacial interaction between hydrophilic biopolymers and hydrophobic thermoplastic matrices. To address this problem, attempt has been made to prepare starch/poly (ε-caprolactone) (PCL) composites using nano-TiO_2 particles [238]. It was observed that mechanical properties and water resistance of these composites were improved significantly with the addition of nano-TiO_2, whereas their transparency got decreased. Formation of an interpenetrating network structure by the addition of nano-TiO_2 into starch/PCL composites has been attributed to the observed properties. Despite unchanged crystal form and crystal size, degree of crystallinity and the crystallization rate of the composites were found to be reduced as observed from the DSC and XRD analysis of the composites.

Another interesting study to improve mechanical properties of starch/polyester biocomposites was carried out through the preparation of native rice starch granules and poly (butylene adipate-*co*-terephthalate) (PBAT) by mechanically blending process using single-screw extruder [239]. The native rice starch was first synthesized as porous material by enzymatic technique through hydrolysis by a mixture of α-amylase and amyloglucosidase to improve the miscibility between the native rice starch granules and PBAT. It was observed that both tensile strength and the elongation at break of the blend increased due to the improved compatibility of the polymers in the blends, which was possible from the pits and holes on the surface of starch granules as observed by SEM.

Interesting composite systems of starch of rice and pea containing citric acid modified respective starch (CAPS and CARS) were prepared by screw extrusion to study the effects of granular CAPS and CARS on structure and various properties using SEM and XRD techniques [240]. They include morphology, thermal stability, and dynamic mechanical thermal analysis, the relationship between the mechanical properties and water content, as well as the water vapor permeability (WVP) of the composite films. No disruption in the thermoplastic process by the granules of CAPS and CARS was observed even though the PS in the matrix was turned into a continuous TPS phase. However, both the granules of CAPS and CARS were found to improve the storage modulus, the glass transition temperature, the tensile strength, and the water vapor barrier, but decreased the thermal stability in the composite. It was also observed that the CARS/TPS composites exhibited a better storage modulus, tensile strength, elongation at break, and water vapor barrier than CAPS/TPS composites. This has been attributed to the smaller size of the CARS granules compared to that of CAPS. Another study of the same system has reported cross-linking of starch films by citric acid [241].

Starch-graft-poly (acrylic acid)/attapulgite system, a superabsorbent composite, has been synthesized by graft copolymerization reaction of starch and acrylic acid (AA) in the presence of attapulgite micropowder in aqueous solution to understand its effect on water absorbency of starch [167]. The parameters to know the water absorbency include weight ratio of AA to starch, initial monomer concentration, neutralization degree of AA, amount of cross-linker, initiator, and attapulgite. It was observed that the superabsorbent composite synthesized under optimal synthesis conditions (weight ratio of AA to starch, 5 : 1) with AA content of 74.68 wt. %, starch content of 14.94 wt. %, initiator content of 0.3 wt. %, and cross-linker content of 0.08 wt. % with an attapulgite content of 10 wt. %. This composite was found to exhibit absorption of 1077 g H_2O/g and 61 g H_2O/g in distilled water and in 0.9 wt. % NaCl solution, respectively. The authors have opined that in view of some of the attributes such as excellent water absorbency and water retention under load, biodegradability in nature, economical and environment friendly, synthesized superabsorbent composite could be useful in various applications, especially in agricultural and horticultural sectors.

8.2.2.2 Starch–Natural Materials-based "Green" Composites

Nonwood-Based Fibers As mentioned earlier, environmental problems encountered with the use of nonrenewable, nondegradable, as well as the nonrecyclability of synthetic polymer (plastic) composites have led to growing interest for the use of biopolymers and biobased reinforcement/fillers to develop eco-friendly green composites. Accordingly, during the last decade or so, there has been focus in the research and development for the production of 100% eco sustainable and "green" composites by replacing nonbiodegradable polymer matrices with biodegradable ones. Several biodegradable natural-derived polymers exist such as polysaccharides (starch, chitin, cellulose, etc.), proteins (collagen, gelatin, casein, albumin, silk, elastin, etc.,), polyesters (polyhydroxyalkanoates), PHB, PLA, lignin, lipids, natural rubber (NR), some polyamides, synthetic polymers such as PVAs, PVAc, and polycaprolactone [39, 113, 161, 184, 240, 242–254]. In majority of cases, these materials degrade through enzymatic reactions in suitable environments. Among the biodegradable polymers, starch belongs to agro-polymers, PLA to chemically synthesized from agro-based monomers, while PHA to microbial derived (PLA). In addition there are also synthetic polyesters, which are processed by conventional polymer route.

These new materials can be used in many application fields of the composites such as panels, packaging, gardening items, etc, which do not require superior mechanical properties [47]. First attempt in this direction was focused on recyclable polymers filled with natural organic fillers such as plant fibers due to the their positive attributes, such as renewability, availability, biodegradability, low specific gravity, and cost in addition to better acoustic properties than the synthetic fibers such as glass fiber, carbon fiber, and aramid fibers [234]. Reported literature shows that in addition to wood flours or saw dust, natural organic fillers such as cellulose, starch granules have also been used in biopolymers such as starch. In general, it has been observed that addition of granular starch to a polymer would lead to a reduction in the elongation at break, while the modulus would increase. Starch being hygroscopic, the water absorption of the composite will be better [234]. In the following paragraphs, some of the "green" microcomposites developed using starch with renewable resource-based materials such as plant- or animal-based fibers will be discussed.

Nonplant Fiber Incorporated Composites One of the natural materials other than the plant fibers used extensively in the composite technology particularly in the development of "green" composite is chitosan. This is because its film has potential applications in agriculture, food, and pharmacy. However, films made of only chitosan exhibit poor mechanical properties and low water resistance. These can be overcome through composite technology by forming miscible, biodegradable composite films from chitosan with other hydrophilic biopolymers. Some such attempts are given below:

In one such study, preparation of a chitosan–starch/keratin fiber (chicken feather) composites by casting/solvent evaporation method has been reported [255]. In this, 5–20 wt. % of three different kinds of keratin fiber reinforcements, namely, short and long biofibers, and rachis particles were used. These constituents are shown schematically in Figure 8.18. Thus obtained composites and the matrix (chitosan–starch) were characterized by their morphology, mechanical, and thermal properties using scanning electron microscopy, thermogravimetry, DSC, and dynamic mechanical analysis, respectively. Enhancement of thermal stability of the composites was reported due to the addition of keratin compared to pure matrix, which was further corroborated by the dynamic mechanical analysis. The latter showed increase in the storage modulus of the composites over that of the pure matrix. These observations have been attributed as uniform dispersion of keratin in the chitosan–starch matrix as observed in their morphology evaluated by scanning electron microscopy suggesting good compatibility between these biopolymers. This was further corroborated by FTIR studies of these materials. It has been reported that the aforementioned results demonstrate that the chicken feathers collected from the poultry farms could be usefully utilized as novel keratin reinforcements to develop new green composites, which exhibited better properties than the pure biopolymer matrix itself.

Two thermally gelatinized corn starches (waxy starch and regular starch with 25% amylose) were mixed with solution of chitosan (deacetylated degree, 90%) to

FIGURE 8.18 Schematic view of the constituents (chitosan–starch–keratin fiber) of the composite obtained. *Source*: Reproduced with the kind permission of the publishers of Cynthia et al. [255]. (*See insert for color representation of the figure.*)

prepare chitosan–starch composites film casting method [57]. For this, first, 2%, w/v of chitosan solution was prepared by dispersing 10 g of chitosan (deacetylated degree of 90%; obtained from Vanson, Redmond, WA) in 500 ml of 1%, v/v lactic acid solution. The solution was filtered with cheese-cloth using vacuum aspiration when the chitosan got dissolved completely in it. Then, 1–4% w/v amounts of starch solutions were prepared by dispersing 25% amylose corn starch or waxy starch obtained from American Maize Products Co., Hammond, in distilled water. This was then heated using hot plates with simultaneous stirring until it gelatinized. Then, the mixture was cooled to 25°C. Both the prepared chitosan and starch were mixed in almost equal proportions (100 ml each of chitosan solution and 1–4% starch solutions). Then, 25% (w/w) of the total solid weight in solution of glycerin was added to the mixture prepared. Finally, the composite mixture was cast onto flat, level Teflon-coated glass plates to obtain composite films and then dried at room temperature for at least 72 h before they were peeled off from the plates. Dried films were conditioned at 50% RH and 25°C for 48 h prior to testing. These films were characterized by various properties. Figure 8.19 shows plots of tensile strength, percent elongation, and water vapor transmission rate (WVTR) of both the matrix and its composite films as functions of ratio of starch to chitosan.

It can be seen that both the tensile strength and percent elongation of composite films show an initial increase followed by decrease with the addition of starches (Fig. 8.19a, b) with regular starch-based composites exhibiting higher values of these two properties. Further, decrease in WVTR of the composite films with the addition of starches was observed (Fig. 8.19c).

Tensile strength and percent elongation of composite films initially increased and then decreased with the addition of starches with regular starch-based composites showing higher values of these two properties. It was also observed that WVTR of the composite films decreased with the addition of starches. On the other hand,

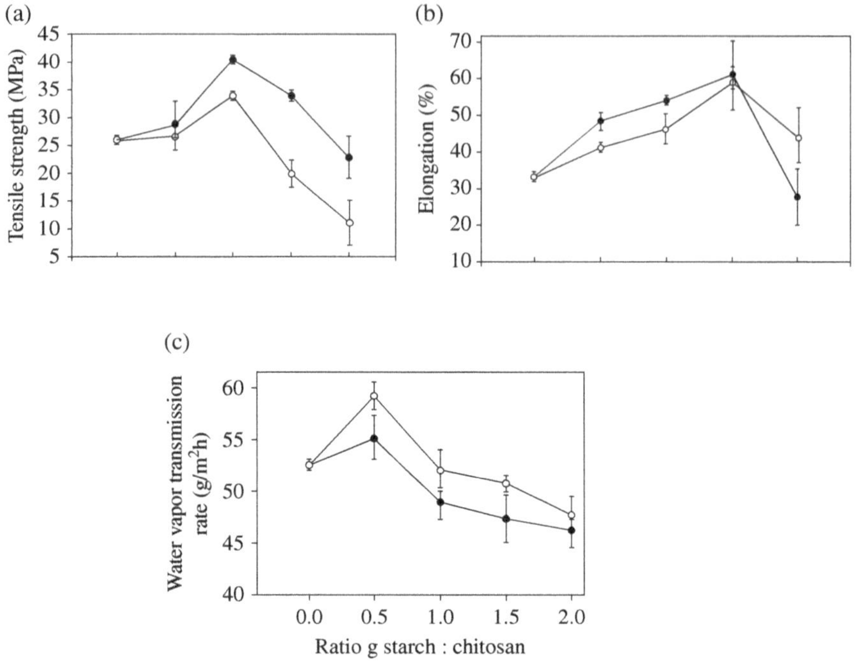

FIGURE 8.19 Plots of (a) tensile strength, (b) % elongation, and (c) water vapor transmission rate (WVTR) of both the matrix and its composite films as functions of ratio of starch to chitosan (solid symbol: regular corn starch—25% amylose; open symbol: waxy corn starch—100% amylopectin, *X*-axis is same for all the figures). *Source*: Xu et al. [57]. Reproduced with permission from Elsevier. Copyright 2005.

addition of gelatinized starch suppressed the crystalline peaks of the chitosan film as well as the amino group band of the chitosan molecule as observed in FTIR study. In fact, the band shifted from 1578 cm^{-1} in the chitosan film to 1584 cm^{-1} in the composite films indicating molecular miscibility existed between these two components.

Another study has reported the preparation of a series of oxidized PS–chitosan (OPS/CS) composite films by a casting and solvent evaporation method [205]. The authors have observed that in addition to hydrogen-bonding enhancement of interactions between OPS and CS molecules due to the formation of electrostatic interaction between the negatively charged carboxyl groups on OPS and the positively charged amino groups on CS. However, OPS/CS blend films exhibited significantly higher tensile strength with lower elongation at break than the PS/chitosan (PS/CS) blend films. Besides, incorporation of CS into the OPS matrix also decreased in moisture uptake by the composite film.

Plant Fiber Incorporated Composites A variety of starches (cassava, castor oil cake, corn, etc.) with a number of plant fibers (banana, coir, flax, jute, sugar cane bagasse, etc.) have been used to develop starch-based "green" composites using different processing techniques. Some of these are listed in Table 8.2. Details of some of these particularly in respect of their structure and properties studied are given in the following pages. It may be worthwhile to note a recent review mentions about the renewed interest on TPS/composites as environmental-friendly materials and as biodegradable renewable resources for a sustainable development [30].

Cassava Starch A "green" biodegradable composite of CaS containing glycerol and potassium sorbate incorporated with wheat bran fiber (contains 40 g of water insoluble fiber per 100 g of bran) has been studied with the aim of understanding the influence of reinforcement content [159]. Casting method of processing was used with three different fractions of wheat bran fiber (1.5 mg, 13.5 mg, and 27.1 mg/g of matrix). Density and dynamic thermal analysis studies carried out at room temperature revealed that density of the matrix was not affected by the increasing fiber content, although it caused the increase in storage modulus, decrease in loss tangent, moisture content, and WVP. However, thermal property studies indicated that the addition of wheat bran shifted the glycerol-rich phase T_g toward higher temperatures, which was attributed to the presence of fiber, which in turn led to an enhancement in the glycerol dispersion. The addition of fibers led to the increase of the yellow index (when a starch is subjected to dry heat, it breaks down to form dextrin also called "pyro dextrin," which are mainly yellow to brown in color and dextrinization is partially responsible for the browning of toasted bread). The improvement in water vapor barrier properties jointly with the enhancement of mechanical properties when fiber was present led to the idea that the composite developed could be used to protect food and extend its shelf life.

CaS incorporated with different amounts (0–30 wt. %) of coir fibers (from green coconuts) composites have been studied by thermal molding using glycerol as plasticizer for the starch [132, 133]. Both the thermoplastic matrix and its composites were given thermal treatment. Structural aspects, tensile properties, moisture absorption, water absorption, and swelling on immersion in water and on exposure to humidity have been evaluated for both untreated and treated matrix and its composites. It was found the composites were free of agglomeration. The authors have observed the following: (i) the tensile strength, YM, and breaking load of treated TPS matrix were higher (3.24 MPa, 59.81 MPa, and 112.68 N, respectively) than those for the untreated matrix (1.56 MPa, 14.56 MPa, and 50.77 N, respectively); in the case of their composites, these values were approximately 373.5 MPa and approximately 176 MPa for YM and 10–11 and 3 MPa for TS for treated and untreated TPS composites containing 0–30% coir fibers, respectively; higher tensile properties for thermoplastic CaS with both the incorporation of fibers up to 15 wt. % of fiber as well as with the thermal treatment; (ii) the water uptake, swelling, and moisture absorption decreased with the incorporation of fibers. For example, values of water uptake of TPS matrix and its composite with 30% coir fiber after 2 h of immersion was about 60% and 38%, respectively, while the effect was more pronounced after 24 h immersion for all the composites; (iii) moisture absorption of TPS matrix and its composites remained constant after about 13 and 10 days, respectively, with the equilibrium moisture for the TPS matrix observed was 15.6%, while it varied

(a) (b)

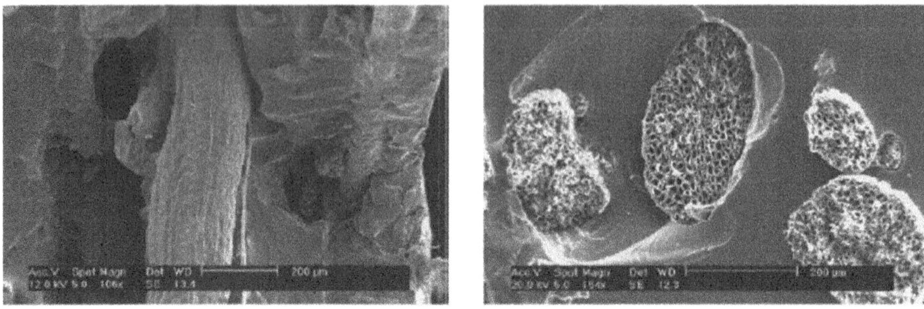

FIGURE 8.20 Fractographs of TPS matrix and filled (a) 30% fiber and (b) detail of fiber–matrix interface. *Source*: Lomelí et al. [132]. Reproduced with permission from Elsevier. Copyright 2011.

between 15.2 and 14% for the TPS composites with 5–30% fiber content; (iv) decrease of linear swelling of TPS matrix with increasing amount of fiber content and immersion time was observed; (v) increase of crystallinity (from 39 to 62%), thermal stability (shown by decreasing degradation temperature), dynamic mechanical properties (storage modulus, glass transition temperature, and lower damping) of the composites were found to increase with increasing amount of fiber incorporation. The authors have opined better interfacial bonding between the matrix and fibers as well as the hindrance to the water absorption caused by the fibers as reported elsewhere [256, 257]. Improved interfacial bonding between the TPS matrix and coir fibers is evident from the SEM pictures showing the fracture surfaces shown in Figure 8.20a, b. On the other hand, lower absorption of water has also been attributed to chemical composition of coir fibers, which has higher lignin content and also due to the presence of wax on the fiber surface.

It was also observed by these researchers that the water absorption of CaS composites was higher compared to that reported earlier for corn starch composites [128]. Furthermore, crystallinity, thermal stability, and storage modulus of the composites were found to increase with increasing amount of fiber incorporation. Furthermore, the properties of the starch–coir fiber composites were found to be not affected after processing suggesting that during the plasticizing process most of the starch remained chemically unmodified although FTIR studies revealed the presence of carbonyl compounds indicating slight decomposition of the starch in the TPS matrix and TPS composites.

Recently, another CaS-based "green" composite containing malt bagasse fiber (by-product of brewing industry) prepared by baking showed good distribution of the fiber throughout the polymeric matrix, which had an amorphous structure [160]. Decreasing water absorption with increase in bagasse fiber content from 5 to 10% was observed in the composite. These composites have been suggested as a possible alternative for packing dry foods by making foam trays using these composites by baking process.

Castor Oil Cake-Based Starch and Its Composites Castor oil is being used in the production of biodiesel, particularly in Brazil, who have been making efforts to increase the production of biodiesel with the aim of meeting the fuel demand as well as recovering the by-products such as castor oil cake and glycerine called as "crude or recovered glycerol" obtained during production [127].

Attempts have been made by some researchers, particularly at the Federal University of Parana (UFPR), Curitiba (PR-Brazil) to make use of these two by-products to develop value-added products considering production of large quantities of these by-products [127, 136, 151, 258–260]. The castor oil cake is an interesting raw material, since it contains about 28–49% of plasticizable component (high protein) and 26–29% of strengthening material (crude fiber) [136, 151, 258], which suggests the potential of cake as a protein source to produce value-added biodegradable material such as thermoplastics or as an attractive matrix for developing "green" composites with suitable reinforcements. This is because many of biodegradable materials have been developed with similar protein fractions [261], and biodegradable films from the castor oil cake can be produced due to ease of their cross-linking in view of their richness in amino acids and good film-genic properties of these proteins [259]. This is justifiable from reported work with CaS containing fibers and sugars (both being capable of strengthening the matrix) producing thermoplastics having the highest YM followed by industrial starch and that derived from cassava roots [147]. In view of the previously mentioned composition of the cake, it has been possible to prepare "green" composites with improved properties [127, 136, 151, 258, 260, 262–264].

Accordingly, attractive propositions have been made by the researchers at (UFPR), Curitiba (PR-Brazil) to develop "green" composites by thermo-molding by the use of castor bean cake either fully as starch or partly substituting conventional starches (in both cases using glycerol as plasticizer) to develop biodegradable composites using lignocellulosic fibers such as banana, sugarcane bagasse, etc. [127, 151] or as reinforcement along with plasticizer in PP, a synthetic polymer, or mainly as matrix along with plasticizer with microfibers of plants or nanocellulose of wood [136].

After assessing the potential of castor oil cake, "recovered glycerol" and sugarcane bagasse fibers for the preparation of "green" (biodegradable) composites, feasibility studies on the development of the composites using factorial design of experiments were carried out to identify the best processing conditions to obtain thermo-molded composite materials and to optimize the best mechanical properties of the composites based on three parameters used, namely, composition, applied load, and temperature of molding [151]. This involved preparation of composites with different amounts of sugarcane bagasse fibers of different length and diameter (length: 10–20 mm and diameter: 0.25–0.85 mm) and evaluation of structural and strength properties and water absorption characteristics. Morphology studies have also been reported to understand the observed properties. Factorial design of experiments revealed the ideal conditions for the "green" composites in terms of fiber content, molding pressure, and temperature. It was observed that increasing temperature and/or pressure resulted in the reduction of water absorption by the composites. This has been attributed to better impermeability existing in the chemical structures and presence of the proteins in the castor oil cake starch together with the sugarcane bagasse fibers, besides the effect of larger compaction of the final products. Statistical analysis of tensile properties of the prepared composites as function of fiber diameter and length of the sugarcane bagasse fibers revealed no significant effect on the YM (an increase of 10 MPa) in both cases, while a decreasing trend with TS with increasing diameter with larger variations in the composites containing the highest fiber diameters compared to the lower diameter fiber composites; higher TS values were observed in the composites with 10 mm length fibers (10% fiber content) than those of 20 mm length fibers (20% fiber content). These results have been attributed to good distribution of the bagasse fibers in the TPS matrix along with the good interaction of the fibers with the matrix as observed in SEM studies, thus demonstrating good adhesion between the constituents of the composites.

Structural studies by XRD revealed no significant changes in its structure, which had distinct domains belonging to the amorphous domains of the starch and the crystalline domains mainly of fiber. On the other hand, significant increase in thermal stability was observed in both the highest and lowest fiber content (15 and 5%) compared to thermoplastic castor oil cake starch. This is attributed to the following: (i) increased resistance to thermal degradation of the components (hemicelluloses, cellulose, and lignin) present in both the matrix and the reinforcement; (ii) highest percentage of starch present in the polymer matrix offering more effective interactions with the plasticizer and (iii) highest adhesion of the molecules by increased adhesion at the matrix-fiber interface leading to a better packing of fibers during processing.

Another study has reported the use of both manual method and controlled method for temperature and pressure during the preparation of "green" composites based on castor oil cake-based starch [265]. Significantly improved tensile properties have been observed for "green" composites of castor oil cake-based starch containing fibers of banana plantation, *Luffa cylindrica*, and sugarcane bagasse (up to 35%) prepared by manual conditions over those of the matrix materials, irrespective of the fiber type and its nature (with or without treatment), with probable ranking of the fibers studied as better reinforcements in the order of sugar cane bagasse fibers, banana fibers, and *Luffa* fibers, respectively. Under controlled conditions of temperature and applied pressure, tensile properties of both the thermoplastic castor oil cake starch and its composites with banana/sugarcane bagasse were found to be significantly higher compared to those obtained in manual conditions. In particular, YM of banana fiber containing composite increased with increasing amount of banana fibers, while it decreased in the case of bagasse fibers. This has been attributed to good interaction of banana fibers with the matrix even with higher fiber content compared to that of bagasse fibers and the presence of fibers in the castor oil cake itself. Observed higher tensile properties of all the composites studied suggested good adhesion between the fibers and the matrix irrespective of the type of fiber and also the fiber content of the castor oil cake itself. In the case of banana fiber composites, improved adhesion of castor oil cake starch with plasticizer used was also attributed to the coating on the surface of the fiber. All these have been corroborated by the fractographic studies carried out by SEM.

Furthermore, based on the observed tensile properties, replacement of castor oil cake starch partially by natural starches such as natural or hydrolyzed corn and pregelatinized or natural CaSs or vice versa could be inferred.

The third study carried out in UFPR/EMBRAPA-Brazil, castor oil cake with pure glycerol was used both as filler in PP and as matrix for nanocellulose of wood [136, 266]. Here the case of using castor oil cake (starch) as filler in PP is discussed while its use as matrix to develop nanocomposites is given later (see Section Starch-Based Bionanocomposites Containing Cellulosic Materials).

In the case of using the castor cake as filler, composites of PP containing different amounts (75–85%) of castor oil cake of two mesh sizes (40 and 60) as reinforcement and glycerin (15–25 wt. %) were prepared by thermo-molding by applying different loads for different times. Apparent density, mechanical properties, and water absorption were evaluated as functions of processing parameters. And obtained results were explained as due to good adhesion between the matrix and the crude fibers in the castor bean cake observed by the fractographic studies. Since cake was used as filler only, more details on this can be seen in Ref. [136].

Based on the aforementioned results of castor oil cake-based starch materials, it may be mentioned that these might provide designers with new alternatives to meet the challenging requirements of new potential applications. These may include substitution of medium-density fiber (MDF) boards in some sectors such as in furniture, automotive (e.g., parcel shelves, door panels, instrument panels, armrests, headrests), and building/construction.

Two patents have also been obtained on the composition and processing of castor oil cake-based composites both as matrix and as filler in conventional thermoplastics, such as high impact polystyrene, LDPE, and PP as well as PHB, a biodegradable biopolymer [260, 267]. The purpose of using castor oil cake with conventional polymers has been reported to aim at increasing the material biodegradability, without compromising its mechanical properties, while use of PHB was to reduce the cost and to improve properties.

Corn Starch One of the early published reports on the corn starch-based "green" composites is by Guimaraes et al. who obtained a patent on the composition and processing of corn starch-based biodegradable composites [268].

Evaluation of the corn starch for its structure and thermal characteristics using XRD and TGA/DTA techniques, processing, and characterization of plasticized corn starch-based composites with banana /sugarcane bagasse fiber have been reported [128]. Thermal analysis studies carried out using TGA/DTA of the starch under study were found to have good thermal stability for composite making. Thermo-molding process with both manual and controlled conditions of temperature and pressure were used to prepare these composites containing both commercial and "recovered" glycerol as plasticizer and banana and sugarcane bagasse fibers (ball milled to 3–5 mm in length) as reinforcements. While the banana fiber was used in three compositions (20, 25 and 35 wt. %), sugarcane bagasse fiber was used with only 25 wt. % with all composites being prepared using both types of glycerol (commercial and recycled one obtained by trans-esterfication of the by-product obtained in the

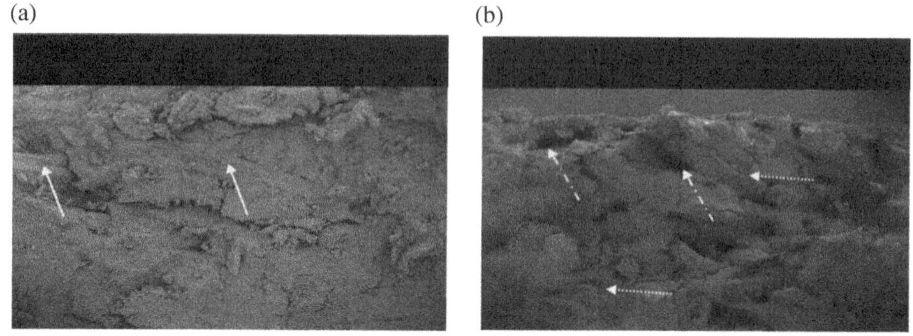

FIGURE 8.21 SEM tensile fractographs of composites of 45% corn starch –30% "recovered" glycerol (a) with 25% banana fiber showing rough fracture surface with lot of cracks and fibers lying parallel to the crack surface; (b) with 25% bagasse fiber showing highly rough fracture surface with lot of small size fibers pullouts and microcavities. *Source*: Adapted from Guimarães [127]. Reproduced with permission from UFPR, Curitiba, Brazil.

biodiesel preparation using castor oil). Tensile properties and fractography were studied as functions of the amount of fiber incorporated, processing conditions (Manual and controlled conditions), and type of glycerol with the same fibers.

Although structural studies by XRD revealed structural changes in both fiber composites, there was not much variation in their crystallinity index (20–21%). On the other hand, marginal improvements in strength properties were observed over those of the plasticized matrix, with tensile properties showing dependence on the processing methods. For example, automatic control of temperature and applied pressure improved the properties for the same amount of fiber content irrespective of the type of glycerol used. "Recovered" glycerol produced more homogeneous composites, which resulted in better properties than those produced with commercial glycerol. Banana fiber containing composites exhibited better properties than those of bagasse fibers. This was understandable in view of the observed morphology of these composites, which seemed to be dictated by the properties of the fibers rather than by the properties of the matrix as evident from the SEM photograph (Fig. 8.21a, b). Also, fractographic studies of tested composite samples supported the observed tensile properties, wherein bagasse fiber composites showed a large number of fiber pullouts and fibers lying perpendicular to the fracture surface compared to those of banana fiber composites (Fig. 8.21a, b).

The composites of this matrix with banana and bagasse fibers prepared by compression molding showed more homogeneous composites with crude glycerin and structural changes in both the fiber composites, while their morphology was dictated by the properties of the fibers rather than by those of the matrix. Improvements in tensile properties were observed in these composites over those of the matrix, which are explained based on fractographic observations.

Another study recommends the corn starch composites with recycled paper (8% w/w) plasticized by 30% w/w glycerol as a viable alternative [201]. On the other

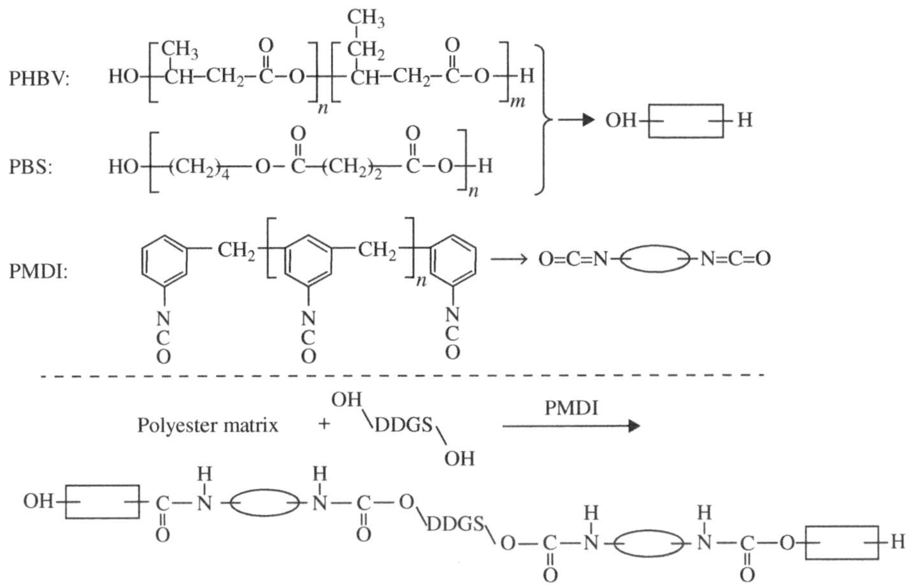

FIGURE 8.22 A schematic of predicted reactions between the polymeric matrix, PMDI, and DDGS. *Source*: Zarrinbakhsh et al. [269, figure 7, p. 1043]. Wiley-VCH Verlag GmbH & Co. KGaA. Copyright 2011.

hand, a biodegradable green composites from distiller's dried grains with solubles (DDGS), a major coproduct of the corn ethanol industry, and a polyhydroxy(butyrate-*co*-valerate) (PHBV)-based bioplastic has been found to exhibit higher thermal stability with enhanced modulus [269]. Figure 8.22 shows predicted reactions between the polymeric matrix, PMDI, and DDGS.

It was observed that DDGS exhibited improved thermal stability on washing by water, which also helped in preventing its decomposition. Using the DDGS with and without water wash and as compatibilizer, composites were prepared by melt processing using the blend of PHBV and 30% PBS as matrix. While the composites containing water-washed DDGS exhibited improved interfacial adhesion between both the PHBV and the matrix PHBV-PBS (as evident from SEM studies), composites containing both water-washed DDGS and compatibilizer DDGS exhibited enhanced tensile and flexural strengths and moduli as well as heat deflection temperature (HDT) compared to the matrix. However, drastic decrease in tensile strength of nonwashed DDGS green composite was observed in composites when water-washed DDGS was used and thus it was found to be compensated by the use of compatibilizer DDGS. Also, existence of some correlation between the storage modulus and HDT of the composites has been observed. In conclusion, based on the observed properties of DDGS containing PHBV and PHBV-PBS composites, the authors have opined that DDGS has the potential to produce biodegradable green composites, which can be used in applications requiring low ductility but high rigidity along with high temperature performance compared to those of the matrix used.

Through the development of models correlating voids and the aspect ratio of fibers with the flexural properties, a dispersion-type starch based an environmental-friendly "green" composite containing cellulose NFs prepared by stirrer mixing has been reported [270]. It was observed that the flexural modulus and flexural strength decreased with increasing void content, but the former increased with the aspect ratio of the fibers used suggesting that both the voids and the aspect ratio greatly influence flexural modulus. The authors have opined that by selection of proper processing method to reduce the void content, flexural properties could be enhanced, even though the processing method used (stir mixing) results in highest flexural strength suggesting sensitiveness of strength to voids generated in the composite. Further, dependence of flexural strength on the average aspect ratio of the fibers could be derived from the regression model whereby the critical aspect ratio was found to be 80. This has been attributed to lowest void content together with uniform dispersion of NFs in the matrix.

Potato Starch "Green" composites of PoS reinforced with alkaline-treated bagasse fiber have been studied to understand the effects of treatment to the fibers, fiber content, and average fiber length on tensile strength, water absorption, and water solubility index [165]. It was observed that (i) water absorption decreased with increasing treated bagasse fiber content and (ii) maximum tensile strength (TS) of 24.5 MPa in composite with 15 wt. % of the treated bagasse fiber. However, at 21.5% of the treated bagasse fiber content, TS was found to decrease to almost the same value as for the matrix (nonbagasse fiber composite), indicating that the average fiber length of this composite was shorter than twice the critical fiber length.

Rice Starch Although biodegradable rice starch films, as is the case with other starches, show potential for use as packaging material, they possess some demerits, such as low mechanical properties and lack of efficient barriers against polarity compounds [271, 272]. However, these have been overcome by modifying its starch properties through chemical and physical treatments and/or blending/incorporating other materials such as other biodegradable polymers or other materials, as mentioned earlier. It is mentioned that selection of improvement should be based on the potential applications in terms of properties and limitations on the use of facilities.

A review by the previous author about the rice starch-based biodegradable films has discussed many relevant points [272]. These are requirements for the formation of rice starch films, factors that affect rice starch-based biodegradable films, methods to enhance the properties of rice starch-based biodegradable films, and various composite systems studied. The requirement for the formation of rice starch films includes one of the constituents such as polymers or compounds that form continuous crystalline or amorphous products, which should be able to form the film as matrix with adequate continuity and cohesion. It can also be coatings, where important factors to be considered are the cohesion between the molecules of the coating material and adhesion between the coating and the support structure. The former dictates barrier and mechanical properties of the film, while the second the ability to withstand the applied pressure and the environment. However, the degree of cohesion depends on many factors such as the chemical structure of the film material, presence of plasticizing and cross-linking agents, final thickness of the film, etc.

Various factors that affect rice starch-based biodegradable films, which may also be applicable to other starches, include amount of amylose and amylopectin present in the starch, types, and amount of plasticizers, lipids, and relative humidity. On the other hand, methods to enhance the properties of rice starch-based biodegradable films include cross-linking, which depends on the type and amount of cross-linker used; irradiation treatment such as ultrasonic, γ or UV, or electron beam radiations; and finally combination/blending including use of reinforcements with other materials such as chitosan, PVA, PLA, or natural fibers or clays. Some of these are briefly mentioned below.

Blending of PVA with rice starch films has been found to increase the tensile strength of the latter and was attributed to the polar interaction between the starch and carboxyl groups of the PLA [273]. Further, possible hydrogen bonding occurring between the carbonyl group (i.e., from ester linkage) in PLA and the hydroxyl group in starch has been the cause for the assumed polar interaction between the two constituents of the composites. The authors have also proposed chemical interactions between starch and PLA indicating possible site for this interaction.

Another study has reported the effect of palm pressed fiber (PPF) surface treatment on the properties of rice starch films [274]. Higher tensile strength, WVP, and thermal properties have been observed in biodegradable rice starch films-based composites with increasing concentration of silane and content of silane treated fiber PPF. The best or the highest mechanical and thermal properties were observed in composites containing 40% of silane-treated PPF. These improved properties have been attributed to filler compatibility resulting in the improved interfacial interactions between the constituents. Besides, increasing concentration of silane resulted in shift of the glass transition temperature (T_g) toward higher temperatures, which has been attributed to possible restriction of the mobility of the starch chain due to the establishment of strong interactions between matrix (rice straw) and the reinforcement (silane-treated PPF).

Bourtoom and Chinnan [275] have studied the effect of incorporation of chitosan fibers and use of different types of plasticizers on the properties of biodegradable rice starch film. They have observed the incorporation of chitosan to biodegradable rice starch films enhanced the tensile strength, WVP, and the crystalline peak structure of starch film but decreased the elongation at the break and film solubility. Improvement of the mechanical properties of the rice starch film chitosan composite has been attributed to the formation of intermolecular hydrogen bonding between NH_3^+ of the chitosan backbone and OH^- of the rice starch, which was corroborated by the FTIR studies, which showed shift of $1541.15\,cm^{-1}$ in the chitosan film to $1621.96\,cm^{-1}$ in the biodegradable blend films. On the other hand, increased WVP of the composite films has been attributed to interactions that were present between the hydroxyl group of rice starch and the amino group of chitosan and the higher hydrophilicity (NH_3^+ groups) of the biodegradable blend films.

On the other hand, the study on the effect of plasticizers used has revealed the highest mechanical resistance with the poorest film flexibility with

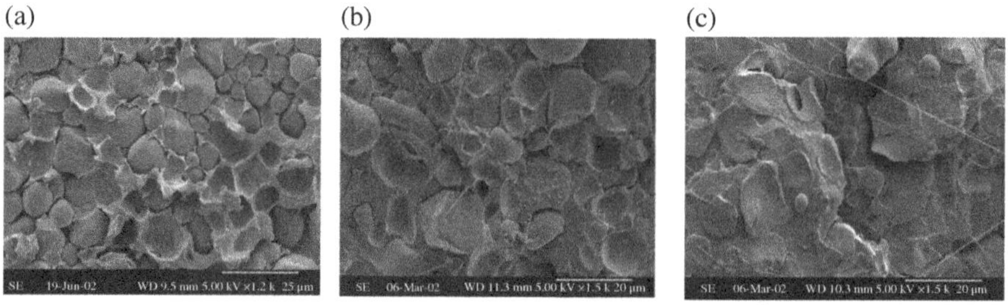

FIGURE 8.23 Scanning electron micrographs of PLA/starch composite showing their microstructures prepared using the one-step method: (a) virgin PLA/starch (55/45); (b) PLA/starch/ MA (55/45)99/1; (c) PLA/starch/MA (55/45)99/1 initiated by 10 wt. % L101 (MA basis). *Source*: Zhang and Sun [276]. Reprinted with permission from American Chemical Society. Copyright 2004.

sorbitol-plasticized films, while the opposite was observed with glycerol and polyethylene glycol-plasticized films. The latter ones also exhibited higher WVP. Observed results with glycerol (more functional groups (—OH)) plasticizer have been attributed to small size along with higher number of glycerol for easy insertion between the polymer chains as well as high hydrophilicity compared to the effect of the larger and higher molecular weight polyethylene glycol molecule.

Wheat Starch Another study has reported a one-step method of preparing PLA/starch composites by blending PLA with wheat starch compatibized by MA using a lab-scale coextruder [276]. In order to improve the compatibility (interfacial adhesion) between PLA, starch, and MA, the researcher has used an initiator, 2,5-bis(*tert*-butylperoxy)-2,5 dimethylhexane (L101). With this, significant improvement in the interfacial adhesion between PLA and starch has been reported. These are shown in SEM pictures of 55 PLA- 45 starch composites with and without MA compatibilizer, revealing good adhesion (Fig. 8.23). This has reflected in the marked increase in the mechanical properties compared to those of virgin composites of PLA/starch with the PLA/starch composites at a constant ratio of 55/45 compatibilized by 1% MA and initiated by 10% L101 (MA basis) showing the highest tensile strength and elongation.

Others Biodegradable composites of different amounts of corn starch and cellulose diacetate (CDA) have been melt processed using epoxidized soya oil (ESO) as a lubricant and triacetin (TA) as a plasticizer and were characterized by their morphology, physical, and thermal properties [277]. The processing of the composites involved direct charging of CDA, starch, TA, and ESO into an extruder at 130–190°C at a screw speed of 100 rpm and was molded under hot pressure to obtain the test specimens. It was observed that the pure CDA exhibited a tensile strength of 62 MPa and an elongation of 4%. However, both the tensile strength and YM got decreased, while the value of elongation increased with increasing starch content in the CDA matrix. On the other hand, dynamic mechanical analysis showed opposite results in storage modulus and tan δ values with the former showing

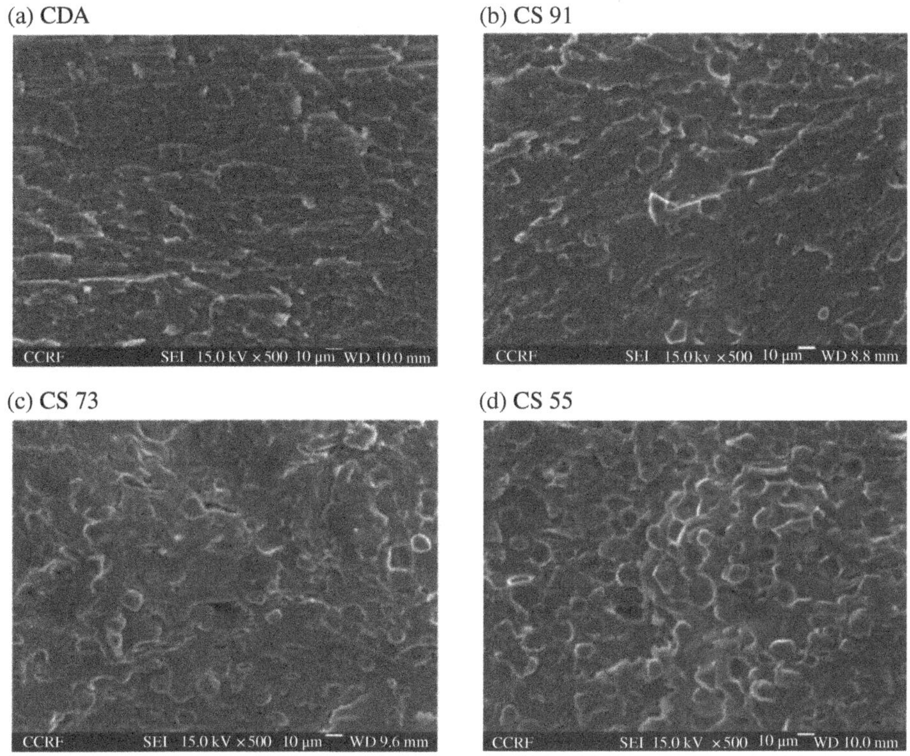

FIGURE 8.24 Fractographs seen in SEM study of (a) CDA/TA/ESO and (b–d) CDA/starch/TA/ESO composites having different amounts of starch (×500). *Source*: Lee et al. [277], figure 2, p.124. Reprinted with permission from Wiley-VCH Verlag GmbH & Co. KGaA. Copyright 2006.

increase and the latter showing decrease with increasing starch level. Improvement in mechanical properties was attributed to very good adhesion between the starch and the CDA matrix as revealed by the SEM study of the fractured surfaces of tested composite. For example, Figure 8.24 shows fractured surfaces of (a) CDA/TA/ESO matrix and those of CDA/starch/TA/ESO composites with different amounts of starch. It can also be seen from the fractographs that incorporation of starch has modified the matrix morphology.

Based on the observation that increasing the amount of starch in the composites resulted in further enhancement of the processability of CDA, the process of preparation of this composite was optimized. Observed results clearly suggested the advantages of using TPS compatibilized with a natural, cheap, and abundant material such as CDA.

Recognizing that a TPS cannot be used directly due to its poor properties in processing, a biodegradable thermoplastic edible corn starch containing ethyl cellulose composite prepared by blending method has been studied to understand the effect of plasticizer (glycerin and glycol), the composition, and the structure on the properties of the composite [46]. It was observed that (i) glycerin was a better plasticizer in processing of the TPS compared to glycol and (ii) ethyl cellulose content of 10% improved mechanical properties, while percent elongation and T_g values increased with increasing amount of starch.

Another study has reported study of corn starch-based biodegradable composites reinforced with date palm and flax fibers, which showed better mechanical properties with 50% fiber content [51].

On the other hand, a very recent study of both mono and hybrid composites of completely biodegradable starch and 20–80% of date palm and 25% of flax fibers prepared by hot pressing the preheated composite mixture with 5 MPa and 160°C for 30 min has been reported [51]. It has been reported that the composites showed very small void fraction (<0.142%) as observed by the density measurements for composites containing up to 50 wt. % fiber content. Further, the composites also exhibited increasing static tensile and flexural mechanical properties (stiffness and strength) with increasing fiber content up to 50 wt. %. Similarly, thermal stability, water uptake, and biodegradation of the composites have also been reported to improve with increasing fiber content. Besides, the results have also indicated that starch-based composites with 50 wt. % fibers content exhibit optimum mechanical properties. The hybrid composite of flax and date palm fibers, 25 wt. % each, has good properties and provides a competitive eco-friendly candidate for various applications.

Vileseca and coworkers have used starch-based commercial biopolymer supplied by Ribawood, S.A (Zaragoza, Spain) as polymer matrix for the composite preparation using jute and hemp fibers strand. In the first case, injection molding method has been used with and without partial delignified jute strands by alkali treatment (NaOH) [27], while in the second case high-shearing mixing process followed by injection molding processing [162]. Increase in strength and stiffness of the jute composites were observed by partial delignification of the jute strands due to the alkali treatment. Influence of the extension of the interfacial adhesion due to the hydrogen bond at the interface on the observed increased mechanical properties was attributed to the application of a blocking reaction to the surface hydroxyl groups of the strands. Higher extension of hydrogen bond at the interface gave higher strength and stiffness.

On the other hand, significant increases in both the UTS and the stiffness were observed for composites containing the highest amount of hemp strands.

Biocomposites of melt-mixed corn starch/sisal fiber composites in an intensive batch mixer connected to a torque rheometer at 130°C with gluten and glycerol have been produced by compression molding [197] with a view to lower water absorption. Increasing YM and UTS were approximately 560 and 162% respectively higher than those of the matrix, while elongation at break decreased. Furthermore, the storage modulus of the composite was found to increase with increasing fiber content whereas tan δ curves showed decreasing trend, confirming the reinforcing effect of the fibers. On the other hand, the water absorption was found to decrease with the addition of sisal fibers, while diffusion coefficient decreased sharply with 5% addition of fiber and further sisal fiber additions showed only small variation. The previous results have been explained as due to good dispersion of the sisal fibers in the matrix as revealed by microscopy observations.

With a view to find a suitable "green" composite to replace the synthetic polymer materials, which are currently being used in some sectors (automobile interior materials, finishing materials of air conditioner, and refrigerator), study of thermo-molded natural soluble starch-based composites with various plasticizers (PVA, polyethylene glycol (PEG), glycerol (G)) (10 wt. % of starch) as matrix and kenaf fiber as filler have been studied [164]. Processing condition used were hot pressing for 30 min at 130°C and about 24 MPa (3500 psi) pressure. The kenaf fibers from kenaf plants of Korea were extracted by retting in 2% NaOH solution. Tensile strength of composites was found to be in the order of composites plasticized with G (12.42 MPa) > PVA (9.72 MPa) > PEG (4.73 MPa) compared to 5.40 MPa of the control. Observed properties have been attributed to good interfacial adhesion between the matrix and their fibers as revealed by the SEM studies.

Studies of injection-molded "green" composites consisting of commercial starch as matrix and short sisal fibers (5–20 wt. %) as reinforcements have been reported [163, 278]. The effect of microstructure and orientation of fibers on the tensile and fracture properties has been looked into in these studies with samples cut in both longitudinal and transverse directions with respect to the melt flow direction. Puncture tests, quasistatic fracture studies, as well as instrumented falling weight impact tests were carried out for all the composites and the plain matrix. Results revealed, irrespective of the sample orientation, an increasing trend in both YM and tensile strength with increasing fiber content with longitudinal cut samples showing more pronounced reinforcing effect due to many fibers mainly oriented longitudinally to the loading direction. Quasistatic fracture tests of single-edge notch bend specimens revealed propagation of cracks in both longitudinal and transverse direction cut samples with stress intensity factor increasing with fiber content. Besides, this was also found to depend on fiber orientation. The authors have also observed increasing trend in fracture toughness, higher values of ductility index, energy at initiation, and total fracture energy, (all these values were higher than the plain matrix) with fiber content and when the impact loading was parallel to the thickness direction as well as higher values of ductility index, energy at initiation, and total fracture energy than the plain matrix. Observed results of fracture toughness with fiber content and microstructure have been explained in terms of the fiber pullout mechanism. Further, injection-molded composites showed higher values of total fracture energy than neat matrix for all compositions in the puncture tests. It has also been observed that due to fiber orientation, damage zone exhibited different patterns besides addition of sisal fibers to the starch matrix to be an efficient means of improving fracture behavior under both quasistatic and impact loading conditions. The authors have also used several theoretical models to predict the elastic properties of these composites.

Recently, preparation of soy proteins plasticized by glycerol-based gelatinized CaS composites containing 20 wt. % bagasse fibers with and without alkali treatment has been reported [198]. Bagasse fiber-based CaS composites exhibited good resistance to retrogradation of starch, flexibility, lightness, and durability, which may be suitable for automotive applications, such as panels, seats, packaging, furnishing, etc. On the other hand, composites containing alkali-treated bagasse fibers exhibited decrease in the water absorption but improvement in the stability of the TPS

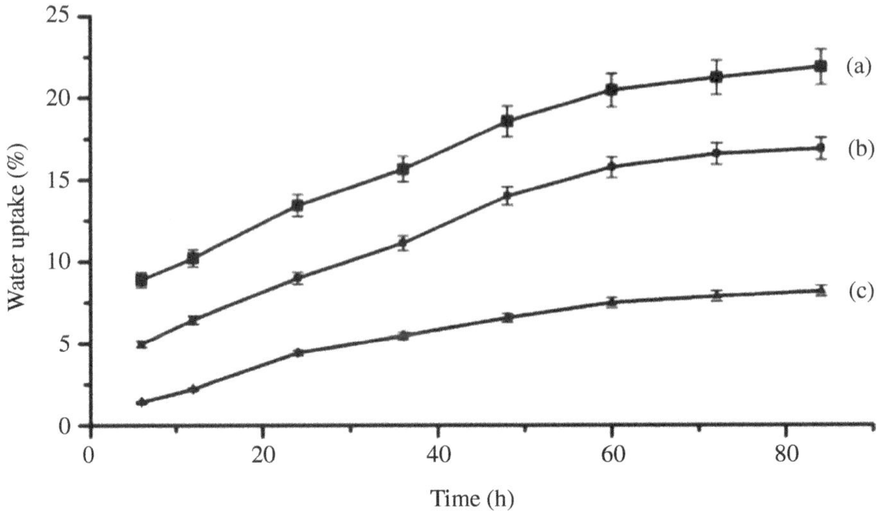

FIGURE 8.25 Water uptake of starch composites investigated: (a) W/MMA-g-starch/GA; (b) W/MMA-g-starch/N; and (c) W/MMA-g-starch/D. Baishya and Maji [279]. *Source*: Reprinted with permission from American Chemical Society. Copyright 2014.

composite during storage for a long period of time. These results have been attributed to improved adhesion between the fiber and the TPS matrix due to the rough surface created as observed in microscopic studies.

Starch–Wood Composites There are not many studies reporting on the water absorption characteristics of starch-based composites. One of the reported studies on this aspect has been of methyl methacrylate-grafted starch-based "green" composite with sawdust (wood flour from *Ipomoea carnea*) using three different cross-linkers, namely, GA, dimethylol-dihydroxy-ethylene urea (DMDHEU), and *N*-methylol acrylamide (NMA) have been carried out [279]. Figure 8.25 shows the plots of water uptake (%) as function of time (h) for all the three types of materials studied. It can be seen that increase in the water uptake capacity with increasing time in all the composites with initial rapid increase, which slowed down later. Further, significant decrease in the water uptake capacity is shown by the composites prepared with MMA-g-starch, while it is the least with the DMDHEU cross-linked composites. It has been reported that these results indicate improved dimensional stability of the wood starch composites. This has been attributed to the interaction of DMDHEU with the functional groups of wood in agreement with observation of earlier researchers [280].

Characterization of these composites for various other properties has been reported the following: (i) DMDHEU-treated wood composites exhibited the overall improvement in all the properties evaluated followed by NMA-treated and GA-treated composites; (ii) the maximum interaction occurred between DMDHEU, wood, and MMA-g-starch; (iii) decrease in crystallinity of wood composites due to incorporation of cross-linker and MMA-g-starch; (iv) maximum improvement of mechanical and viscoelastic properties observed in DMDHEU composites; (v) decrease in the surface roughness of the composites due to cross-linking, which also improved flame retardancy of the composites.

8.2.2.3 Starch-based Nanocomposites Starch is well known to have multilevel structures from macro to molecular scales from starch granules to semicrystalline and crystalline regions. Following the generally accepted definition of bionanocomposites as "a hybrid nano structured material consisting of naturally occurring materials for both matrix and the reinforcing phases, with one of these having nano size and the product showing nano size at least in one dimension," starch-based bionanocomposites can be defined as a mixture of starch-based biopolymers with nanosized fillers/reinforcements with designed size, shape, and surface chemistry properties [281]. Bionanocomposites are not new concepts since a number of materials occurring in nature such as enamel and dentine in teeth, bone, ivory, etc. are natural bionanocomposites. In the case of starch-based bionanocomposites, nanofillers/reinforcements include plant-based NFs/whiskers/nanocellulose, synthetic polymer NFs, cellulose nanowhiskers, solid-layered clays, carbon nanotubes, and graphene. Starch nanocomposites are complex systems because they involve interactions at multiple length scales brought about by the inherent multilevel structure of starch and nanoparticle interactions. Since starchy materials can be used either as the continuous polymeric phase (matrix) or the nanodispersed phase (filler), or both, which result in better properties than the original starch, a number of reports have been published on the starch-based nanocomposites, which cover many of the aforementioned nanomaterials and are briefly discussed under appropriate headings in the following pages. In fact, a comparison of reinforcing effect of the different fillers such as inorganic clays, carbon nanotubes, and polysaccharide-based (cellulose, starch) nanoparticles used to improve the mechanical properties of TPS has been recently reported [17, 190, 269, 282]. Since there are many published papers on the preparation techniques to obtain NFs/cellulose/whiskers, etc., which are reviewed recently, these are not included here. Interested readers may see [23] for more details on this.

Processing of starch-based nanocomposites has already been explained (see Section 8.2.1.2). As regards the characterization of starch-based bionanocomposites, this has been carried out using various techniques as in the case of microcomposites. These include (i) optical and SEM for morphology studies to understand the homogeneity of dispersion, presence of voids, presence of aggregates, orientation of whiskers, etc.; (ii) wide-angle XRD for structural analysis; (iii) mechanical testing; and (iv) TGA/DTA/DMTA for evaluation of thermal properties.

Starch-Based Bionanocomposites Containing Cellulosic Materials

PLANT-BASED CELLULOSE NANOCOMPOSITES Preparation of thin films of TPS such as corn starch (TPS) matrix reinforced with 0–10 wt. % kenaf cellulose NF loadings (0–10 wt. %) by casting method by evaporating the mixtures of aqueous suspension of NFs, starch, and glycerol has been reported [16, 43, 45, 47, 200].

The composite mixture was found to have undergone gelatinization process at the same time. A homogeneous dispersion of NFs within the TPS matrix was observed along with strong adhesion between the filler and matrix. Besides, significant decrease in the moisture sensitivity of the TPS film significantly was observed with the addition of nanoreinforcements, while about 20% decrease in moisture content at equilibrium was observed for the addition of 10 wt.% NFs compared to 5.7% for the raw fiber.

Another study undertaken to assess the effect of different processing methods and amount of cellulose microfibrils in the behavior of the composites has reported the development of composite films based on TPS of PoS and cellulose microfibril (CMF) from Colombian agro-industrial wastes, namely, from vascular bundles of banana rachis and the BC microfibrils produced by *Gluconacetobacter* genus bacteria with the culture media of pineapple peel juice [283, 284]. It was observed that when cellulose microfibrils were added before gelatinization, the mechanical properties showed improvement with significant increases in both the YM and tensile strength of both the composites compared to those of starch matrix.

A biodegradable composite with Mater-biR-Y104U/C starch-based polymer matrix incorporated with alpha grass as the reinforcement has been reported [199]. It was observed that the value of YM of the prepared composite was similar to that of PP-based composites, while it increased significantly with increasing reinforcing amount of fibers, reaching values up to 7.2 GPa, comparable to those of PP–glass fiber composites. This indicated that these biodegradable composites can replace PP–glass fiber composites in some applications where stiffness is an important criterion. Further, it was also observed when the alpha grass fibers were given treatment, the fibers exhibited a high value of about 34 GPa showing enough stiffening possibilities. Although the glass fiber has higher stiffening power by volume content, modulus efficiency factor of the developed biodegradable composites was found to be about 0.5, similar to those of polyolefin-based natural fiber composites.

A cost-effective and fully biodegradable "green" composite containing 1–5 wt. % nanocellulose fibers (Celish, obtained from Daicel Chemical Industries Co. Ltd., Japan) having a length of 1 μm and diameter of 20–30 nm mixed with corn starch resin using ball mill with ceramic balls has been reported [157]. Two layers of ramie fabric was pasted with the previous composite mixture and hot compression molded at 150°C under the pressure of 13.1 MPa (after evaporating water in the resin and stopping the heating process) to obtain fully biodegradable nano "Green" textile composites. In fact, before the compression molding, the authors have used hand layup for laying the fabric properly. It was observed that cellulose NFs were well dispersed resulting in higher interfacial area and good fiber networking within the matrix resin, which would lead to better barrier properties. This makes safe use of this "textile" composite in various environments having slight moisture and oil. Besides, good interactions at the fiber–matrix interface and the compatibility between cellulose and corn starch resin molecules resulted in improved properties in the prepared composite. Analysis of sorption characteristics of water, lubricating oil, and diesel in the textile composites revealed that (i) with increasing fiber content in the composite, sorption increased till a maximum was observed in the case of water and diesel with the uptake was in the order of maximum with water followed by diesel and finally by lubricating oil. While the sorption in the case of water was faster, it was steady and slowest in the case of diesel. Besides, the water sorption mechanism in the nanocomposites was found to exhibit slight deviation from Fickian mode. Observed sorption characteristics by water and the lubricating oil have been attributed to the nature and properties of these solvents, such as low molecular weight, small-sized polar molecule in the case of water, and high viscosity along with high molecular weight and larger size of the hydrocarbon molecules in the case of oil.

Zhen [48] has reported development of a TPS polymer (TPS supplied by Chengdu Shudu Pharmaceuticals Co. Ltd.) composite by solution casting method using appropriate solvent systems. This composite contained nanocellulose fiber extracted from cotton by chemical treatment followed by mechanical separation and was intended for the sports industry with the objective of overcoming the disposal problem of waste sports equipment. It was interesting to see that this biological material exhibited special properties of nanomaterials, rigidity of cellulose, dimensional stability, and biodegradability. The tensile strength and modulus of the nanocomposite films were found to be significantly enhanced, while the degradation and water absorption increased significantly in comparison to the pure TPS matrix.

Studies have also been made to prepare and characterize nanocomposites using WMS plasticized with sorbitol as the matrix and tunicin whiskers as the reinforcement acting as a stable aqueous suspension [83]. Obtained nanocomposites were characterized for their structure by wide-angle X-ray scattering, scanning electron microscopy, and DSC. Water absorption characteristics have also been studied. The authors have observed a single glass-rubber transition with no evidence of transcrystallization of amylopectin on cellulose whisker surfaces along with resultant antiplasticizing effects. Also, initial increase in the glass-rubber transition temperature of the plasticized amylopectin matrix for about 10–15 wt. % whiskers content was observed, which then found to decrease. Further, increasing either moisture content or whiskers content resulted in a significant increase in crystallinity of the composites.

With a view to develop maize starch-based nanocomposites, first an environment friendly, sustainable, and water-soluble cross-linker, 1,2,3,4-butane *tetra*carboxylic acid (BTCA), was used to cross-link WMS, which was blended with micro/nanofibrillated cellulose (MFC) to obtain bionanocomposite [161]. Smooth, transparent, and defect-free films could be obtained with industrially pregelatinized WMS without any plasticizer with this starch. Various steps involved in obtaining the nanocomposite are shown in Figure 8.26 showing the stability of cross-linked composite of maize starch/MFC in water.

It has been reported that this method would provide a benign and convenient way to produce cross-linked starch-based composite films (≈300 μm in thickness), comparable to commercially available plastic sheets and as well be easily scaled up for commercial production. It was observed that cross-linking brought down the moisture absorption as well as made the films and composites insoluble in water. Due to the incorporation of MFC, enhanced tensile properties with a YM of over 2.3 GPa, fracture strain of 3.1%, and fracture stress of 39 MPa, along with significantly high toughness were obtained in cross-linked WMS composite films containing 15% MFC.

FIGURE 8.26 Various steps involved in obtaining the nanocomposite. *Source*: Dastidar and Netravali [161]. Reprinted with permission from American Chemical Society. Copyright 2013. (*See insert for color representation of the figure.*)

With a view to develop higher flexural properties, a study on the preparation and characterization of starch-based dispersion-type biodegradable resin and cellulose nanofibers has been reported [270]. It was observed that voids and the aspect ratio of the fibers used have large effect on the flexural modulus in that flexural modulus increased with aspect ratio of the fiber, while it decreased with increasing void content. Also, the critical value of aspect ratio of the fiber used was found to be approximately 80 for reinforcing the matrix to obtain higher flexural strength value of the composite. Furthermore, flexural strength of the composite was found to improve when the composite was processed by stirrer mixing technique due to the enhanced uniform dispersion of nanofibers along with the lowest void content.

Another study has reported development of PS–bamboo cellulose crystals using casting method [158]. Good and comparable reinforcing effect was observed even at low a fiber content of 8 wt. % compared to that of microcomposites with 30–50 wt. % cellulosic fibers [170, 171].

Bionanocomposites of CaS–CW (0–20 wt. %) with sorbitol and glycerol mixture as plasticizers has been studied [172]. For this purpose, a homogeneous mixture of starch and the plasticizer was first made by mixing it for a fixed time and speed in Rheomixer. Then the mixture was subjected to the ultrasonic treatment for about 5 min, followed by the addition of required amount of aqueous suspension of whiskers. This was then followed by hot pressing the mixture poured into suitable mold at an appropriate temperature so that no volatilization of glycerol took place as well as no thermal degradation of the whiskers. Hot pressed sample was then dried at 60–70°C to constant weight before determining tensile, thermal, and other properties. A very high increase in elongation and increase in TS up to 5 wt. % nanoreinforcement loading, which then decreased were observed, both of which could be attributed to reinforcing effect by carbohydrate present in nanowhiskers. On the other hand, YM decreased significantly for glycerol-containing starch compared to that of sorbitol TPS, while it increased significantly for the composites. Besides, decreased crystallinity (as determined by XRD), thermal stability (as determined by TGA), and water absorption were observed with the incorporation of both bagasse whiskers and nanocellulose. These have been attributed to hindering of the reinforcing effect or swelling due to possible transcrystallization (as seen by XRD), which may induce lower interactions between nanofibrils and starch.

Castor oil cake-based starch nanocomposites containing 2, 5, and 10 wt. % of nanocellulose of raft wood (*Ochroma pyramidale* (Cav.)) and 10 wt. % glycerol as plasticizer have been prepared by chemical and mechanical methods by thermo-molding process [266]. Increase in flexural modulus, water absorption, and swelling thickness, but no significant change in the apparent density, were observed with the increasing amount of incorporation of nanocellulose indicating appropriateness of processing method used. These observed results have been explained on the basis of fractographic studies. These revealed presence of agglomerated nanocellulose fibers along with the short fibers present in castor oil cake particularly in 5 wt. % fiber content nanocomposites. Probably, in the absence of agglomeration of NFs, the properties would have been much better than the observed ones.

"Green" composites of 6–50 wt. % cellulose NFs incorporated into various starches such as corn, wheat, potato, and waxy maize have been reported [285]. It was observed that tensile properties of PoS-based composites showed the highest increase in tensile strength (~400%) followed by that of corn starch composites (~300%) for the same amount of fiber content (40 wt. %) compared to up to 250% increase in tensile strength with 5–10% loading of cellulose NFs in different matrices (Prof. M. Saini, Private Communication, University of Toronto, Canada).

TPS–PCL blend and its composite with nanosilica (nano-SiO_2) have been prepared by melt extrusion technique using a twin extruder and characterized for mechanical properties [286]. It was observed that in general, mechanical properties of composites were better than the starch–PCL blend of 50:50. In fact, the tensile strength of the SiO_2/TPS/PCL composite with 1.8% SiO_2 increased by 20%, while percent elongation by 33%.

In one of the rare combinations of matrix and reinforcement of the same material, rice starch films reinforced with rice nanocrystal composites prepared by solution casting have been reported recently [86]. It has been observed that the composites containing 20% SNCs exhibited the best mechanical properties presenting a high tensile strength of about 16.43 MPa and elongation at break of 5.76%. Further, it has also been found that the water barrier properties and thermal stability of rice starch films are enhanced with increasing amount of SNC incorporation.

Chen et al. [92] have reported development of a new environment-friendly high-strength nanocomposite consisting of elastic polyurethane (WPU)-WMS nanocrystals by casting and evaporating techniques. The crystalline platelets of nanostarch crystals having an average equivalent diameter of 25–40 nm were produced by the sulfuric acid hydrolysis of WMS granules. It was observed that dispersion of SNCs into the WPU matrix as small aggregates was achieved by the predispersing process. Efficient strengthening of the composites could be achieved due to strong interactions between filler and matrix. Significant increases in YM (from 0.6 to 3.2 MPa) and tensile strength (from 10.4 to 24.1 MPa) with high elongation at break

(1148–1136%) were observed for composite films with 1 wt. % SNCs and kept at 75% RH for 1 week. Besides, these composites exhibited good thermal stability.

Studies on the rice starch films-microcrystalline cellulose with average length of about 0.480 μm obtained from PPF have been reported [271]. Characterization of the composite by tensile strength, elongation, and water-resistance properties revealed increase in tensile strength (from 5.16 MPa for pure rice starch film to 44.23 MPa for the composite) and water resistance, but decrease in elongation at the break of the composites. These results clearly demonstrated reinforcement effect by the fibers indicating the strong interactions between microcrystalline cellulose from PPF as well as between microcrystalline cellulose from PPF and rice starch film matrix.

BC NANOCOMPOSITES Interesting studies of development of CaS-based composites with the incorporation of BC (with and without enzymatic hydrolysis by *Trichoderma reesei* endoglucanases) composite prepared by casting method have been reported [61, 152]. The following results have been reported:

1. Freshly prepared TPS films show only an amorphous halo in the X-Ray diffraction pattern indicating complete gelatinization of the starch. However, after conditioning for 10 days at 43% RH, a crystalline pattern developed.
2. Increasing slope with increasing time of treatment of cellulose fibers compared to an almost flat nature of the matrix was observed in stress–strain curves of the bionanocomposites. Values of YM, UTS, and percent strain evaluated from these curves showed an initial slow increase in both YM and UTS with increasing time of hydrolysis in the case of partially hydrolyzed BC containing nanocomposites, suggesting substantial reinforcing effect in both these values, although the extent of this effect depended upon the duration of reaching a maximum for about 60 min in both cases. However, with further increase time of hydrolysis, both values decreased and finally reached almost the same values as those of TPS matrix itself after 120 min. On the other hand, the values of percent strain at break were found to be higher for short enzymatic treatment times but more scattered than those at the optimum duration of hydrolysis (60 min), which was attributed to poor dispersion of the NFs into the TPS matrix.
3. Values of YM, UTS of bionanocomposites containing hydrolyzed BC fibers for 60 min, were found to be four times and two times, respectively, the values of those without treatment to fibers (140.6 ± 40.3 MPa; 4.15 ± 0.66 MPa), while they were 17 times and 8 times, respectively, higher than the plasticized matrix (33.4 ± 4.3 MPa; 1.09 ± 0.39 MPa). Thus the effectiveness of enzymatic treatment of the nanocellulose fibers in enhancing significantly their reinforcement capacity could be seen, which was attributed to the disruption of amorphous regions within the cellulose NF. On the other hand, reversal of the above gains was observed in the composites containing cellulose NFs treated for longer time of hydrolysis.
4. Cellulose NF content had affected the tensile properties such that no change in both YM and UTS (~175 and ~3.5 MPa, respectively) up to 2 wt. % of their addition, while these increased dramatically (~550–570 MPa and ~9.6 MPa,

respectively) with 2.5–5 wt. % fiber, where after they decreased. On the other hand, percent strain at break decreased up to 80% of TPS matrix with fiber addition of 0.25–1.25 wt. %, but it further decreased to 7% for 2.5 wt. % NF addition. These results are explained as due to formation of a percolation network [287, 288]. Poor dispersion of the fibers in the TPS matrix resulted in large scatter in the values of both YM and UTS at the higher cellulose fiber contents.

5. Morphology studies of bionanocomposites containing hydrolyzed nanocellulose fibers revealed fractures with a granular aspect due to partial crystallization of the starch induced by the cellulose fibers, but not due to the crystalline structure of the cellulose itself as observed earlier in TPS–cellulose whisker composites [65, 289], as evident from the holes and protuberances due to the pullout of fiber bundles.

In another study of using BC in the development of starch-based composites, PoS (32.8% amylose and 67.1% amylopectin) plasticized by glycerin was incorporated with nanoribbons of BC, which were synthesized by *in situ* fermentation and chemical cross-linking by citric acid [290]. Enhanced mechanical properties, improved thermal behavior, as well as water stability have been reported for these nanocomposites. For example, YM of cross-linked nanocomposite (715.82±56.84 MPa) was 400 times that of TPS matrix (1.85±0.15 MPa), while fourfold increase in tensile strength (5.65±1.36 MPa) over the TPS matrix (1.26± 0.05 MPa) compared to threefold increase (3.54±0.39 MPa) of noncross-linked nanocomposites have been observed. Uniform, continuous, and straight nanoscale network-like BC elements oriented in plane via the compression of BC pellicles have been reported to contribute to the observed very high YM values of nanocomposites. However, not much change in strain at break was observed in all these composites, but these values were very much lower (125%) than those of TPS matrix as has been observed by others in similar systems [61, 291]. It may also be observed from the observed YM and tensile strength values of cross-linked composites were higher than those of noncross-linked composites. This suggests that the effectiveness of cross-linking resulting in network formation by the nanoribbons along with the formation of covalent bonds whereby effective stress transfer through the specimen has taken place. All these results have been corroborated by fractographic studies. It has been concluded that the aforementioned results confirm about the use of starch in the packaging industry.

Starch–Graphene Composites Potato starch incorporated with grapheme nanosheets composites prepared by a novel method has been reported recently [202]. In this method, graphene oxide was reduced with hydrazine hydrate in the presence of starch. Figure 8.27a–d shows the morphology of graphene and starch–graphene nanocomposite as observed in SEM and TEM.

Figure 8.28a, b shows the plot of water vapor permeation (WVP) and tensile properties as functions of filler content. It can be seen from the Figure 8.28a that the tensile strength reached to the maximum of about 25 MPa from about 8 MPa for PS, while percent elongation at break decreased from about 25% to about 4% for TPS containing 1.774 wt. % of graphene compared to the plasticized starch (PS). On the other hand, water absorption decreased drastically with the incorporation of about

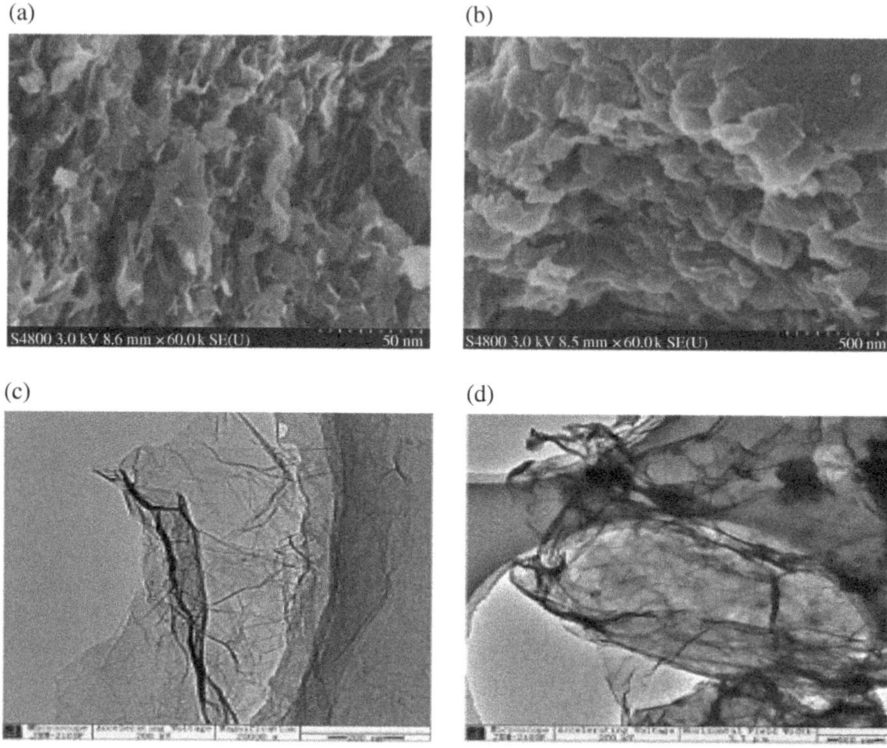

FIGURE 8.27 Electron micrographs of graphene and starch–graphene composites: (a) SEM of graphene; (b) SEM of starch–graphene composite; (c) TEM of graphene (d) TEM of starch–graphene composite. *Source*: Zheng et al. [202]. Reprinted with permission from American Chemical Society. Copyright 2013.

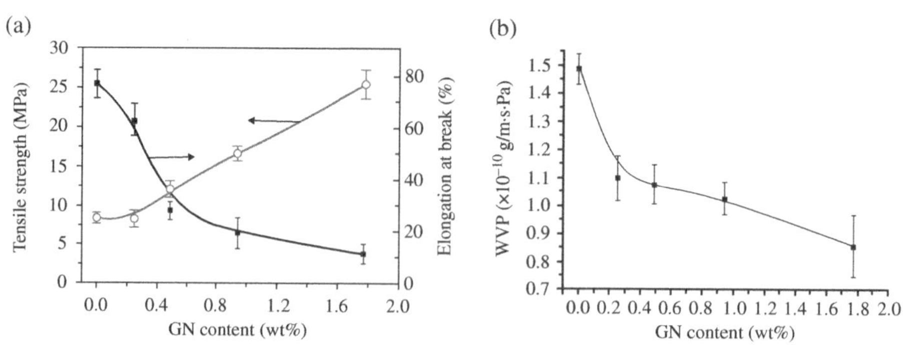

FIGURE 8.28 Plots showing the effect of GN–starch content on (a) TS and elongation at break of PS/GN–starch composites and (b) WVP of PS/GN–starch composites. *Source*: Zheng et al. [202]. Reprinted with permission from American Chemical Society. Copyright 2013.

0.2 wt. % grapheme and then slowly till up to about 1.774 wt. %. This suggested moisture barrier even at a very low loading (0.248 wt. %) of GN–starch fillers. The thermal conductivity of the composite was found to be 9.7×10^{-4} S/cm for the PS composite with 1.774 wt. %.GN–starch content. This has been attributed to the strong interaction between starch in GN–starch and the PS matrix. The authors have also observed high solubility and stability of the composite in water.

Another study has reported the fabrication of plasticized (PS) composites containing raw and reduced graphene oxides (GO & RGO) as fillers [292]. Flat and transparent sheets of both RG and RGO were first prepared by oxidizing graphite to get GO and then it was reduced using glucose to get RGO. Higher values of tensile strength, elongation at break, and moisture barrier, but lower thermal stability have been observed in PS-GO composites compared to those of PS-RGO composites. These could be attributed to the formation of stronger interaction of GO with the PS matrix than the RGO. Similarly, PS-GO composites were found to be UV light resistant while PS-RGO composites showed different thermal conductivity values at different environments (1.07×10^{-4}, 6.92×10^{-4} and 0.01 S/cm, respectively, at RH of 50, 75, and 100%).

Starch/Inorganic Nanoparticle Composites Considering the application of polymer–metal nanoparticle composites in the areas of optoelectronics, nonlinear optical devices, and color filters, in recent years much attention has been shown in research on these materials. Some of these materials include PoS (gel)-based nanocomposites containing nanoparticles of gold (Au) and Silver (Ag). These composite materials have been prepared by an environmentally benign "green" method using mild reducing conditions [293]. This method is reported to be advantageous and clean. The method consists of preparing the metal nanoparticles by reduction of metal ions in solution, while the PoS gel was being dried on flat surface to obtain foils containing nanometal structure. Taking into account the problem of aggregation of nanoparticles induced by their high surface energy, surface passivating reagents and capping reagents have been usually used in the reaction system.

Besides, using an environmentally friendly method, the authors have also prepared dialdehyde starch (DAS)-based composites DAS/Ag and DAS/Au with nanosilver and nanogold as reducing and protecting agents. Formation of about 10 nm ball shaped Ag and Au nanoparticles situated within the polysaccharide template have been reported based on UV–vis spectroscopy, TEM, and XRD studies. Significant rise of molecular weight of polysaccharide chains constituting DAS/Ag and DAS/Au composite template have led to the formation of nanosilver- and nanogold-containing composites leading to induction of the polymerization and/or cross-linking reactions of DAS molecules.

Recently, another nanocomposite films of PoS containing Ag nanoparticles by an environmentally friendly *ex situ* method (green chemistry process) have been reported [294]. The method consists of two steps. In the first, colloidal silver dispersion was prepared using water, glucose, and soluble starch as solvent, reducing agent, and stabilizing agent, respectively. In the second step, the as-prepared nanoparticles were dispersed in a PoS–glycerol matrix. While studying the influence of amount of glucose and reaction time on the prepared particle size and distribution of the silver nanoparticles, two distinct silver nanoparticle populations in the nanocomposite films with 5 and 20–50 nm size were observed. Both TEM and UV–vis studies gave evidences of lower nanoparticles in the nanocomposites compared to the colloidal silver dispersions. The authors have attributed potential interactions developed between the nanoparticles and the starch matrix to the observed reduced particle size. These nanocomposites have been reported to be promising candidates for many applications in antimicrobial packaging, biomedicines, and sensors.

Starch–Clay (Layered) Nanocomposites The recent increasing interest toward "green"/biodegradable polymers has favored the continued investigations on the systems mentioned previously, showing also their limitations. On the other hand, the success achieved by nanocomposites has fostered the search for new systems where the polymer matrix such as starch is biodegradable. The final properties of such nanocomposites will depend on a number of factors, including the biodegradable type of starch used as well as the nanosized filler, their mutual compatibility, the filler dispersion, and the processing conditions. Accordingly, other interesting starch-based "green" composite systems are the bionanocomposites of starch filled by layered clays including modified montmorillonite and chitosan intercalation. A number of published reports are available on this subject [95, 166, 187, 203, 295–298], which are briefly discussed in the following pages. Considering low hydrophobicity of modified clays (e.g., Cloisite Na^+ or 30B) and their compatibility with starches, layered silicates such as montmorillonite nanoclays have been incorporated in to different types of starches for synthesizing starch nanocomposite systems [17]. It has been observed that the basal spacing of d_{001} plane in the clay has increased from 1.17 to 1.85 nm by modification in the case of Cloisite 30B. Further, it was found that the clay particles were mainly stacked and aggregated in microcomposites of starch polymers, where the polymer chains have not diffused by reptation (this means thermal motion of very long linear, entangled macromolecules in polymer melts or concentrated polymer solutions) into the clay interlayers spacing. On the other hand, alternating layered silicates and polymer chains have been observed in the structure of intercalated nanocomposites.

This polymer diffusion into the clay interlayer spacing leads to an increase in the basal spacing (d_{001}).In exfoliated nanocomposites the clay layers (nanofillers) are individually delaminated and homogeneously and fully dispersed into the 3D matrix (Fig. 8.29).

It has been observed that the properties of starch-layered clay nanocomposites are closely related to their structure and leading to large improvements in their performance, which can be mainly attributed to the high interface area between nanoclay and matrix as well as to the corresponding dispersion state [95, 295, 296].

There is also recently published report about preparation of nanocomposites based on a starch-derived matrix and three different lamellar silicates [300]. This was characterized for rheological, morphological, and tensile properties through the effects of the rheological flow on the dispersion, improvement of intercalation/exfoliation, and possible aligning of the nanosilicate particles along the flow direction. It was observed that studied properties (rheological, morphological, and tensile properties) were significantly different depending on the nature of compatibility between the

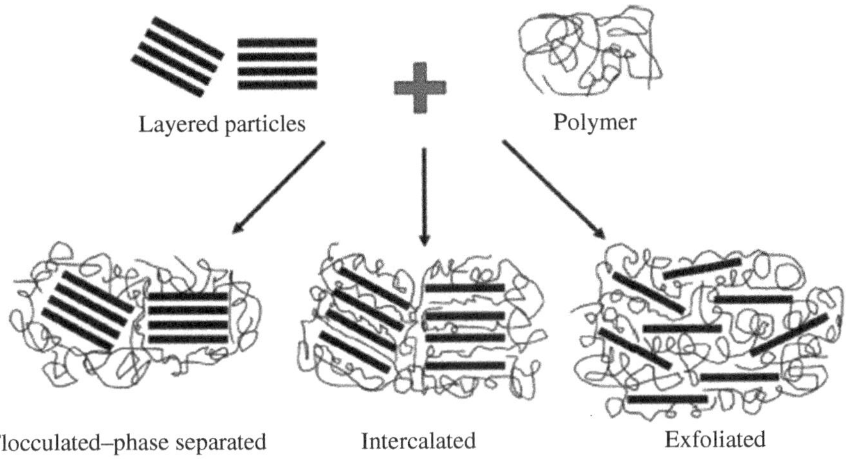

FIGURE 8.29 Dispersion of layered particles in polymer matrix. *Source*: Alexandre and Dubois [299]. Reproduced with permission from Elsevier.

polymer matrix and the nanofiller, as well as the draw ratio and therefore the orientation degree.

With a view to improve the properties of prepared biodegradable glycerol-plasticized sago thermoplastic starch (TPSS), montmorillonite (MMT) was incorporated as filler to prepare nanocomposites by hot-pressing method [166]. The C–O groups of the starch were found to be shifted to higher wave number, while the reactive hydroxyl groups of MMT shifted to lower wave numbers as observed from FTIR spectrum. Further, the crystallization of the nanocomposites was restrained by the nanofiller and intercalated in TPSS as observed in XRD spectrum. It was also observed that the thermal stability of the nanocomposites was better than that of the plasticized matrix used (TPSS) as observed in TGA/DTA studies, while uniform dispersion of nanofiller (MMT) was observed in SEM studies.

It is reported that nanocomposites of biopolymer such as starch incorporated with clay, prepared by coprecipitation in a chemical such as ethanol through the addition of low amounts of clay, form a new class of materials exhibiting improved mechanical properties [301]. However, major problems remaining as challenges with the starch-based clay nanocomposites have been reported to be the dispersion of clay in the matrix and the miscibility between the biopolymer (starch) and clay [297, 302–304]. In one recent study, no significant influence by clay on the type of crystalline structure of starch molecules, but low crystallinity was observed in the native normal corn starch (plasticized with glycerine)-based nanocomposites containing three different kinds of clay sodium montmorillonite (Na^+MMT), chitosan-modified montmorillonite, and laponite [305]. Also, improved YM (65%) and tensile strength (30%) without any decrease in elongation at break have been observed for the TPS-5 wt. % clay nanocomposite compared to those of TPS matrix. However, these properties decreased with further increase in clay content. It has also been mentioned that 65%

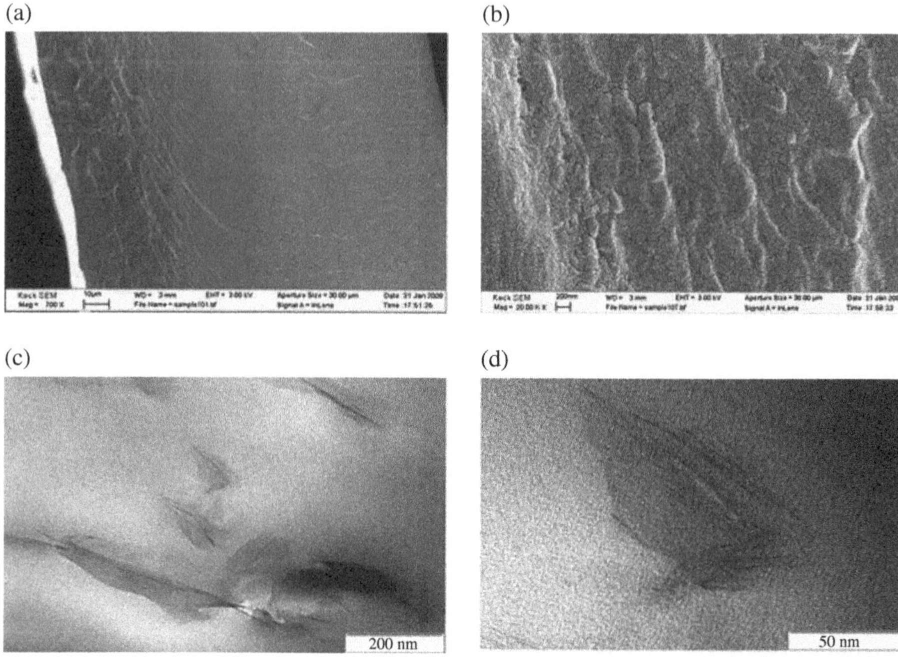

FIGURE 8.30 SEM micrographs of fractured surfaces starch–5% montmorillonite (clay) nanocomposites; (a) unfilled starch sample, (b) starch–clay nanocomposites, (c and d) TEM images of the same nanocomposites at two different magnifications. *Source*: Chung et al. [305]. Reproduced with permission with Elsevier. Copyright 2010.

improvement of YM of the composites is a significantly better compared to that observed by other researchers for starch-based nanocomposites containing the same amount of nanoparticles [306–312]. Observed mechanical properties have been attributed to poor clay dispersion and low crystallinity of the matrix material. Stability of matrix with respect to water during transportation and storage was found to improve with the addition of clay in the materials. These results have been discussed through fractography studied by SEM. Fractured surfaces of TPS matrix and TPS-5 wt. % clay nanocomposite shown in Figure 8.30a, b reveal a homogeneous surface in both the matrix and its nanocomposite suggesting even low amount of clay being fairly compatible and miscible with starches resulting in well-dispersed nanocomposites. These figures also reveal that the starch granules have been disrupted completely with the clay being dispersed in the polymer matrix. No aggregation of clay was observed even at higher magnifications. On the other hand, TEM studies of these composites revealed exfoliated clay layers coexisting with some clay agglomerates (Fig. 8.30c) and internally disordered but somewhat aligned clay agglomerates(Fig. 8.30d).

Another study has reported the preparation and characterization of urea- and ethanolamine-plasticized corn starch "green" composites incorporated with activated montmorillonites [203]. While montmorillonites were activated using ethanolamine as the activated solvent, the composites were prepared by blend extrusion method. Mechanical properties of composites were found to be improved after being cooled by liquid nitrogen compared to those obtained at room temperature. Morphology studies using SEM and TEM techniques revealed that (i) the melting urea was crystallized on the MMT sheets in the inducement of ethanolamine together with the overlapping of the column-shape crystal whisker, (ii) exfoliated MMT sheets in the composites acting as the inhomogeneous nucleation agent, and finally (iii) reticulating fiber structure in the liquid nitrogen-cooled composites.

Yet another study has been carried out on the nanocomposites of TPS (potato/wheat) incorporated with 0–7 vol. % of organophilized sodium montmorillonites and silicate clays (layered silicates) [297]. The organophilation of sodium montmorillonites and silicate clays was done using aminododecanoic acid, stearyl dihydroxyethyl ammonium chloride, and distearyl dimethyl ammonium chloride surfactants, respectively. Only limited exfoliation was observed during the homogenization of the composites irrespective of using a wide variety of organophilizers. Further, competitive dissolution and adsorption of the surfactant and the plasticizer were observed in the composites. Besides, the properties were found to be dependent on the type and strength of interaction. For example, the amino acid-modified clay-containing composites showed almost 10 times the tensile strength compared to that containing aliphatic amine surfactant-treated silicate thus suggesting wide variation of the reinforcing effect of the silicates. Besides, other properties such as viscosity, color, and water adsorption were found to change in the same way as those observed in strength and are dominated by interactions.

8.2.2.4 Starch Nanoparticles in Composites Just as in the case of micro- and nanocomposites, where incorporation of reinforcements of micro- or nanoreinforcements, nanostarch particles when incorporated into any biopolymer would not only enhance the mechanical properties but also physical properties such as permeability or fire retardancy. Of course, the extent of enhancement in properties of nanocomposites with starch nanofillers depend on the nature and effectiveness of interactions at the interfacial region, that is, on both the surface area and the dispersion of the particles. In the case of these type of nanocomposites, the surface area depends on the dimensions of the dispersed particles from 0.5 to 250 g/m^2 for natural fibers and up to 1000 g/m^2 for cellulose nanofibrils, exfoliated clays, and carbon nanotubes [17]. Starch is the second most studied organic material for producing nanocrystals after nanocellulose.

Rice starch films incorporated with about 15% SNCs revealed improvement of mechanical properties of the composite over that of the matrix rice starch film [313] suggesting increased crystalline peak structure of rice starch film by the incorporation of nanostarch crystals. Besides, increasing amount of nanocrystals into the rice starch films have been found to improve the water barrier properties and the char yield as well as shifting of the glass transition temperature (T_g) toward higher temperatures with increasing amount of nanocrystals in the rice starch films. The latter indicated enhanced thermal stability of rice starch films containing starch nanocrystals.

TABLE 8.6 Relative tensile test data for starch nanocrystal-filled nanocomposites[a]

Samples	Starch	RH (%)	YM (MPa)	TS (MPa)	ε_b (%)
NR-5%	Waxy	00	1.6	1.95	0.91
NR-10%	Waxy	00	2.5	2.97	0.90
NR-20%	Waxy	00	5.0	3.89	0.76
NR-5%	Waxy	43	2.5	2.00	0.90
NR-10%	Waxy	43	7.5	3.00	0.90
NR-20%	Waxy	43	19.2	3.10	0.80
TPS-G25-5%	Waxy	43	7.3	3.60	0.30
TPS-G25-10%	Waxy	43	7.5	4.20	0.20
TPS-G25-20%	Waxy	43	22.0	9.80	0.007
TPS-S25-6%	Waxy	11	1.08	0.96	0.73
TPS-S25-20%	Waxy	11	1.08	0.58	0.34
TPS-S25-40%	Waxy	11	0.98	0.42	0.24
TPS-S25-6%	Waxy	43	1.30	1.11	0.94
TPS-S25-20%	Waxy	43	2.18	1.75	0.66
TPS-S25-40%	Waxy	43	2.99	1.65	0.04
TPS-S25-6%	Waxy	53	2.51	1.68	0.50
TPS-S25-20%	Waxy	53	8.58	2.90	0.21
TPS-S25-40%	Waxy	53	15.52	5.10	0.06
TPS-S25-20%	Waxy	75	7.74	2.12	0.56
TPS-S25-40%	Waxy	75	35.46	4.01	0.26
TPS-S25-5%	Waxy	43	2.13	2.61	0.90
TPS-S25-10%	Waxy	43	2.23	3.61	0.92
TPS-S25-15%	Waxy	43	2.69	4.18	0.65
TPS-S35-5%	Waxy	43	1.65	2.61	0.86
TPS-S35-10%	Waxy	43	3.12	6.06	0.60
TPS-S35-15%	Waxy	43	7.06	12.33	0.42
PVA-G2-5%	Pea	43	n.a	1.11	1.04
PVA-G2-10%	Pea	43	n.a	1.08	1.03
PVA-G2-15%	Pea	43	n.a	0.97	1.00
SPI-G30pph-1%	Pea	RT	2.36	1.37	0.38
SPI-G30pph-2%	Pea	RT	2.91	1.52	n.a

ε_b(%), strain at break; b, recalculated from original data; c, recalculated data; G, glycerol; NR, natural rubber; pph, parts per hundred; PVA, polyvinyl alcohol; RH, relative humidity; SPI, soya protein isolate; TPS, thermoplastic starch; TS, tensile strength; YM, Young's modulus.

[a] Le Corre et al. [17]. Reprinted with permission from American Chemical Society. Copyright 2010.

It has been reported that the reinforcing effect of SNCs is more significant in TPS (polymers of PVA and soy protein isolate (SPI)-starch) blend than in NR [17, 91, 240, 248, 282, 314–317]. This is evident from Table 8.6, which shows values of YM, tensile strength, and percent strain at break obtained at different relative humidities in TPS and natural fiber composites containing starch nanoparticles. Strong interactions between the filler and amylopectin chains and a possible crystallization at the filler/matrix interface have been attributed to the higher reinforcing effect observed with the former matrix [282]. An immediate use of this has been taken advantage through the development of commercial biotires in collaboration with Goodyear, who used Mater-Bi (Novamont), a nanoparticle from starch by partially replacing conventionally used lamp black and silica in the tires [91].

A synergistic reinforcement of waterborne polyurethane(WPU) using both SNCs and CW has been reported to be an interesting and innovative approach for the use of nanocrystals as fillers [314]. Thus prepared WPU/1% SNC/0.4% CW nanocomposites exhibited significant reinforcing effect than other composites without SNC. This has been attributed to the formation of a new type of network permitted by strong hydrogen-bonding interactions both between the nanofillers themselves as well as between the nanofillers and the WPU matrix.

Generally, mechanical properties along with WVP (barrier) properties are also important for polymer packaging materials for their use in food sector applications [17, 317]. Considering current tendency to use eco-friendly polymers such as PLA-based polymeric materials for such applications and that SNCs are better than clays in flexible food packaging applications together with the fact that platelet-shaped particles are preferred, since they alter the diffusion path of penetrant molecules and improve the barrier properties, attempts have been made to use exfoliated starch nanoparticles as fillers in PLA in place of microcrystalline cellulose, which has low oxygen permeability due to higher degree of crystallinity of PLA [17, 316].

A comparison of WVP values of polymeric matrices filled with mineral fillers and organic fillers is shown in Table 8.7 [240, 248, 315, 318, 319]. More filler content of MCC is required for getting comparable value to that of ZnO.

Table 8.8 shows comparison of WVP of nanocomposites of different matrices (plasticized pea and chitosan, NR) filled with either mineral fillers or organic fillers including maize SNCs [17, 320–323]. It can be seen from the Table 8.8 that more MCC filler content is required for getting comparable value to that of ZnO.

PLA matrix incorporated with starch-grafted PCL using PSNCs prepared by H_2SO_4 hydrolysis of native PS granules has been reported [204]. It has been observed that the percent elongation at break of the composite increased up to 25%, while the YM decreased by the addition of nanostarch crystals. Introduction of PCL component, which provided flexibility to nanocomposites and the formation of an interfacial layer, which facilitated stress transfer to rigid nanocrystals were reported to be the causes for the observed properties. It was also found that barrier properties could also be modified by using SNC, while optimum filler content was only 5%.

With a view to find potential applications for various biodegradable polymer matrices, a study on water absorption characteristics was carried out [17]. A rapid increase of the diffusion coefficient was observed in nanocomposites of a nondegradable polymer matrix containing about 10 wt. % wax maize SNCs, which was attributed to the geometrical percolation effect leading to the creation of a diffusion pathway. On the other hand, no change in water uptake was observed in nanocomposites of soy protein matrix containing PS nanocrystal, due to the strong interfacial interaction between the filler and the matrix [89]. However, at equilibrium, a decrease in water uptake from 69.7 to 50.6% has been reported for the pullulan matrix-based nanocomposites with the increase in the SNCs content up to 40% [324].

TABLE 8.7 Water vapor permeability (WVP) of biopolymer materials used in packaging (conditions: 23°C, 50% RH), and mild conditions (25°C, 75% RH)[a]

Material (film thickness)	T (°C)	RH (%)	WVP (10^{11}g/m s Pa)	PO_2 ((cm^3 μm)/(day m^2 kPa))
Amylose+40% glycerol	23	50	119	7
Amylopectin+40% glycerol	23	50	144	14
Chitosan+40% glycerol	20	75		0.1–04
Pure chitosan (15.2 μm/2 μm)	20/23	75/0	4.5	0.5203 cm^3/m^2 day
Corn starch (63.1 μm)	25	75	17.7	
Wheat starch (500 μm)	20	n.a.		1.2096
Cassava (300–4000 μm)	25	58	45	
Cassava+20% glycerol (100 μm)	23	50	240	
Waxy maize (300–4000 μm)	25	58	38	
Pea starch+6.5% sorbitol	22	10	99.4	
Pea starch+10.87% sorbitol	22	10	186.1	
High amylose	25	50	34	
Zein-coated high amylose	25	50	1170	
PLA	23	50	1.34	160
PHA	23	50		150
PHB(61 μm)	25	50	0.245	
Ecovio		ASTM F1249	0.92	1426
Ecoflex		ASTM F1249	1.7	3329
LDPE		ASTM D1434	0.01	6897
PET	23	0/95	0.28	153.2
NR	23	50	0.0107	1.67

b, differing conditions for O_2 permeability; n.a., data not available in original paper; PO_2, permeability of oxygen recalculated from the original paper; RH, relative humidity.
[a]Le Corre et al. [17]. Reprinted with permission from American Chemical Society.

Similar results of decrease in water uptake from 78 to 62% have been reported for nanocomposites of PVA matrix containing 40 wt. % SNCs, cellulose microfibrils [47], and tunicin whiskers [26, 91].

8.3 APPLICATIONS

Considering the variety of starch-based “green” composites developed, their processing methods used, structure, and properties obtained, a number of applications for the starch-based “green” composites have been attempted. Some of these including on their products applicable to various sectors, especially in biomedical and automotive uses [325–333], are patented.

One such patented system is starch/nanosilver biodegradable composite films produced by preprocessing starch, degrading and oxidizing starch, and preparing starch solution [327, 328]. The steps involved include blow molding of the

TABLE 8.8 Water vapor permeability (WVP) of polymer matrix filled with either mineral fillers or organic fillers[a]

Matrix	Filler	Filler content (%)	T (°C)	RH (%)	WVP (10^{11}g/m s Pa)	Relative WVP
Mineral fillers glycerol plasticized pea starch	Nano ZnO	0	25	75	47.6	1.00
		0.5	25	75	37.5[b]	0.79
		1	25	75	27.5[b]	0.58
		2	25	75	25.0[b]	0.53
		3	25	75	22.1	0.46
		4	25	75	21.8	0.46
Glycerol-plasticized chitosan	Unfilled		27.1	76.2	131	1.00
	Na-MMT	5	26.4	78.8	98	0.75
	Cloisite 30B	5	24.3	78.2	92	0.70
	Nanosilver	5	24.5	78.1	95	0.73
	Ag-ion	5	22.3	77.3	96	0.73
PLA	Unfilled		25	50	1.8	1.00
	Cloisite Na+	5	25	50	2.08	1.16
	Cloisite 30B	5	25	50	1.7	0.94
	Cloisite 20A	5	25	50	1.15	0.64
Organic fillers glycerol-plasticized pea starch	Microcrystalline cellulose	0	25	75	50.1	1.00
		2.5	25	75	33.0[b]	0.66
		6	25	75	27.5[c]	0.55
		8.5	25	75	25.5	0.51
		12	25	75	30.0[b]	0.60
Xylan	Sulfonated cellulose whiskers (gmil/hm^2 (WVTR. film thickness))	0	25	75	304	1.00
		10	25	75	174	0.57
Natural rubber	Waxy maize starch nanocrystals	0	25	50	3.41	1.00
		5	25	50	2.41[d]	0.71
		10	25	50	2.36[d]	0.69
		20	25	50	1.88[d]	0.55

[a] Le Corre et al. [17]. Reprinted with permission from American Chemical Society.
[b] Recalculated from original data.
[c] Relative WVP refers to ratio of WVP values to those obtained for the unfilled matrix.
[d] Recalculated data.

preprocessed starch into film to dryness using nitrogen, and obtaining double-layer starch self-assembly film, and then immersing this film into 1–2% silver ammonia solution. This was then heated to 68–72°C for reacting to finally obtain the composite film. The above process has been reported to be simple having low equipment requirements. The product is reported for using in medical antibacterial materials particularly for dressing.

A starch-acrylic hydrophilic monomer and/or vinyl monomer was copolymerized by reactive extrusion in the presence of water, layered silicates, and a radical initiator to obtain biodegradable nanocomposites [331].

Liu [329] has prepared a completely biodegradable starch (maize/potato/Maniocca-Cassava)-based composite material. This composite material comprises of 10–25 wt. % of starch, 20–50 wt. % aliphatic polyester, 10–20 wt. % of a starch compatibilizer, and 10–60 wt. % of a natural plant fiber (China grass textile fiber/sisal fiber/wood). First, the natural plant fibers was crushed followed by mixing them with 2% NaOH solution and oven dried at 100–130°C for about 60–90 min. till moisture content was below 2%. This dried material was then mixed for 5–20 min with starch, aliphatic polyester, and starch compatibilizer in a preheated internal mixer to obtain a blended natural plant fiber-reinforced starch-based composite material. This was then made into tablets. Thus obtained material exhibited high strength and modulus with excellent plasticity, processability, and resistance to water, acids, and alkalis besides completely naturally biodegradable. The author has claimed this to be inexpensive and would have scope for broad applications.

Another starch-based fully biodegradable plastic composite containing biological catalyst, plasticizer, and PBS has been patented [333]. The patent claims the process is based on biochemical treatment on the natural starch using various fully decomposable additives, with a view to change the structure of the starch. This can be processed by various methods such as extrusion molding, injection molding, and blow molding. Besides, the additions result in 100% degradation effect after the starch-based fully biodegradable plastic enters into soil for 30 days. It also claims the advantages of good film-forming property, strength, and thermoplastic-processing property having great hydrophobicity besides of low cost. The intended use for this composite has been reported to be disposable living goods and medical (sanitary) products, packaging materials, and agricultural thin films.

Starch–carboxymethyl cellulose composites containing a plurality of nonpermanent intrafiber metal (aluminum, boron, bismuth, titanium, and zirconium ions) cross-links have been patented [332]. These are intended in personal care products, while its mixed polymer composite fiber as super absorbents for adults and as infant diapers with only carboxyalkyl cellulose instead of carboxymethyl cellulose and several nonpermanent intrafiber metal cross-links.

Considering poor mechanical properties, weak aging resistance and lack of processing performance, and not completely biodegradable existing PLA composite, a fully biodegradable composite of low cost that can be produced in large scale has been patented by Zhejiang Haizheng Biological Material Company [325]. This composite has a preset amount of PLA, 10% of TPS, 5–30% of inorganic filler,

0.5–2% antioxidizer, and 0.1–2% lubricant (all in wt. % with respect to 100 wt. % of PLA). It has been processed in steps comprising of mixing, extruding, granulating, and finally, hot press forming.

Another PLA/starch composite with compatibilizer has been patented for possible applications in molding cutlery, cups, plates, bowls, or packaging [330]. This biodegradable composite exhibited excellent stability toward thermal distortion, improved thermal deformation tolerance of above 120°C, improved molding or other thermal forming properties, and increased biodegradable rates relative to PLA matrix, no surface shine, and improved thermotolerance as well as workability. It is reported to be molded into disposable items for heat-related applications.

Another patent has been reported with starch–isocyanate group containing biodegradable material prepared by heat mixing of 5–95% starch or modified starch, 95–5% biodegradable resin, and 0.1–50% isocyanate group material [327]. This material is reported to have improved interfacial adhesion between starch and a biodegradable resin and therefore improved mechanical properties. The method of preparation of this composite is reported to be efficient, low cost, and allow producing the mechanically excellent starch-based composite material on a large scale. This composite material is intended for products such as artificial timber and film.

8.4 PERSPECTIVES

Preparation of *in situ* composites by bottom-up approach is expected to result in composites with defined properties, which can fine-tune the processability and degradability. In fact, there are some recent reports of preparation of PLA composites containing nanocellulose whiskers modified with lactide and polymerized using stannous octoate [215]. Similar studies using SNCs may become a very fertile area in the days to come. The other material in future could be exfoliated clay-MMT including kaolinite-based layered silicates incorporated starch inorganic–organic hybrid composites, which are expected to be predominant field.

Furthermore, with increased environmental and ecological awareness all over the world, the use of synthetic plastics is expected to become almost extinct. Therefore, complete attention with renewed interest would be directed toward use of biodegradable systems particularly starch-based ones, which would have ease of processability at low cost and exhibit improved mechanical properties coupled with lower water absorption. This may call for intensive research and development of starch-based systems or even nanostarch in combination with nanocellulose systems incorporated bionanocomposites to tune the moisture absorption and other properties.

8.5 CONCLUDING REMARKS

Considering the growing environmental concerns and the economical reasons, the need to replace conventional materials including composites based on dwindling fossil resources by new set of materials prepared from renewable resources seems to be on the rise. Within this context, starch will certainly play a very important role as a source of viable alternatives particularly for its exploitation in nonfood applications, which is not new. However, due to the boom of fossil-based petroleum and the development of polymer materials associated with it, considerable decline took place after the Second World War. Fortunately, with the emergence of new successful commercial products from starch-based plastics evoked interest at the end of the last millennium, which has been vigorously pursued in recent times. Accordingly, recognizing the fact that starch, a versatile raw material with possibility of both chemical and physical modifications, is a cheap raw material, readily available from different species, various sources of starch and its structure and properties are discussed. Use of starch in the synthesis of biodegradable "green" composites including recent trends in nanocomposites are reviewed to underline the usage of starch from various sources to convert into nanocrystals and modification of starch to use in conjunction with other degradable matrices in addition to nondegradable plastics to become degradable along with the use of natural raw materials such as plant fibers and clays. Further, research and development is being directed toward the "green" composites with modified starch in nanodimensions as well as the composites with modifiers with a view to have improved processability using common polymer-processing techniques. In addition, preparation of *in situ* composites by bottom-up approach is expected to result in composites with defined properties, which can fine-tune the processability and degradability. PLA composites containing nanocellulose whiskers modified with lactide and polymerized using stannous octoate to prepare the composites are reported. Following this, there is possibility of similar studies using SNCs being taken up in future. Also, exfoliated clay including kaolinite-based layered silicates incorporated starch inorganic–organic hybrid composites, which are expected to reign the field of "green" composites. Furthermore, with increased environmental and ecological awareness, the use of synthetic plastics is expected to be directed toward biodegradable systems with improved mechanical and processability coupled with lower water absorption. The research and development of these systems will result in increased use of nanostarch in combination with nanocellulose systems to tune the moisture absorption. All these would finally show a bright future to starch.

ACKNOWLEDGMENTS

The authors are grateful to Prof. Susheel Kalia, editor of this book entitled *Biodegradable Green Composites*, for inviting them to contribute this chapter and to the publisher John Wiley & Sons, Inc. They also express their sincere thanks to all the publishers (American Chemical Society, Elsevier, Express Polymer Letters, IOP Publishing, Wiley-VCH Verlag GmbH & Co., World Journal of Nano Science and Engineering), contact persons of these, and RightsLinks, who have all helped in getting the necessary permissions, as well as all the authors who have given their kind permission to reproduce the figures and tables used from their published work free of cost. Dr. V.S. Prasad would like to thank the Director of the National Institute for Interdisciplinary Science and Technology (NIIST-CSIR), Thiruvananthapuram-695019

(Kerala, India), for the encouragement for research. He also sincerely acknowledges the KSCSTE for funding in Nanocellulose Research. The authors also thank Mrs. S. Mini, Principal Technical Officer in NIIST-CSIR, Thiruvananthapuram (Kerala), for helping in providing necessary literature whenever it was sought during the preparation of this chapter. Dr. K.G. Satyanarayana would like to express sincere thanks to Poornaprajna Institute for Scientific Research (PPISR, Bengaluru), with whom he has been associated, for their encouragement.

REFERENCES

[1] Baillie, C. (2004). (Ed.) *Green Composites*, Woodhead Publishing Series in Composites Science and Engineering, Cambridge, 320.

[2] Chabba, S., Matthews, G. F., and Netravali, A. N. (2005). "Green" composites using cross-linked soy flour and flax yarns. *Green Chem.*, **7**, 576–581.

[3] La Mantia, F. P. and Morreale, M. (2011). Green composites: a brief review. *Compos. Part A*, **42**, 579–588.

[4] Shanmugasundaram, O. L. (2009). Green composites: manufacturing techniques and applications. *Ind. Text. J.*, **120**, 63–69.

[5] Ball, S. G. and Morell, M. K. (2003). From bacterial glycogen to starch: understanding the biogenesis of the plant starch granule. *Annu. Rev. Plant Biol.*, **54**, 207–233.

[6] Blennow, A., Engelsen, S. B., Nielsen, T. H., Baunsgaard, L., and Mikkelsen, R. (2002). Starch phosphorylation: a new front line in starch research. *Trends Plant Sci.*, **7**(10), 445–450.

[7] Buleon, A., Colonna, P., Planchot, V., and Ball, S. (1998). Starch granules: structure and biosynthesis. *Int. J. Biol. Macromol.*, **23**(2), 85–112.

[8] Carvalho, A. J. F. (2008). Starch: major sources, properties and applications as thermoplastic materials. In: *Monomers, Polymers and Composites from Renewable Resources*, (Eds. Belgacem, M. N. and Gandini, A.), Elsevier, Oxford, 1st Edition. pp. 321–342.

[9] Damager, I., Engelsen, S. B., Blennow, A., Moller, B. L., and Motawia, M. S. (2010). First principles insight into the alpha glucan structures of starch: their synthesis, conformation and hydration. *Chem. Rev.*, **110**, 2049–2080.

[10] Daniel, J., Whistler, R. L., Voragen, A. C. J., and Pilnik, W. (1994). Starch and other polysaccharides. In: *Ullmann's Encyclopedia of Industrial Chemistry*, (Eds. Elvers, B., Hawkins, S., and Russey, W.), 5th Edition. **A25**, VCH Verlagsgesellschaft mbH, Weinheim, pp. 1–62.

[11] Dastidar, T. G., and Netravali, A. N. (2012). Improving resin and film forming properties of native starches by chemical and physical modification, *J. Biobased Mater. Bioenergy*, **6**(1), 1–24.

[12] Dufresne, A., Meideiros, E. S., and Orts, W. J. (2009). Starch based nanocomposites. In: *Starch: Characterisation Properties and Applications*, (Ed. Bertolini, A.), CRC Press, Rio de Janeiro, pp. 205–251.

[13] Galliard, T. (1987). Starch availability and utilization. In: *Starch: Properties and Potential* (Ed. Galliard, T.), 1st Edition. John Wiley & Sons, Inc., New York, pp. 1–15.

[14] Getner, E. and Steller, R. (2012). Blends of natural and synthetic polymers as degradable materials, Part 1. Blends with starch. *Przem. Chem.*, **91**(12), 2407–2412.

[15] Avérous, L. (2014). Formulation and development of biodegradable and bio-based multiphase materials: plasticized starch-based materials. In: *Environmental Impact of Polymers*, (pp. 155–199), (Ed. Hamaide, T., Deterre, R., and Feller, J.-F.), John Wiley & Sons, Inc., Hoboken, NJ, DOI: 10.1002/ 9781118827116. ch9.

[16] Kalia, S., Kaith, B. S., Kaur, I., and Njuguna, J. (2013). Biofiber-reinforced thermoplastic composites. In: *Polymer Composites* (Eds. Thomas, S., Joseph, K., Malhotra, S. K., Goda, K., and Sreekala, M. S.), Wiley-VCH Verlag GmbH & Co. KGaA, Weinheim, pp. 239–288.

[17] Le Corre, D., Bras, J., and Dufresne, A. (2010). Starch nanoparticles: a review. *Biomacromolecules*, **11**, 1139–1153.

[18] Lin, N., Huang, J., Chang, P. R., Anderson, D. P., and Yu, J. (2011). Preparation, modification, and application of starch nanocrystals in nanomaterials: a review. *J. Nanomater.*, **2011**, Article ID 573687, 13 pages http://dx.doi.org/10.1155/2011/573687.

[19] Mosab, K., Kotiba, H., and Fawaz, D. (2012). Thermoplastic starch blends: a review of recent works. *Polym. Sci. Ser. A*, **54**(2), 165–176.

[20] Netravali, A. N., and Chabba, S. (2003). Composites get greener. *Mater. Today*, **6**, 22–26.

[21] Pérez, S., Baldwin, P. M., and Gallant, D. J. (2009). *Starch: Chemistry and Technology*, 3rd Edition. Elsevier, Inc., New York, p. 140.

[22] Satyanarayana, K. G., Gregorio G. C. A., and Wypych, F. (2009). Biodegradable composites based on lignocellulosic fibers—an overview. *Prog. Polym. Sci.*, **34**(9), 982–1021.

[23] Satyanarayana, K. G., Wypych, F., Woehl, M. A., Ramos, L. P., and Marangoni, R. (2011). Nanocomposites based on starch and fibers of natural origin. In: *Handbook of Bioplastics and Biocomposites Engineering Applications*, (Ed. Pilla, Srikanth), **1**, Scrivener Publishing LLC, Salem, MA pp. 469–508.

[24] Shogren, R. L. (1998). Starch, properties and material applications. In: *Biopolymers from Renewable Resources* (Ed. Kaplan, D. L.), Springer, Berlin Heidelberg, New York, p. 30.

[25] Tester, R. F. and Karkalas, J. (2002). *Starch in Biopolymers*, Wiley-VCH, Weinheim, p. 381.

[26] Tomasik, P. and Schilling, C. H. (2004). Chemical modification of starch. *Adv. Carbohydr. Chem. Biochem.*, **59**, 175–403.

[27] Vilaseca, F., Mendez, J. A., Pelach, A., Llop, M., Canigueral, N., Girones, J., Turon, X., and Mutje, P. (2007). Composite materials derived from biodegradable starch polymer and jute strands. *Process Biochem.*, **42**(3), 329–334.

[28] Visakh, P. M., Mathew, A. P., and Thomas, S. (2013). Natural polymers: their blends, composites and nanocomposites: state of art, new challenges and opportunities. In: Sabu Thomas, Aji J. Mathew, PM Visakh (Eds.). *Advances in Natural Polymers*, (pp. 1–20) Springer, Berlin.

[29] Visinescu, D., Tirsoaga, A., Patrinoiu, G., Tudose, M., Paraschiv, C., Ianculescu, A., and Carp, O. (2010). Green synthetic strategies of oxide materials: polysaccharides-assisted synthesis part I. Starch-assisted synthesis of nanosized metal-oxides. *Rev. Roum. Chim.*, **55**(11–12), 1017–1026.

[30] Wattanakornsiri, A. and Tongnunui, S. (2014). Sustainable green composites of thermoplastic starch and cellulose fibers. *Songklanakarin J. Sci. Technol.*, **36**(2), 149–161.

[31] Xie, F., Halley, P. J., and Averous, L. (2011). Bionanocomposites based on starch. In: *Nanocomposites with Biodegradable Polymers* (Ed. Mittal, V.), Oxford University Press, Oxford, p. 234.

[32] Xie, F., Halley, P. J., and Averous, L. (2012). Rheology to understand and optimize processability, structures and properties of starch polymeric materials. *Prog. Polym. Sci.*, **37**(4), 595–623.

[33] Yu, L., Dean, K., and Li, L. (2006). Polymer blends and composites from renewable resources. *Prog. Polym. Sci.*, **31**, 576–602.

[34] Yu, L., Liu, X., Petinakis, E., Dean, K., and Bateman, S. (2013). Starch based blends, composites and nanocomposites. In: *Advances in Natural Polymers: Composites and Nanocomposites*, vol. **18**, (Eds. Thomas, S., Visakh, P. M., and Mathew, A. P.). Springer, Berlin, pp. 121–154.

[35] Zhu, F. (2014). Structure, physicochemical properties, modifications, and uses of sorghum starch. *Compr. Rev. Food Sci. Food Saf.*, **13**(4), 597–610.

[36] Cao, X. D., Chen, Y., Chang, P. R., Stumborg, M., and Huneault, M. A. (2008). Green composites reinforced with hemp nanocrystals in plasticized starch. *J. Appl. Polym. Sci.*, **109**, 6, 3804–3810.

[37] Zhang, Y. and Han, J. H. (2006). Plasticization of pea starch films with monosaccharides and polyols. *J. Food Sci.*, **71**(6), E253–E261.

[38] Gallant, D. J., Bouchet, B., and Baldwin, P. M. (1997). Microscopy of starch: evidence of a new level of granule organization. *Carbohydr. Polym.*, **32**(3–4), 177–191.

[39] Ma, X., Chang, P. R., and Yu, J.-G. (2008). Properties of biodegradable thermoplastic pea starch/carboxymethyl cellulose and pea starch/microcrystalline cellulose composites. *Carbohydr. Polym.*, **72**(3), 369–375.

[40] Ma, X.-F., Chang, P. R., Yu, J.-G. (2008). Characterizations of glycerol plasticized-starch (GPS)/carbon black (CB) membranes prepared by melt extrusion and microwave radiation, *Carbohydr. Polym.*, **74**(4), 895–900.

[41] Ma, X.-F., Chang, P. R., Yu, J.-G., and Wang, N. (2008). Preparation and properties of biodegradable poly (propylene carbonate)/thermoplastic dried starch composites. *Carbohydr. Polym.*, **71**(2), 229–234.

[42] Ma, X.-F., Jian, R., Chang, P. R., and Yu, J. (2008). Fabrication and characterization of citric acid-modified starch nanoparticles/plasticized-starch composites. *Biomacromolecules*, **9**(11), 3314–3320.

[43] Cho, D., Kim, H.-J., and Drzal, L. T. (2013). Surface treatment and characterization of natural fibers: effects on the properties of biocomposites. In: *Polymer Composites* (Eds. Thomas, S., Joesph, S. K., Malhotra, K. G, and Sreekala, M. S.), Wiley-VCH Verlag GmbH & Co. KGaA, Weinheim, pp. 133–177.

[44] Dufresne, A. (2007). Polymer nanocomposite from biological sources. In: *Biopolymers Technology*, (Ed. Bertolini, A. C.), Cultura Academia, Sao Paulo, pp. 59–83.

[45] Nair, A. B., Sivasubramanian, P., Balakrishnan, P., Nair, K. A., and Sreekala, M. S. (2013). Environmental effects, biodegradation and life cycle analysis of fully biodegradable "green" composites. In: *Polymer Composites*, vol. **3**, (Eds. Thomas, S.;

Joseph, S. K., Malhotra, K., Goda, K., and Sreekala, M. S.), Wiley-VCH Verlag GmbH & Co. KGaA, Weinheim, pp. 515–568.

[46] Tao, J., Cui, Y. H., Ji, X. L., Ma, L., and Wo, D. Z. (2007). Advances in Composite Materials and Structures Part 1&2. In: *Key Engineering Materials*. (Kim, J. K., Wo, D. Z., Zhou, L. M., Huang, H. T., Lau, K. T., Wang, M. Eds.) **334–335**, pp. 345–348.

[47] Thomas, S., Joseph, K., Malhotra, S. K., Goda, K., and Sreekala, M. S. (2013). Fully biodegradable "green" composites. In: *Polymer Composites*, (Eds. Thomas, S. and Joseph, K., Malhotra, S. K., Goda, K., and Sreekala, M. S.), Wiley-VCH Verlag GmbH & Co. KGaA, Weinheim, pp. 431–463.

[48] Zhen, W. (2011). Study on nano cellulose/starch composite, sports materials, modeling and simulation. In: *Advances in Materials Research*, vol. **187** (Ed. Wu, Y. W.), Trans. Tech. Publications, Inc., Dürnten, pp. 544–547.

[49] Heinze, T., Liebert, T., Heublein, B., and Hornig, S. (2006). Functional polymers based on dextran. In: *Advances in Polymer Science* (Ed. Klemn, D.), **205**, Springer-Verlag, Berlin/Heidelberg, pp. 199–291.

[50] Akhila, R., Prasad, V. S., and Emilia, T. A. (2006). Enzymatic esterification of starch using recovered coconut oil. *Int. J. Biol. Macromol.*, **39**(4–5), 265–272.

[51] Ibrahim, H., Farag, M., Megahed, H., and Mehanny, S. (2014). Characteristics of starch-based biodegradable composites reinforced with date palm and flax fibers. *Carbohydr. Polym.*, **101**, 11– 19.

[52] Labet, M., Thielemans, W., and Dufresne, A. (2007). Polymer grafting onto starch nanocrystals, *Biomacromolecules*, **8**(9), 2916–2927.

[53] Namazi, H. and Dadkhah, A. (2008). Surface modification of starch, *J. Appl. Polym. Sci.*, **110**(4), 2405–2412.

[54] Mali, S., Grossmann, M. V. E., Garcia, M. A., Martino, M. N., and Zaritzky, N. E. (2008). Antiplasticizing effect of glycerol and sorbitol on the properties of cassava starch Films. *Braz. J. Food Technol.*, **11**(3), 194–200.

[55] Dufresne, A. and Vignon, M. R. (1998). Improvement of starch film performances using cellulose microfibrils. *Macromolecules*, **31**, 2693–2696.

[56] He, R., Wang, X. L., Wang, Y. Z., Yang, K. K., Zeng, H. B., and Ding, S. D. (2006). A study on grafting poly (1, 4-dioxan-2-one) onto starch via 2, 4-tolylene diisocyanate. *Carbohydr. Polym.*, **65**(1), 28–34.

[57] Xu, Y. X., Kim, K. M., Hanna, M. A., and Nag, D. (2005). Chitosan- starch composite film: preparation and characterization. *Ind. Crop. Prod.*, **21**(2), 185–192.

[58] Romero-Bastida, C. A., Bello- Pérez, L. A., Garcia, M. A., Martino, M. N., Solorza-Feria, J., and Zaritzky, N. A. (2005). Physicochemical and microstructural characterization of films prepared by thermal and cold gelatinization from non-conventional sources of starches*Carbohydr. Polym.*, **60**(2), 235–244.

[59] Mo, X. Z., Zhong, Y. X., Liang, CH. Q., and Yu, S. J. (2010). Studies on the properties of banana fibers-reinforced thermoplastic cassava starch composites: preliminary results. *Adv. Mater. Res.*, **87–88**, 439–444.

[60] Fakirov, S., and Bhattacharayya, D. (2007). *Handbook of Engineering Biopolymers*, Cincinatti, OH, Hanser Publishers.

[61] Woehl, M. A., Canestraro, C. D., Mikowski, A., Sierakowski, M. R., Ramos, L. P., and Wypych, F. (2010). Bionanocomposites of thermoplastic starch reinforced with bacterial cellulose nanofibres: Effect of enzymatic treatment on mechanical properties. *Carbohydr. Polym.*, **80**, 866–873.

[62] Corradini, E., Texeira, E. M., Agnelli, J. A. M., and Mattoso, L. H. C. (2007). *Amido Termoplástico*, Embrapa Instrumentação Agropecuária, São Carlos, SP, p. 27.

[63] Gáspár, M., Benko, Z., Dogossy, G., Reczey, k., and Czigany, T. (2005). Reducing water absorption in compostable starch-based plastics. *Polym. Degrad. Stab.*, **90**, 563–569.

[64] Carraher Jr., C. H. (1996). *Polymer Chemistry: An Introduction*, 3rd Edition. Marcel Dekker, New York.

[65] Angles, M. N. and Dufresne, A. (2000). Plasticized starch/tunicin whiskers nanocomposites: 1. Structural analysis. *Macromolecules*, **33**, 8344–8353.

[66] Angles, M. N. and Dufresne, A. (2001). Plasticized starch/tunicin whiskers nanocomposites: 2. Mechanical behavior. *Macromolecules*, **34**, 2921–2931.

[67] Avérous, L., Fringant, A., and Moro, L., (2001). Plasticized starch–cellulose interactions in polysaccharide composites. *Polymer*, **42**(15), 6565–6572.

[68] Lopez-Rubio, A., Lagaron, J. M., Ankerfors, M., Lindstrom, T., Nordqvist, D., Mattozzi, A., and Hedenqvist, M. S. (2007). *Carbohydr. Polym.*, **68**, 718.

[69] Talja, R. A., Helen, H., Roos, Y. H., and Jouppila, J. (2007). Effect of various polyols and polyol contents on physical and mechanical properties of potato starch-based films. *Carbohydr. Polym.*, **67**(3), 288–295.

[70] Wilhelm, H. M., Sierakowski, M. R., Souza, G. P., and Wypych, F. (2003). Starch films reinforced with mineral clay. *Carbohydr. Polym.*, **52**, 101–110.

[71] Cervera, M. F., Karjalainen, M., Airaksinen, S., Rantanen, J., Krogars, K., Heinämäki, J., ColarteA. I., YliruusiJ. (2004). Physical stability and moisture sorption of aqueous chitosan–amylase starch films plasticized with polyols. *Eur. J. Pharm. Biopharm.*, **58**(1), 69–76.

[72] Matveev, Y. I., Grinberg, V. Y., and Tolstoguzov, V. B. (2000). The plasticizing effect of water on proteins, polysaccharides and their mixtures. Glassy state of biopolymers, food and seed. *Food Hydrocoll.*, **14**, 425–437.

[73] Blanshard, J. M. V. (1987). Starch granule structure and function: a physicochemical approach. In: *Starch: Properties and Potential*, (Ed. Galliard, T.), John Wiley & Sons, Inc., New York.

[74] Knutson, C. A. (1990). Annealing of maize starches at elevated temperatures. *Cereal Chem.*, **67**(4), 376–384.

[75] Zavareze, E. R. and Dias, A. R. (2011). Impact of heat-moisture treatment and annealing in starches: a review. *Carbohydr. Polym.*, **83**, 317–328.

[76] Angellier, H., Molina-Boisseau, S., Belgacem, M. N., and Dufresne, A. (2005). Surface chemical modification of waxy maize starch nanocrystals. *Langmuir*, **21**(6), 2425–2433.

[77] Jayakody, L. and Hoover, R. (2002). The effect of lintnerization on cereal starch granules. *Food Res. Int.*, **35**(7), 665–680.

[78] Li, W., Corke, H., and Beta, T. (2007). Kinetics of hydrolysis and changes in amylose content during preparation of microcrystalline starch from amylose maize starches. *Carbohydr. Polym.*, **69**(2), 398–405.

[79] Robin, J. P., Mercier, C., Charbonniere, R., and Guilbot, A. (1974). Lintnerized starches, gel filtration and enzymatic studies of insoluble residues from prolonged acid treatment of potato starch. *Cereal Chem.*, **51**, 389–406.

[80] Dufresne, A., Cavaille, J. Y., and Helbert, W. (1996). New nanocomposite materials: microcrystalline starch reinforced thermoplastic, *Macromolecules*, **29**(23), 7624–7626.

[81] Namazi, H. and Dadkhah, A. (2010). Convenient method for preparation of hydrophobically modified starch nanocrystals with using fatty acids. *Carbohydr. Polym.* **79**(3), 731–737.

[82] Namazi, H., Fathi, F., and Dadkhaha, A. (2011). Hydrophobically modified starch using long-chain fatty acids for preparation of nanosized starch particles. *Scientia Iranica*, **18**(3), 439–445.

[83] Mathew, A. P. and Dufresne, A. (2002). Morphological investigation of nanocomposites from sorbitol plasticized starch and tunicin whiskers. *Biomacromolecules*, **3**(3), 609–617.

[84] Siqueira, G., Abdillahi, H., Bras, J., and Dufresne, A. (2010) High reinforcing capability cellulose nanocrystals extracted from syngonanthus nitens (capim dourado). *Cellulose*, **17**, 289–298.

[85] Figueira, J. A., Carvalho, P. H., Sato, H. H. (2011). Sugarcane starch: quantitative determination and characterization. *Ciênc. Tecnol. Aliment*, **31**(3), 806–815.

[86] Piyada, K., Waranyou, S. and Thawien, W. (2013). Mechanical, thermal and structural properties of rice starch films reinforced with rice starch nanocrystals. *Int. Food Res. J.*, **20**(1), 439–449.

[87] Putaux, J. L., Potocki-Véronèse, G., Remaud-Simeon, M., Buléon, A. (2006). α-D-glucan-based dendritic nanoparticles prepared by in vitro enzymatic chain extension of glycogen. *Biomacromolecules*, **7**(6) 1720–1728.

[88] Bhattacharya, T., Bandyopadhyay, N. R., Roy, D., and Bhattacharyya, D. K. (2013). Novel approach to prepare biopolymers based nanocrystals. *Adv. Appl. Sci. Res.*, **4**(4), 251–254.

[89] Zheng, H., Ai, F., Chang, P. R., Huang, J., Dufresne, A. (2009). Starch nano crystal-reinforced soy protein plastics. *Polym. Compos.*, **30**(4), 474–480.

[90] Putaux, J. L., Molina-Boisseau, S., Momaur, T., and Dufresne, A. (2003). Platelet nanocrystals resulting from the disruption of waxy maize starch granules by acid hydrolysis. *Biomacromolecules*, **4**(5), 1198–1202.

[91] Chen, Y., Cao, X., Chang, P. R., and Huneault, M. A. (2008). Comparative study on the films of poly(vinyl alcohol)/pea starch nanocrystals and poly(vinyl alcohol)/native pea starch. *Carbohydr. Polym.*, **73**(1), 8–17.

[92] Chen, G., Wei, M., Chen, J., Huang, J., Dufresne, A., and Chang, P. R. (2008). Simultaneous reinforcing and toughening: new composites of waterborne PU filled with low loading level of starch nanocrystals. *Polymer*, **49**(7), 1860–1870.

[93] Wang, Y. and Zhang, L. (2008). High-strength waterborne polyurethane reinforced with waxy maize starch nanocrystals. *J. Nanosci. Nanotechnol.*, **8**(11), 5831–5838.

[94] Le Corre, D., Bras, J., and Dufresne, A. (2009). Nanoparticles for eco-efficient packaging: influence of botanic. In: *2nd International Conference on Biodegradable Polymers and Sustainable Composites (BIOPOL 2009)*, (Ed. Jimenez, A.), Alicante, Spain (6 pages).

[95] Asha, S. C. and Prasad, V. S. (2013). Synthesis and characterisation of self assembling hybrid copolymer templates based on aminopropyl tri–(2-carboxy-3-(pentadeca-8,11,14-trienyl)phenoxy silane modified layered silicate for micromolding of polymer vesicles. 23rd FAPS Polymer Congress And Macro-2013, Indian Institute of Science, Bangalore, May 13–15, 2013.

[96] Dufresne, A., Kellerhals, M. B., and Wilholt, B. (1999). Transcrystallization in mcl-PHAs/cellulose whiskers composites. *Macromolecules*, **32**, 7396–7401.

[97] Raquez, J.-M., Goffina, A.-L., Duquesnea, E., Habibiahb, Y., and Duboisa, Ph. (2012). Novel nanocomposites reinforced with polysaccharide (starch) nanocrystals from interfacial ring opening polymerization to melt processing implementation. Biobased Monomers, Polymers, and Materials ACS Symposium Series, Chapter 16, vol. **1105**, 257–268.

[98] Angellier, H., Putaux, J.-L., Molina-Boisseau, S., Dupeyre, D., and Dufresne, A. (2005). Starch nanocrystal fillers in an acrylic polymer matrix. *Macromol. Symp.*, **221**(1), 95–104.

[99] Thielemans, W., Belgacem, M. N., and Dufresne, A. (2006). Starch nanocrystals with large chain surface modifications. *Langmuir*, **22**(10), 4804–4810.

[100] Pérez, S. and Bertof, E. (2010). The molecular structures of starch components and their contribution to the architecture of starch granules: a comprehensive review. *Starch-Stärke*. **62**(8), 389–420.

[101] Imberty, A. and Pérez, S. (1988). A revisit to the three-dimensional structure of B-type starch. *Biopolymers*, **27**, 1205–1221.

[102] Ptak, S., Roczkowska, M., Zarski, A., and Kapusniak, J. (2014). Esterification of starch with fatty acids. New opportunities and challenges, *Przem. Chem.*, **93**(4), 472–479.

[103] Thakur, V. K., Thakur, M. K., Gupta, R. K., Prasanth, R., and Kessler, M. R. (2013). Green composites: an introduction. In: *Green Composites from Natural Resources*, (Ed. Thakur, V. K.), CRC Press. (Taylor & Francis), Boca Raton, **419** pp. 1–10.

[104] Coppin, A., Varre, J. S., Lienard, L., Guerardel, Y., Soyer-Gobillard, M. O., Buleon, A., Ball, S., and Tomavo, S. (2005). Evolution of plant-like crystalline storage polysaccharide in the protozoan parasite *Toxoplasma gondii* argues for a red alga ancestry. *J. Mol. Evol.*, **60**, 257–267.

[105] Englyst, H. N., Veenstra, J., and Hudson, G. J. (1996). Measurement of resistant starch in vitro and in vivo. *Br. J. Nutr.*, **75**, 327–337.

[106] Damager, I., Olsen, C. E., Blennow, A., Denyer, K., Møller, B. L., and Motawia, M. S. (2003). Chemical synthesis of methyl 6′-alpha-maltosyl-alpha-maltotrioside and its use for investigation of the action of starch synthase II. *Carbohydr. Res.*, **338**(2), 189–197.

[107] Engelsen, S. B., Madsen, A. O., Blennow, A., Motawia, M. S., Møller, B. L., and Larsen, S. (2003). The phosphorylation site in double helical amylopectin as investigated by a combined approach using chemical synthesis, crystallography and molecular modeling. *FEBS Lett.*, **541**, 137–144.

[108] O'Sullivan, A. C. and Pérez, S. (1999). The relationship between internal chain length of amylopectin and crystallinity in starch. *Biopolymers*, **50**, 381–390.

[109] Griffin, B. E., Jarman, M., and Reese, C. B. (1988). The synthesis of oligoribonucleotides—IV: preparation of dinucleoside phosphates from 2′,5′-protected ribonucleoside derivatives. *Tetrahedron*, **24**, 639–662.

[110] Bergo, P. V. A., Sobral, P. J. A., and Prison, J. M. (2009). Physical properties of cassava starch films containing glycerol. *Gene Conserve*, **8**, 727–734.

[111] Avérous, L. and Boquillon, N. (2004). Biocomposites based on plasticized starch: thermal and mechanical behaviors. *Carbohydr. Polym.*, **56**(2), 111–122.

[112] Stepto, R.F.T. (2006). Understanding the processing of thermoplastic starch. In: World polymer congress and 41st international symposium on macromolecules, MACRO-2006, Rio de Janeiro. 2006 [in CD-ROM. paper no. 1862].

[113] Zhang, X., Do, M. D., and BilykA. (2006). Enhancement of the mechanical performance of wheat proteins-based biodegradable polymer materials via chemical modification. In: World polymer congress and 41st international symposium on macromolecules, MACRO-2006, Rio de Janeiro. [in CD-ROM, paper no. 1863].

[114] Bourban, C., Karamuk, E., deFondaumiere, M. J., Ruffi eux, K., Mayer, J., and Wintermante, E. (1997). Processing and characterization of a new biodegradable composite made of a PHB/V matrix and regenerated cellulosic fibers. *J. Environ. Polym. Degrad.*, **5**, 159–166.

[115] Curvelo, A. A. S., de' Carvalho, A. J. F., and Agnelli, J. A. M. (2001). Thermoplastic starch–cellulosic fibers composites: preliminary results. *Carbohydr. Polym.*, **45**, 183–188.

[116] Fowler, P. A., Hughes, J. M., and Elias, R. M. (2006). Biocomposites: technology, environmental credentials and market forces. *J. Sci. Food Agric.*, **86**, 1781–1789.

[117] Kuciel, S. and Liber-Knec, A. (2009). Biocomposites on the base of thermoplastic starch filled by wood and kenaf fiber. *J. Biobased Mater. Bioenergy*, **3**, 1–6.

[118] Avella, M., Martuscelli, E., and Raimo, M. (2000). Review—Properties of blends and composites based on poly (3-hydroxy) butyrate (PHB) and poly (3-hydroxybutyrate-hydroxyvalerate) (PHBV) copolymers. *J. Mater. Sci.*, **35**, 523–545.

[119] Cao, Y., Shibata, S., and Fukumoto, I. (2006). Mechanical properties of biodegradable composites reinforced with bagasse fiber before and after alkali treatments. *Compos. Part A*, **37**, 423–427.

[120] Carvalho, A. J. F., Curvelo, A. A. S., and Agnelli, J. A. M. (2001). A first insight on composites of thermoplastic starch and kaolin. *Carbohydr. Polym.*, **45**, 189–195.

[121] Chen, L., Qiu, X. Y., Xi, Z, Hong, Z, Sun, J, Chen, X. S., Jing, X. (2006). Poly (L-lactide)/starch blends compatibilized with poly (l-lactide)-g-starch copolymer. *Carbohydr. Polym.*, **65**, 75–80.

[122] Choi, J. S., Lim, S. T., Choi, H. J., Hong, S. M., Mohanty, A. K., DrzaL, L. T., Misra, M., and Wibowo, A. C. (2005). Rheological, thermal, and morphological characteristics of plasticized cellulose acetate composite with natural fibers. *Macromol. Symp. Special Issue Bio-Based Polym. Recent Prog.*, **224**(1), 297–308.

[123] Choi, J. S., Lim, S. T., Choi, H. J., Hong, A. K., Drzal, L. T., Misra, M. and Wibowo, A. (2004). Preparation and characterization of plasticized cellulose acetate biocomposite with natural fiber. *J. Mater. Sci.*, **39**(21), 6631–6633.

[124] Cunha, A. M., Liu, Z. Q., Feng, Y., Yi, X.-S., and Bernardo, C. A. (2001). Preparation, processing and characterization of biodegradable wood flour/starch–cellulose acetate compounds. *J. Mater. Sci.*, **36**, 4903–4909.

[125] Fang, J. and Fowler, P. (2003). The use of starch and its derivatives as biopolymer sources of packaging materials. *J. Food Agric. Environ.*, **1**, 82–84.

[126] Gousse, C., Chanzy, H., Exoffier, G., Soubeyand, L., and Fleury, E. (2002). Stable suspensions of partially silylated cellulose whiskers dispersed in organic solvents. *Polymer*, **43**(9), 2645–2651.

[127] Guimarães, J. L. 2010. Preparação e Caracterização de Compósitos de Amido Plastificado com Glicerol e reforçados com Fibras Naturais. Ph.D. Thesis, Engineering and Materials Science. UFPR. 190 p. Curitiba (PR-Brasil). [In Portuguese].

[128] Guimarães, J. L., Satyanarayana, K. G., Wypych, F., Saul, C. K., and Ramos, L. P., (2010). Studies on the processing and characterization of corn starch and its composites with banana and sugarcane fibers of Brazil. *Carbohydr. Polym.*, **80**, 130–138.

[129] Huang, C. Y., Kuo, J. C., and Yang, S. Y. (2006). Effect of plasticizer on the properties of biodegradable PVA/starch blends. In: World polymer congress and 41st international symposium on macromolecules, MACRO-2006. [CD-ROM, paper No.0707]. Brasil: Rio de Janeiro.

[130] Kumar, A. P. and Singh, R. P. (2008). Biocomposites of cellulose reinforced starch: improvement of properties by photo-induced cross linking. *Bioresour. Technol.*, **99**(18), 8803–8809.

[131] Lim, S. T., Chang, E. H., and Chung, H. J. (2001). Thermal transition characteristics of heat-moisture treated corn and potato starches. *Carbohydr. Polym.*, **46**, 107–115.

[132] Lomelí, R. M. G., Satyanarayana, K. G., Iwakiri, S., Bolzon, M. G., Tanobe, V., and Syden-stricker, F. T. (2011). Study of the properties of biocomposites: part I—Cassava starch-green coir fibers of Brazil. *Carbohydr. Polym.*, **86**(4), 1712–1722.

[133] Lomelí, R. M. G., Satyanarayana, K. G., Iwakiri, S., Bolzon, M. G., Tanobe, V., and Sydenstricker, F. T. (2014). Bio-composites of cassava starch-green coconut fiber: part II—structure and properties. *Carbohydr. Polym.*, **102**, 576–583.

[134] Luna, G., Villada, H., and Velasco, R. (2009). Thermoplastic cassava starch of reinforced with Fique's fiber: preliminary studies (Almidón termoplástico de yuca reforzado con fibra de fique preliminares). *Dyna*, **76**(159), 145–151.

[135] Ma, X.-F., Yu, J.-G., and Kennedy, J. F. (2005). Studies on the properties of natural fibers-reinforced thermoplastic starch composites. *Carbohydr. Polym.*, **62**, 19–24.

[136] Marina, M. N. K.2013. Potencial do Uso da a Torta de Mamona (*Ricinus communis* L.) Como Matriz Para Produção de Compósitos e Nanocompósitos. Tese de Doutorado Ciências Florestais e da Madeira, Universidade Federal do Paraná (UFPR), Curitiba (PR-Brasil). [In Portuguese].

[137] Mattoso, L. H. C., Leão, A. L., and Frollini, E. (Eds.) (2002). Natural polymers and composites. In *Proceedings of 3rd International Symposium on Natural Polymers and Composites—ISNaPol 2002*, Embrapa Agricultural Instrumentation, São Carlos, SP.

[138] Mattoso, L. H. C., Leão, A. L., and Frollini, E. (Eds.). (2004). Natural polymers and composites. In: *Proceedings of 5th International Symposium on Natural Polymers and Composites—ISNaPol 2004 (paper No.90)*, Embrapa Agricultural Instrumentation, São Pedro.

[139] Misra, S., Tripathy, S. S., Misra, M., Mohanty, A. K., and Nayak, S. K. (2002). Novel eco-friendly biocomposites-fabrication and processing evaluation. *J. Reinf. Plast. Compos.*, **21**, 55–70.

[140] Mohanty, A. K., Misra, M., and Hinrichsen, G. (2000). Biofibres, biodegradable polymers and biocomposites: an overview. *Macromol. Mater. Eng.*, **276**, 1–24.

[141] Müller, C. M. O., Laurindo, J. B., and Yamashita, F. (2009). Effect of cellulose fibers on the crystallinity and mechanical properties of starch-based films at different relative humidity values. *Carbohydr. Polym.*, **77**, 293–299.

[142] Nishino, T., Matsuda, I., and Hirao, K. (2004). All-cellulose composite. *Macromolecules*, **37**, 7683–7687.

[143] Ochi, S. (2006). Development of high strength biodegradable composites using Manila hemp fiber and starch-based biodegradable resin. *Compos. Part A*, **37**, 1879–1883.

[144] Rodriguez-Gonzalez, F. J., Ramsay, B. A., and Favis, B. D. (2003). High performance LDPE/thermoplastic starch blends: a sustainable alternate to pure polyethylene. *Polymer*, **44**, 1517–1526.

[145] Shamai, K., Bianco-Peled, H., and Shimoni, E. (2003). Polymorphism of resistant starch type III. *Carbohydr. Polym.*, **54**, 363–369.

[146] Singh, S., Mohanty, A. K., Sugie, T., Takai, Y., and Hamada, H. (2008). Renewable resource based biocomposites from natural fiber and polyhydroxybutyrate-co-valerate (PHBV) bioplastic. *Compos. Part A.*, **39**, 875–886.

[147] Teixeira, E. M. (2007). Utilização de amido de mandioca na preparação de novos materiais termoplasticos (Use of cassava starch for the preparation of new thermoplastic materials), Doctoral thesis in Portuguese, University of São Paulo, São Carlos, Brazil.

[148] Torres, F. G., Arroyo, O. H., and Gomez, C. (2007). Processing and mechanical properties of natural fiber reinforced thermoplastic starch biocomposites. *J. Thermoplast. Compos. Mater.*, **20**, 207–223.

[149] Tran, P., Gravier, D., and Narayan, R. J. (2006). Biocomposites synthesized from chemically modified soy oil and biofibers. *Appl. Polym. Sci.*, **102**, 69–75.

[150] Tserki, V., Matzinos, P., and Panayiotou, C. (2006). Novel biodegradable composites based on treated lignocellulosic waste flour as filler part II: development of biodegradable composites using treated and compatibilized waste flour. *Compos. Part A.*, **37**, 1231–1238.

[151] Vanderleia, A, (2009), "Obtaining of thermolded composites through the use of castor oil cake plastified with glycerol, derivative of the transesterification of oils and fats" (Obtenção de compósitos termomoldados a partir da torta de mamona plastificada com glicerol, derivado do processo de transesterificação de óleos e gorduras). Masters' Thesis, Federal University of Parana (UFPR), Curitiba (PR-Brasil). (In Portuguese).

[152] Woehl, M. A. (2009). Bionanocompósitos de amido plastificado com glicerol reforçados com nanofibras de celulose bacteriana, Master's Thesis, Federal University of Parana (UFPR), Curitiba, Brazil (in Portuguese).

[153] Wool, R. P., Khot, S. N., LaScala, J. J., Bunker, S. P., Lu, J., Thielemans, W., Can, E., Morye, S. S., Williams, G. I., Lankey, R. L., Anastas, P. T. (2002). Affordable composites and plastics from renewable resources: part II: manufacture of composites. In: *Advancing Sustainability Through Green Chemistry and Engineering*, (Eds. Lankey, R. L. and Anastas, P. T.), ACS Symposium Series, American Chemical Society, Washington, DC, **823**, pp. 205–224.

[154] Drzal, L. T., Mohanty, A. K., and Mishra, M. (2002). Opportunities for value added bio-based materials. Oral Presentation at the "Creating Value for Biobased Resources: Moving Beyond Petroleum" Conference, Kansas City, MO, USA, November 11–13, 2002.

[155] Liu, W., Thayer, K., Misra, M., Drzal, L. T., and Mohanty, A. K. (2007). Processing and physical properties of native grass-reinforced biocomposites. *Polym. Eng. Sci.*, **47**(7), 969–976.

[156] Lundquist, L., Marque, B., Hagstrand, P. O., Leterrier, Y., and Minson, J. A. E. (2003). Novel pulp fiber reinforced thermoplastic composites. *Compos. Technol.*, **63**, 137–152.

[157] Sreekala, M. S., Goda, K., and Devi, P. V. (2008). Sorption of water, oil and diesel, in: cellulose nanofiber reinforced corn starch resin/ramie fabric. *Compos. Interfaces*, **15**(2–3), 281–299.

[158] Liu, D., Zhong, T., Chang, P. R., Li, K., and Wu, Q. (2010). Starch composite reinforced by bamboo cellulosic crystals. *Bioresour. Technol.*, **101**, 2529–2536.

[159] Fama, L., Gerschenson, L., and Goyanes, S. (2009). Starch-vegetable fibre composites to protect food products. *Carbohydr. Polym.*, **75**(2), 230–235.

[160] MelloR. P. F. and Mali, S. (2014). Use of malt bagasse to produce biodegradable baked foams made from cassava starch. *Ind. Crop. Prod.*, **55**, 187–193.

[161] Dastidar, T. G. and Netravali, A. (2013). Cross-linked waxy maize starch-based "green" composites. *ACS Sustainable Chem. Eng.*, **1**(12), 1537–1544.

[162] Vilaseca, F., Mendez, J. A., Canigueral, N., Turon, X., Barbera, L., Girones, J., Pimenta, M. T. B., Pelach, M. A., and Mutje, P. (2007). Biodegradable composite materials from starch-based biopolymer and hemp strands. *Afinidad*, **64**(527), 22–24.

[163] Alvarez, V., Vazquez, A., and Bernal, C. (2006). Effect of microstructure on the tensile and fracture properties of sisal fiber/starch-based composites. *J. Compos. Mater.*, **40**(1), 21–35.

[164] Song, K. H. and Kim, I. S. (2013). Effects of plasticizer on the mechanical properties of kenaf/starch biocomposites. *Fibers Polym.*, **14**(12), 2135–2140.

[165] Chen, Y., Ishikawa, Y., Maekawa, T., and Zhang, Z. (2006). Preparation of acetylated starch/bagasse fiber composites by extrusion. *Trans. ASABE*, **49**(1), 85–90.

[166] Zuraida, A., Yusliza, Y., Nurizan, O., Anuar, H., Zahurin, H., and Noorasikin, S. (2012). MMT to modify the thermal properties of thermoplastic starch (glycerol plasticized), properties of montmorillonite-reinforced thermoplastic sago starch composites. In: *Materials and Manufacturing Technologies XIV*, (Eds. Yigit, F. and Hashmi, M. S. J.), Advances in Materials Research, **445**, pp. 469–474.

[167] Li, A., Zhang, J., and Wang, A. (2007). Utilization of starch and clay for the preparation of superabsorbent composite. *Bioresour. Technol.*, **98**(2), 327–332.

[168] Heux, L., Chauve, G., and Bonini, C. (2000). Nonflocculating and chiral-nematic self-ordering of cellulose microcrystals suspensions in nonpolar solvents. *Langmuir*, **16**(21), 8210–8212.

[169] Oksman, K., Mathew, A. P., Bondeson, D., and Kvien, I. (2006). Manufacturing process of cellulose whiskers/polylactic acid nanocomposites. *Compos. Sci. Technol.*, **66**, 2776–2784.

[170] Bogoeva-Gaceva, G., Avella, M., Malinconico, M., Buzarovska, A., Grozdnov, A., Gentile, A., and Errico, M. (2007). Natural fiber eco-composites. *Polym. Compos.*, **28**(1), 98–107.

[171] Eichhorn, S. J., Baillie, C. A., Zafeiropoulos, N., Mwaikambo, L. P.Ansell, M. P., Dufresne, A., Entwistle, K. M., Herrera-Franco, P. J., Escamilla, G. C., Groom, L., Hughes, M., Hill, C., Rials, T. G., and Wild, P. M. (2001). Review: current international research into cellulosic fibers and composites. *J. Mater. Sci.*, **36**(9), 2107–2131.

[172] Teixeira, E. D., Pasquini, D., Curvelo, A. A. S., Corradini, E., Belgacem, M. N., and Dufresne, A. (2009). Cassava bagasse cellulose nanofibrils reinforced thermoplastic cassava starch. *Carbohydr. Polym.*, **78**, 422–4311.

[173] Gindl, W. and Keckes, J. (2004). Tensile properties of cellulose acetate butyrate composites reinforced with bacterial cellulose. *Compos. Sci. Technol.*, **64**, 2407–2413.

[174] Brown, E. E. and Laborje, M. P. G. (2007). Bioengineering bacterial cellulose/poly(ethylene oxide) nanocomposites. *Biomacromolecules*, **8**, 3074–3081.

[175] Ljungberg, N., Bonini, C., Bortolusi, F., Boisson, C., Heux, L., and Cavaillé, J.Y. (2005). New nanocomposite materials reinforced with cellulose whiskers in atactic

polypropylene: effect of surface and dispersion characteristics. *Biomacromolecules*, **6**(5), 2732–2739.

[176] Araki, J., Wada, M., Kuga, S., and Okano, T. (1998). Flow properties of microcrystalline cellulose suspension prepared by acid treatment of native cellulose. *Colloids Surf. A.*, **142**, 75–82.

[177] Bonini, C. and Heux, L. (2005). Dispersion of microfibrils and/or microcrystals, notably of cellulose, in an organic solvent. European Patent Application number: EP20000949531, (Published Date: November 9, 2005) [Original French Patent. FR 99.07493-(1999/2000)].

[178] Liu, H., Xie, F., Yu, L., Chen, L., and Li, L. (2009). Thermal processing of starch-based polymers. *Prog. Polym. Sci.*, **34**(12), 1348–1368.

[179] Hunealt, M. A., Li, H., Chapleau, N., and Favis, B. D. (2006) Processing and properties of poly (lactic acid)/plasticized starch blends. In: World polymer congress and 41st international symposium on macromolecules, MACRO-2006, Rio de Janeiro. 2006 [in CD-ROM. Paper No. 841].

[180] Spychaj, T., Schmidt, B., Ulfig, K., and Markowska-Szcupak, A. (2012). Starch-grafted-*N*-vinylformamide copolymers manufactured by reactive extrusion: synthesis and characterization. *Polimery*, **57**(2), 95–100.

[181] Chen, L., Qiu, X. Y., Deng, M. X., Hong, Z. K., Luo, R., Chen, X. S., and Jing, X. B. (2005). The starch grafted poly (l-lactide) and the physical properties of its blending composites. *Polymer*, **46**(15), 5723–5729.

[182] Lönnberg, H., Fogelström, L., Samir, M.A.S.A., Berglund, L., Malmstöm, E., and Hult, A. (2008). Surface grafting of microfibrillated cellulose with poly (-caprolactone)—synthesis and characterization. *Eur. Polym. J.*, **44**(9), 2991–2997.

[183] Hubbe, M. A., Rojas, O. A., Lucia, L. A., and Saini, M. (2008). Cellulosic nanocomposites: a review, *Bioresource*, **3**(3), 929–980.

[184] García, N. L., Famá, L., Dufresne, A., Aranguren, M., and Goyanes, S. (2009). A comparison between the physico-chemical properties of tuber and cereal starches. *Food Res. Int.*, **42**(8), 976–982.

[185] Tan, Y., Xu, K., Li, L., Liu, C., Song, C., and Wang, P. (2009). Fabrication of size controlled starch based nanospheres by nanoprecipitation. *ACS Appl. Mater. Interfaces*, **1**(4), 956–959.

[186] Liu, D., Wu, Q., Chen, H., and Chang, P. R. (2009). Transitional properties of starch colloid with particle size reduction from micro- to nanometer. *J. Colloid Interface Sci.*, **339**(1), 117–124.

[187] Nair, B. P., Pavithran, C., Sudha, J. D., and Prasad, V. S. (2010). Microvesicles through self assembly of PS-clay nanocomposite. *Langmuir*, **26**(3), 1431–1434.

[188] Kim, J.-Y., Park, D.-J., and Lim, S.-T. (2008). Fragmentation of waxy rice starch granules by enzymatic hydrolysis. *Cereal Chem.*, **85**(2), 182–187.

[189] Kim, J.-Y. and Lim, S.-T. (2009). Preparation of nano sized starch particles by complex formation with *n*-butanol. *Carbohydr. Polym.*, **76**(1), 110–116.

[190] Angellier-Coussy, H. and Dufresne, A. (2013). Mechanical properties of starch-based nanocomposites. In: *Biopolymer Nanocomposites: Processing, Properties, and Applications*, (pp. 261–292), (Eds. Dufresne, A., Thomas, S., Pothen, L. A.), John Wiley & Sons, Inc., Hoboken, NJ.

[191] Niu, Y. and Li, H. (2012). Controlled release of urea encapsulated by starch-g-poly (vinyl acetate). *Ind. Eng. Chem. Res.*, **51**(38), 12173–12177.

[192] Ohkita, T. and Lee, S. H. (2005). Crystallization behavior of poly (butylene succinate) corn starch biodegradable composite. *J. Appl. Polym. Sci.*, **97**(3), 1107–1114.

[193] Wu, H. X., Liu, C. H., Chen, J. G., Chen, Y., Anderson, D. P., and Chang, P. R. (2010). Oxidized pea starch/chitosan composite films: structural characterization and properties. *J. Appl. Polym. Sci.*, **118**(5), 3082–3088.

[194] Inceoglu, F. and Menceloglu, Y. Z. (2013) Transparent low-density polyethylene/starch nanocomposite films. *J. Appl. Polym. Sci.*, **129**(4), 1907–1914.

[195] Gupta, A. P., Sharma, M., and Kumar, V. (2008). Preparation and characterization of potato starch based low density polyethylene/low density polyethylene grafted maleic anhydride biodegradable polymer composite. *Polym. Plast. Technol. Eng.*, **47**(9), 953–959.

[196] Ismail, A. and Gamal, M. A. B. (2010). Water resistance, mechanical properties, and biodegradability of poly(3-hydroxybutyrate)/starch composite. *J. Appl. Polym. Sci.*, **115**(5), 2813–2819.

[197] Corradini, E., Agnelli, J. A. M., De Morais, L. C., and Mattoso, L. H. C. (2008). Study of properties of biodegradable composites of starch/gluten/glycerol reinforced with sisal fibers. *Polimeros-Ciencia E Tecnologia*, **18**(4), 353–358.

[198] Kaewtatip, K. and Thongmee, J. (2014). Preparation of thermoplastic starch/treated bagasse fiber composites. *Starch-Stärke*, **66**(7–8), 724–728.

[199] Espinach, F. X., Julian, F., Alcala, M., Tresserras, J., and Mutje, P. (2014). High stiffness performance alpha-grass pulp fiber reinforced thermoplastic starch-based fully biodegradable composite, *Bioresource*, **9**(1), 738–755.

[200] Karimi, S., Dufresne, A., Tahir, P. M., Karimi, A., and Abdulkhani, A. (2014). Biodegradable starch based composites: effect of micro and nanoreinforcements on composite properties. *J. Mater. Sci.*, **49**(13), 4513–4521.

[201] Wattanakornsiri, A., Pachana, K., Kaewpirom, S., Sawangwong, P., and Migliaresi, C. (2011). Green composites of thermoplastic corn starch and recycled paper cellulose fibers. *Songklanakarin J. Sci. Technol.*, **33**(4), 461–467.

[202] Zheng, P., Ma, T., and Ma, X. (2013). Fabrication and properties of starch-grafted graphene nano sheet/plasticized-starch composites. *Ind. Eng. Chem. Res.*, **52**(39), 14201–14207.

[203] Huang, M. F. and Yu, J. G. (2006). Structure and properties of thermoplastic corn starch/montmorillonite biodegradable composites. *J. Appl. Polym. Sci.*, **99**, 170–176.

[204] Yu, J., Ai, F., Dufresne, A., Gao, S., Huang, J., and Chang, P. R. (2008). Structure and mechanical properties of poly (lactic acid) filled with (starch nanocrystal)-graft-poly (ε-caprolactone). *Macromol. Mater. Eng.*, **293**(9), 763–770.

[205] Wu, K., Hu, Y., Song, L., Lu, H. D., and Wang, Z. Z. (2009). Flame retardancy and thermal degradation of intumescent flame retardant starch-based biodegradable composites. *Ind. Eng. Chem. Res.*, **48**(6), 3150–3157.

[206] Battegazzore, D., Alongi, J., and Frache, P. (2014). LA based composites containing natural fillers, thermal, mechanical and barrier properties, *J. Polym. Environ.*, **22**(1), 88–98.

[207] Sun, X. S., Acioli-Moura, R., and Li, W. (2007). Effects of starch types on mechanical properties of poly(lactic acid)/starch composites. *Int. Polym. Proc.*, **22**(5), 410–417.

[208] Liu, X. G., Zhao, N. P., Yang, K. K., Wang, Y. Z., Zheng, C. Y., Li, S. Z., and Zhang, Z. P. (2008). Preparation of poly (lactic acid)/etherified starch composites, *Iran. Polym. J.*, **17**(12), 947–952.

[209] Xiong, Z., Dai, X., Zhang, R., Tang, Z., Na, H and Zhu, J. (2014). Preparation of biobased monofunctional compatibilizer from cardanol to fabricate polylactide/starch blends with superior tensile properties. *Ind. Eng. Chem. Res.*, **53**, 10653–10659.

[210] Ke, T. and Sun, X. (2000). Physical properties of poly (lactic acid) and starch composites with various blending ratios. *Cereal Chem.*, **77**, 761–768.

[211] Wang, H., Sun, X., and Seib, P. (2001). Strengthening blends of poly (lactic acid) and starch with methylenediphenyl diisocyanate. *J. Appl. Polym. Sci.*, **82**, 1761–1767.

[212] Kiangkitiwan, N. and Srikulkit, K. (2013). Poly (lactic acid) filled with cassava starch-g-soybean oil maleate. *Sci. World J.*, **2013**, 7.

[213] Ghosh, S., Viana, J. C., Reis, R. L., Mano, J. F., Nakamura, T., Yamashita, K., and Neo, M. (2006). *Bioceramics-18, Parts 1&2*, Key Engineering Materials, Trans Tech Publications Inc., Switzerland and USA, **309–311**, pp. 1109–1112.

[214] Fabunmi, O. O., Tabil, L. G., Chang, P. R., Panigrahi, S. (2007). Developing biodegradable plastics from starch. Paper number RRV07130, ASABE/CSBE North Central Intersectional Meeting, Published by the American Society of Agricultural and Biological Engineers, St. Joseph, Michigan; [From:www.asabe.org].

[215] Li, S., Xiong, Z., Fei, P., Cai, J., Xiong, H., Tan, J., and Yu, Y. (2013). Parameters characterizing the kinetics of the nonisothermal crystallization of thermoplastic starch/poly (lactic acid) composites as determined by differential scanning calorimetry. *J. Appl. Polym. Sci.*, **129**(6), 3566–3573.

[216] Jukola, H., Nikkola, L., Gomes, M. E., Chiellini, F., Tukiainen, M., Kellomaki, M., Chiellini, E., Reis, R. L., and Ashammakhi, N. (2008). Development of a bioactive glass fiber reinforced starch-polycaprolactone composite. *J. Biomed. Mater. Res. B Appl. Biomat.*, **87**(1), 197–203.

[217] Kim, E. G., Kim, B. S., and Kim, D. S. (2007). Physical properties and morphology of polycaprolactone/starch/pine-leaf composites. *J. Appl. Polym. Sci.*, **103**(2), 928–934.

[218] Ge, X. C., Xu, Y., Meng, Y. Z., and Li, R. K. Y. (2005). Thermal and mechanical properties of biodegradable composites of poly(propylene carbonate) and starch-poly(methyl acrylate) graft copolymer, *Compos. Sci. Technol.*, **65**(14), 2219–2225.

[219] Huang, C. Y., Yang, S. Y., and KuoJ. C. (2006) Effect of PVA content on the biodegradation and mechanical properties of PVA/starch blends. In: World polymer congress and 41st International Symposium on macromolecules, MACRO-2006, Rio de Janeiro. [in CD-ROM; paper no. 0714].

[220] Nievas, M., Solanas, V., Tripoli, S. B., Bucala, V., and Villar, M. A. (2006). Evaluation of biodegradability of starch-based polyethylene composite. In: World polymer congress and 41st international symposium on macromolecules, MACRO-2006, Rio de Janeiro, Brazil. [in CD-ROM, paper no. 1410].

[221] Priya, B., Gupta, V. K., Pathania, D., and Singha, A. S. (2014). Synthesis, characterization and antibacterial activity of biodegradable starch/PVA composite films reinforced with cellulosic fibre, *Carbohydr. Polym.*, **109**, 171–179.

[222] Ramaraj, B. (2007). Crosslinked poly(vinyl alcohol) and starch composite films. II. Physicomechanical, thermal properties and swelling studies. *J. Appl. Polym. Sci.*, **103**(2), 909–916.

[223] Riyajan, S. A., Jitdaphon, W., and Leejarkpai, T. (2014). Effect of additives on the physical properties of a biopolymer hydrogel from epoxidized natural rubber, poly (vinyl alcohol), and starch. *Kautsch. Gummi Kunstst.*, **67**(3), 34–40.

[224] Simão, R. A., Martins, M., Silva, M. L., Thire, M. S. M., and Andrade, C. T. (2006). Hydrophobicity of CVDa-C:H coatings produced on corn starch plasticized films: contact angle and AFM adhesive forces measurements. In: World polymer congress and 41st international symposium on macromolecules, MACRO-2006, Rio de Janeiro. 2006 [in CD-ROM; paper no. 1844].

[225] Singha, A. S. and Kapoor, H. (2014). Effect of silk fibroin reinforcement on the properties of potato starch-polyvinyl alcohol blend films. *Int. J. Polym. Anal. Charact.*, **19**(3), 212–221.

[226] Leja, K. and Lewandowicz, G. (2010). Polymer biodegradation and biodegradable polymers—a review. *Pol. J. Environ. Stud.*, **19**(2), 255–266.

[227] Chang, P. R., Ai, F., Chen, Y., Dufresne, A., and Huang, J. (2008). The effect of starch nanocrystal-graft-PCL on mechanical properties of water borne PU based nanocomposites. *J. Appl. Polym. Sci.*, **111**(2), 619–627.

[228] Chang, P. R., Jian, R. J., Zheng, P. W., Yu, J. G., and Ma, X. F. (2010). Preparation and properties of glycerol plasticized-starch (GPS)/cellulose nanoparticle (CN) composites. *Carbohydr. Polym.*, **79**, 301–305.

[229] Liu, X., Gao, C., Sangwan, P., Yu, L., and Zhen, T. (2014). Accelerating the degradation of polyolefins through additives and blending. *J. Appl. Polym. Sci.*, **131**(18), 2282–2287.

[230] Gisha, E. L., Lissy, J., Prasad, V. S., and Pillai, C. K. S. (2006). Sugar end-capped polyethylene: ceric ammonium nitrate initiated oxidation and melt phase grafting of glucose onto polyethylene and its microbial degradation. *Polym. Degrad. Stab.*, **91**(7), 1484–1494.

[231] Gupta, A. P., Kumar, V., Sharma, M., and Shukla, S. K. (2009). Physicochemical studies on interaction behaviour of potato starch filled low density polyethylene grafted maleic anhydride and low density polyethylene biodegradable composite sheets. *Polym. Plast. Technol. Eng.*, **48**(6), 587–594.

[232] Ali, R. R., Rahman, W., Kasmani, R. M., Ibrahim, N., Mustapha, S. N. M., and Hasbullah, H. (2013). Tapioca starch biocomposite for disposable packaging ware. *Chem. Eng. Trans.*, **32**, 1711–1716.

[233] Godbole, S., Gote, S., Latkar, M., and Chakrabarti, T. (2003). Preparation and characterization of biodegradable poly-3-hydroxybutyrate–starch blend films. *Bioresour. Technol.*, **86**, 33–37.

[234] Bagheri, R. (1999). Effect of processing on the melt degradation of starch filled polypropylene, *Polym Int.*, **48**(12), 1257–1263.

[235] Prachayawarakorn, J., Sangnitidej, P., and Boonpasith, P. (2010). Properties of thermoplastic rice starch composites reinforced by cotton fiber or low-density polyethylene. *Carbohydr. Polym.*, **81**(2), 425–433.

[236] Brebu, M. and Vasile, C. (2010). Thermal degradation of lignin—a review. *Cellul. Chem. Technol.*, **44**(9), 353–363.

[237] Apopei, D. F., Dinu, M. V., Trochimczuk, A. W., and Dragan, E. S. (2012). Sorption isotherms of heavy metal ions onto semi-interpenetrating polymer network cryogels based on polyacrylamide and anionically modified potato starch. *Ind. Eng. Chem. Res.*, **51**(31), 10462–10471.

[238] Fei, P., Shi, Y., Zhou, M., Cai, J., Tang, S., and Xiong, H. (2013). Effects of nano-TiO_2 on the properties and structures of starch/poly(ε-caprolactone) composites. *J. Appl. Polym. Sci.*, **130**, 4129–4136.

[239] Thothong, S., Sriroth, K., Tantatherdtam, R., and Jarerat, A. (2012). Production and characterization of porous starch granule and poly(butylene adipate-co-terephthalate) blend as a bio-composite. In: *Advances in Chemical Engineering II*, (Eds. Liu, Z., Peng, F. and Liu, X.), Parts 1–4, Advanced Materials Research, **550–553**, pp. 1513–1521.

[240] Ma, X.-F., Chang, P. R., Yu, J. G., and Stumborg, M. (2009) Properties of biodegradable citric acid-modified granular starch/thermoplastic pea starch composites. *Carbohydr. Polym.*, **75**(1), 1–8.

[241] Reddy, N. and Yang, Y.Q. (2010). Citric acid cross linking of starch films. *Food Chem.*, **118**(3), 702–711.

[242] Alves, V. D., Mali, S., Beleia, A., and Grossmann, M. V. E. (2007). Effect of glycerol and amylose enrichment on cassava starch film properties. *J. Food Eng.*, **78**(3), 941–946.

[243] Arvanitoyannis, I., Biliaderis, C. G., Ogawa, H., and Kawasaki, N. (1998). Biodegradable films made from low-density polyethylene (LDPE), rice starch and potato starch for food—packaging applications—part 1. *Carbohydr. Polym.*, **36**(2–3), 89–104.

[244] Auras, R. A., Harte, B., Selke, S., and Hernandez, R. (2003). Mechanical, physical, and barrier properties of poly (lactide) films. *J. Plast. Film Sheeting*, **19**(2), 123–135.

[245] BASF, Biodegradable Polymers—Inspired by Nature Ecoflex, Ecovio, October, 2008. [From: www.bioplastics.**basf**.com Accessed on October 24, 2014].

[246] Bertuzzi, M. A., Vidaurre Castro, E. F., Armada, M., and Gottifredi, J. C. (2007). Water vapor permeability of edible starch based films. *J. Food Eng.*, **80**(3), 972–978.

[247] Erkskea, D., Viskerea, I., Dzenea, A., Tupureinaa, V., and Savenkovab, L. (2006). Biobased polymer composites for films and coatings. *Proc. Est. Acad. Sci. Chem.*, **55**(2), 70–77.

[248] Ma, X.-F., Chang, P. R., Yang, J., and Yu, J. (2009). Preparation and properties of glycerol plasticized-pea starch/zinc oxide-starch bionanocomposites. *Carbohydr. Polym.*, **75**(3), 472–478.

[249] Rhim, J.-W. and Ng, P. K. W. (2007). Natural biopolymer-based nanocomposite films for packaging applications. *Crit. Rev. Food Sci. Nutr.*, **47**(4), 411–433.

[250] Rhim, J.-W., Hong, S.-I., and Ha, C.-S. (2009). Tensile, water vapor barrier and antimicrobial properties of PLA/nanoclay composite films. *LWT-Food Sci. Technol.*, **42**(2), 612–617.

[251] Ryu, S. Y., Rhim, J. W., Roh, H. J., and Kim, S. S. (2002). Preparation and physical properties of zein-coated high-amylose corn starch film. *LTW-Food Sci. Technol.*, **35**(8), 680–686.

[252] Saxena, A. and Ragauskas, A. J. (2009). Water transmission barrier properties of biodegradable films based on cellulosic whiskers and xylan. *Carbohydr. Polym.*, **78**(2), 357–360.

[253] Sumesh, S., Surya, B. S., and Prasad, V. S. (2014). Proceedings National Conference on Materials Science and Technology, (NCMST-14), July 28–30, 2014, IIST, Thiruvananthapuram, Kerala, India.

[254] Van Tuil, R., Fowler, P., Lawther, M., and Weber, C. J. (2000). Properties of biobased packaging materials. In: *Biobased Packaging Materials for the Food Industry: Status and Perspectives*, (Ed. Weber, C. J.), KVL, Department of Dairy and Food Science, Frederiksberg, pp. 13–44.

[255] Cynthia, G. F.-H., Arturo, C.-C., Carlos, V.-S., Víctor, M. C., José, L. R.-A., Armando, A.-C., Perla, E. G.-C., and Ana, L. M.-H. (2014). All green composites from fully renewable biopolymers: chitosan-starch reinforced with keratin from feathers. *Polymer*, **6**, 686–705.

[256] Alvarez, V. A., Ruseckaite, R. A., and Vazquez, A. (2003). Mechanical properties and water absorption behavior of composites made from a biodegradable matrix and alkaline treated sisal fibers. *J. Compos. Mater.*, **37**(17), 1575–1588.

[257] Sreekumar, P. A., Saiter, J. M., and Thomas, S. (2009). Water absorption and ageing behavior of natural fiber composites. In: *Natural Fiber Reinforced Polymer Composites From Macro to Nanoscale*, (Eds. Thomas, Sabu and Pothan, Laly A.), Old City publishing Inc., Philadelphia, pp. 446–472.

[258] Lacerda, R. S., Bittante, A. M. Q. B., Chambi, H., Gomide, C. A., Moraes, I. C. F., Carvalho, R. A., and Sobral, P. J. A. (2011). Castor bean cake as raw material for biodegradable material. In: *Proceedings of 11th International Congress on Engineering and Food (ICEF11), 2011*, (Eds. Taoukis, Petros S., Shoforos, Nikolaos G., Karathanos, Voios T., and Saravacos, George D.), Cosmosware, Athens, pp. 1–6.

[259] Makishi, G. L. A., Lacerda, R. S., Bittante, A. M. Q. B., Chambi, H. N. M., Costa, P. A., Gomide, C. A., Carvalho, R. A., and Sobral, P. J. A. (2013). Films based on castor bean (*Ricinus communis* L.) proteins cross linked with glutaraldehyde and glyoxal. *Ind. Crop. Prod.*, **50**, 375–382.

[260] Machado, A. C. O. M., Ribeiro, C. M. S., Freire, D. M. G., Cunha, L. M., Castilho, L. R., Dias, M. L., and Nazerth, N. J. (2010). Material compósito contendo material termoplástico e material protéico e processo de obtenção. BR PI0804101-6 A2. [From: http://www.patentesonline.com.br/material-comp-sito-contendo-material-termopl-stico-e-material-prot-ico-e-processo-229188.html. Accessed on July 16, 2014].

[261] Tharanathan, R. N. (2003). Biodegradable films and composite coatings: past, present and future. *Trends Food Sci. Technol.*, **14**, 71–78.

[262] Alves, E. E. N., Souza, C. F., Inoue, K. R. A., and Leite, T. S. (2010). Potential of production of biogas from the castor bean (*Ricinus communis* L.) cake. Paper No. 10. Annual International Meeting of American Society of Agricultural and Biological Engineers (ASABE) Sponsored by ASABE David L. Lawrence Convention Center, Pittsburgh, PA, USA, June 20–23, 2010.

[263] Bittante, A. M. Q. B., Lacerda, R. S., Oliveira, T., Makishi, G. L. A., Costa, P. A., Chambi, H. N. M., Gomide, C. A., and Sobral, P. J. A. (2014). Properties of biodegradable films made with proteins extracted from castor bean (*Ricinus communis*) cake: effect of protein extraction pH. *Chem. Eng. Trans.*, **37**, 751–756.

[264] Stork, R. R. and Rocha, M.C.G.R. (2010). Composites of low density polyethylene and castor press cake. *Polym. Plast. Technol. Eng.*, **49**(3), 1352–1355.

[265] Guimarães, J. L., Cursino, A. C. T., Saul, C. K., Sierrakowski, M. R.Ramos, L. P., Wypych, F., Satyanarayana, K. G. (2015). Effect of processing conditions on tensile

properties of green composites of castor oil cake starch-plant fibers. *J. Bioprocess. Biotechn.*, **5**(1):198. doi: 10.4172/2155-9821.1000198 (8 pp).

[266] Marina, M. N. K., Washington, L. E. M., Satyanarayana, K. G., and Elaine, C. L. (2013). Evaluation of physical and mechanical properties of nanocomposites of castor oil cake based starch reinforced with nanocellulose. VII Workshop on Applied Nanotechnology to Agrobusiness. Embrapa Instrumentação, São Carlos, 10 on June 13, 2013, pp. 241–243.

[267] Guimarães, J. L., Satyanarayana, K. G., Wypych, F., Ramos, L. P., and Cursino, A. C. T. (2006) Method for thermal molding for deriving and extracting vegetable oil from castor bean cake for using these materials in packaging industry, involves using plastic with glycerol derivative. International patent, BR200605742-A, 2006a, (Publication Date: July 29, 2009).

[268] Guimarães, J. L., Satyanarayana, K. G., Wypych, F., and Ramos, L. P. (2006) Thermally molded direct-rolled plastic composites is based on starch with glycerol hydrate reinforced with fibers of vegetable or mineral and use of degradable polymer materials. International Patent, BR200602428-A, 2006, (Publication Date: November 27, 2007).

[269] Zarrinbakhsh, N., Misra, M., and Mohanty, A. K. (2011). Biodegradable green composites from distiller's dried grains with solubles (DDGS) and a polyhydroxy (butyrate-*co*-valerate) (PHBV)-based bioplastic. *Macromol. Mater. Eng.*, **296**(11), 1035–1045.

[270] Dong, C. S. and Takagi, H. (2014). Flexural properties of cellulose nanofibre reinforced green composites. *Compos. Part B*, **58**, 418–421.

[271] Wittaya, T. (2009). Microcomposites of rice starch film reinforced with microcrystalline cellulose from palm pressed fiber. *Int. Food Res. J.*, **16**, 493–500.

[272] Wittaya, T. (2012). Rice starch-based biodegradable films: properties enhancement. In Structure and Function of Food Engineering, InTech. [From: 10.5772/ 47751 Accessed on October 24, 2014].

[273] Yew, G. H., Mohd Yusof, A. M., Mohd Ishak, U. S., and Ishiaku, Z. A. (2005). Water absorption and enzymatic degradation of poly (lactic acid)/rice starch composites. *Polym. Degrad. Stab.*, **90**(3), 488–500.

[274] Phattaraporn, T., Waranyou, S., and Thawien, W. (2011). Effect of palm pressed fiber (PPF) surface treatment on the properties of rice starch films. *Int. Food Res. J.*, **18**(1), 535–547.

[275] Bourtoom, T. and Chinnan, M. S. (2008). Preparation and properties of rice starch-chitosan blend biodegradable film. *Lebensm. Wiss. Technol.*, **41**(9), 1633–1641.

[276] Zhang, J.-F. and Sun, X. (2004). Mechanical properties of poly (lactic acid)/ starch composites compatibilized by maleic anhydride, *Biomacromolecules*, **5**, 1446–1451.

[277] Lee, S. Y., Cho, M. S., Nam, J. D., and Lee, Y. (2006). Melting processing of biodegradable cellulose diacetate/starch composites. *Macromol. Symp.*, **242**, 126–130.

[278] Alvarez, V., Vazquez, A., and Bernal, C. L. (2005). Fracture behavior of sisal fiber-reinforced starch-based composites. *Polym. Compos.*, **26**(3), 316–323.

[279] Baishya, P. and Maji, T. K. (2014). Studies on effects of different cross-linkers on the properties of starch-based wood composites. *ACS Sustainable Chem. Eng.*, **2**(7), 1760–1768.

[280] Yuan, J., Hu, Y., Li, L., and Cheng, F. (2013). The mechanical strength change of wood modified with DMDHEU. *Bioresources*, **8**, 1076–1088.

[281] Darder, M., Aranda, P., and Ruiz-Hitzky, E. (2007). Bionanocomposites: a New concept of ecological, bioinspired, and functional hybrid materials. *Adv. Mater.*, **19**(10), 1309–1319.

[282] Angellier, H., Molina-Boisseau, S., Dole, P., and Dufresne, A. (2006). Thermoplastic starch- waxy maize starch nanocrystal nanocomposites. *Biomacromolecules*, **7**(2), 531–539.

[283] Montoya, U., Rojas, O., Gañán, P. F., Marind, D., and Castroe, C. I. (2014). Synthesis of thermoplastic starch-bacterial cellulose nanocomposites via *in situ* fermentation. *J. Braz. Chem. Soc.*, **25**(9), 1607–1613.

[284] Montoya, U., Zuluaga, R., Castro, C., Goyanes, S., and Ganan, P. (2014). Development of composite films based on thermoplastic starch and cellulose microfibrils from Colombian agro-industrial wastes. *J. Thermoplast. Compos. Mater.*, **27**(3), 413–426.

[285] Vazquez, A. and Alvarez, V. A. (2008). Starch–cellulose fiber composites. In: *Biodegradable Polymer Blends and Composites from Renewable Resources*, (Ed. Yu, Long), John Wiley & Sons, Inc., Hoboken, NJ pp. 241–286, DOI: 10.1002/9780470391501. ch11.

[286] Liu, W. Y., Chen, Y., Ouyang, L., Liu, Y. J., and Hao, X. H. (2012). Processing and mechanical properties of starch and PCL composite reinforced by Nano-SiO_2, functional materials and nanotechnology. In *Advanced Materials Research* (Eds. Xu, B., and Li, H. Y.), Trans Tech Publications, Switzerland, **496**, pp. 134–137.

[287] Favier, V., Dendievel, R., Canova, G., Cavaille, J.-Y., and Gilormini, P. (1997). Simulation and modeling of three-dimensional percolating structures: case of a latex matrix reinforced by a network of cellulose fibers. *Acta Mater.*, **45**(4), 1557–1565.

[288] Habibi, Y., Lucia, L. A., Rojas, O. J. (2010). Cellulose nanocrystals: chemistry, self-assembly, and applications. *Chem. Rev.*, **110**, 3479–3500.

[289] Dalmas, F., Chazeau, L., Gauthier, C., Cavaille, J.-Y., and Dendievel, R. (2006). *Polymer*, **47**, 2802–2812.

[290] Osorio, M. A., Restrepo, D., Velásquez-Cock, J. A.Zuluaga, R. O., Montoya, U., Rojas, O., Gañán, P. F., Marin, D., and Castro, C. I. (2014). Synthesis of thermoplastic starch-bacterial cellulose nanocomposites via *in situ* fermentation. *J. Braz. Chem. Soc.*, **25**(9), 1607–1613.

[291] Martins, I. M. G., Magina, S. P., Oliveira, L., Freire, C., Silvestre, A., Neto, C., and Gandini, A. (2009). New biocomposites based on thermoplastic starch and bacterial cellulose. *Compos. Sci. Technol.* 2009, **69**(13), 2163–2168.

[292] Ma, T. T., Chang, P. R., Zheng, P. W., and Ma, X.-F. (2013). The composites based on plasticized starch and graphene oxide/reduced graphene oxide. *Carbohydr. Polym.*, **94**, 63–70.

[293] Khachatryan, K., Khachatryan, G., Fiedorowicz, M., Para, A., and Tomasik, P. (2013). Formation of nanometal particles in the dialdehyde starch matrix. *Carbohydr. Polym.*, **98**, 568–573.

[294] Cheviro, P., Gouanve, F., and Espuche, E. (2014). Green synthesis of colloid silver nanoparticles and resulting biodegradable starch/silver nanocomposites. *Carbohydr. Polym.*, **108**(8), 291–298.

[295] Asha, S. C., Ansira, P. M., and Prasad, V. S. (2014). Synthesis, Characterisation and Self Assembly of Oleyl Amine Modified MMT Based Hybrid Copolymers of Styrene

and Acrylic Acid, *ICAFM 2014*, February 19–21, 2014, CSIR-NIIST, Thiruvananthapuram.

[296] Asha, S. C., Aswathy, M. J., and Prasad, V. S. (2013). Synthesis Characterization and Self Assembling Properties of Polymer Clay Nanocomposites by in-situ Intercalative Copolymerisation of Cinnamic Acid-CTAB Adduct Modified MMT with Styrene, *Nano India 2013*, Feburary19–20, NIIST, Thiruvananthapuram.

[297] Bagdi, K., Muller, P., and Pukanszky, B. (2006). Thermoplastic starch/layered silicate composites: structure, interaction, properties. *Compos. Interfaces*, **13**(1), 1–17.

[298] Darder, M., Colilla, M., and Ruiz-Hitzky, E. (2003). Biopolymer–clay nanocomposites based on chitosan intercalated in montmorillonite. *Chem. Mater.*, **15**, 3774–3780.

[299] Alexandre, M. and Dubois, P. (2000). Polymer-layered silicate nanocomposites: preparation, properties and uses of a new class of materials. *Mater. Sci. Eng.*, **28**(1–2), 1–63.

[300] La Mantia, F. P., Arrigo, R., and Morreale, M. (2014). Effect of the orientation and rheological behaviour of biodegradable polymer nanocomposites. *Eur. Polym. J.*, **54**, 11–17.

[301] Zhao, R., Torley, P., and Halley, P. J. (2008). Emerging biodegradable materials: starch and protein-based bio-nanocomposites. *J. Mater. Sci.*, **43**, 3058–3071.

[302] Chiou, B. S., Yee, E., Wood, D., Shey, J., Glenn, G., and Orts, W. (2006). Effects of processing conditions on nanoclay dispersion in starch–clay nanocomposites. *Cereal Chem.*, **83**, 300–305.

[303] Park, H. M., Lee, W. K., Park, C. Y., Cho, W. J., and Ha, C. S. (2003). Environmentally friendly polymer hybrids. Part I. Mechanical, thermal, and barrier properties of thermoplastic starch/clay nanocomposites. *J. Mater. Sci.*, **38**, 909–915.

[304] Park, H. M., Li, X., Jin, C. Z., Park, C. Y., Cho, W. J., and Ha, C. S. (2002). Preparation and properties of biodegradable thermoplastic starch/clay hybrids. *Macromol. Mater. Eng.*, **287**, 553–558.

[305] ChungY.-L., Ansari, S., Estevez, L., Hayrapetyan, S., Giannelis, E. P., and Lai, H.-M. (2010). Preparation and properties of biodegradable starch–clay nanocomposites. *Carbohydr. Polym.*, **79**, 391–396.

[306] Chivrac, F., Gueguen, O., Pollet, E., Ahzi, S., Makradi, A., and Averous, L. (2008). Micromechanical modeling and characterization of the effective properties in starch-based nano-biocomposites. *Acta Biomater.*, **4**, 1707–1714.

[307] Chivrac, F., Pollet, E., Schmutz, M., and Averous, L. (2008). New approach to elaborate exfoliated starch-based nanobiocomposites. *Biomacromolecules*, **9**, 896–900.

[308] Kampeerapappun, P., Aht-ong, D., Pentrakoon, D., and Srikulkit, K. (2007). Preparation of cassava starch/montmorillonite composite film. *Carbohydr. Polym.*, **67**, 155–163.

[309] Kvien, I., Sugiyama, J., Votrubec, M., and Oksman, K. (2007). Characterization of starch based nanocomposites. *J. Mater. Sci.*, **42**, 8163–8171.

[310] Lu, Y. S., Weng, L. H., and Cao, X. D. (2006). Morphological, thermal and mechanical properties of ramie crystallites-reinforced plasticized starch biocomposites.*Carbohydr. Polym.*, **63**, 198–204.

[311] Mondragon, M., Mancilla, J. E., and Rodriguez-Gonzalez, F. J. (2008). Nanocomposites from plasticized high-amylopectin, normal and high-amylose maize starches. *Polym. Eng. Sci.*, **48**, 1261–1267.

[312] Pandey, J. K. and Singh, R. P. (2005). Green nanocomposites from renewable resources: effect of plasticizer on the structure and material properties of clay-filled starch. *Starch-Starke*, **57**, 8–15.

[313] Piyada, K. and Witthaya, T. (2010). Properties and characteristics of rice starch films reinforced with rice starch nanocrystals. Proceeding of International Conference on Applied Science Engineering and Technology, Conference Venue River View Hotel, Singapore, August 25–27, 2010.

[314] Novamont, environmental product declaration (epd) mater – bi® pe type. http://www.materbi.com/ing/html/prodotto/tecnologie/additivi.html (accessed September 23, 2015). [From Ref. 160]

[315] Petersson, L. and Oksman, K. (2006). Biopolymer based nanocomposites: comparing layered silicates and microcrystalline cellulose as nano reinforcement. *Compos. Sci. Technol.*, **66**(13), 2187–2196.

[316] Vandeputte, G. E. and Delcour, J. A. (2004). From sucrose to starch granule to starch physical behaviour: a focus on rice starch. *Carbohydr. Polym.*, **58**(3), 245–266.

[317] Wang, Y., Tian, H., and Zhang, L. (2010). Role of starch nanocrystals and cellulose whiskers in synergistic reinforcement of waterborne polyurethane. *Carbohydr. Polym.*, **80**(3), 665–671.

[318] Rindlav- Westling, A., Stading, M., and Gatenholm, P. (2002). Crystallinity and morphology in films of starch, amylase and amylopectin blends. *Biomacromolecules*, **3**(1), 84–91.

[319] Rindlav-Westling, A. and Gatenholm, P. (2002). Surface composition and morphology of starch, amylose, and amylopectin. *Biomacromolecules*, **4**(1), 166–172.

[320] Butler, B. L., Vergano, P. L., Testin, R. F., Bunn, J. M., and Wiles, J. L. (1996). Effect of plasticizers and fatty acids on mechanical and permeability. *J. Food Sci.*, **61**, 953–961.

[321] García, M. A., Pinotti, A., and Zaritzky, N. E. (2006). Physicochemical, water vapor barrier and mechanical properties of corn starch and chitosan composite films. *Starch-Starke*, **58**(9), 453–463.

[322] Gaudin, S., Lourdin, D., Forssell, P. M., and Colonna, P. (2000). *Carbohydr. Polym.*, **43**(1), 33–37.

[323] Salleh, E., Muhamad, I., and Khairuddin, N. (2007). Preparation, characterization and antimicrobial analysis of antimicrobial starch-based film incorporated with chitosan and lauric acid. *Asian Chitin J.*, **3**, 55–68.

[324] Kristo, E. and Biliaderis, C. G. (2007). Physical properties of starch nanocrystal-reinforced pullulan films. *Carbohydr. Polym.*, **68**(1), 146–158.

[325] Han, C., Chen, Z., Bai, Z., and Dong, L. (2009). Polylactic acid composite material contains preset amount of polylactic acid, thermoplastic starch, plasticizer, inorganic filler, antioxidizer and lubricant. Patent No. CN101157792-A, 2009, Zhejiang Haizheng Biological Material Co.

[326] Lee, S.-H., Okita, T., and Kimura, H. (2005). Composite material of starch and biodegradable resin. Japanese Patent No. JP 2005154706-A, Assignee: Green Bio: KK.

[327] Li, L., Cui, Q., Tang, F., and Xia, B. (2005). Preparation of starch/nano silver antibacterial composite film used for biodegradable medical antibacterial bandage or antibacterial dressings by processing starch, degrading and oxidizing starch and preparing starch solution. Chinese Patent Nos. CN102552963-A & CN102552963-B.

[328] Lidong, L., Hua, X., Ma, N., Tang, F., and Cui, Q. (2013). Preparation method of starch/nano-silver antibacterial composite film. Chinese Patent Nos. CN102552963-A and also B (Publication Date: September 25, 2013).

[329] Liu, S. (2012). Natural plant fiber reinforced starch based composite material comprises natural plant fiber, starch aliphatic polyester, and starch compatibilizer. Patent Number: CN102477170-A, Shanghai Genius Advanced Material Co Ltd.

[330] Nie, L., Parker, M. D., Bassi, S. D., and Maningat, C. C. (2007). Composite product, E.G. Cutlery, cups, plates, bowls, or packaging, includes semi crystalline polylactic acid, starch, and compatibilizer. Patent No.: US2007092745-A1, MGP Ingredients Inc.

[331] Spychaj, T., Krala, G., and Schmidt, B. (2013). Biodegradable nanocomposite hydrogel sorbent and a method for preparing a biodegradable hydrogel nanocomposite sorbent. Patent No. PL393869-A1 (Publication Date: February 15, 2011), Univ Zachodniopomorski Technologiczny.

[332] Weerawarna, S. A. (2009). Biodegradable superabsorbent particles. Patent No.US 2009325797-A1 (Publication Date: December 31, 2009), Assignee: Weyerhaeuser Co. Federal Way, USA.

[333] Zhong, R. (2012). Sanitary products based on PBS, starch-based fully biodegradable plastic used for disposable sanitary products, comprises starch, composite biological catalyst, composite plasticizer, and polybutylene succinate, in parts weight of preset range. Patent No. CN102585299-A, Guangdong Zhongtairun Biotechnology Co.

9

GREEN COMPOSITE MATERIALS BASED ON BIODEGRADABLE POLYESTERS

Pramendra Kumar Bajpai

Division of Manufacturing Processes and Automation Engineering, Netaji Subhas Institute of Technology, Dwarka, New Delhi, India

9.1 INTRODUCTION

Composite processing industries have catalyzed their activities in the development of new composite materials, which involve low or zero harm to environment and human health. Specifically, new polymer composites known as natural fiber-reinforced polymer composites have been processed, which are environmentally friendly and

Biodegradable Green Composites, First Edition. Edited by Susheel Kalia.

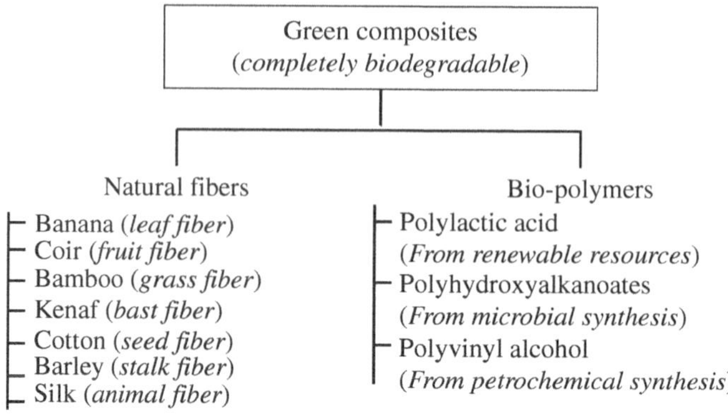

FIGURE 9.1 Example of natural fibers and biopolymers.

biodegradable in nature. Earlier, polymer composites incorporate synthetic fibers (such as glass, carbon, and aramid fibers) into thermoset (such as epoxy) and thermoplastic (such as polypropylene (PP), polyvinyl chloride, polyethylene, etc.) synthetic polymers. These polymer composites have excellent structural properties and are used in various engineering applications. The main drawback associated with these composites is their nonbiodegradable nature when these are disposed into the environment after its usable life. Natural fiber-reinforced polymer composites incorporate natural fibers into synthetic polymers or biodegradable polymers. When natural fibers are reinforced with synthetic polymer such as epoxy and PP, the resulting composite is known as partially biodegradable composite. When natural fibers are reinforced with biopolymers such as polylactic acid (PLA), the resulting composite is known as fully biodegradable or green composite [1–6]. There are a variety of natural fibers and biopolymers that are used to process different kinds of green composites. Few examples of green composite constituents are shown in Figure 9.1.

Apart from their full biodegradability, green composites offer a number of advantages over synthetic polymer composites such as lightweight characteristics, acoustic insulation, and corrosion resistance [7, 8]. Though green composites have many superior properties, there are several challenges in the processing of green composites such as selection of suitable fabrication scheme and optimization of processing parameters to produce high-quality composites, compatibility of natural fibers with biopolymers, and their poor adhesion efficiency. The varying properties of natural fibers and its dependency over topographical regions are another major challenge in standardizing the green composite processing [9].

This chapter mainly covers various primary processing techniques to develop green composites. Effect of different processing techniques on mechanical properties (tensile strength, flexural strength, and impact strength) of the developed green composite has also been discussed in this chapter. The chapter also includes a critical analysis on various processing parameters to have a strong adhesion between natural fibers and biopolymers and subsequently to develop high-quality green composites.

9.2 FABRICATION TECHNIQUES FOR GREEN COMPOSITES

Green composites are developed using the same fabrication techniques that are used to develop synthetic polymer composites with little or no change in the fabrication scheme. Hand lay-up, compression molding, film stacking, prepregging technique, injection molding, extrusion process and extrusion with injection or compression molding, pultrusion, and resin transfer molding are some of the fabrication techniques that are widely used to process natural fiber-reinforced polymer composites.

9.2.1 Hand Lay-Up Fabrication Technique

The simplest fabrication technique to produce polymer composites is the hand lay-up process. This technique is generally preferred when polymer is in liquid form and natural fibers are in mat form. In this technique, polymer is laid up alternatively on fiber mats in a metallic mold. Extra resin is squeezed out because of applied pressure. The number of layers depends on the final thickness of the composite required. After specific curing time, mold is opened and composite laminate is taken out. Care should be taken while applying pressure because nonuniform pressure will lead to the development of composite of uneven thickness. Resin-rich area, fiber displacement, and void contents are some of the defects that are generally observed in composites fabricated through hand lay-up technique. To minimize the void content, a roller is rolled over back and forth on fiber mat at every lay-up to release the trapped air. A release gel is sprayed on metallic mold plate to avoid sticking of polymer to mold plate and to allow easy removal. A thin plastic sheet can also be used at top and bottom mold plate to avoid sticking of polymer to the mold plate. The schematic of hand lay-up fabrication technique is shown in Figure 9.2. Though this technique involves less capital investment, the production rate is slow and also high reinforcement is a little difficult to achieve through this process.

Green composites comprising denim fabric (a twilled cotton fabric) into PLA were fabricated by using hand lay-up method. Composites were developed with different numbers of denim layers. Evaluation of mechanical properties (impact strength and tensile strength) and thermal properties of the developed green composites showed that the three-layered denim-reinforced composite resulted in best properties among all specimens [10]. Cylindrical composite pins (normally used for tribological testing)

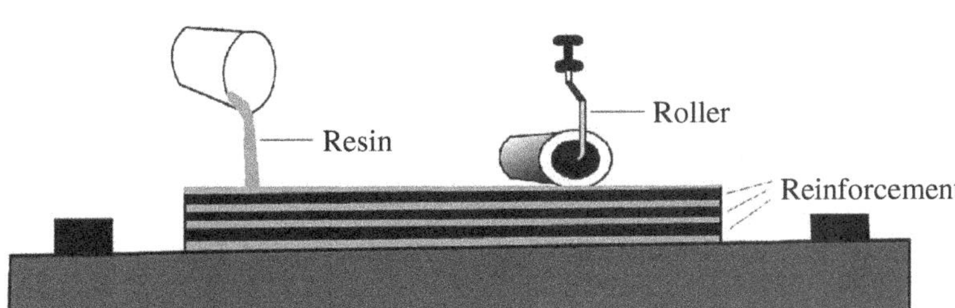

FIGURE 9.2 Schematic of hand lay-up technique.

can also be developed using hand lay-up principle by modifying the mold design. *Lantana camara* fiber (a type of natural fiber)-reinforced epoxy biocomposite pins of length 35 mm and a diameter of 10 mm were developed successfully for abrasive wear testing. Chopped *L. camara* fibers in the desired proportion were added to the mixture of epoxy resin and hardener at room temperature. The mixture was poured into cylindrical mold after proper stirring (to obtain a uniform dispersion of fibers). After that, the upper and lower part of the mold was fixed properly with nuts and bolts, and cured cylindrical specimens were taken out after curing at room temperature for wear testing [11].

9.2.2 Compression Molding

Compression molding technique is one of the widely used fabrication techniques to develop green composite laminates. The basic principle behind this technique involves application of predefined pressure and prespecified temperature, which converts the compound (mixture of fiber and polymer) to be molded into the shape of mold cavity, and desired product is obtained. Both thermoset and thermoplastic polymers can be processed in compression molding. Natural fibers can be used either in mat form (woven mat, unidirectional mat, randomly oriented fiber mat) or in short and chopped fiber form.

When natural fibers are used in mat form to be reinforced with thermoplastic polymer, which is in pellets form, then compression molding is performed through film stacking method to fabricate green composites. In this technique initially, polymer pellets are converted into thin polymer sheets of predefined thickness using compression molding process at specified temperature and pressure. These polymer sheets and natural fiber mats are alternatively stacked in a metallic mold to the desired thickness. At the top and bottom of the inner surface of mold, some spray-like silica release gel or plastic sheet of higher melting point (enough higher than the working temperature of polymer matrix to be used) is used to avoid sticking of polymer matrix to the mold surface and to allow easy removal of processed composite after cooling. The schematic of compression molding method is shown in Figure 9.3. Time of processing, working temperature, and application of pressure should be critically analyzed and should be optimized to fabricate the high-performance green composites. Excess temperature and pressure will lead to the deterioration of natural fiber surface and subsequently its properties. Lower temperature and pressure will lead to improper adhesion between fiber and polymer matrix, which in turn will result in poor composite properties.

Press molding has been used to fabricate green composites comprised of bamboo fibers (BF) as reinforcements and polyvinyl alcohol (PVA) and cornstarch-based biodegradable resin (CPR). For BF/PVA composites, fibers were dipped into the PVA solution, and the preform fiber sheets were obtained after drying. For the BF/CPR composites, BF were arranged parallel to the longitudinal direction in the mold and sandwiched between two CPR films and then pressed under light pressure. Several preforming (PF) sheets were laminated in the mold and then hot pressed under high pressure and at temperature of 150°C for 30 min to develop green composites [12].

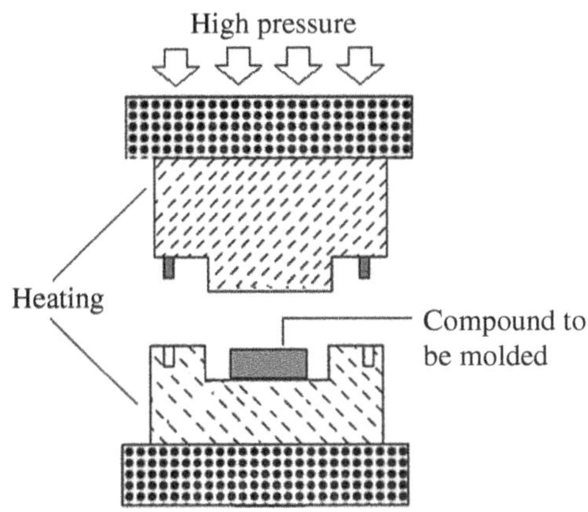

FIGURE 9.3 Compression molding fabrication technique.

Fully green composites were developed by reinforcing a CPR with curaua fibers through three fabrication methods, which are (i) direct method (DM), (ii) PF methods, and (iii) prepreg sheet (PS) method. In DM, a sliver of curaua fibers was inserted into a metallic mold with the resin, which was poured directly into them. Initially, low pressure was applied on the material at 150°C for 1 h. Then, pressure was increased and consistently applied during cooling till room temperature. In PF, the composite was fabricated by hot-pressing preforms of resin-pasted fiber slivers. In PS, slivers were placed on a metallic plate and stretched, and resin was applied into them, and thin PS were obtained. A set of five sheets, each with identical fiber orientation, was inserted in the mold and pressed to fabricate the desired green composite [13]. In a study, the manufacturing of PLLA/flax fiber green composites by film stacking process was optimized to attain desired qualities of processed composite. The studied parameters were temperature, time, compressibility, permeability, and the viscosity of the molten polymer. The degradation of the flax fibers with respect to the applied compaction load and the applied temperature was also studied. The results of the study show that higher load is required for compaction of flax mats and the transverse permeability of flax mats is lower as compared with glass mats. It has been concluded from the study that film stacking fabrication technique is more suitable than in-plane impregnation processes such as resin transfer molding due to the limitation on the flow ability of resin [14].

The film stacking process has been successfully used in the fabrication of various laminated green composites. The natural fibers used were nettle, *Grewia optiva*, and sisal fibers in woven mat form with PLA (which is a biopolymer) as a matrix material. In this fabrication technique, first of all, PLA pellets were converted into PLA films of approximately 1 mm in thickness. Natural fiber mats and PLA films were stacked alternatively in a metallic mold. At the top and bottom of the metallic mold, thin Teflon sheets were used to avoid sticking of polymer with metallic mold. The whole assembly was hot pressed at around 180°C (melting point of used PLA polymer) and at high MPa pressure for few minutes. The cooling was being carried out under

(a) (b) (c)

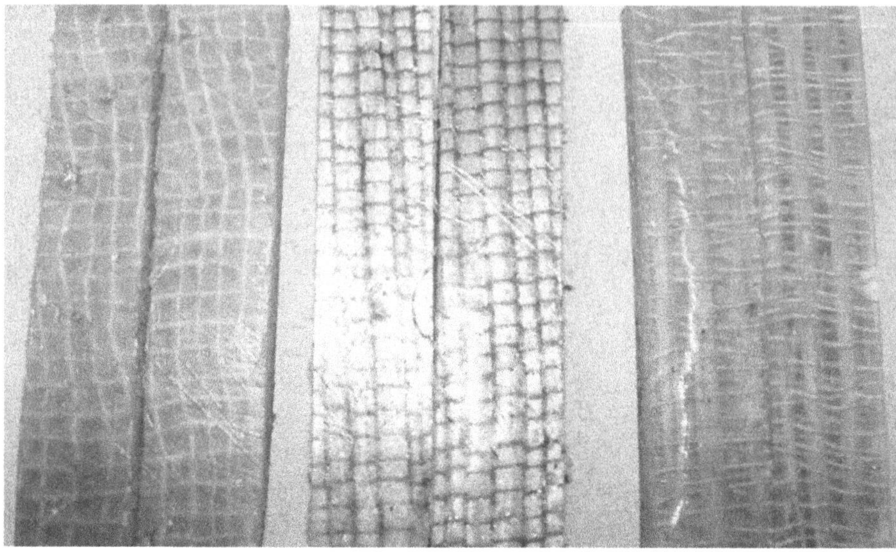

FIGURE 9.4 Green composite laminates processed through film stacking technique. (a) PLA/nettle. (b) PLA/sisal. (c) PLA/*Grewia optiva*. (*See insert for color representation of the figure.*)

pressure, and finally composite laminate was taken out. The developed green composite specimens are shown in Figure 9.4. The fabricated green composites have shown favorable mechanical properties and enhanced tribological and machining characteristics [15–17].

The SEM micrographs of the developed composites are shown in Figure 9.5. Figure 9.5a shows the SEM image of the surface of the composite after its fabrication through film stacking technique. It is clear from the image that the surface is smooth, clear, and free from any impression. Figure 9.5b and c shows the fractured surface after tensile testing and wear-out surface due to sliding the composite specimen using pin on disc apparatus during sliding wear testing of the developed composite. SEM images clearly depicts that fibers are strongly bonded with polymer matrix. The adhesion efficiency of natural fibers with biodegradable matrix depends upon the fabrication technique to great extent.

9.2.3 Injection Molding Fabrication Technique

Extrusion and injection molding processes are commonly used fabrication techniques to develop short fiber-reinforced green composites. Through extrusion process, pellets of natural fiber and polymer blends are developed. These pellets are further processed through either compression molding or injection molding to fabricate the green composite in desired shape. In injection molding process, short natural fibers and polymer pellets are fed into the barrel where polymer softens due to heating in the barrel chamber. Mixing action takes place due to the rotation of screw. Complete

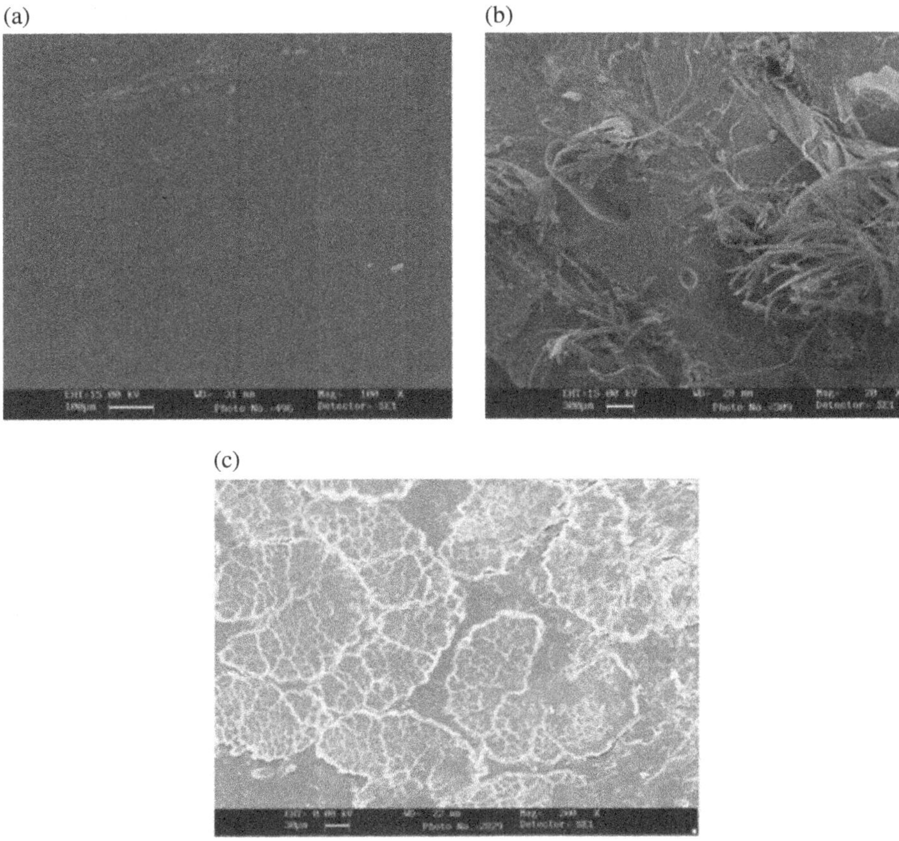

FIGURE 9.5 SEM images of the (a) surface of the developed composite, (b) fractured surface after tensile testing, and (c) surface after sliding wear testing.

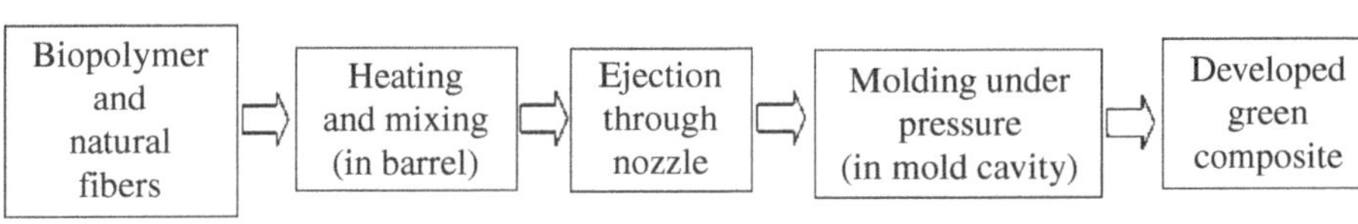

FIGURE 9.6 Fabrication scheme for injection molding.

mixing and uniform distribution of fibers into the polymer are ensured through rotation of screw, and any agglomeration of fibers is avoided. The blend is injected into the mold cavity through the nozzle attached at the end of barrel. The specified pressure is applied on the mold for a definite period of time, and desired composite product is taken out of the mold cavity after cooling. Screw speed, temperature in the barrel chamber, injection pressure, nozzle cross-sectional area, time, and amount of pressure applied on the mold are some of the governing parameters to decide the quality and performance of fabricated green composite. The fabrication scheme is shown in Figure 9.6.

Twin-screw extruder has been employed to develop pellets of recycled newspaper cellulose fiber-reinforced PLA polymer. The pelletized composites were injection molded after drying in a convection oven. The injection molding was carried out on injection molding machine having four temperature zones with the nozzle temperature at 185°C and the zones at 183°C. The hold, pack, and fill pressures were maintained at different predefined pressures. The mechanical and thermo-mechanical properties of the fabricated composites were found to be comparable with the corresponding properties of PP composites. The results have shown that recycled newspaper cellulose fiber-reinforced PLA composites have a potential to replace glass in many applications. The SEM investigations have confirmed the uniform dispersion of the recycled newspaper cellulose fibers in the PLA matrix [18]. Hybrid biocomposites comprising maple wood fiber in polyhydroxy(butyrate-*co*-valerate) (PHBV) biopolymer and filled with microsized talc can be processed effectively with the injection molding process. The wood fibers and PHBV were dried to remove the moisture. PHBV pellets and wood fibers were physically mixed and fed to the microextruder, and then talc was added to the melt mix in a definite ratio. Finally, all the three components were properly mixed in the microextruder for an appropriate and uniform dispersion of the talc and wood fibers in the melt. The melt from the extruder was collected in a preheated mini-injection molder, and this was transferred to the molding machine at an optimized injection pressure for specimen fabrication [19].

9.2.4 Resin Transfer Fabrication Technique

In resin transfer molding technique, polymer in liquid form is injected through ports into the mold cavity where fibers in the form of mat (woven fabric, unidirectional mat, and randomly oriented fiber mat) are already placed. The viscosity of the polymer plays a very important role for proper uniform flow and distribution of resin over the fiber mat. High viscosity results in fiber displacement, fiber washout, improper wetting, and poor adhesion between fiber and matrix and deficient resin areas at some places in the mold. Sometimes, a catalyst is also used in the polymer as an accelerator. The surface finish of the composite product developed through this technique is very good on both sides. The other parameters that affect the quality of the composite product include injection speed of the polymer and applied pressure. The fabrication scheme is shown in Figure 9.7.

It has been established that jute fibers can be processed as successfully as synthetic fibers such as glass fibers using resin transfer molding. Jute fiber nonwoven mats were reinforced with unsaturated polyester (PET)/styrene to develop biocomposite

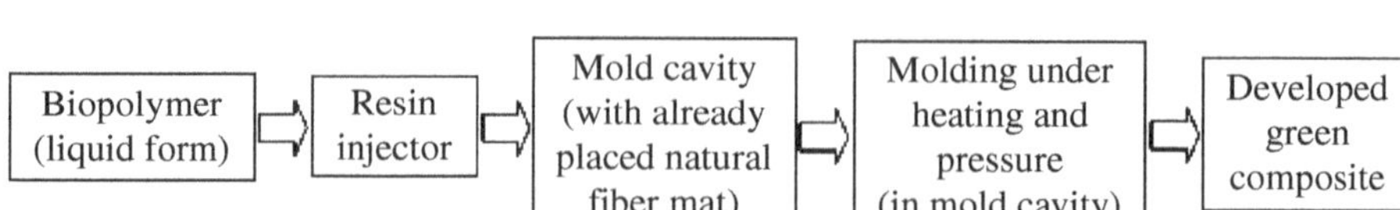

FIGURE 9.7 Fabrication scheme for resin transfer molding.

through resin transfer molding technique same as used for glass fiber-reinforced composites. At the beginning of resin injection, the mold temperature was 55°C, and injection pressure was arbitrary, but did not exceed 240 kPa. Mechanical properties of the developed composites were evaluated and found to be comparable with commercial glass-reinforced specimens [20].

9.2.5 Pultrusion Fabrication Technique

To fabricate composite products having constant cross section continuously, generally, pultrusion process is recommended. In this technique, continuous fiber roving or mat is passed into a hot die through a resin tank positioned in between. Resin-dipped fibers acquire the shape of mold cavity in the hot die and at the same time get cured due to heating. The cured composite product is pulled through a pulling mechanism and trimmed with a cutter in desired size. The speed with which fibers are passed into resin tank (to properly wet the reinforcement) and die temperature (to achieve proper curing) should be accurately optimized to fabricate good quality composite products. Pulling mechanism also keeps the composite product straight and in horizontal position. The fabrication scheme is shown in Figure 9.8.

High-quality hybrid polymer composites have been successfully developed using E-glass with unidirectional hemp yarn or woven hemp fabric reinforcements into PET resin through pultrusion process. It was observed that hemp fibers could be successfully pulled through a typical open bath setup without damage and could withstand the processing temperature of 177°C. The aesthetic appearance of the surface of the developed part was excellent having a natural look [21]. Lightweight, corrosion-resistant, and energy-efficient core-filled pultruded panel has been fabricated using a bio-based PET resin through pultrusion process. The core-filled panel was made using the one-step process in which the foam core enters the heated die along with the glass reinforcements. Four fiber preforms and two resin baths were used during the manufacturing of panels, and the temperature of the die during the manufacturing was maintained between 250 and 300°F [22]. It has been shown that pultrusion process is a suitable process for fabrication of jute and kenaf fiber-reinforced PET composites and their hybrids with glass fibers. In all hybrids, natural fiber content was 40 wt.%, while glass fiber content was 25 wt.%. The continuous fiber rovings were first impregnated with unsaturated PET resin in a resin impregnation tank. There was a pulling device to pull the jute fibers impregnated with resin through a steel die to obtain the desired shape. Curing process was carried out in a precision machined forming and curing die [23].

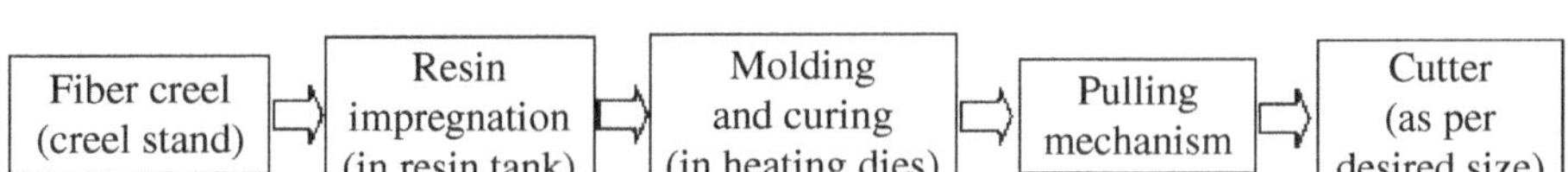

FIGURE 9.8 Fabrication scheme for pultrusion process.

9.3 PROCESSING OF GREEN COMPOSITES THROUGH MICROWAVE HEATING

Curing and processing of natural fiber-reinforced polymer composites through microwave energy are an unconventional way of composite processing, which has received considerable attention in recent years due to a number of favorable reasons. Curing through microwave heating is a cost-effective and green approach (pollution-free) for composite processing. Microwave heating is an energy-efficient process than thermal heating such as used in compression molding and injection molding techniques.

Various unidirectional natural fiber-reinforced PP composites have been fabricated through microwave heating process. The natural fibers used for reinforcement were coir, bamboo, and rattan fibers. Evaluation of mechanical properties has shown favorable results in support of microwave processing of green composites with some modification and improvements in the setup [24]. In another study, natural fiber-reinforced thermoplastic composite laminates have been fabricated through microwave curing technique. Sisal and *G. optiva* natural fibers in woven mat form were reinforced with two types of thermoplastic polymers (PP and ethylene vinyl acetate). Results of mechanical properties (tensile, flexural, and impact strength) of the fabricated composites showed improvement in the properties as compared to neat polymers. Also, a simulation software was used to simulate microwave curing process and to optimize microwave wattage and exposure time [25]. Joining of green composite specimen has also been carried out successfully through microwave heating technique [26, 27]. Nettle and *G. optiva* natural fiber-reinforced PLA composites were developed through conventional film stacking technique. Lap joints of the developed green composites were prepared through microwave heating process and adhesive bonding technique. The results have shown that the joint strength of microwave-processed specimens was higher as compared to the joint strength of adhesively bonded specimens [28].

9.4 APPLICATION OF GREEN COMPOSITE

As the awareness to save the environment is increasing, synthetic polymer composites are being discarded wherever possible, and green composites are finding its place in many application areas because of its biodegradable nature. Natural fiber-reinforced PET composites have been used for the engine and transmission cover of automobiles [29]. The spur gear and gib–cotter joint have been developed using roselle fiber-reinforced PET composite [30]. Natural fiber-based composites have got its substantial usage in automotive interiors and exterior parts because of its lightweight characteristics, which has direct impact on fuel consumption. The application of PLA fabric has been studied, and the study shows that the PLA fabric has cleared almost all specifications formulated by an original equipment manufacturer except few properties. Also, PLA fabric was found comparable with PET automotive fabric [31]. Deep drawability of ramie woven fabric-reinforced green composite has been explored, and the results show that the products based on deep drawing operation are expected for green composites [32].

Wood plastic green composites are used in various fields of applications such as automobiles, construction, electronic parts, consumer products, and so on, with maximum usage in automobile sector for making components such as door panels, dashboard, cabin linings, and so on [33]. Various natural fiber-reinforced thermoplastic and thermoset composites (such as cotton, flax, coconut, and sisal fibers reinforced with PP, PET, or some thermoset set resin) have been used for the processing of automotive parts like glove box, door panels, floor panels, backrest, trunk floor, and trunk panel [4]. Bio-based composite roof structures have been developed using cellulose fibers reinforced with soy oil-based resin. The developed bio-based composite structures have shown good mechanical performance and found suitable for roof structures [34]. In a study, life cycle assessment was performed to analyze the replacement of glass fiber-based composites by jute fiber composites to produce automotive structural frontal bonnet of an off-road vehicle (Buggy). The case study has concluded that application jute fiber composites are advantageous in Buggy enclosures [35].

9.5 CONCLUDING REMARK

Natural fiber-reinforced biopolymer-based green composites are those newly emerging materials that are completely biodegradable in nature and thus are environment-loving materials. Green composites have received attention from many commercial application areas specifically for nonstructural components. The performance of green composites mainly depends upon interfacial characteristics between natural fiber and biopolymer. Therefore, it becomes imperative to select the right fabrication scheme for the development of high-quality green composites. The present chapter has covered various fabrication techniques to develop green composites. The chapter has also highlighted the emerging fabrication techniques such as microwave curing to process green composites.

REFERENCES

[1] Huda, M.S., Drzal, L.T., Mohanty, A.K., Misra, M. (2007). The effect of silane treated and untreated talc on the mechanical and physicomechanical properties of polylactic acid/newspaper fibers/talc hybrid composites. *Composites: Part B*, **38**, 367–379.

[2] Singh, I., Bajpai, P.K., Malik, D., Madan J., Bhatnagar, N. (2012). Microwave joining of natural fiber reinforced green composites. *Advanced Material Research*, **410**, 102–105.

[3] Oksman, K., Skrifvars, M., Selin, J.F. (2003). Natural fibers as reinforcement in polylactic acid (PLA) composites. *Composite Science and Technology*, **63**, 1317–1324.

[4] Holbery, J., Houston, D. (2006). Natural-fiber-reinforced polymer composites in automotive applications. *JOM: The Journal of the Minerals, Metals and Materials Society*, **58**, 80–86.

[5] Bajpai, P.K., Singh, I., Madaan J. (2013). Frictional and adhesive wear performance of natural fibre reinforced polypropylene composites. *Proceedings of the Institution of Mechanical Engineers, Part J: Journal of Engineering Tribology*, **227**, 385–392.

[6] Torres, F.G., Arroyo, O.H., Gomez, C. (2007). Processing and mechanical properties of natural fiber reinforced thermoplastic starch biocomposites. *Journal Thermoplastic Composite Materials*, **20**, 207–223.

[7] Bajpai, P.K., Singh, I., Madaan, J. (2014). Development and characterization of PLA-based green composites: A review. *Journal of Thermoplastic Composite Materials*, **27**, 52–81.

[8] Satyanarayana, K.G., Arizaga, G.G.C., Wypych, F. (2009). Biodegradable composites based on lignocellulosic fibers: An overview. *Progress in Polymer Science*, **34**, 982–1021.

[9] Singh, I., Bajpai, P.K., Dhawan, V. (2013). Primary and secondary processing of biocomposites. In: Thakur, V.K., Singh, A.K., editors. *Biomass-Based Biocomposites*. Smithers Rapra Publishing, Shawbury. pp. 61–83.

[10] Lee, J.T., Kim, M.W., Song, Y.S., Kang, T.J., Youn, J.R. (2010). Mechanical properties of denim fabric reinforced poly(lactic acid). *Fibers Polymers*, **11**, 60–66.

[11] Deo, C., Acharya, S.K. (2010). Effect of fiber content on abrasive wear on lantana camara fiber reinforced polymer matrix composite. *Indian Journal of Engineering and Materials Science*, **17**, 219–223.

[12] Yong, C.A.O., Yi-qiang, W.U. (2008). Evaluation of statistical strength of bamboo fiber and mechanical properties of fiber reinforced green composites. *Journal of Central South University of Technology*, **15**, 564–567.

[13] Gomes, A., Matsuo, T., Goda, K., Ohgi, J. (2007). Development and effect of alkali treatment on tensile properties of curaua fiber green composites. *Composites: Part A*, **38**, 1811–1820.

[14] Ouagne, P., Bizet, L., Baley, C., Bréard, J. (2010). Analysis of the film stacking processing parameters for PLLA/flax fibre biocomposites. *Journal of Composite Materials*, **44**, 1201–1215.

[15] Bajpai, P.K., Singh, I., Madaan, J. (2012). Comparative studies of mechanical and morphological properties of polylactic acid and polypropylene based natural fiber composites. *Journal of Reinforced Plastics and Composites*, **31**, 1712–1724.

[16] Bajpai, P.K., Singh, I., Madaan, J. (2013). Tribological behavior of natural fiber reinforced PLA Composites. *Wear*, **297**, 829–840.

[17] Bajpai, P.K., Singh, I. (2013). Drilling behavior of sisal fiber reinforced polypropylene composite laminates. *Journal of Reinforced Plastics and Composites*, **32**, 1569–1576.

[18] Huda, M.S., Drzal, L.T., Mohanty, A.K., Misra, M. (2006). Chopped glass and recycled newspaper as reinforcement fibers in injection molded poly(lactic acid) (PLA) composites: A comparative study. *Composites Science and Technology*, **66**, 1813–1824.

[19] Singh, S., Mohanty, A.K., Misra, M. (2010). Hybrid bio-composite from talc, wood fiber and bioplastic: Fabrication and characterization. *Composites: Part A*, **41**, 304–312.

[20] O'dell, J.L. (1997). Natural fibers in resin transfer molded composites, in *Proceedings of the Forth International Conference on Wood fiber–Plastic Composites*, May 12–14, Madison, USA, pp. 280–285.

[21] Lackey, E., Vaughan, J.G., Inamdar, K., Hancock, B. (2007). Statistical characterization of pultruded composites with natural fiber reinforcements–Part A: Fabrication. *Journal of Natural Fibers*, **4**, 73–87.

[22] Vuppalapati, R.R., Chandrashekhara, K., Schuman, T.P. (2009). Manufacturing and performance evaluation of core-filled, pultruded bio-composite panels, in *Proceedings of the 3rd Annual ISC Research Symposium, ISCRS 2009*, April 14, Rolla, Missouri.

[23] Akil, H.M., De Rosa, I.M., Santulli, C., Sarasini, F. (2010). Flexural behaviour of pultruded jute/glass and kenaf/glass hybrid composites monitored using acoustic emission. *Materials Science and Engineering A*, **527**, 2942–2950.

[24] George, T.J, Jacob, N., Francis, F.K., Joseph, M. (2013). Investigation on mechanical properties of various microwave cured natural fiber reinforced polymer composites. *International Journal of Engineering Research and Technology*, **2**, 2027–2032.

[25] Ali, S., Bajpai, P.K, Singh, I., Sharma, A.K. (2014). Curing of natural fibre-reinforced thermoplastic composites using microwave energy. *Journal of Reinforced Plastics and Composites*, **33**, 993–999.

[26] Bajpai, P.K., Singh, I., Madaan J. (2012). Finite element model for microwave heating of thermoplastic composites. *International Journal of Materials Engineering Innovation*, **3**, 247–258.

[27] Singh, I., Bajpai, P.K., Malik, D., Sharma, A.K., Kumar, P. (2011). Feasibility study on microwave joining of green composites. *Akademeia*, **1**, ea0101.

[28] Bajpai, P.K., Singh, I., Madaan, J. (2012). Joining of natural fiber reinforced composites using microwave energy: Experimental and finite element study. *Materials and Design*, **35**, 596–602.

[29] Drzal, L.T., Mohanty, A.K., Mehta, G., Misra, M. (2002). Low-cost, bio-based composite materials for housing applications. *Advances in Building Technology*, **1**, 225–232.

[30] Thiruchitrambalam, M., Athijayamani, A., Sathiyamurthy, S., Syed Abu Thaheer, A. (2010). Review on the natural fiber-reinforced polymer composites for the development of roselle fiber-reinforced polyester composite. *Journal of Natural Fibers*, **7**, 307–323.

[31] Ghosh, S., Krishnan, S. (2007). Application of poly (lactic acid) fibres in automotive interior. *Indian Journal of Fibre and Textile Research*, **32**, 119–121.

[32] Nakamura, R., Goda K., Noda, J., Ohgi, J. (2009). High temperature tensile properties and deep drawing of fully green composites. *eXPRESS Polymer Letters*, **3**, 19–24.

[33] Ashori, A. (2008). Wood–plastic composites as promising green-composites for automotive industries. *Bioresource Technology*, **99**, 4661–4667.

[34] Dweib, M.A., Hu, B., Shenton, H.W., Wool, R.P. (2006). Bio-based composite roof structure: Manufacturing and processing issues. *Composite Structures*, **74**, 379–388.

[35] Alves, C., Ferrão, P.M.C., Silva, A.J., Reis, L.G., Freitas, M., Rodrigues, L.B., Alves, D.E. (2010). Ecodesign of automotive components making use of natural jute fiber composites. *Journal of Cleaner Production*, **18**, 313–327.

10

APPLICATIONS OF GREEN COMPOSITE MATERIALS

KORONIS GEORGIOS, ARLINDO SILVA, AND SAMUEL FURTADO

Department of Mechanical Engineering, Instituto Superior Tecnico, University of Lisbon, Lisbon, Portugal

Biodegradable Green Composites, First Edition. Edited by Susheel Kalia.

10.1 INTRODUCTION

The topic of green composites has gathered an enormous momentum in both the academic and industrial sectors due to the need to render new products more environmentally friendly. Several major reports and journal papers have been produced that try to get the most important data in one place. Of these, the report from the UK Knowledge Transfer Network initiative on materials [1] and the review papers on natural fiber composites from Fuqua et al. and Thakur et al. [2, 3] stand as important literature for anyone interested in this topic. Of the myriad combinations of possible fibers and polymers to form composites, there is a need to select those that maximize the performance for a given application. A methodology has been proposed [4], with a specific example for automotive applications that enables a designer to select the appropriate combination of fiber and matrix. However, the use of natural fibers in industrial applications is not without issues. Leão et al. discuss the technical and social issues in the generalized use of natural fibers in the particular case of the automotive industry in South America [5].

Several areas of intervention for the use of green composites will be discussed. Applications of green composites in consumer products open up the chapter with a number of recent examples of success. Biomedical applications and packaging are discussed afterwards. The field of transportation is then tackled at length, given the enormous range of products that have found use for green composites. Construction is also an area of increasing demand for materials like green composites. Energy has been using CFRP and glass fiber-reinforced plastic (GFRP) for some years and is currently turning its attention to green composites as well. Finally, sports and leisure contribute to a few more applications of green composites. All of these applications for green composites have found improvements of some kind when compared to other solutions: these improvements can be in terms of economic and functional performance, by either reducing manufacturing costs, reducing the weight, or enhancing the functional performance of the products; or they can contribute to a more environmentally friendly product; or in other situations they can improve both the environmental and functional performances.

10.2 GREEN COMPOSITE MATERIALS

Broadly speaking, a green composite is a product being manufactured entirely by renewables or a mix of synthetic and natural-based products at a discretionary percentage. Furthermore, by the addition of annually renewable materials in the composite's constitution, the overall composite itself can be also rendered "green" and therefore to a certain extent will contribute to a more sustainable application in terms of material usage and energy savings. All this goes to show that by bringing sustainable products to the market, provided that their constituents derived from renewable resources, may well show very promising potential and benefits to companies, natural environment, and end customers when compared to materials made of dwindling petroleum resources. Such innovative green composites have been tested in numerous studies in an attempt to explore their performance on several applications for future developments. Table 10.1 illustrates a number of studies that examined different types of reinforced bioresins with several kinds of natural fibers.

10.2.1 Reinforcement

Natural plant fibers that come from harvesting plants grow in crop fields and can be integrated as filaments or reinforcements in composite manufacturing in the same way as ceramic or glass fibers, for instance. They exhibit numerous distinct advantages over synthetic ones, which have been extensively acknowledged by the scientific community in recent years [14–17]. Among the advantages as regards the natural fiber adoption, the most important is their superior environmental performance. They are an organic renewable product that require little energy, and they are also carbon dioxide neutral, that is, they do not return excess carbon dioxide to the atmosphere when they are composted or combusted and besides they provide carbon dioxide sequestration and they are biodegradable [18, 19]. Additionally, they are less abrasive to processing equipment; they need less energy in their production, resulting in lower pollutant emissions; and they are cheaper to buy and of lower density while additionally having high thermal conductivity and acoustic insulating properties. Table 10.2 lists data from several authors about fiber properties.

The reinforcing efficiency of plant fibers is related to the nature of cellulose and its crystallinity. Also the microfibril angle of the cellulose in the layers of the cell walls is important in the process of understanding the fiber elastic, creep, and strength properties [28]. High cellulose content and low microfibril angle are desirable properties of fibers, which are intended for use as reinforcement in polymer composites.

In general, natural fibers display an inability to provide a consistent pattern of physical properties in a given year; those properties may vary from every harvesting season and/or from harvesting region based on interchangeable sun, rain, and soil conditions. Moreover essential parameters are the type of ground on which the plant grows, the amount of water the plant receives during its growth, the year of the harvest, and most importantly the kind of processing and production route. An approach to address this problem is to mix batches of fibers from different harvests. Blending fibers provides a hedge against variability in any single fiber crop. By having multiple suppliers of fiber and harvests, the ratio of fibers ensures relatively consistent performance in the finished part [29]. Alternatively, it is introduced to the market that a genetic transformed fiber variety may guarantee reinforced composite of consistent performance [30].

Another main drawback of the use of fibers is the low processing temperature required (limited thermal stability). An increase in temperature can create irreversible damage and leads to porosity, which reduces the mechanical properties of the fibers [31]. In order to avoid this processing risk, the ranges of temperatures as well as the processing time have to be limited accordingly [32].

TABLE 10.1 Mechanical properties of several green composites

Resin reinforcement	Elongation to break (%)	Tensile strength (MPa)	Young's modulus (GPa)	Processing	Reference
1. Starch + 30% jute	2 ± 0.2	26.3 ± 0.55	2.5 ± 0.23	Thermoplastic injection molding	[6]
2. PLA + 30% ramie	4.8 ± 0.2	66.8 ± 1.7	n.s.	Hot-pressing sheet molding	[7]
3. PLA + 30% jute	1.8 ± 0	81.9 ± 2.9	9.6 ± 0.36	Thermoplastic injection molding	[8]
4. PTP® + 25% hemp	n.s.	≈62 ± 2	≈7.2 ± 0.3	Compression molding	[9]
5. PHBV + 30% jute	0.8 ± 0	35.2 ± 1.3	7 ± 0.26	Thermoplastic injection molding	[10]
6. PLLA + 30% flax	≈2.3 ± 0.2	≈98 ± 12	≈9.5 ± 0.5	Film stacking compression molding	[11]
7. PHB + 30% flax	≈7 ± 1.5	≈40 ± 2.5	≈4.7 ± 0.3	Film stacking compression molding	[11]
8. PLA + 30% flax	1 ± 0.2	53 ± 3.1	8.3 ± 0.6	Twin-screw extruder + compression molding[a]	[12]
9. PP + 30% flax	2.7 ± 1.5	29.1 ± 4.2	5 ± 0.4	Twin-screw extruder + compression molding[a]	[12]
10. PP + 30% jute	1.4 ± 0.1	47.9 ± 2.7	5.8 ± 0.47	Thermoplastic injection molding	[8]
11. PP + 30% fiberglass	3.01 ± 0.22	82.8 ± 4.0	4.62 ± 0.11	Compression molding	[13]

n.s., non studied.

[a] Long fiber composite.

TABLE 10.2 Properties of several natural fibers in relation with E-glass

Fibers	Density (g/cm^3)	Diameter (μm)	Tensile strength (MPa)	Young's modulus (GPa)	Elongation at brake (%)
Flax	1.5	40–600	345–1500	27–39	2.7–3.2
Hemp	1.47	25–250	550–900	38–70	1.6–4
Jute	1.3–1.49	25–250	393–800	13–26.5	1.16–1.5
Kenaf	1.5–1.6	2.6–4	350–930	40–53	1.6
Ramie	1.5–1.6	10–80	400–938	61.4–128	1.2–3.8
Sisal	1.45	50–200	468–700	9.4–22	3–7
Curaua	1.4	7–10	500–1100	11.8–30	3.7–4.3
Abaca	1.5	10–30	430–813	31.1–33.6	2.9

Note: The values are adopted from the studies and databases [8, 20–27].

TABLE 10.3 Properties of natural polymers

Polymer	Density (g/cm^3)	Melting point (T_m °C)	Tensile strength (MPa)	Young's modulus (GPa)	Elongation at brake (%)
Thermoplastic starch	1.00–1.39	110–115	5–6	0.125–0.85	31–44
PLA	1.21–1.25	150–162	21–60	0.35–3.5	2.5–6.0
PLLA	1.25–1.29	170–190	15.5–65.5	0.83–2.7	3.0–4.0
PHB	1.18–1.26	168–182	24–40	3.5–4.0	5.0–8.0
PHBV	1.23–1.25	144–172	20–25	0.5–1.5	17.5–25.0

Note: The values marked down are adopted by the studies of Refs. [20, 37–44].

10.2.2 The Matrix

Bioresin is a term used to describe a resin or a resin formulation derived from a biological source [33]. Sometimes, it is also used to describe resin formulations only partially derived from biological source. The most widely applied renewable resources include plant oils, polysaccharides (mainly cellulose and starch), and proteins [34, 35]. In the last century, the cultivation of oilseed crops, primarily for food use, led to the investigation of plant oils as precursors for resin systems [33]. Examples of plant oil used are from rapeseed, soybean, castor, pine, etc. [34, 36]. Currently several other commercial bioresin formulations can be found in the acronyms listed in Table 10.3: PLA represents the polylactic acid and PLLA connotes the poly-L-lactide. PHB stands for polyhydroxybutyrate, another aliphatic polyester, and PHBV is the copolymer poly(3-hydroxybutyrate-co-3-hydroxyvalerate). Contrary to the fibers obtained from bio-based resources, these resins have an important advantage—reproducible properties—since they are industrialized products designed specifically for a number of applications in the consumer market [45].

The major barriers for widespread acceptance of biopolymers, as substitutes for traditional nonbiodegradable polymers, are performance limitations and the high cost. For example, PLA is sensitive to hydrolytic degradation under melt processing condition in the presence of small amounts of moisture; thus, the hygroscopic nature of the natural fibers has a negative effect on the adhesion mechanism as well as on the biodegradability of bio-based composites [46]. Also, other problems like brittleness, low heat distortion temperature, high gas permeability, low melt viscosity for further processing, delicate treatments, and/or coating requirements in long-life applications also restrict their use in a wide range of applications [45, 47]. The high cost of biopolymers compared to traditional plastics is not due to the raw material costs for biopolymer synthesis; rather it is mainly attributed to the low volume of production. Another challenge for the development of biodegradable polymers lies in the fact that such biopolymers should be stable during storage or usage and then degrade once disposed of after their intended lifetime [48].

10.3 CONSUMER PRODUCTS

Increasing consumer awareness for the adoption of greener consumer goods over the past few years has been recorded, in parallel with a growing demand for a more sustainable, environmentally friendlier lifestyle, especially when bearing in mind the informed buyers that may be willing to pay more as they also turn out to be more ecologically conscious on green products [49]. As a result of being sensitized, they want to purchase eco-friendly solutions and services, preferring firms that are in favor of ecological practices. From a corporate perspective, traditional firms and start-ups have been putting quite a lot of effort into releasing series of products aiming to show their green credentials and claim a reduced impact on the environment.

On June 2005, NEC Corporation launched a novel developed plastic by combining biomass-based additives such as kenaf fiber and PLA. This new composite material made of kenaf fiber-reinforced PLA was integrating the highest percentage of renewable content applied of any bioplastic for electronic equipment (90% of resin content, excluding inorganics). It was commercialized jointly with Unitika Ltd. in September 2004, when the material was used to make the dummy card for a personal computer. NEC in collaboration with engineering labs tested this composite for its appropriateness of use on other electronic device housing applications. These composites (PLA/kenaf fiber and PLA/kenaf fiber/flexibilizer) show good practical characteristics for housing materials of electronic products in comparison with petroleum-based plastics used in housings such as glass fiber-reinforced acrylonitrile butadiene styrene (ABS) resin [50]. Following that, the material was adopted for the casing of an "eco" mobile phone launched on the Japanese market on March 10, 2006 [51]. In 2010, NEC commercialized a business personal computer that in 90% of its plastic housing uses NuCycle™, which is a flame-retardant bioplastic of 75% plant-based material applied on the model—NEC Mate Type ME (see Fig. 10.1, right) [52].

NEC is targeting high thermal conductivity PLA composites and further investigated cardanol-bonded cellulose thermoplastics [53], which are synthesized by chemically bonding cellulose, a primary component of materials such as wood and straw, with the oily component cardanol, which is derived from the agricultural

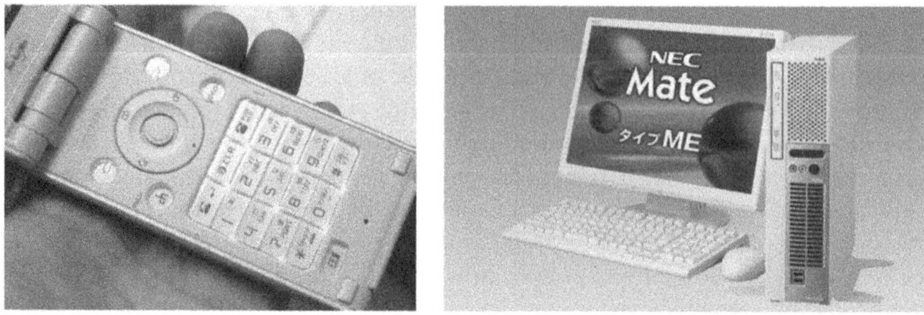

FIGURE 10.1 Mobile phone (FOMA N701iECO) casing made of kenaf fiber-reinforced polylactic acid (left). Mate Type ME desktop business personal computer that in 90% of its plastic housing uses NuCycle™ (right). *Source*: Courtesy of NEC Corporation. (*See insert for color representation of the figure.*)

by-product of cashew nut shells. In the interim, NEC Corporation aims to benefit from this technology and start producing in mass volumes a cellulose-based high-functionality bioplastic during the fiscal year of 2016. Their expectation is to deploy the upcoming material in electronic devices and various other durable products.

In an effort to satisfy the upcoming market of green composites, Netcomposites and Design Research Centre, Sheffield Hallam University Art developed the BioFurniture Cabinet. This furniture is a range of modular compostable storage cabinets that is a combination of molded biocomposite panels and wooden structures. The panels are made from Biotex, a commercially available biocomposite material made of a flax and PLA textile, which is thermoformed into shape from a woven textile [54]. The resultant developments have been shown in a solo exhibition, BioFurniture, London Design Festival on September 2013, London, as well as in the trade exhibition of JEC Europe, Paris, 2013.

As indicated by the study of Shih et al. [55], the modified fiber disposable chopsticks can be reused along with PLA to form novel fiber-reinforced green composites exploiting this to reduce waste in chopstick-using countries. Other recent green composite applications include a biodegradable paper wristwatch with the commercial name Patch [56], novel composites from modified fiber of disposable chopsticks [55], and carbon fiber from lignin [57]. Particularly for the latter, Innventia and Swerea SICOMP launch a road map for carbon fiber from lignin, introducing a cost-effective lignin-based carbon fiber derived from wood-based products. The target is to have developed by 2025 a new high-value Swedish product for use primarily in advanced composite materials.

Despite subjective interest from the green composite producer path on bio-based products, several independent organizations and national institutions are taking actions in order to verify the amount of biodegradability on whether the final product is compostable. These bodies are offering custom certifications; these certifications allow firms to provide credible compostable claims for their products in the marketplace. The Biodegradable Products Institute (BPI) [58] is a nonprofit organization that educates manufacturers and the public about biodegradability as it relates to composting bioplastic materials and other coatings and packaging that may go into commercial compost facilities. ASTM standards 6400 [59] and 6868 [60] and its compostable label may be found on products such as garbage bags, service ware, and packaging. Scientific Certification Systems (SCS) [61], an independent certifier, has developed certification standards for biodegradable soaps, detergents, and cleaners. BioPreferred [62] is a national program in the United States managed by the US Department of Agriculture that certifies and awards labels to qualifying products to increase consumer recognition of bio-based products. The purpose is to increase the purchase and use of such products while creating new jobs and providing new markets for farm commodities.

10.4 BIOMEDICAL APPLICATIONS

Chemistry for modern medicine linked along with the addition of green composites is expected to be a major theme of chemical activities in the present century. Consequently, a great attention is paid to biopolymers and their potential importance in biomedical applications. Following that, green composites have begun to appear in the form of implants in different segments of the broad medical application field.

Tissue engineering is one of the areas where novel composites are being developed for a number of applications, showing remarkable promise in the biomedical field [63–66]. It is a field that has been vaguely explored as a very imperative research topic in order to regenerate and/or improve not only tissues but organs and biological functions as well. As a result biodegradable materials like PLLA and PDLLA have been often combined with other degradable polymers such as PLGA, PEG, and chitosan to create composites with desirable material properties [67]. Chitosan, the N-deacetylated derivative of chitin, is one of the most promising polymers that combine bioresorption, absence of cytotoxicity, and low environmental impact during processing [68, 69]. Accordingly, chitosan and its composites have generated enormous scientific interest for biomedical applications owing to its various advantages such as biocompatibility, biodegradability, and osteoconductive properties as for their use of three-dimensional (3D) scaffolds in which cells are seeded before implantation. The most common concept underlying tissue engineering is to combine a biodegradable matrix (scaffold), living cells, and/or biologically active molecules to form a construct promoting the repair and regeneration of the targeted tissue (Fig. 10.2).

Another segment for green composite application within the biomedical field is the implants and medical devices that thereafter are able to dissolve and be absorbed into the human body. The most common materials used in these implants for orthopedic or medical applications are generally metals (e.g., stainless steel, cobalt-chrome and titanium alloys), polymers (e.g., ultrahigh molecular weight polyethylene, poly(ether-ether-ketone)), and ceramics (e.g., hydroxyapatite). However, the last two decades of the twentieth century saw a paradigm shift from biostable biomaterials to bioabsorbable or biodegradable (hydrolytically and enzymatically degradable) biomaterials for medical devices that could help the body to repair and regenerate the damaged tissues [70].

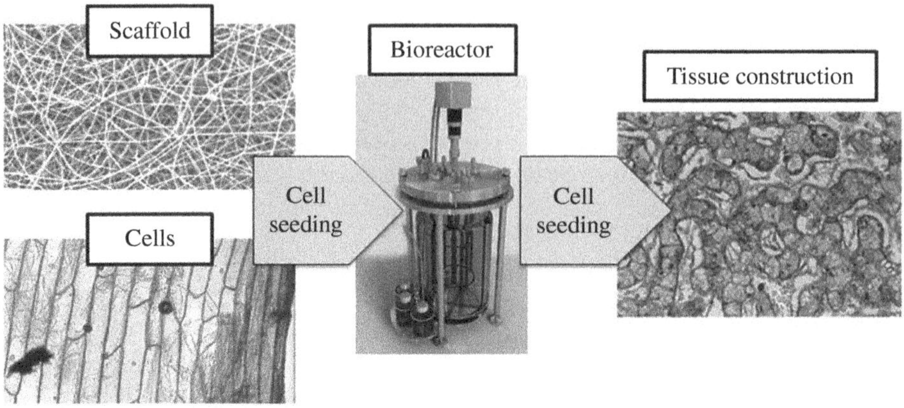

FIGURE 10.2 Scaffold-guided approach for tissue engineering. Cells are expanded *ex vivo* and seeded onto a bioactive scaffold. The tissue-engineered construct is then dynamically cultured into a bioreactor system allowing for improvement of culture conditions.

Recently, more research is focusing on composite nanofibrous scaffolds as they possess both the physical properties of synthetic polymers and bioactivity of natural polymers. The scaffold, made of composites, must possess specific characteristics including high porosity, high surface area, structural strength, specific 3D shape, and biodegradability. 3D hydroxyapatite/chitosan nanocomposite rods with layer-by-layer structures have been constructed via an *in situ* precipitation method, resulting in bioabsorbable parts that can be used for bone fracture internal fixation [71, 72]. 3D biocompatible scaffolds with novel zirconium–chitin-based composites have been introduced by Ehrlich et al. [73]. Such specimens were prepared for the first time using hydrothermal synthesis. These zirconium–chitin composites have a high potential for use in a broad range of applications ranging from synthetic catalysis to biocompatible materials for bone and dental repair [73].

In another kind of application, green composites can be applied for wound dressing. Different types of chitin- and chitosan-based dressing materials are commercially available. Chitin and chitosan in the form of composites, gels, nanofibers, films, nonwovens, and scaffolds have been used in order to regenerate wounded tissues [70, 74, 75] due to their excellent biological and physicochemical properties such as biodegradability, biocompatibility, nontoxicity, antimicrobial, and hydrating features.

Bouillat Terrier Orthopédie has developed their first "ecologic orthosis" made of Twinflax 100% flax fabric (235 g/m^2) and an SR Greenpoxy 55 (Sicomin) matrix. The orthosis is produced via the infusion process and trimmed using a jigsaw with a diamond blade [76].

More recently, Naseri et al. [77] have developed chitosan-/polyethylene oxide-based randomly oriented fiber mats reinforced with chitin nanocrystals (ChNC) for use in wound dressing application. Randomly oriented fiber mats were prepared by electrospinning, and the effect of ChNC as reinforcement/functional additive as well as genipin as cross-linking reagent was evaluated, which is composed of chitosan blended with PEO in a 1:1 mass ratio as matrix phase and ChNC as reinforcing phase.

10.5 PACKAGING

Packaging has been one of the major fields of application of green polymers and composites in recent years. Especially in what concerns food packaging, research and legislation have been intense. Biodegradability and composting of food packaging have been a legislation concern in the European Union [78]. Srinivasa and Tharanathan discuss at length the issues of nonbiodegradable packaging materials, classify the different biopolymers currently used for this application (see Fig. 10.3), and propose chitin-/chitosan-based materials as particularly adequate biopolymers [79].

Tang et al. reviewed biopolymers and biopolymer-based nanocomposites to be used as materials for packaging of food products [80]. They divide biopolymers for packaging in six different groups: polysaccharide, starch based; polysaccharide, cellulose based; other polysaccharide based; protein based; polyhydroxyalkanoate (PHA) based; and PLA based. Biopolymer blends are also covered, in a total of six further types. For bionanocomposites, seven more material types are identified: starch based, cellulose based, pectin based, gluten based, gelatin based, soy protein based, PLA based, and PHB based. The paper discusses the various techniques that have been used for developing cost-effective biodegradable packaging materials with optimum mechanical strength and oxygen and moisture barrier properties. The particular case of nanocomposite films based on natural polymers is discussed,

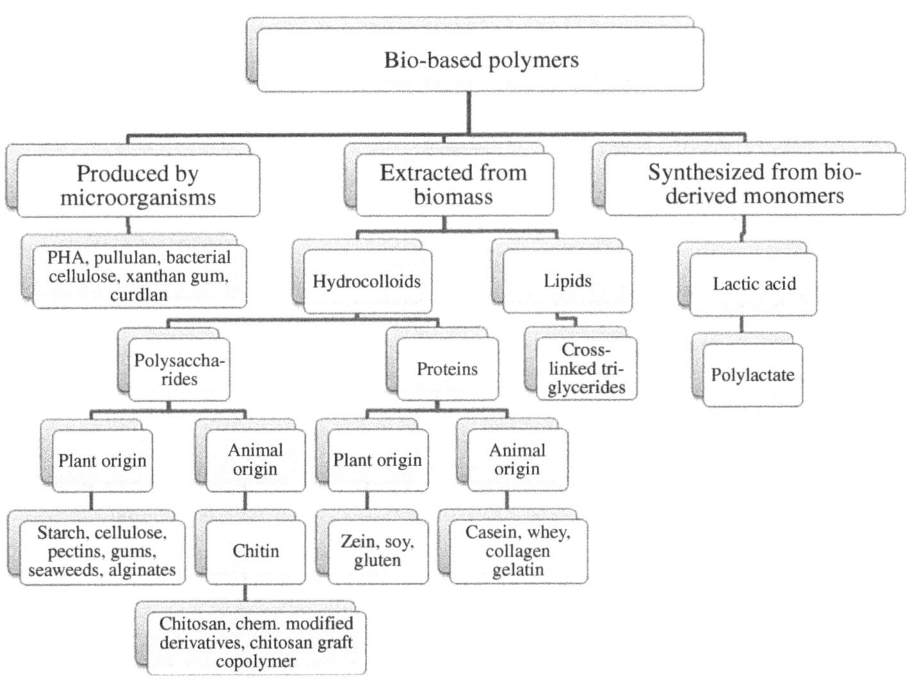

FIGURE 10.3 Different types of bio-based polymers for biopackaging. *Source*: Adapted from Srinivasa and Tharanathan [79].

together with applications, by Ref. [81]. The authors discuss the mechanisms of strengthening biopolymer films with nanoparticles and other treatments. Youssef [82] further makes the case for nanocomposites with biofunctional properties for food packaging applications. One of the most important properties of biodegradable packaging materials is their reduced environmental impact. Davis and Song assessed the impact on waste management of biodegradable packaging based on raw materials from crops and compared them with oil-based polymer packaging materials. The authors concluded that biodegradable packaging materials are most suitable for single-use disposable packaging applications where the postconsumer use packaging can be locally composted as a means of recycling the materials. However, for the widespread application of these biodegradable materials to be environmentally successful, appropriate collection, transportation, and treatment technologies are considered crucial [83]. Guana et al. proposed acetylated starch/PLA as a material for loose-fill packaging and discussed different compositions and preparation schemes to obtain the desired packaging properties [84].

10.6 TRANSPORTATION INDUSTRY

The largest segments for all natural fiber composite applications are the automotive and construction. The automotive sector is very important for the natural fibers (such as flax, kenaf, hemp, jute, and so on), accounting for over 96% of plant fiber-reinforced plastics in use for nonstructural automotive interior components, while wood plastic composites (WPCs) are the preferred in the building and construction sector [85, 86]. The use of natural composites in the automotive industry started with Henry Ford. Around 1942, he began experimenting, initially by using compressed soybeans in order to produce composite plastic-like components. During that period the petroleum-based chemicals were very cheap, and therefore soy-based plastics could not find economic importance [87]. Now, due to the new environmental regulations and depletion and uncertainty of petroleum sources, scientists and manufacturers have continuously revived their interest to derive new generations of composite materials from renewable sources, offering a new impetus for biocomposite material applications [87].

Many major automakers have now a real interest in global sustainability. Virtually all the major car manufacturers, such as Daimler Chrysler, Mercedes, Volkswagen, Audi group, BMW, Ford, Opel, Peugeot, Renault, and Volvo, currently use natural fiber composites in various applications. Daimler Chrysler identifies bio-based materials as one of the two key parts of its plan to create a global sustainability network [88, 89].

Daimler-Benz has been exploring the idea of replacing glass fibers with plant fibers in automotive components since 1991. Mercedes used jute-based door panels in its E-class vehicles in 1996. In September 2000, Daimler Chrysler began using plant fibers for their vehicle production. The bast fibers are primarily used in automotive applications because they exhibit the greatest strength. The other advantages of using bast fibers in the automotive industry include weight savings of 10–30% and the corresponding cost savings [89]. Volvo is targeting to reduce weight and the material and production costs of its car door panels by replacing a GFRC with a natural composite, wood fiber (a WPC). The decrease in production costs derives from less wear on machines, shorter cycle times, and less energy use. At the same time, the increase in use of renewable materials will also increase the material recyclability possibilities [90].

Natural fiber composites play an important role in the automotive companies' search for new materials because they combine different properties, like sound and vibration abatement capability and reduced weight (it is estimated that approximately 75% of a vehicle's energy consumption is directly related to factors associated with the vehicle's weight [48, 91]), and simultaneously fulfill the very stringent environmental legislations in different countries. These environmental legislations (e.g., the European Directive 2000/53/EC) are pushing the automotive companies to evaluate the environmental impact of the entire vehicle's life cycle, from raw materials to manufacturing to disposal [92, 93].

Natural fibers are currently used in significant quantities for automotive components, and they are a good replacement of glass fibers as well as an adequate reinforcement of synthetic polymers. They are used in trim parts, dashboard parts, door panels, parcel shelves, seat cushions, backrests and headrests, cabin linings, trunk liners, and so on. The plant fibers' superior damping characteristics are increasing their use for thermoacoustic insulation purposes as well. Another example is the use of flax fibers in car disk brakes to replace asbestos fibers. However, there are only very few exterior parts made from plant fiber composites so far [86, 89, 90, 94]. An overview of some automotive usage of plant fibers by the major automakers is presented in Table 10.4.

Typical amounts of plant fibers used for different applications in the automotive industry are front door linens (1.2–1.8 kg), rear door linens (0.8–1.5 kg), boot linens (1.5–2.5 kg), parcel shelves (up to 2.0 kg), seat backs (1.6–2.0 kg), sunroof sliders (up to 0.4 kg), and headliners (average 2.5 kg).

The most common natural fiber composites used in the last decades are WPCs, short natural fiber injection-molded compounds, and nonwoven natural fiber mats. However, continuous woven natural fiber-reinforced mats have also become a viable option for more demanding applications, extending the range of potential applications. Petrochemical polymers are more frequently used at present, although, as expected in years to come, bio-based polymer composite parts are the emergent constituent materials in such applications. The main manufacturing processes used in bio-based polymers are compression and injection molding [86].

Some recent examples of these new successful applications of plant fiber-reinforced composites are presented below:

- Volkswagen incorporates Faurecia's flax-reinforced polypropylene (PP) technology in its 2015 Volkswagen Golf model. This bio-based composite will be incorporated in the production of several components, namely, seating, door panels, and front-end modules [96].

TABLE 10.4 Plant fiber usage in the automotive industry

Automotive company	Vehicle	Part
Audi	A2, A3, A4, A6, A8, Avant, Roadster, Coupe	Seat back, side and back door panel, boot lining, hat rack, spare tire lining
BMW	3, 5, and 7 series and others	Door panels, headliner panel, boot lining, seat back
Daimler-Benz	A, C, E, and S class	Door panels, windshield/dashboard, business table, pillar cover panel
Fiat	Punto, Brava, Marea, Alfa Romeo 146 and 156	—
Ford	Mondeo CD 162 and Focus	Door panels, B-pillar, boot liner
Opel	Astra, Vectra, and Zafira	Headliner panel, door panels, pillar cover panel, instrument panel
Peugeot	New model 406	—
Renault	Clio	—
Rover	Rover 2000 and others	Insulation, rear storage shelf/panel
Volkswagen	Golf A4, Passat Variant, and Bora	Door panel, seat back, boot lid finish panel, boot liner
Volvo	C70 and V70	—
Mitsubishi	Space Star, Colt	Door panels, instrument panels

From Ref. [95], http://biggani.com/files_of_biggani/mashiur/interview/omar_faruk.pdf

- To ideally satisfy the strict requirements of the vehicle manufacturing industry, Naroplastic and Narodur were developed. Naroplastic is a thermoplastic natural fiber plastic composite presenting high strength resistance combined with low weight benefits and manufactured by compression molding. Compared to conventional materials, this biocomposite plays an important role in conserving natural resources, storing CO^2, saving fuel, increasing passive safety, and reducing noise levels. In turn, Narodur is a thermosetting natural fiber plastic composite with high strength and especially high stiffness combined with a relatively low weight. It is also manufactured to suit the automotive but also the construction, shipbuilding, or furniture industries [90].
- Fibrolon®, a WPC, developed by FKuR Kunststoff GmbH, with the important advantage of being formable into complex profiles, panels, and hollow profiles and thus with a particular interest for components for automotive interior production [90].
- In the United Kingdom, different natural fiber compounds, produced by hand lay-up, vacuum bagging, and RTM, have been incorporated in the sports car Lotus Eco Elise [97]. Lotus swapped out its typical fiberglass reinforcements for hemp fibers in the composite body panels, the double-curvature fixed hardtop, and the spoiler [98]. Exposed hemp fibers in an unpainted stripe from the bumper to the spoiler made a striking ecocontrast to the metallic finish, which signals immediately that this car is different.

The development of novel technologies, like long fiber "twistless" yarns, is also creating new possibilities for high-performance biocomposites. It is the case of Biotex company that has been developing a range of prototype components from new high-performance biocomposite materials including a marine headliner panel from Fairline Boats and two interior automotive parts from Jaguar Land Rover. Due to its combination of good mechanical properties and environmental performance, Jaguar Land Rover selected Biotex flax/PP "twistless" fabric to produce prototype door modules for both the Land Rover Defender and Jaguar XF models. It was found that the Defender Front Door Module made by biocomposite was approximately 60% lighter than the current steel part at the same stiffness, while the Jaguar XF Rear Door Module was 35% lighter than the current glass-filled PP component for the same thickness [99].

Toyota has developed automotive parts from press-molded kenaf/PLA nonwoven mats, including the Toyota Raum spare wheel cover, which is claimed to be the first 100% natural automotive product in the world (although Henry Ford was experimenting with hemp-reinforced soy resin panels for cars as early as the 1940s). Soybean–natural fiber composites have been used commercially by John Deere, under the trade name Harvest Form®, to produce body panels and cab roofs for hay balers. The parts were produced by the RIM process and 25% weight savings are claimed. More recently, EcoTechnilin (United Kingdom) has supplied a composite sandwich panel consisting of nonwoven flax/bioresin facings and a paper honeycomb core (its commercial name is "FibriCard") for load floor in the Jaguar F-type convertible [86]. The Ford Motor Company claims that they are the first automaker to develop and use an environmentally friendly wheat straw-reinforced plastic in a vehicle. The wheat straw is the waste by-product of wheat, meaning that this application can contribute to a significant waste reduction. Their first application of the natural fiber-based plastic—containing 20% wheat straw biofiller—was on the 2010 Ford Flex's third-row interior storage bins. They claim that by this substitution alone there was a recorded reduction in petroleum usage by 9 metric tons (Mt) and in CO^2 emissions by 13.5 Mt per year.

In order to achieve high-performance materials, Faurecia introduced BioMat, a 100% bio-based fiber (hemp) and polybutylene succinate (PBS) using the Natural Fiber for Lean Injected Design (NAFILean) process, for injected instrument panels, center consoles, and door panels. NAFILean is used for invisible components such as trim parts, top covers, and air ducts. In collaboration with Mitsubishi Chemical and other partners, they created new processes for extracting substances from natural sources for the matrix of the injection material. The company is expecting that BioMat will appear in vehicles on the roads as early as 2016 [100].

A successful example of replacement of the majority of parts traditionally made from plastics by high-quality, safe, and fully durable biomaterials can be found in the Biofore Concept Car designed and manufactured by students from the Helsinki Metropolia University of Applied Sciences. UPM Formi (a durable, high-quality biocomposite for injection molding, extrusion, and thermoforming production) and UPM Grada (thermoformable wood material) were used to manufacture the passenger compartment floor, center console, display panel cover, door panels, front mask, side skirts, dashboard, and interior panels, presenting a significant improvement in the overall environmental performance of car manufacturing. By incorporating these UPM's biomaterials, the concept car is approximately 150 kg lighter than its equivalents, resulting in lower fuel consumption as stated by the manufacturers [101].

The natural fiber composites have been offering some great potential in other sectors of the transportation industry. Some examples have been already presented by electric scooter and train manufacturers.

In the aeronautic sector, the natural fiber composites are not yet fulfilling the extremely stringent regulations. However, a flax/phenolic composite was used to replace aluminum sheets for building the structural members of Spitfire fuselages for British military aircrafts during the Second World War due to aluminum shortage [85].

10.7 CONSTRUCTION

The use of natural fibers by the construction industry will help achieve a more sustainable consumption pattern of building materials [102].

As mentioned before, WPCs are the most preferred in the construction sector. The most common applications are indoor panels, decking, railing, window frames, outdoor furniture, picnic tables, garden benches, pallets, boards, bridges, and so on (manufactured through extrusion, compression molding, and injection molding) [85, 86]. In the specific case of bamboo fiber-reinforced composites, applications are more extensive: indoor and outdoor floors, trestle bridge floor, bathroom floor, container soleplate, train and car floor, indoor and outdoor furniture, garden furniture, outdoor, wood structure, doors and windows, and facade materials [103].

Some further examples are presented in the coming paragraphs.

In 2009 the Louisiana Museum of Modern Art in Denmark opened a pavilion made from a range of eco-friendly technologies including natural fibers and bio-based polymers, solar cells, and self-cleaning coatings [86].

In 2013 a full biocomposite building facade, clad with bio-based panels made from hemp fiber and a bio-based resin, was installed in a gas receiving station in the Netherlands [86].

BamDeck™ composite decking is making its production using a 100% recycled material, reinforced with bamboo fibers, two times stronger and four times harder than the competition [104].

The British company Tech-Wood International Ltd. is offering a modular house construction kit made of WPCs (with about 75% fiber content) named "simply housing" [105].

ThyssenKrupp Plastics is delivering an easy-to-install decking (TimberTech) made of WPC. TimberTech presents several advantages like the possibility for the boards to be fastened directly to a wooden subframe without predrilling, floating installation, stain resistance, low maintenance, and weather resistance (capable of withstanding persistent rain and hot sun without warping [106]).

Natural fiber-reinforced composites (an example of which is the kenaf-reinforced polymer) can be also used to replace some use of unsustainable wood sourcing in building, thus lessening local environmental degradation [90].

Natural fibers are also used as stand-alone materials in insulation, thatch roofing, and geotextiles. For example, hemp slivers, a waste product of hemp fiber production, can be used in the form of hemp lime as blocks or sprayed onto walls to provide structure and excellent insulation properties. Hemp lime was used in the construction of the Adnams Brewery warehouse in Suffolk, United Kingdom, contributing to savings of approximately £40,000/year in energy costs by enabling the building to remain at a constant 16°C without the use of heating or air conditioning [86]. Also, an application of hemp fiber fleece as insulation material, in Germany, has shown a high carbon capture capability [97]. Thirty-five thousand protective blocks made from GreenGran granules (natural fiber-reinforced biopolymers PLA and PHB) were placed on riverbanks and dams to prevent riverbank erosion and have shown encouraging results [107].

10.8 ENERGY INDUSTRY

The energy industry shows an important potential for the application of natural fiber composites. Due to the higher environmental protection awareness of modern society, the demand for cleaner energy resources is considered a vital necessity. The most important application in the energy industry can be found in the construction of wind turbine blades. There are several examples where natural fibers successfully replace E-glass in wind turbine blade construction.

In China, since 2010, 100 sets of 800KW bamboo wind turbine blades have been set in operation due to their high wind strength, high hardness, and good toughness. The final cost is 15% lower when compared to a conventional glass fiber wind turbine blade [103].

In the University of Nottingham, a flax/polyester composite 3.5 m-long wind turbine blade was successfully manufactured using vacuum infusion. The blade successfully passed the IEC/British Standard 61400 test for acoustic noise and presented a weight reduction of 10% while in parallel having a lower environmental impact when compared with similar blades made of synthetic fibers. The flax composite blade presented also showed improved fatigue test results when compared with the corresponding blades made of E-glass [85].

Biotex (flax/PP composite) has been used to produce natural fiber-reinforced blades for a rooftop wind turbine by the University of Stuttgart [108].

10.9 SPORTS AND LEISURE INDUSTRY

The sports and leisure industries also present great opportunities for biocomposite applications. Plant fiber-reinforced composites have indeed been successfully used in sports and leisure outdoor applications. Mainly due to their excellent vibration damping and health safety properties, they are being applied in boat hulls and canoes, snowboards, surfboards, bike frames, tennis rackets, toys, musical instruments, and so on [85, 86, 94].

10.9.1 Boat Hulls and Canoes

One of the most important fields of application for green composites within this industry section is in the marine sector. A flax-reinforced catamaran hull is manufactured using vacuum infusion [109]. On another project, a 4 m-long canoe was built entirely of flax-reinforced PLA to demonstrate the feasibility of manufacturing boat structures from these 100% bio-based materials [110].

In 2011, a prototype boat with a composite structure incorporates up to 50% flax fiber for the Mini Transat 6.50 transatlantic yacht race. For the particular case, an improved impregnation process was used, which was limiting the humidity absorption by the use of natural fibers [111].

Hiendl NFC is a natural fiber composite developed by H. Hiendl GmbH & Co. KG, consisting of 70% wood fibers and 30% PP. It was used to manufacture a new 12.5 m-long mahogany deck plank for the dragon boat, an Asian heritage. Lighter profiles than traditional wooden ones were achieved, and the annual maintenance work required was reduced to practically zero [112].

In order to create more environmentally friendly hulls for electric-powered boats, Lavender Composites successfully tested a combination of Sicomin's Greenpoxy 55 and Amorim's CoreCork core. The end result was to manufacture a hull with minimal environmental impact while maintaining the highest performance possible. Through the use of the infusion manufacturing process, it was also possible to reduce volatile organic compound releases [113].

Plasmor used flax fiber mats and PLA thermoplastic films along with a vacuum-bagged film stacking process in order to successfully produce a prototype of a small pleasure canoe made entirely of biocomposites. Since the material can be composted at the end of its intended use, the canoe will have a reduced environmental impact and simultaneously will contribute in fulfilling the demanding waste disposal regulations. The company advocates that the prototype is a proof that the materials and process developed can be used to build large parts; therefore the process is ready to be transferred to a shipyard [114].

VE Paddles shaped a high-performance paddle made of 100% flax-reinforced blade mounted on a standard reinforced fiberglass. The paddle was presented at the 2013 PADDLEexpo in Nürnberg where it received a lot of interest and enthusiasm from the attendees [115].

10.9.2 Snowboards/Skis and Surfboards

The use of natural composites in snowboards/skis and surfboards is already a reality, and several examples can be found. For example, Magine Snowboards and Skis has developed a biocomposite snowboard using Composites Evolution's Biotex flax fabric. The construction is a sandwich structure composed by a wooden laminate core between two layers of Biotex flax fabric and a top and bottom plastic layer, all bonded together using an eco-epoxy resin system [116].

Another promising application has been launched by Northern California-based Xboards, Inc., when it introduced an eco-friendly snowboard made from flax fiber composite under the commercial name BioBoard™ [117]. This novel composite board is reported to combine powerful shock-absorbing properties with improved resiliency and strength. The freestyle FlaxJacket™ board has twofold vibration dampening power compared to Kevlar and three times that of fiberglass while reducing board chattering and leg fatigue for the rider.

10.9.3 Toys

In several cases, toys made of biocomposites present technical, cost-related, or ecological advantages when compared with traditional synthetic solutions. That is the case of babel pico, a strategy game for two players, and Kipp Kipp Ahoi, a skill game from Ravensburger made by a wood injection-molded material [118]. The Dutch company presented toy cars made of natural fiber and PP using the injection molding process [97].

10.9.4 Musical Instruments

The excellent sound behavior of plant fibers makes them a promising material for musical applications. Some applications are presented in the following.

A special instance of resource conservation was introduced by Herbert Neureiter in the Austrian Tyrol. Clarinets, which are often made from endangered African blackwood (ABW), can be produced by injection molding using Fasal material. The risk of splitting and the absorption of moisture are virtually null. "In spite of the lower weight of about 10% the material shows a very good sound behavior and there are hardly any differences to massive wood." Customers are given a 5-year warranty for dimensional stability. The high density of Fasal results in similar sound characteristics to those of solid wood [118].

Alpaca used a combination of flax fabric, carbon fiber, and bio-derived resins to manufacture a travel guitar, resulting in a lightweight and strong instrument with a good, resonant sound. They also claim that their guitar can handle anything a good adventure brings: dirt, water, bumps, and bruises [119].

Blackbird Guitars claims to have produced the world's first concert-level plant fiber composite musical instrument, offering warm, big sound with lightweight strength of aerospace-inspired composite construction. The guitar (Clara ukulele) is made from Ekoa—a proprietary, renewable material, with the tone of vintage wood and the durability of cutting-edge, eco-friendly construction [120].

High-quality flax fibers, enclosed by two thin layers of glass fibers, were used as a core material for speaker manufacturing. The final bio-based composite satisfied the key criteria of a high-performance diaphragm: high internal damping, high velocity of sound, and high flexural rigidity.

10.10 CONCLUSIONS

Green composites have been finding their way into everyday life, substituting all sorts of other materials, from metals to petroleum-based plastics and composites, with tremendous environmental benefits. The trend will continue for years to come, as more and more bio-based composites are developed and incorporated into the products we design, use, and dispose of.

REFERENCES

[1] Knowledge Transfer Network. (2014). Technology Overview—Biocomposites. UK, Materials KTN, NetComposites **40**.

[2] Fuqua, M. A., Huo, S., Ulven, C. A. (2012). Natural fiber reinforced composites. *Polymer Reviews*, **52**, 259–320.

[3] Thakur, V. K., Thakur, M. K., Gupta, R. K. (2014). Review: Raw natural fiber-based polymer composites. *International Journal of Polymer Analysis and Characterization*, **19**, 256–271.

[4] Koronis, G., Silva, A., Fontul, M. (2013). Green composites: A review of adequate materials for automotive applications. *Composites: Part B*, **44**, 120–127.

[5] Leão, A., Sartor, S. M., Caraschi, J. C. (2006). Natural fibers based composites: Technical and social issues. *Molecular Crystals and Liquid Crystals*, **448**, 161/[763]–177/[779].

[6] Vilaseca, F., Mendez, J., Pelach, A., Llop, M., Canigueral, N., Girones, J., Turon, X., Mutje, P. (2007). Composite materials derived from biodegradable starch polymer and jute strands. *Process Biochemistry*, **42**, 329–334.

[7] Yu, T., Ren, J., Li, S., Yuan, H., Li, Y. (2010). Effect of fiber surface-treatments on the properties of poly(lactic acid)/ramie composites. *Composites Part A: Applied Science and Manufacturing*, **41**, 499–505.

[8] Bledzki, A. K., Jaszkiewicz, A., Scherzer, D. (2009). Mechanical properties of PLA composites with man-made cellulose and abaca fibres. *Composites Part A: Applied Science and Manufacturing*, **40**, 404–412.

[9] Müssig, J., Schmehl, M., von Buttlar, H. B., Schönfeld, U., Arndt, K. (2006). Exterior components based on renewable resources produced with SMC technology: Considering a bus component as example. *Industrial Crops and Products*, **24**, 132–145.

[10] Bledzki, A. K., Jaszkiewicz, A. (2010). Mechanical performance of biocomposites based on PLA and PHBV reinforced with natural fibres: A comparative study to PP. *Composites Science and Technology*, **70**, 1687–1696.

[11] Bodros, E., Pillin, I., Montrelay, N., Baley, C. (2007). Could biopolymers reinforced by randomly scattered flax fibre be used in structural applications? *Composites Science and Technology*, **67**, 462–470.

[12] Oksman, K., Skrifvarsb, M., Selinc, J. F. (2003). Natural fibres as reinforcement in polylactic acid (PLA) composites. *Composites Science and Technology*, **63**, 1317–1324.

[13] Design, G. (2010). *CES EduPack materials selection software*. Granta Design Limited, Cambridge, MA.

[14] Wool, R., Khot, S. N. (2000). *Bio-based resins and natural fibers*. ASM International, Materials Park. Available at http://www.asminternational.org/pdf/XXspotlightsXX/BiobasedResins.pdf. Accessed September 8, 2010.

[15] Zah, R., Hischier, R., Leao, A., Braun, I. (2007). Curauá fibers in the automobile industry: A sustainability assessment. *Journal of Cleaner Production*, **15**, 1032–1040.

[16] Joshi, S., Drzal, L., Mohanty, A., Arora, S. (2004). Are natural fiber composites environmentally superior to glass fiber reinforced composites? *Composites Part A: Applied Science and Manufacturing*, **35**, 371–376.

[17] Bismarck, A., Mishra, S., Lampke, T., Plant fibers as reinforcement for green composites. In Mohanty, A. K., Misra, M., Drzal, L. T., (Eds.), *Natural fibers, biopolymers, and biocomposites*. CRC Press, Taylor & Francis Group, Boca Raton, 2005.

[18] Williams, G. I., Wool, R. P. (2000). Composites from natural fibers and soy oil resins. *Applied Composite Materials*, **7**, 421–432.

[19] John, M. J., Thomas, S. (2008). Biofibres and biocomposites. *Carbohydrate Polymers*, **71**, 343–364.

[20] Mohanty, A. K., Misra, M., Drzal, T. L., Selke, S. E., Harte, B. R., Hinrichsen, G., Natural fibers, biopolymers, and biocomposites. In Mohanty, A. K., Misra, M., Drzal, L. T., (Eds.), *Natural fibers, biopolymers, and biocomposites: An introduction*. CRC Press, Taylor & Francis Group, Boca Raton, 2005, pp. 1–36.

[21] Wambua P. (2003). Natural fibres: Can they replace glass in fibre reinforced plastics? *Composites Science and Technology*, **63**, 1259–1264.

[22] Sedan, D., Pagnoux, C., Smith, A., Chotard, T. (2008). Mechanical properties of hemp fibre reinforced cement: Influence of the fibre/matrix interaction. *Journal of the European Ceramic Society*, **28**, 183–192.

[23] Arbelaiz, A., Fernández, B., Cantero, G., Llano-Ponte, R., Valea, A., Mondragon, I. (2005). Mechanical properties of flax fibre/polypropylene composites. Influence of fibre/matrix modification and glass fibre hybridization. *Composites Part A: Applied Science and Manufacturing*, **36**, 1637–1644.

[24] George, J., Sreekala, M. S., Thomas, S. (2001). A review on interfacial modification and characterization of natural fiber reinforced plastic composites. *Polymer Engineering and Science*, **41**, 1471–1485.

[25] Bledzki, A. K., Mamun, A. A., Jaszkiewicz, A., Erdmann, K. (2010). Polypropylene composites with enzyme modified abaca fibre. *Composites Science and Technology*, **70**, 854–860.

[26] Goda, K., Sreekala, M. S., Gomes, A., Kaji, T., Ohgi, J. (2006). Improvement of plant based natural fibers for toughening green composites: Effect of load application during mercerization of ramie fibers. *Composites Part A: Applied Science and Manufacturing*, **37**, 2213–2220.

[27] Kenaf-fiber.com. (2009). Technical data and information about thermal insulation and soundproofing of kenaf natural fibers. Available at http://www.kenaf-fiber.com/en/infotec-tabella2.asp. Accessed February 22, 2010.

[28] Eichhorn, S. J., Baillie, C. A., Zafeiropoulos, N., Mwaikambo, L. Y., Ansell, M. P., Dufresne, A., Entwistle, K. M., Herrera-Franco, P. J., Escamilla, G. C., Groom, L., Hughes, M., Hill, C., Rials, T. G., Wild, P. M. (2001). Review: Current international research into cellulosic fibres and composites. *Journal of Materials Science*, **36**, 2107–2131.

[29] Brosius, D. (2006). Natural fiber composites slowly take root. *Composite Technology*, February, 32–37.

[30] Mougin, G. (2006). Natural fibre composites—Problems and solutions. JEC Composites Magazine. Paris, JEC, 25. Available from: http://www.jeccomposites.com/news/composites-news/natural-fibre-composites-problemsand-solutions. Accessed October 25, 2011.

[31] Gassan, J., Bledzki, A. K. (2001). Thermal degradation of flax and jute fibers. *Journal of Applied Polymer Science*, **82**, 1417–1422.

[32] Wielage, B., Lampke, T., Marx, G., Nestler, K., Starke, D. (1999). Thermogravimetric and differential scanning calorimetric analysis of natural fibres and polypropylene. *Thermochimica Acta*, **337**, 169–177.

[33] Fowler, P. A., Tverezovskiy, V. V., Elias, R. M., Chappell, C. G., Fitchett, C. S., Laughton, N. G., Seefeld, J. F. (2009). Vegetable oil: Starting material for a new bio-resin system. COST Action E49 Conference "Management and Control of VOC Emissions from Wood-Based Panels," 2009, Nantes, France.

[34] Laine, L., Rozite, L. (2010). Eco-efficient composite materials- State of the Art. Anacompo. Available at http://www.ketek.fi/anacompo/STATE%20OF%20THE%20ART.pdf. Accessed March 13, 2012.

[35] Raquez, J. M., Deléglise, M., Lacrampe, M. F., Krawczak, P. (2010). Thermosetting (bio)materials derived from renewable resources: A critical review. *Progress in Polymer Science*, **35**, 487–509.

[36] Unknown. (2006). Clean, Green technology makes sustainable resins from vegetable oil. Available at http://www.bc.bangor.ac.uk/news-and-resources/news/documents/Replantpressrelease.pdf. Accessed August 25, 2015.

[37] Van de Velde, K., Kiekens, P. (2002). Biopolymers: Overview of several properties and consequences on their applications. *Polymer Testing*, **21**, 433–442.

[38] Averous, L., Moro, L., Dole, P., Fringant, C. (2000). Properties of thermoplastic blends: Starch-polycaprolactone. *Polymer*, **41**, 4157–4167.

[39] Clarinval, A. M. (2002). Classification and comparison of thermal and mechanical properties of commercialized polymers. International Congress & Trade Show, Industrial Applications of Bioplastics. IENICA. York, UK, February 3–5, 2002.

[40] Tsuji, H. (2000). In vitro hydrolysis of blends from enantiomeric poly(lactide)s Part 1. Well-stereo-complexed blend and non-blended films. *Polymer*, **41**, 3621–3630.

[41] Phong, L., Han, E. S. C., Xiong, S., Pan, J., Loo, S. C. J. (2010). Properties and hydrolysis of PLGA and PLLA cross-linked with electron beam radiation. *Polymer Degradation and Stability*, **95**, 771–777.

[42] Ahankari, S. S., Mohanty, A. K. Misra, M. (2011). Mechanical behaviour of agro-residue reinforced poly(3-hydroxybutyrate-co-3-hydroxyvalerate), (PHBV) green composites: A comparison with traditional polypropylene composites. *Composites Science and Technology*, **71**, 653–657.

[43] Li, Y., Shimizu, H. (2009). Improvement in toughness of poly(l-lactide) (PLLA) through reactive blending with acrylonitrile-butadiene-styrene copolymer (ABS): Morphology and properties. *European Polymer Journal*, **45**, 738–746.

[44] Curvelo, A. A. S., de Carvalho, A. J. F., Agnelli, J. A. M. (2001). Thermoplastic starch-cellulosic fibers composites: Preliminary results. *Carbohydrate Polymers*, **45**, 183–188.

[45] Koronis, G., Silva, A., Fontul, M. (2011). Green composites for an electric vehicle body: A review of adequate materials' combination. Proceedings of the 16th International Conference on Composite Structures. Porto, Portugal, June 28–30, 2011.

[46] Mukherjee, T., Kao, N. (2011). PLA based biopolymer reinforced with natural fibre: A review. *Journal of Polymers and the Environment*, **19**, 714–725.

[47] Koster, R., Cardon, L., Ragaert, K., Harten, P. (2010). Bio-moulding and design. *Bioplastic Magazine*, **5**, 13–14.

[48] Mohanty, A. K., Misra, M., Drzal, L. T. (2002). Sustainable bio-composites from renewable resources: Opportunities and challenges in the green materials world. *Journal of Polymers and the Environment*, **10**, 19–26.

[49] Han, H., Hsu, L. T., Lee, J. S. (2009). Empirical investigation of the roles of attitudes toward green behaviors, overall image, gender, and age in hotel customers' eco-friendly decision-making process. *International Journal of Hospitality Management*, **28**, 519–528.

[50] Serizawa, S., Inoue, K., Iji, M. (2006). Kenaf-fiber-reinforced poly(lactic acid) used for electronic products. *Journal of Applied Polymer Science*, **100**, 618–624.

[51] Nec.com. Bioplastics for Electronic Equipment. Available at http://www.nec.com/en/global/rd/innovative/bioplastics/top.html. Accessed March 12, 2010.

[52] Puppi, D., Chiellini, F., Piras, A. M., Chiellini, E. (2010). Polymeric materials for bone and cartilage repair. *Progress in Polymer Science*, **35**, 403–440.

[53] Tanaka, S., Honzawa, H., Iji, M. (2013). Development of cardanol-bonded cellulose thermoplastics: High productivity achieved by using isocyanate-modified cardanol. *Journal of Applied Polymer Science*, **130**, 1578–1587.

[54] Anonymous. Biofurniture. Available at http://biofurniture.weebly.com/. Accessed June 7, 2014.

[55] Shih, Y.-F., Huang, C.-C., Chen, P.-W. (2010). Biodegradable green composites reinforced by the fiber recycling from disposable chopsticks. *Materials Science and Engineering A*, **527**, 1516–1521.

[56] Anonymous. Patch: The original paper watch. Available at http://www.patchwatches.it/. Accessed June 5, 2014.

[57] Anonymous Innventia AB Research Institute. Swedish lignin-based carbon fibre in composite materials of the future. Available at http://www.innventia.com/carbonfibre. Accessed May 29, 2014.

[58] BPI. The Biodegradable Products Institute. Available at http://www.bpiworld.org/. Accessed May 29, 2014.

[59] ASTM. 6400-12. (2012). *Standard specification for labeling of plastics designed to be aerobically composted in municipal or industrial facilities*. ASTM International, West Conshohocken, PA.

[60] ASTM. D6868-11. (2011). *Standard specification for labeling of end items that incorporate plastics and polymers as coatings or additives with paper and other substrates designed to be aerobically composted in municipal or industrial facilities*. ASTM International, West Conshohocken, PA.

[61] Anonymous SCS Global Services. Scientific Certification Systems: Setting the standard for sustainability. Available at http://www.scsglobalservices.com/. Accessed June 18, 2014.

[62] Holvikivi, J. (2007). Culture and cognition in information technology education. *European Journal of Engineering Education*, **32**, 73–82.

[63] Santo, V. E., Duarte, A. R. C., Gomes, M. E., Mano, J. F., Reis, R. L. (2010). Hybrid 3D structure of poly(d,l-lactic acid) loaded with chitosan/chondroitin sulfate nanoparticles to be used as carriers for biomacromolecules in tissue engineering. *The Journal of Supercritical Fluids*, **54**, 320–327.

[64] Ngiam, M., Liao, S., Patil, A. J., Cheng, Z., Yang, F., Gubler, M. J., Ramakrishna, S., Chan, C. K. (2009). Fabrication of mineralized polymeric nanofibrous composites for bone graft materials. *Tissue Engineering, Part A*, **15**, 535–546.

[65] Yazdimamaghani, M., Vashaee, D., Assefa, S., Shabrangharehdasht, M., Rad, A. T., Eastman, M. A., Walker, K. J., Madihally, S. V., Köhler, G. A., Tayebi, L. (2014). Green synthesis of a new gelatin-based antimicrobial scaffold for tissue engineering. *Materials Science and Engineering: C*, **39**, 235–244.

[66] Qian, J., Xu, W., Yong, X., Jin, X., Zhang, W. (2014). Fabrication and in vitro biocompatibility of biomorphic PLGA/nHA composite scaffolds for bone tissue engineering. *Materials Science and Engineering: C*, **36**, 95–101.

[67] Cai, K., Yao, K., Yang, Z., Qu, Y., Li, X. (2007). Histological study of surface modified three dimensional poly (d, l-lactic acid) scaffolds with chitosan in vivo. *Journal of Materials Science. Materials in Medicine*, **18**, 2017–2024.

[68] Muzzarelli, R. A. A. (1983). Chitin and its derivatives: New trends of applied research. *Carbohydrate Polymers*, **3**, 53–75.

[69] Guitian Oliveira, N., Sirgado, T., Reis, L., Pinto, L. F. V., da Silva, C. L., Ferreira, F. C., Rodrigues, A. (2014). In vitro assessment of three dimensional dense chitosan-based structures to be used as bioabsorbable implants. *Journal of the Mechanical Behavior of Biomedical Materials*, **40**, 413–425.

[70] Nair, L. S., Laurencin, C. T. (2007). Biodegradable polymers as biomaterials. *Progress in Polymer Science*, **32**, 762–798.

[71] Wang, Z., Hu, Q., Cai, L. (2010). Chitin fiber and chitosan 3D composite rods. *International Journal of Polymer Science*, **2010**, 7.

[72] Oh, D. X., Hwang, D. S. (2013). A biomimetic chitosan composite with improved mechanical properties in wet conditions. *Biotechnology Progress*, **29**, 505–512.

[73] Ehrlich, H., Simon, P., Motylenko, M., Wysokowski, M., Bazhenov, V. V., Galli, R., Stelling, A. L., Stawski, D., Ilan, M., Stocker, H., Abendroth, B., Born, R., Jesionowski, T., Kurzydlowski, K. J., Meyer, D. C. (2013). Extreme Biomimetics: Formation of zirconium dioxide nanophase using chitinous scaffolds under hydrothermal conditions. *Journal of Materials Chemistry B*, **1**, 5092–5099.

[74] Jayakumar, R., Tamura, H., Prabaharan, M., Sudheesh Kumar, P. T., Nair, S. V., Furuike, T., Novel chitin and chitosan materials in wound dressing. In Laskovski, A., (Ed.), *Biomedical engineering: Trends in material science*. InTech, Rijeka, Croatia, 2011, pp. 3–24.

[75] Muzzarelli, R. A. A., Nanochitins and nanochitosans, paving the way to eco-friendly and energy-saving exploitation of marine resources. In Hoefer R., (Ed.), *Polymer science: A comprehensive reference*. Elsevier, Amsterdam, 2012, pp. 153–164.

[76] Anonymous. (2012). Advanced composite Solutions. *JEC Composite*, **72**, 15–19.

[77] Naseri, N., Algan, C., Jacobs, V., John, M., Oksman, K., Mathew, A. P. (2014). Electrospun chitosan-based nanocomposite mats reinforced with chitin nanocrystals for wound dressing. *Carbohydrate Polymers*, **109**, 7–15.

[78] Avella, M., Bonadies, E., Martuscelli, E., Rimedio, R. (2001). European current standardization for plastic packaging recoverable through composting and biodegradation. *Polymer Testing*, **20**, 517–521.

[79] Srinivasa, P. C., Tharanathan, R. N. (2007). Chitin/chitosan—Safe, ecofriendly packaging materials with multiple potential uses. *Food Reviews International*, **23**, 53–72.

[80] Tang, X. Z., Kumar, P., Alavi, S., Sandeep, K. P. (2012). Recent advances in biopolymers and biopolymer-based nanocomposites for food packaging materials. *Critical Reviews in Food Science and Nutrition*, **52**, 426–442.

[81] Rhim, J.-W., Ng, P. K. W. (2007). Natural biopolymer-based nanocomposite films for packaging applications. *Critical Reviews in Food Science and Nutrition*, **47**, 411–433.

[82] Youssef, A. M. (2013). Polymer nanocomposites as a new trend for packaging applications. *Polymer-Plastics Technology and Engineering*, **52**, 635–660.

[83] Davis, G., Song, J. H. (2005). Biodegradable packaging based on raw materials from crops and their impact on waste management. *Industrial Crops and Products*, **23**, 147–161.

[84] Guana, J., Eskridgeb, K. M., Hanna, M. A. (2005). Acetylated starch-polylactic acid loose-fill packaging materials. *Industrial Crops and Products*, **22**, 109–123.

[85] Shah, D. U., Schubel, P. J., Clifford, M. J. (2013). Can flax replace E-glass in structural composites? A small wind turbine blade case study. *Composites Part B: Engineering*, **52**, 172–181.

[86] Lucintel. (2011). Opportunities in Natural Fiber Composites. Lucintel, USA.

[87] Drzal, L. T., Mohanty, A. K., Misra, M. (2001). Bio-composite materials as alternatives to petroleum-based composites for automotive applications. Automotive Composites Conference. A. C. D. Society of plastic engineers. Troy, MI, 2008.

[88] Unknown. (2009). Special report: Cars made of plants?. Available at http://www.edmunds.com/fuel-economy/special-report-cars-made-of-plants.html. Accessed November 25, 2011.

[89] Ashori, A. (2008). Wood–plastic composites as promising green-composites for automotive industries! *Bioresource Technology*, **99**, 4661–4667.

[90] Anonymous. (2013). IBIB2012-13 International Business directory for innovative bio-based plastics and composites. Available at http://www.bio-based.eu/iBIB. Accessed August 25, 2015.

[91] The Budd Company, D. C. (2001). Composite materials promise increased fuel efficiency. *Advanced Materials and Chemicals*, 26–33. Available at http://www.atp.nist.gov/eao/sp950-3/budd_co.pdf. Accessed September 24, 2015.

[92] Holbery, J., Houston, D. (2006). Natural-fiber-reinforced polymer composites applications in automotive. *The Journal of the Minerals, Metals and Materials Society*, **58**, 80–86.

[93] Silva, C. A. (2010). Sustainable Design of automotive components through jute fiber composites: an integrated approach Lisbon, PhD Thesis, Universidade Técnica de Lisboa, Instituto Superior Técnico, Portugal.

[94] Quarshie, R., Carruthers, J. (2014). *Technology overview: Biocomposites*. Materials KTN and NetComposites. Available at http://www.netcomposites.com/downloads/Biocomposites%20Guide.pdf. Accessed September 24, 2015.

[95] Faruk, O. Cars from Jute and Other Bio-Fibers. Available at http://biggani.com/files_of_biggani/mashiur/interview/omar_faruk.pdf. Accessed August 8, 2014.

[96] Anonymous. (2014). JEC Composites: Flax reinforced polypropylene in 2015 Volkswagen Golf. Available at http://www.jeccomposites.com/news/composites-news/flax-reinforced-polypropylene-2015-volkswagen-golf. Accessed August 8, 2014.

[97] The European Industrial Hemp Association (EIHA). (2010). European Hemp fibres for diverse bio-based products, *Bio-based News*, July 15, 2009, The European Industrial Hemp Association (EIHA).

[98] Malnati, P. (2009). ECO Elise concept: Lean, speedy and green. *Composites Technology*, **15**, 46–48.

[99] Weager, B. (2010). High-performance biocomposites: Novel aligned natural fibre reinforcements. *JEC Composites*, **55**, 31–32.

[100] Anonymous. (2013). JEC Composites: Faurecia presents advancements in use of natural composites. Available at http://www.jeccomposites.com/news/composites-news/faurecia-presents-advancements-use-natural-composites. Accessed August 8, 2014.

[101] Anonymous. (2014). JEC Composites: UPM and metropolia present the biofore concept car. Available at http://www.jeccomposites.com/news/features/biocomposites/upm-and-metropolia-present-biofore-concept-car. Accessed August 8, 2014.

[102] Yan, L., Chouw, N., Jayaraman, K. (2014). Flax fibre and its composites: A review. *Composites Part B: Engineering*, **56**, 296–317.

[103] Anonymous. (2014). JEC Composites: The development of bamboo fiber composites in China. Available at http://www.jeccomposites.com/news/features/biocomposites/development-bamboo-fiber-composites-china. Accessed August 8, 2014.

[104] Anderson, J. P., Altan, M. C. (2012). Properties of composite cylinders fabricated by bladder assisted composite manufacturing. *Journal of Engineering Materials and Technology*, **134**, 044501–044501.

[105] Anonymous. (2011). JEC composites: Complete houses made of wood-plastic composites. Available at http://www.jeccomposites.com/news/composites-news/complete-houses-made-wood-plastic-composites. Accessed August 8, 2014.

[106] Anonymous. (2014). JEC composites: ThyssenKrupp plastics extends its range of wood-plastic composites. Available at http://www.jeccomposites.com/news/features/biocomposites/thyssenkrupp-plastics-extends-its-range-wood-plastic-composites. Accessed August 8, 2014.

[107] Snidjer, M. (2010). NF Bioplastics to save coral reefs. *Bioplastic Magazine*, **5**, 34–35.

[108] Anonymous. (2014). JEC composites: Blades for rooftop wind turbine manufactured with Biotex flax. Available at http://www.jeccomposites.com/news/features/biocompos ites/blades-rooftop-wind-turbine-manufactured-biotex-flax. Accessed August 8, 2014.

[109] Rijswijk, K., Brouwer, W. D., Beukers, A. (2003). *Application of natural fibre composites in the development of rural societies*. FAO, Rome.

[110] Bourçois, D., Baley, C., Grohens, Y., LeMaguer, G., LeDuigou, A., Deux, J.-M., Davies, P., Finot, J.-M., Chabaud, C. (2011). Environment-friendly composites for marine applications: The Navecomat project. *JEC Composite*, **67**, 35–37.

[111] Anonymous. (2011). A flax-composite boat for transatlantic races. *JEC Composites*, **67**, 33–34.

[112] Anonymous. (2010). Biomaterial for dragon boat decking. *Bioplastic Magazine*, **5**, 30.

[113] Anonymous. (2011). Using the infusion process to produce a hull only with green materials reducing overall weight. *JEC Composite*, **68**, 41.

[114] Anonymous. (2012). First small watercraft that is compostable at end of life. *JEC Composite*, **71**, 25.

[115] Anonymous. (2013). JEC composites: Flax composite paddle by VE Paddles using Bcomp's flax composite. Available at http://www.jeccomposites.com/news/composites-news/flax-composite-paddle-ve-paddles-using-bcomps-flax-composite. Accessed August 8, 2014.

[116] Anonymous. (2012). Biocomposite snowboard using Biotex flax fabric. Available at http://www.jeccomposites.com/news/composites-news/biocomposite-snowboard-using-biotex-flax-fabric. Accessed August 8, 2014.

[117] XBoards. (2012). Bioboard TM: Natural materials plus increased vibration dampening. Available at http://www.x-boards.com/home.html. Accessed June 5, 2014.

[118] Bittermann, R. (2010). Wood composites for toys. *Bioplastic Magazine*, **5**, 32–33.

[119] Anonymous. (2014). JEC Composites: Alpaca carbon fiber travel guitar. Available at http://www.jeccomposites.com/news/composites-news/alpaca-carbon-fiber-travel-guitar. Accessed August 8, 2014.

[120] Anonymous. (2013). JEC Composites: Blackbird guitar introduces an ukulele made with natural fibers. Available at http://www.jeccomposites.com/news/composites-news/blackbird-guitar-introduces-ukulele-made-natural-fibers. Accessed August 8, 2014.

INDEX

Biodegradable Green Composites, First Edition. Edited by Susheel Kalia.